Directions in Energy Policy: A Comprehensive Approach to Energy Resource Decision-Making

Directions in Energy Policy:

A Comprehensive Approach to Energy Resource Decision-Making

Second in a Series
Proceedings of the International Scientific Forum
on an Acceptable World Energy Future
November 27–December 1, 1978

Forum Chairman
Behram Kurşunoğlu

Edited by
Behram Kurşunoğlu
Arnold Perlmutter

Center for Theoretical Studies
University of Miami

Ballinger Publishing Company • Cambridge, Massachusetts
A Subsidiary of Harper & Row, Publishers, Inc.

This book is printed on recycled paper.

International Standard Book Number: ISBN 0-88410-089-8

Library of Congress Catalog Card Number: 79-21524

Printed in the United States of America

Library of Congress Cataloging in Publication Data

International Scientific Forum on an Acceptable World Energy Future, University of Miami, 1978.
Economic aspects of energy production.

Bibliography: p.
Includes index.
1. Power resources—Congresses. 2. Energy industries—Congresses. 3. Energy policy—Congresses. I. Kurşunoğlu, Behram, 1922- II. Perlmutter, Arnold, 1928- III. Miami, University of, Coral Gables, Fla. Center for Theoretical Studies. IV. Title.
HD9502.A2I59 1978 333.7 79-21524
ISBN 0-88410-089-8

Contents

List of Tables

List of Figures

Preface

In light of worldwide energy events triggered in part by the uneven supply of conventional fuel, hysteria on the part of some against nuclear energy, and the appearance of dark clouds on the economic horizons, the time span between successive International Scientific Forums seems to be quite long compared to the ever-shortening periods between significant energy induced changes. At this writing, we hope that by the time we convene our Third International Scientific Forum in Nice, France, the world will not have experienced further energy-related shocks of the type that we have recently witnessed.

This volume, comprising the proceedings of the second energy forum, supported in part by the Exxon Education Foundation, discusses the world energy future. It does not, of course, and could not, in fact, predict all that has happened and will happen in 1979 that was "energy oriented." Even two of the Forum's most illustrious members, Professor Edward Teller and Professor Eugene Wigner, could not have foreseen the 1979 nuclear incident at Three Mile Island, the gas lines in American cities, and the recent impetus in the U.S. Congress and in U.S. industry toward the development of synthetic fuels, arising from shortages of energy and price escalation.

This volume differs from the first in that greater emphasis is placed on near- and long-term economic and legal considerations of energy-related questions. Yet, as befits a gathering of eminent scientists, scholars, engineers, and industrialists, there are several key discussions of new sources of energy that could profoundly and positively affect the global picture. In all this we cannot hope to pass through the coming transitions in an economically meaningful way without the contribution of nuclear energy.

As an introduction to these proceedings, we believe that the overview of the Forum prepared by the eminent nuclear pioneer Dr. Karl Cohen with the assistance of the session moderators serves as a concise statement of the deliberations and conclusions.

The Editors

Overview

Karl Cohen

ENERGY DEMAND

In spite of higher energy prices and considerable improvement in the efficiency of energy use, energy demand will continue to grow, though not at the rates of the last twenty years. World demand is likely to have doubled early in the next century and to have quadrupled by the middle of the century. This demand increase estimate is contingent upon a much needed economic improvement in the developing world.

ENERGY MODELING

Energy models project alternative futures of energy demand, supply, and price movement. Differences between projections and between models reflect intrinsic uncertainties about future economic behavior and about the success and cost of future technologies. There is, however, a growing consensus that changes in energy demand or of modes of energy supply occur slowly. Public policy must recognize this characteristic. To be effective, a public energy policy must be timely and consistent over long periods. Standards for environmental protection should be stable, rational, and consistent between alternative energy sources. Further, in view of the uncertainties mentioned, public policy should be based on a range of estimates that reflects different contingencies.

ENERGY AND ECONOMIC GROWTH

Although energy use is pervasive throughout industrialized economies, substantial reductions in the growth of energy consumption are possible with only relatively small reductions in the growth of economic output (GNP). The precise GNP growth reductions associated with an energy growth restriction depend upon the elasticity of energy demand, the value share of energy in the economy, the economic policies undertaken, changes in the rate of capital formation, and whether the reductions are motivated by cost increases or by other policy measures.

NUCLEAR

The Forum saw no new evidence to change its previous conclusion that a significant contribution to world energy supply by nuclear fission is essential over at least the next five decades. All prudent scenarios for such energy production rely on a link between present and advanced nuclear systems by reprocessing of nuclear fuels. We recommend that the international issues inhibiting the development of reprocessing and fuel recovery, in both the thorium and uranium cycles, be promptly resolved.

CONSERVATION

Due in large part to the success of OPEC, energy demand grows less than it otherwise would have and is paid for at higher prices. Therefore, more efficient use of energy in many instances becomes cost-effective. The time scale for introduction of these improvements is usually long, being related to the turnover time of huge existing stocks of buildings, transportation equipment, and industrial process equipment. Consistent long-range public policies will help to accelerate the reduction of energy demand growth by more efficient use of energy. Such policies should include an R & D strategy for improved energy use commensurate with the expected high future prices of new energy supplies.

NORTH AMERICA REGION

A freer exchange of technologies and resources between Canada, Mexico, and the United States would be mutually advantageous.

OIL AND GAS

The Forum expects that considerable oil and gas resources exist that can be developed into reserves at prices between today's OPEC prices and the probable much higher price of synthetic oil and gas from coal. On economic grounds, the OPEC price level is not expected to increase in real terms (excluding, of course, short-term political perturbations).

LEGAL

There is a tendency in some countries to resort to increasingly labyrinthine trial type procedures, sometimes before several agencies, to resolve the social and political issues involved in energy decisions. While the growth of such complicated and duplicative procedures is intended to compensate for both a basic fear of complex technology and a lack of confidence in our institutions to govern properly, such procedures have enormous costs for society in delay, uncertainty, and in some instances, frustration of public policy without any demonstrable benefit over alternative processes. Various forms of less complex decisionmaking have produced not only excellent technical analyses, but also final decisions on a timely basis and with wide public acceptance—for example, the Windscale proceedings in Great Britain. Therefore, these less ritualistic modes of decisionmaking, sometimes styled "legislative," are recommended as being more appropriate to governmental energy decisions.

PUBLIC SAFETY: RISKS AND BENEFITS

Standards for environmental protection—for example, against carcinogens in foods or air pollution from plant emissions—should be stable, rational, and applied across the board to all risks. Ad hoc application of standards to some hazards and neglect of others results in unbalanced application of resources and unjustifiable public anxieties.

STANDARDIZATION

Standardization of complex industrial equipment is highly desirable. Reliability, safety, and economics are all improved by the learning process. Changes in standards must be rationed carefully to avoid unexpectedly negative effects because of interruption of the learning process. Plant-unique requirements imposed by local governments should be avoided for the same reason. Scaling up plant sizes to obtain economies of scale must also be done with great circumspection.

PROLIFERATION OF NUCLEAR WEAPONS

Reprocessing and breeder reactor development are economically unattractive for small-scale nuclear energy systems. Widespread economically motivated use of these technologies is distant. On the other hand, policies designed to prevent the spread of nuclear weapons that interfere with a nation's legitimate economic goals will not be successful. Multilateral accords between weapons and non-weapons states are necessary to manage this problem. Unilateral actions by weapons states will be counterproductive.

ATMOSPHERIC CARBON DIOXIDE

The possibility of serious climatic changes from the accumulation of anthropogenic CO_2 in the atmosphere, as from burning of fossil fuels and deforestation, is widely recognized. However, our knowledge both of the global CO_2 balance and of global climatic effects of CO_2 accumualtion is rudimentary. During the next decades, a sustained international scientific study, incorporating data gathering and theoretical analysis, should be implemented. Regulations against the use of fossil fuels in advance of fuller understanding of the phenomena and of possible countermeasures are premature.

Session A

Finding an Evolving Balance of Energy Technologies

Part I

Evolving Global Energy Balances and Constraints

Chapter 1

Political Fusion: Scientific Research and the Legislative Process*

Carl D. Pursell
Congress of the United States, Michigan

I come to this meeting as a student, knowing that there is much I can learn from you and nothing I can tell you about science or energy that you do not already know. But I also come with the hope that I can share some knowledge with you from my special field—the legislative process. Since this rather unscientific world of politics and human chemistry often has such a profound impact on your scientific endeavors, I think it is only right that the leaders of the scientific community be encouraged to have an equally important impact on legislative decisionmaking.

For the past two years, since being elected to Congress from my district in Michigan, one of my primary goals has been to serve as a catalyst in the effort to accelerate research and development for alternative energy sources. While I have introduced legislation on a broad range of new energy sources, my particular emphasis has been on fusion energy.

I believe we need a worldwide partnership for energy emancipation, to free us from the threat of an energy shortage and the conflicts that that would certainly cause. I see this as the greatest threat to world peace and stability in the coming decades.

Solving this dilemma requires a cooperative international effort by science, government, and industry to meet a goal that transcends the personal interests of any individual, group, or nation. That is the central theme and focus of my efforts. And since the ability of science and industry to respond to this challenge depends on what government does, the people in this audience must assume a

*I would like to recognize, and express my gratitude for, the excellent assistance of the staff of the Laser Fusion Department of Lawrence Livermore Laboratories and the staff of the Department of Energy Division of Laser Function in the preparation of graphic illustrations and transparencies used in the original presentation of this paper.

more active role than ever in assisting and influencing government decisions. I believe that you can be the crucial factor in the equation we are seeking.

The legislative decisionmaking process on laser fusion research provides an excellent example of the complications and the opportunities we face. When I began working on the laser fusion issue in the Science and Technology Committee as a new member of Congress, several things became immediately apparent: one, there were important potential applications of laser fusion for both national security and civilian energy production; two, the military aspects of the program had by far the greatest emphasis; three, private sector involvement was essential to assure the most rapid possible development of fusion energy; four, the history of the program showed only a limited commitment on the part of the government to develop the full potential of laser fusion; and five, individual ambitions and competitions were, in some cases, getting in the way of progress toward the goals all seemed to share. Clearly there was a need to seek new leadership on all sides, and a more cooperative attitude was essential.

Working with scientists, congressional staff people, and other members of Congress, we developed a new strategy based on intensive personal work designed to generate a more cooperative approach. There can be no claims of complete success. And we fully realize there may be serious obstacles along the road. But I believe the last two years have yielded considerable progress. There is now a much greater awareness in the Congress and in the media of the promise of fusion energy.

I have received excellent cooperation from the laser fusion people in the Department of Energy and see them as allies in developing the energy potential of fusion and in encouraging wider participation by private and academic research groups. But I believe the most significant result of our effort has been the addition of a new dimension in the national laser fusion program—the dimension of separately identified civilian energy objectives.

This past year, for the first time, we achieved approval in the Science and Technology Committee for separate funding for the civilian aspects of the program, which was then approved by the Appropriations Committee and subsequently by the full House. The eventual conference report between House and Senate also recognized the civil energy potential of laser fusion and the important role that the private and academic sectors have in the ultimate success of fusion as a basic energy source.

I believe we have the ship turned around and headed out to sea. We hope we are on a good course; and we are actively enlisting others, like the influential leaders gathered here, to join us in nurturing this cooperative approach and in moving the partnership forward in a positive manner.

Many of you are very sophisticated in the workings of government. For those who are not already knee-deep in this process, I think it is vital that you wade in. In the Congress, we have a fifteen stage process to obtain funding for a program like laser fusion. And success usually requires a perfect batting average. It is possible to hit fourteen of fifteen and still end up with the same net result as striking out completely.

A proposal must clear three subcommittees and three full committees and must then be approved by the full House. The same number of hurdles must be surmounted in the Senate. And the fifteenth—and perhaps the most critical—hurdle of all is the final conference between the House and the Senate.

Even if you are successful in reaching all these plateaus, the entire program can be shot down in the executive branch, either by the White House or by the Office of Management and Budget (OMB). OMB is a somewhat hidden group with a great deal of muscle. It is often overlooked as a primary force in government decisions.

The congressional decisionmaking process is very much like the laser fusion process. Just as powerful laser beams are focused on the target from various angles in an experiment, at every stage of the legislative process tremendous pressures are focused from all angles to influence the congressional decision.

I discovered early in my work on this issue the conflicting pressures from the executive agencies such as OMB, federal departments such as Energy Research and Development Administration (ERDA) (which we restructured as the Department of Energy), national scientific laboratories, private research groups, independent analysts, a variety of congressional committees and staffs with conflicting program and jurisdictional goals, and numerous members of the House and Senate and their personal staffs. But I also found that at many stages of the process, the influence of independent scientists was not as strong a factor as it could be.

Because I want to encourage such influence, I am going to dramatize the critical pressure points, and introduce you to some of the leading players, by reviewing the key stages involved in our work on laser fusion legislation.

First I met with my personal congressional staff, to set a strategy to be followed in trying to get our amendment for additional funding through the Congress. This is a very important phase that is often overlooked by people outside Congress who are trying to influence decisionmaking. Often your greatest influence can be generated by working closely with the staff of an interested member of Congress, because members' actions are often the product of meetings such as this one.

We then take our idea to the committee level, working methodically to win over support one member at a time. A man who will be very important to the entire scientific energy research community in coming years is Congressman Don Fuqua of Florida, who is taking over as chairman of the House and Technology Committee. Another extremely important voice in decisions affecting scientific research is Congressman John Wydler of New York, who is the ranking minority member on Science and Technology.

I find it is generally much more effective to work one on one with individual members than to come before the committee cold and with a big proposal and a big speech. Programs that are pushed in that way are often very big failures.

It is also extremely important to work closely with the committee staffs at every step of the legislative process. There are so many complex proposals before Congress that it is often the staff people who make the life or death decisions on individual portions of a major program. The staff of the Fossil and Nuclear Subcommittee is the key subcommittee of Science and Technology in decisions on fusion and many other energy programs.

When I develop a program like this, I also go out of my way to encourage scientists in the field to offer their advice. I conferred, for example, with Dr. Henry Gomberg of KMS Fusion, which is located in my home state of Michigan.

An important message that I want to convey to you today is that my interest in fusion energy came about largely because of the efforts of Dr. Gomberg, Dr. Robert Hofstadter, and other active scientists, who spent many hours explaining the potential of the program and answering my questions. My thesis is that each of you could have a similar impact with other members of Congress.

This kind of outside consultation is especially important to me as I develop a proposal with the committee staff, my staff, and other members. During this development stage, I also took the time to visit the Laser Fusion Division at the Department of Energy for discussions with Dr. Martin Stickley and his staff.

The congressional staff people and I also made intensive personal inspections of such leading federal laboratories as Sandia, Los Alamos, and Lawrence Livermore. Many hours were spent in personal discussions with their scientists, and we continued to confer with all these people throughout the legislative process.

The Armed Services Committee and its staff have a very powerful influence on laser fusion decisions in the House. Another part of our effort to decrease the historical tensions surrounding the laser fusion program and to generate a more cooperative approach was to meet with Chairman Mel Price of Illinois, other Armed Services Committee members, and their committee staff people. Our approach has been to show the Armed Services Committee people that we recognize the importance of their national security concerns and that the increased emphasis we seek for fusion energy does not necessarily have to conflict with their responsibilities.

Now we come to the really critical test in the House—the Appropriations Committee, and especially the Public Works Subcommittee, which has responsibility for laser fusion funding. The Public Works Chairman is Tom Bevill of Alabama, and another powerful member of the Public Works Subcommittee is John Myers of Indiana. If any of the programs envisioned by an authorizing committee, such as Science and Technology, are to be translated into real money, they must win the approval of the proper subcommittee and of the full Appropriations Committee. And perhaps nowhere in the legislative process is the staff more important than on the Appropriations Committee, which must consider the funding levels for a tremendous variety of programs approved by the other substantive committees.

Another important development in our laser fusion effort involved the General Accounting Office. The GAO–not to be confused with the scandal-prone GSA (General Services Administration)–is an outstanding analytical arm of the government. My office was the scene of a critical meeting during the struggle to increase civilian laser fusion research funding. The Armed Services Committee asked GAO to evaluate the performance of KMS Fusion. Since KMS is such a significant part of the civilian laser fusion program, this report would obviously have a major impact on our proposals.

When I heard that the GAO team was ready to report to the Armed Services Committee, I requested to sit in on the briefing, and the report turned out to be very positive. To assure that these results would be known to all concerned, I secured permission for a repeat briefing in my office for the Science and Technology and the Appropriations Committee staffs. That was a major factor in our success in the House.

All the while we were working with the committees, I was taking our case to individual members of the House, because in addition to the extensive work off the floor, the floor debate is critical. Even the most meticulously prepared proposal can be stopped cold at this crucial point. An eloquent and respected congressman can make or break a bill in debate, and this kind of homework is essential before bringing a proposal to the floor.

The floor debate is an often overlooked stage where leading scientists can also have a real impact. All members of Congress, whether they are on one of the key committees or not, have a voice in the decision at this point. So if you have laid the groundwork by meeting with members from your part of the country, they can help during the formulation of a proposal and also offer crucial voting support to assure passage on the floor.

There is an equally exhaustive process involved in the Senate. Thus, it is extremely important that our scientific leaders communicate the importance of intensive energy research to the members and staffs of the Senate as well.

Then comes what is often the biggest hurdle of them all, the conference committee that settles the differences between the House and Senate versions of a bill. The committee staffs again play a critical role during this phase of the process, as we work out language and strategy that we present at the next meeting of the conference committee.

Even after all of the fifteen steps of the legislative process, after all the staff work, outside consultations, and research, there are still major external pitfalls. The president can unravel an entire year's work with one stroke of his pen by vetoing the bill. In fact, that is exactly what happened this year to the bill containing laser fusion funding. It was vetoed and sent back for Congress to work out a new compromoise acceptable to the White House. And even now we're still doing follow-up work with OMB and the Department of Energy (DOE) to assure that the congressional intent to fund civilian energy applications of laser fusion is carried out.

There are many parallels between the fusion process and the legislative process. I have described how the great pressures from many sources must be delicately coordinated to affect the decision at the right time and in the right way to achieve the desired result, just as laser beams must strike the target properly to obtain a useful fusion reaction. When the process works, it can be a beautiful thing to see, just like an image of a successful laser fusion target experiment. To obtain such a successful result, a scientist must be concerned with such precise factors as time, temperature, and density. In my political laboratory, I guess you could say we are concerned with very imprecise factors, such as tone, temperament, and intensity.

In either case, all the elements must come together in just the right way or the effort fails. If the laser pulses in a fusion experiment are not synchronized, the fuel pellet is not compressed properly, and we do not achieve a useful reaction. If we don't have the proper balance in a legislative decision, the pressure from one direction overpowers other influences. It may look good from one angle, but we do not achieve a result that advances the overall goals of the program.

The legislative budget process often works like a laser chain. You can visualize it as an illustration of a laser pulse as it proceeds through a series of amplifiers of increasing diameter. Similarly, we can visualize the budget process as starting with a proposal in the executive branch and proceeding through a series of subcommittees and committees. As the proposal moves along this legislative laser chain, the dollar figure often expands as each step brings new ideas, additions, and changes in the program.

This underscores my point that the shape of the program can be influenced at any stage of the legislative process—right up to and including the final conference report—by the right information presented by the right person at the right time. And it is equally important to remember that the dollar figure can be reduced just as easily as it can be increased.

The Department of Energy has supplied me with charts that track the changes approved by various committees in laser fusion operating funds over the past three years. These charts cover only operating funds and do not include such items as construction and capital acquisition.

The chart for fiscal year 1977 (Figure 1-1) shows a considerably simplified process compared to present conditions. The Joint Committee on Atomic Energy had the major role and added about $15 million as it considered the bill. The House and Senate Appropriations Committees added funds to bring their bill into line with the joint committee, and operation funds totaled $75.8 million.

The fiscal year 1978 budget chart (Figure 1-2), which corresponds to the year I entered Congress, shows the major committee shakeup that had been undertaken. Authorizing decisions were split in the House between the Science and Technology and Armed Services Committees. The jurisdictional battle that resulted caused the authorization bill to lag behind the appropriations measure,

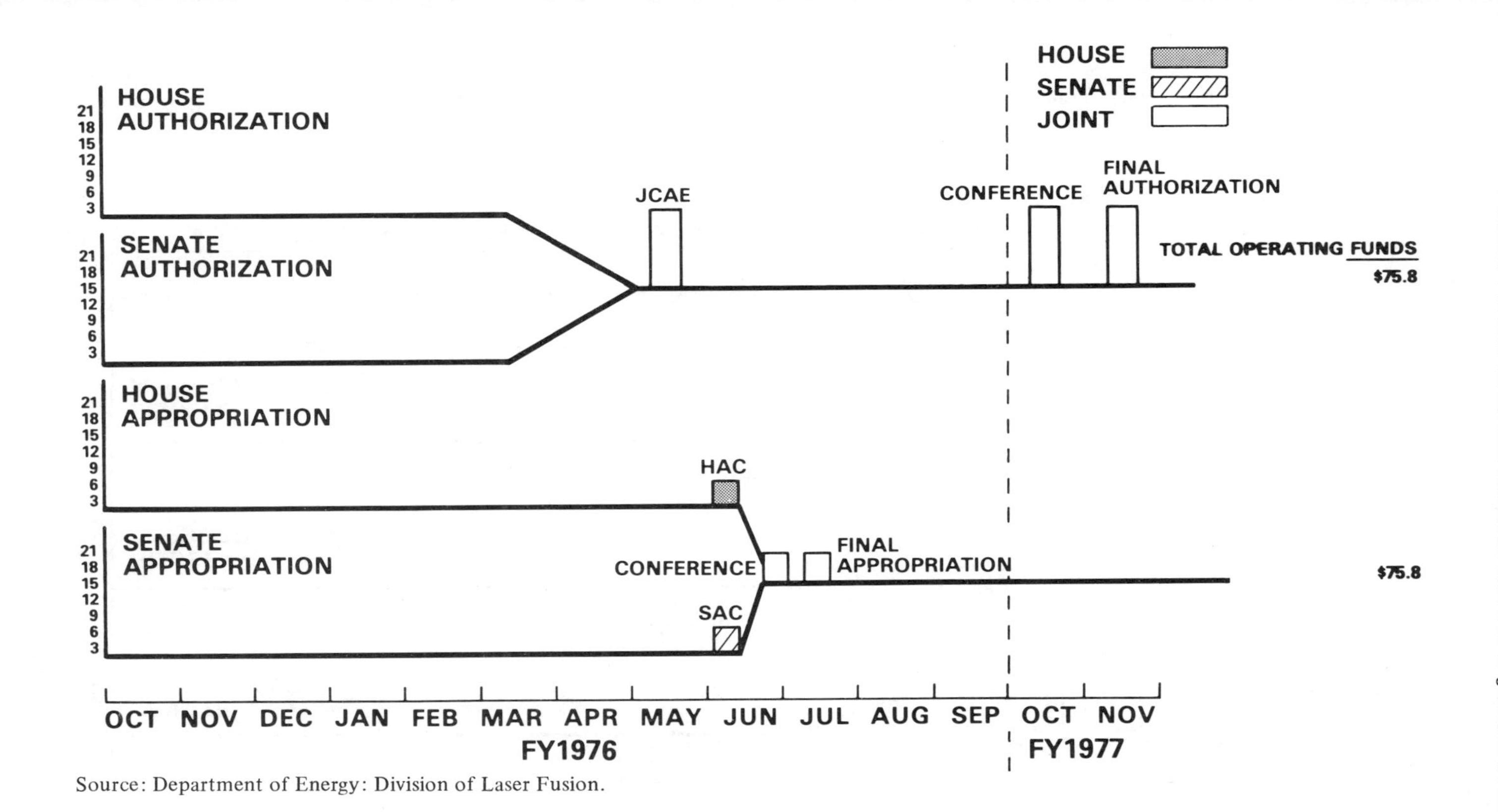

Source: Department of Energy: Division of Laser Fusion.

Figure 1–1. Budget Outlays: Congressional Actions, Inertial Confinement Fusion Program, Operating Expenses FY 1977 Add-ons (In Millions).

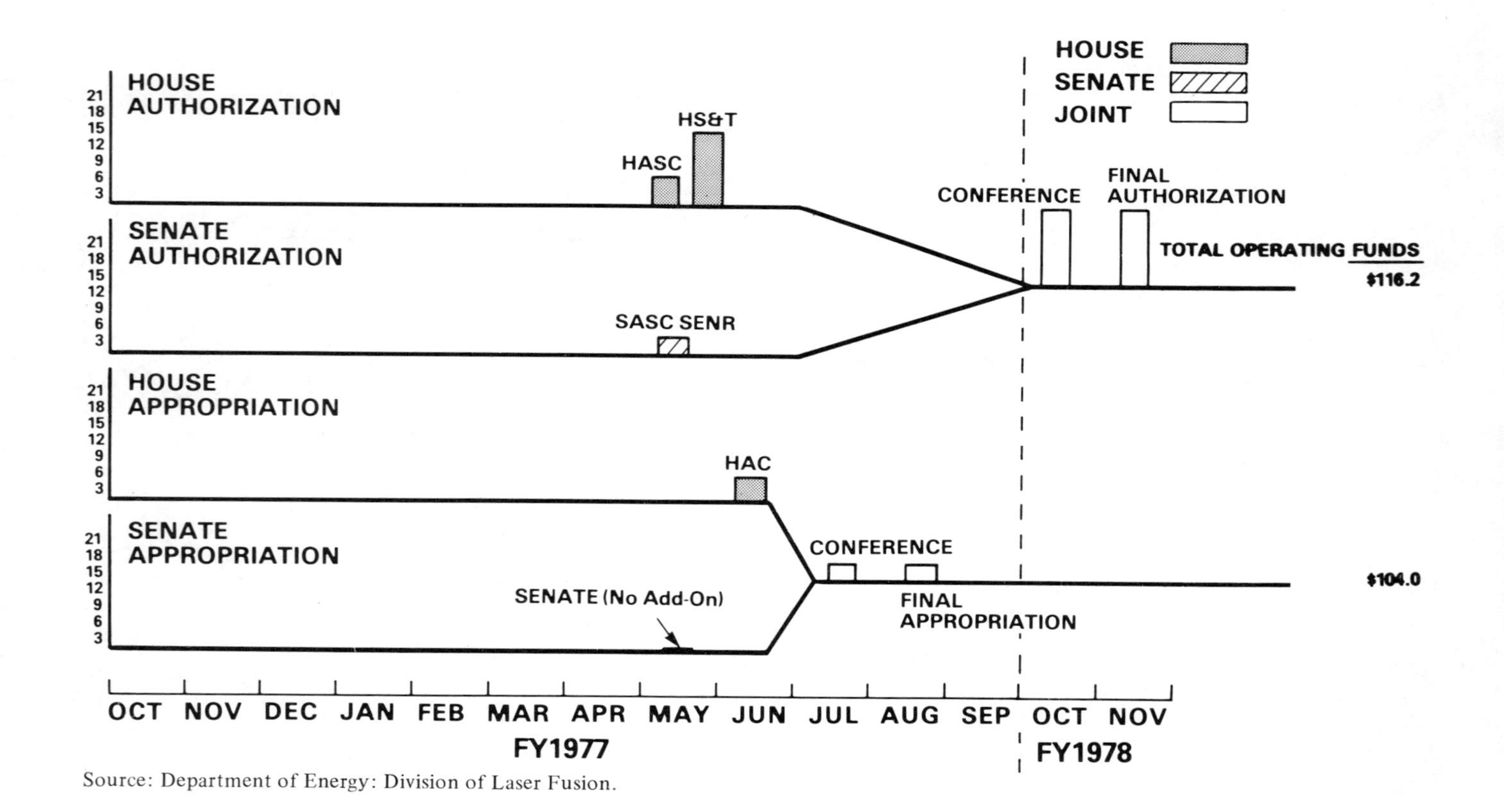

Source: Department of Energy: Division of Laser Fusion.

Figure 1–2. Budget Authority: Congressional Actions, Inertial Confinement Fusion Program, Operating Expenses FY 1978 Add-ons (In Millions).

which set operating funds at $104 million. The authorizing bill ended up significantly higher, at $116.2 million, primarily due to a $9.2 million amendment that I sponsored.

In the fiscal year 1979 budget (Figure 1-3) we faced the possibility of reduced operating funds for the laser fusion program, but both the Science and Technology and the Armed Services Committees added significant funding. We approved $8.8 million in the Science and Technology Committee, and the Armed Services Committee added $12 million. Not all of the increase survived the conference committee, but we kept operating funds at $104 million and designated part of the total for civilian applications.

In the coming years we are facing an increasingly difficult task in achieving adequate research budgets. There is no doubt that economic considerations and the mood of the country are going to result in tighter and tighter federal budgets.

Adding money for new or expanded programs is one kind of challenge. But we face the considerably more difficult challenge of shifting priorities and reallocating funds within a budget with much more severe restraints. The intensity of the battle lines will be far tougher, as sharply competing interests fight for existing funds in a context of scarce financial resources. Thus I am asking for assistance from this audience to develop, in every way possible, support from the scientific community to give energy research the funding priority it deserves.

For instance, I have introduced legislation to establish an Energy Research Trust Fund to underwrite an accelerated energy R&D program similar to our early space program. This is a positive plan to assure that fusion and a wide variety of other potential new forms of energy receive the intense emphasis that they do not have today.

We are going to need your influence as never before to see that programs like this have a chance. Because of the budget restrictions we face, it is entirely possible that more heavily organized groups, such as those pushing for development of programs like the B-1 bomber, will win the fight for priorities by the sheer force of their lobbying effort.

The world's scientists already do a great deal for humanity. But more and more the needs of this world compel you to also work outside the laboratory. The times require an international team of scientist-statesmen who can help all of us to see the desperately needed priorities of coming generations.

The work of a legislator, when it is done right, can be very much like that of a scientist. Both of us require exhaustive preparation, extreme patience, long man hours, and tremendous lab development time before we can produce results that we can offer to the world.

To get the desired results, we have to achieve the proper chemistry through careful planning and experimentation. And we must be willing to suffer many setbacks and partial successes, from which we can learn what we need to do to have our efforts succeed.

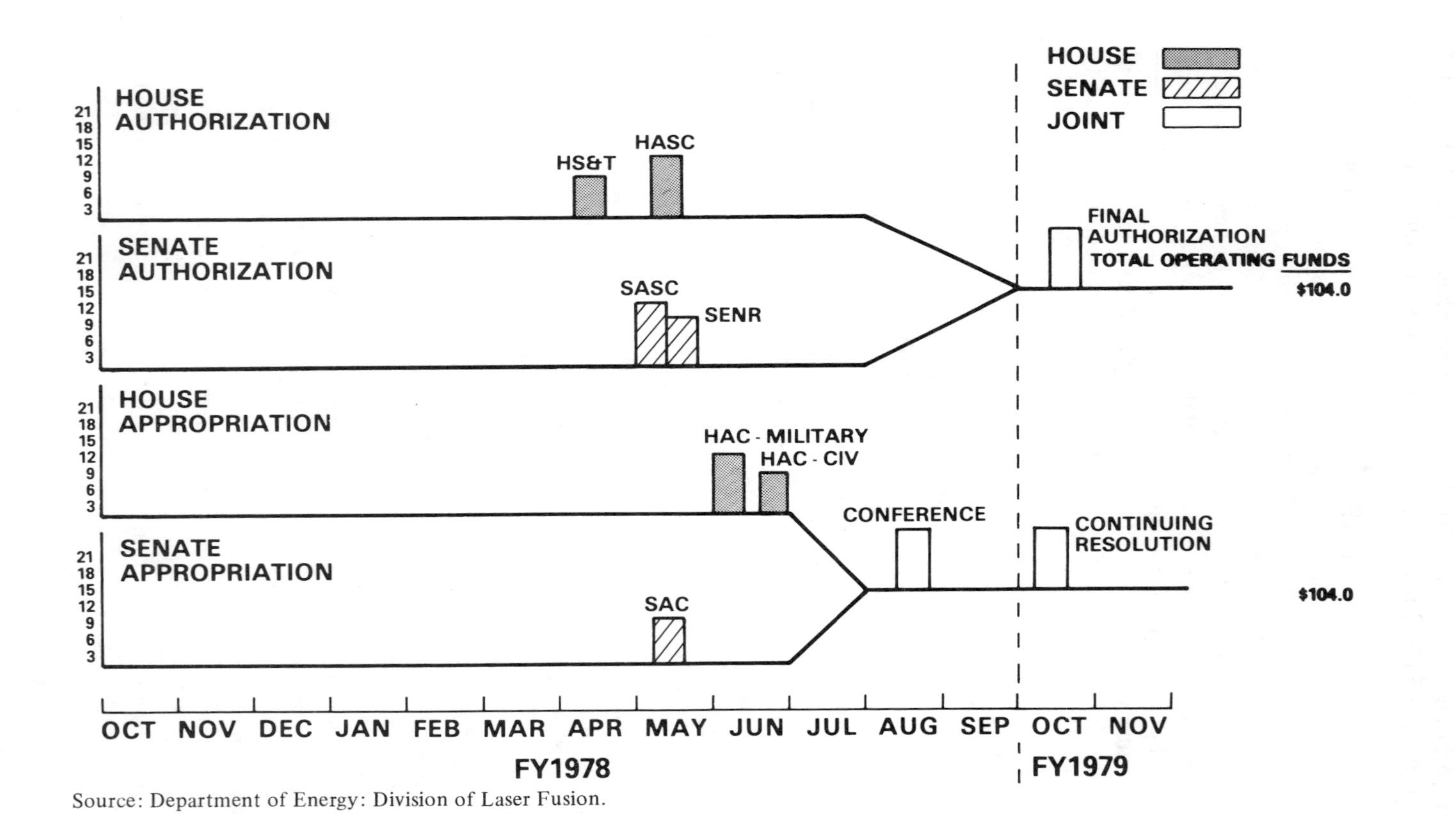

Source: Department of Energy: Division of Laser Fusion.

Figure 1–3. Budget Authority: Congressional Actions, Inertial Confinement Fusion Program, Operating Expenses FY 1979 Add-ons (In Millions).

The time constraints of our annual congressional budget process and our two year terms between elections often cause our political decisions to be rushed. Without proper development time, we often fail to reach optimum results. But like a careful scientist, I am willing to learn by experience. I also have the patience to know that a laudable goal is worth trying for again and again until success is achieved.

But the congressional aspect of the fusion energy program is only one part of the overall effort. You could visualize it as one chain in a multichain laser, with the work of scientists, governments, and industries in many countries focusing their efforts toward the single goal of fusion energy.

It is the responsibility of all of us at this conference, and of our legislative and scientific colleagues throughout the world, to join in the international strategy I mentioned—a worldwide partnership to make fusion and other new sources of energy a reality. If we do this, and if we really put the ultimate goal ahead of personal or national interests, then our efforts will converge in a reaction that we can be proud of as scientists and statesmen.

You and your colleagues have a leadership capability which is probably greater than that of any other group in our society. In many ways our world suffers from a lack of leadership toward the higher goal of life, toward an improvement in the quality of life for all people. Everything we do can be part of a larger effort to benefit mankind. We can do nothing greater, and mankind deserves nothing less. We have it within our power to light the fire of a whole new range of energy sources. And as President John Kennedy so eloquently said: "The glow from that fire, can truly light the world."

❋ *Chapter 2*

World Energy Outlook and Options

Bent Elbek
Niels Bohr Institute

ENERGY FUTURES: NEAR, TRANSITIONAL, AND DISTANT

One of the most important aspects of the energy problem is that of *time*. In this Chapter I shall distinguish between three different parts of the energy future: the *near future*, extending, say, until twenty years from now; the *transitional future,* being the period between twenty and forty years from now; and the *distant future*, being more than forty years removed. In each of these periods we can select a typical year–say, 1990, 2010, and 2030. My thesis–that these periods are distinct from an energy point of view–presents widely different problems and, although more difficult to see, also widely different elements of resolution.

In the near future in a technical sense, there are few energy problems. The world can continue to thrive on oil. However, the supply margin is becoming narrower, and political disturbances might well generate shortages and price increases. On the whole, relatively constant oil prices are likely in the near future and worldwide demand growth could be moderate, perhaps 3 percent per annum for primary energy, even with some recovery of the world economy. On the supply side coal and nuclear would be stronger, but not much stronger because of the inability of most governments to make the necessary decisions. So by 1990 oil would still be by far the most important energy source (see Table 2-1), in spite of all statements as to the necessity of saving oil and using something else.

The state of indecision in the energy field and the general sluggishness of the world economy could make us enter the transitional future, the twenty to forty year period, largely unprepared for the changes expected on the energy scene.

Table 2-1. World Primary Energy Supply Structure (low growth case).

	1977 (percent)	*1990 (percent)*	*2010 (percent)*	*2030 (percent)*	*2050 (percent)*
Oil	44	39	26	12	5
Gas	18	18	12	8	7
Coal	30	26	23	22	22
Nuclear	2	10	31[a]	48[a]	55[a]
Hydro	6	6	5	5	5
Solar	0	1	3	5	6
	100	100	100	100	100
Total EJ/y	280	400	760	810	900

[a] Includes possible contribution by fusion.

Source: *Energy: Global Prospects 1985-2000,* Report of the Workshop on Alternative Energy Strategies (New York: McGraw-Hill Book Company, 1977); *Energy Supply to the year 2000,* Second Technical Report of WAES (Cambridge, Mass.: MIT Press, 1977). *Energy Supply-Demand Integrations to the year 2000,* Third Technical Report of WAES (Cambridge, Mass.: MIT Press, 1977); A Future Energy Scenario, Wolf Häfele and Wolfgang Sassin, Tenth World Energy Conference, Istanbul 1977.

Even with a modest economic development, oil demand could reach technical and political production ceilings and hit them fairly hard because of the lack of anticipatory actions in the preceding period. The result could be nations rushing into action trying to catch up, a period when nuclear energy is rehabilitated and coal and gas projects executed. However, the impact of such delayed actions could be insufficient to avert an energy crisis. Energy prices are likely to increase significantly, in real terms perhaps to levels between 150 and 200 percent of those of today. Although the energy bill might still amount to only a few percent of the GNP, it would constitute a much larger fraction of foreign trade and cause severe balance of payments problems for energy-importing regions, notably Japan and Europe, but also North America and the developing world.

There are many uncertainties connected to forecasts for this period. However, barring major catastrophies and war, I find it highly improbable that the global energy problem would turn out to be a specter. There are some very real foundations for an extended energy crisis toward the end of the century. At the same time, there is hope that the apparent need for new patterns of consumption and for new sources of supply will initiate the necessary changes in the world's energy system. Some of these changes, such as combustion of coal and thermal nuclear reactors, might be of a preliminary nature, but still pointing forward to more viable solutions.

In the distant future, beyond forty years, we should be able to see the outline of a more permanent world energy balance. In the industrial countries the demand growth could be quite modest, in principle zero. For the developing countries, however, all reason points to continued growth, making these countries all-decisive for world energy demand. Some of the measures that could be of help in the transitional period, would in the long run be grossly inadequate. Energy efficiency improvements, the transition to coal and nuclear energy, and supplementing solar sources, will no longer suffice. Very drastic changes are necessary in the patterns of energy use, and major new sources of supply must be found. It is of course the main purpose of this forum to assess what these patterns and sources might be.

There is a very clear difference between the problems of the periods I have called the near, the transitional, and the distant future. Nevertheless, this distant future is only forty years away. Add another twenty years and the world must generate about 50 percent of its energy, an amount corresponding to about twice that of all energy consumed today, from sources that presently are largely unknown. This is the challenge that perhaps not we, but most certainly our children, will be facing.

In the following sections I shall present some–in part highly personal–views on important aspects of the three periods of the future enumerated above. I shall consider whether one can hope for a reasonably smooth transition from the type of energy system we have today, through the kind of system we are reasonably sure that we can have if we want, to a distant system of which we can scarcely see an outline. In this discussion I shall consider first some trends in the demand for energy that in the long run could be important and then, more briefly, the types of supply that might meet the demand.

ENERGY AND ECONOMIC DEVELOPMENT

The relationship between energy and economy has been much discussed and has been analyzed from different points of view. A simple plot or a tabulation (Table 2-2) clearly shows a correlation between GNP and energy consumption.

Table 2-2. Energy-GNP Relation for Major Regions (1975).

Region	*Energy* *EJ/y*	*GNP* *10^9 \$/y*	*GNP/capita* *\$*	*Energy/GNP* *MJ/\$*
North America	80.5	1678	7100	48
Latin America	12.5	327	1025	38
Japan	14.2	496	4450	29
West Europe	50.3	1695	4645	30
USSR, East Europe	59.6	925	2550	64
China	16.3	315	380	52
Asia (excluding China)	10.9	210	210	52
Africa	5.0	163	414	31
Middle East	4.1	153	1990	27
Oceania	3.1	94	4480	33
World	256.6	6056	1550	42

Source: *Energy: Global Prospects 1985-2000,* Report of the Workshop on Alternative Energy Strategies (New York: McGraw-Hill Book Company, 1977).

But as is often pointed out, the ratio of energy use to GNP is not a constant, either in time or from one world region to another. Even when averages for major regions are considered, the 1975 ratio E/GNP (the energy intensity) varies by almost a factor of two, being lowest in Western Europe and Japan and highest in North America and the USSR. There seems to be no correlation between per capita income and energy intensity.

The fluctuations in energy intensity probably reflect for the most part the local availability of energy. Where energy in the past has been cheap and abundant, the energy intensity is high. If energy has been scarce, the energy intensity is low. In view of the impending energy shortage on a global scale, it can be expected that future energy intensities will scatter less and be more closely related to the stage of economic development.

In order to discover possible long-term energy intensity trends of a society, it is of interest to study various indicators other than energy. Figure 2-1 illustrates trends in the composition of the GNP for a number of countries. The main

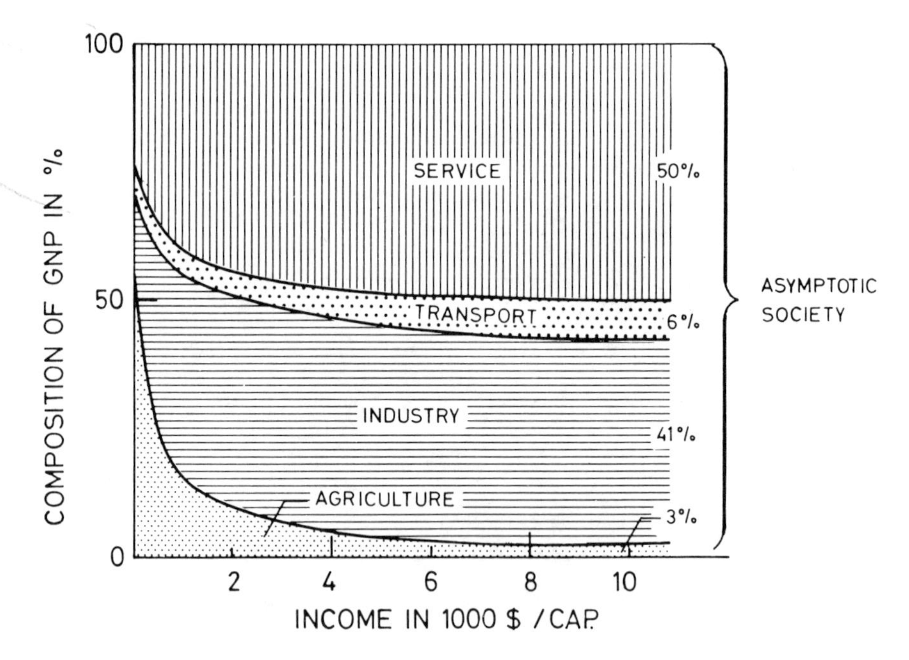

Source: Based on data for twelve countries from *United Nations Statistical Yearbook* (New York: 1977).

Figure 2-1. Percentage Composition of GNP as a Function of per Capita Income. It should be noted that the USA shows a somewhat larger service sector (58 percent) than shown here as asymptotic value. Industry includes construction.

trend is a sharp decline in the contribution from agriculture and increases for industry, transport, and services as the GNP per capita grows. For the high income countries, the relative contributions are fairly constant and point to an asymptotic society where services account for approximately 50 percent of the GNP.

Table 2-3 shows, corresponding to Figure 2-2, typical GNP compositions for high, medium, and low income regions. The changing composition alone implies a declining energy intensity as an economy grows. However, more importantly, the energy intensity of each sector also seems to be strongly dependent on the stage of development. This dependence is not easy to quantify by reference to the present situation alone because, as already mentioned, some regions have, for historical reasons, excessive energy intensities. Nevertheless, the trends illustrated for the four economic sectors in Table 2-4 are reasonable averages of the situation today and reproduce, when combined with the GNP data in Table 2-3, today's world energy demand.

Table 2-3 also gives an estimate of the energy intensity of the same economic sectors in an "asymptotic" society. The basis for the asymptotic values are

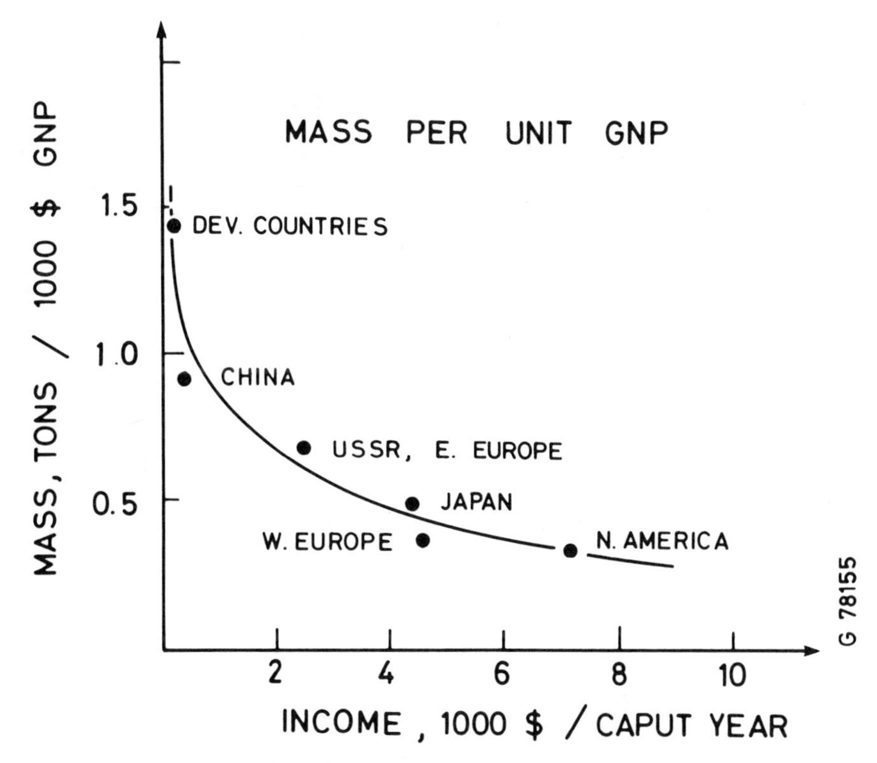

Figure 2-2. Mass Intensity (mass per unit GNP) of Some Selected Economies. For definition of mass see text and Table 2-5.

Table 2-3. Typical Composition of GNP.

Country	*Per capita 1975$/year*	*Population 10^6*	*Agriculture (percent)*	*Industry (percent)*	*Transport (percent)*	*Service (percent)*	*Total (percent)*
Low Income	300	2371	40	30	5	25	100
Middle income	1700	650	12	40	5	43	100
High income	4800	884	3	41	6	50	100

Sources: *Energy: Global Prospects 1985–2000*, Report on the Workshop on Alternative Energy Strategies (New York: McGraw-Hill, 1977); *Energy Supply to the year 2000*, Second Technical Report of WAES (Cambridge, Mass.: MIT Press, 1977).

Table 2-4. Typical Primary Energy Intensity of Sectors (1975).

Country	*Agriculture MJ/$*	*Industry MJ/$*	*Transport MJ/$*	*Service MJ/$*	*Households (percent)*	*Total MJ/$*
Low income	5	100	300	15	9	55
Middle income	10	150	150	10	10	66
High income	20	40	100	7	25	33
Asymptotic	16	26	60	6	25	22

Sources: Average estimates based on *Energy Supply-Demand Integrations to the year 2000,* Third Technical Report of WAES (Cambridge, Mass.: MIT Press, 1977); BP Statistical Review of the Oil Industry (London: British Petroleum Company Ltd., 1976). See also *World Bank Atlas* (Washington, D.C.: The World Bank, 1977), table 2.

present values in the most energy-efficient regions (Europe and Japan) reduced by 20 percent. This takes into account further efficiency improvements and further maturing of the economy in the direction of quality products and services rather than production of bulk commodities. The 20 percent efficiency improvement is relatively modest. However, as it applies to primary energy, it must also take into account increasing losses connected to, for example, a larger share of electricity and synthetic fuels.

One can use the asymptotic society values in Table 2-3 and 2-4 to estimate the possible long-term energy demand for the world. As an ultimate case one can consider a world with 10^{10} people each having an income of \$6,000 (1975). The resulting energy demand turns out to be 1300 EJ per year (1 EJ = 10^{18} J) compared to today's world demand (cf. Table 2-2) of 260 EJ per year. Per capita consumption would be 130 GJ/y, corresponding to 4.1 kW.

Now, this very aggregated example just goes to show that even a high level of income (West German standard) for an ultimate world population of 10^{10} does not necessarily lead to an excessive energy demand. However, to what extent an even distribution of wealth could be accomplished in the long run is another matter, and it might take considerable time to make the energy intensities of all regions converge.

MATERIAL CONTENT OF GNP

One can extend these considerations concerning the economic structure and energy demand of present and future societies and look into not only the "energy" intensity but also the "matter" intensity of the GNP. How much steel, aluminum, paper, cement, chemicals, and food are required for each \$1,000 of GNP? Now this is perhaps a physicist's view of GNP, and I hope economists will forgive me this transgression. However, in these times where we worry about resources of energy and materials, I think it is of some interest to find out what is the mass of the GNP.

Table 2-5. Material Content of GNP in kg/$1000 (1975).

	North America	*West Europe*	*Japan*	*USSR*	*China*	*Asia*
Steel	71	88	205	217	92	88
Aluminum	2.6	1.9	2.1	2.3	0.5	1.0
Paper	33	20	27	13	19	10
Cement	43	109	132	187	95	199
Chemicals[a]	60	52	82	49	26	33
Grain[b]	129	78	35	206	667	1145
"Total mass"[c]	339	349	483	674	900	1430
Energy[d]	1147	717	693	1529	1242	1242

Source: *Energy: Global Prospects 1985-2000,* Report of the Workshop on Alternative Energy Strategies (New York: McGraw-Hill Book Company, 1977).

[a]Oil equivalent of feedstock.

[b]Rice + maize + wheat + barley.

[c]Sum of the six entries above, with the exception of aluminum. These commodities are those produced in greatest quantity.

[d]Oil equivalent of energy consumption in kg/$1000 of GNP (cf. Table 2-2).

Table 2-5 illustrates this for some selected regions at different stages of development. The materials listed are, with the exception of aluminum, those handled in greatest quantity in all societies. They are also the most energy demanding. The materials in Table 2-5 can be estimated to require about 50 percent of the gross energy demand for their production and handling.

At first sight it is perhaps surprising that the differences among regions are not larger. Apart from grain, the amounts of material entering each unit of the GNP in most cases vary by less than a factor of four—this in spite of GNP per capita variations of up to a factor of thirty-five. This illustrates a considerable stability in GNP composition (see also Figure 2-1) irrespective of the size of the GNP. This stability is also reflected in the near constancy of the E/GNP ratios given in Table 2-2.

The regions considered in Table 2-5 are large and to a considerable extent self-sufficient in most of the materials considered. The numbers do, however, refer to production and are not corrected for import and export. This should therefore be interpreted with some caution as far as level of consumption is concerned.

The finer details of the material composition of the GNP could be of some interest for long-term energy projections. In general the more mature economies, such as those of North America and Europe, require aluminum, chemicals, and paper (!). The upsurging economies of Japan and the USSR are extremely steel intensive, whereas the developing countries' mass budget is dominated by grain. For the energy transition period it could be of significance that the developing countries might then pass through a period of heavy industrial development as a necessary step toward a more diversified economy.

If we add up all the materials listed in Table 2-5, we get a primitive measure of the "matter intensity" of an economy, the "mass per unit GNP." This mass is much larger in the developing economies than in the industrial countries. A dollar's worth of GNP weighs about 1.5 kg in India but only 300 g in the United States. This difference is to some extent due to the inclusion of food (grain) in the mass. However, even without food some of this trend is conserved, but with the "steel and concrete" economies of the USSR and Japan put in prominence. With food included, as in Figure 2-2, the mass intensity declines rapidly as an economy develops.

One should of course keep in mind that the picture is reversed if we look at the mass produced by each individual. This is illustrated in Table 2-6 and Figure 2-3, showing the total mass per capita of the six commodities studied here. Clearly the mass per individual is much larger in the developed than in the developing economies. Each North American producces about 2.4 t of materials per year, whereas the average for a South Asian is 0.3 t. Thus wealth is still to a considerable extent associated with mass.

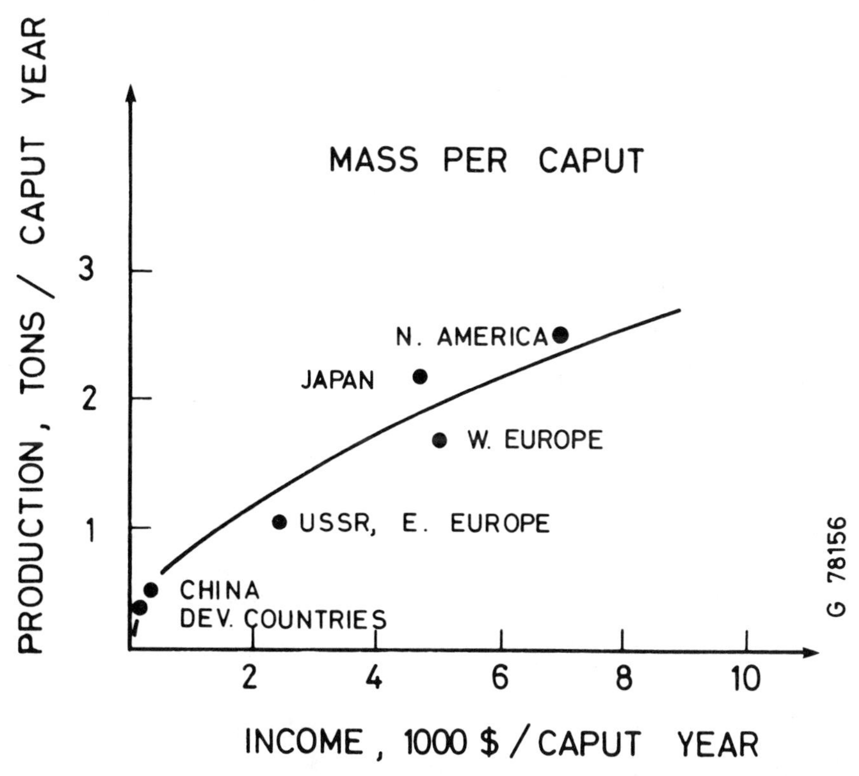

Figure 2-3. Mass per capita Year for Some Selected Economies. See also Table 2-6.

Table 2-6. Material Content of GNP in kg/capita.

	North America	*West Europe*	*Japan*	*USSR*	*China*	*South Asia*
Steel	504	400	912	217	92	9
Aluminum	18	9	9	2	0.5	0.2
Paper	234	98	120	13	19	2
Cement	305	495	587	187	95	41
Chemicals	426	236	365	49	26	7
Grain	915	355	155	525	253	240
"Total mass" (kg/cap.y)	2402	1593	2148	993	485	299
GNP ($/cap.y)	7120	4645	4450	2550	380	210
Energy (kgoe/cap.y)	8144	3211	3035	4024	472	259
Population (10^6)	236	365	112	254	822	1002

See notes for Table 2-5.

THE QUESTION OF FOOD

There has been much concern over the world's ability to feed its growing population in the future. At the same time there has been criticism of the methods employed in modern agriculture, which are also considered wasteful in terms of energy.

As one can deduce from Tables 2-3 and 2-4, food production accounts for only a few percent of the world's energy consumption. This, however, refers only to energy directly used in agriculture for the running of tractors, harvesters, irrigation, and so on. But in addition, agriculture indirectly requires substantial quantities of energy feedstocks for fertilizers and other materials. Also, one must remember that in developed economies, agriculture is only a first part of an extensive food system. It can be estimated that a modern food system requires about 12 percent of the total energy.

Most productions also show an economy of scale when judged on an energy basis. This is not true for agriculture where high yields require a disproportionately high energy input. A primitive system can easily deliver a nutritional energy ten times higher than the input of energy for work and fertilizer. If we look at an advanced, high-yielding system, the return is much less. Perhaps only about 50 percent of the input energy is recovered as food, as the combined result of energy-intensive cropping and large conversion to animal foodstuffs.

I think that such calculations are extremely misleading and, if acted upon, that they might lead to a food disaster that the world otherwise could easily avoid by the use of energy. Figures 2-4 and 2-5 illustrate the grain yield as it depends on two important energy inputs to agriculture, fertilizer and tractors, based on the actual performance of agriculture in different parts of the world. Without exception, high yields are associated with large energy inputs, but with the law of diminishing returns quite evident. The yields are approximately proportional to the square root of the energy input. A doubling of the yield would therefore require four times as much energy for agriculture.

One could ask how much energy would be needed to feed an ultimate world population of 10^{10}. If we take as a basis a minimum daily diet of 11 MJ, of which 20 percent is of animal origin, we can estimate the energy requirements if high yield agriculture is used throughout. For plant production we can assume an energy return (energy ratio) of a factor of four. For animal matter the return is assumed to be ten times less. Thus the total yearly gross energy requirements for food delivered at the farm gate to 10^{10} people would be:

$$\text{Vegetable:} \quad 10^{10} \times 8.8 \times 365 \times \frac{1}{4} \text{ MJ/y} = 8 \text{ EJ/y}$$

$$\text{Animal:} \quad 10^{10} \times 2.2 \times 365 \times \frac{1}{0.4} \text{ MJ/y} = \underline{20 \text{ EJ/y}}$$

$$\text{Total} \quad 28 \text{ EJ/y}$$

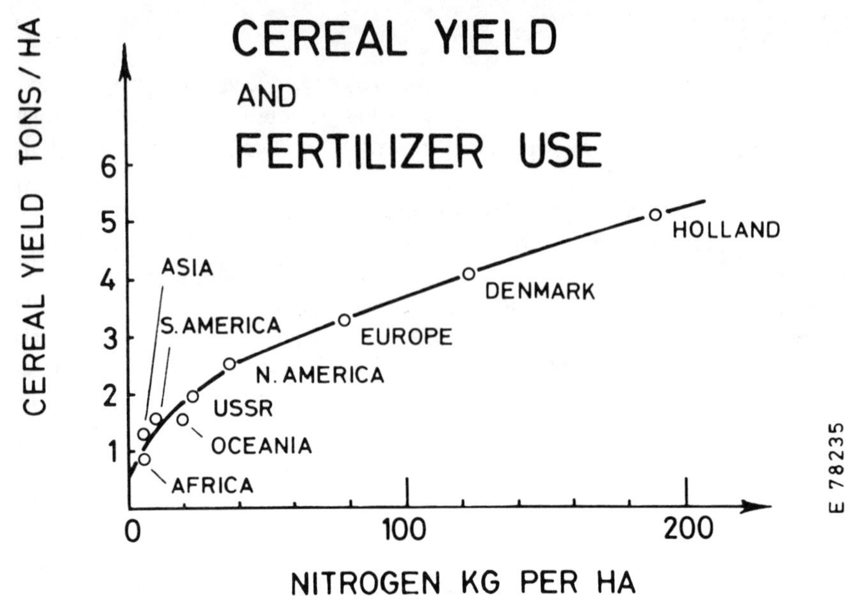

Source: *FAO Production Yearbook* (New York: 1974).

Figure 2-4. Yield of Cereals in Different Regions as a Function of Nitrogen Fertilizer Use.

This is certainly a considerable rate of energy (about 11 percent of present world energy use). However, if compared to a projected asymptotic energy demand of 1,300 EJ/y, the energy required for future food production does not seem excessive. Although agricultural energy for a long time would have to be largely oil and gas, this simple estimate underlines that energy is not a severely limiting factor for providing an adequate diet for a growing population.

ENERGY DEMAND PROJECTIONS

At last year's Energy Forum, I presented an energy demand projection (Figure 2-6) until the year 2000. I need not repeat the assumptions behind these projections but just mention that they give, for moderate world economic growth and large energy savings, a year 2000 energy demand of 630 EJ/y—or two and a half times that of today. Now, one year later, I see no compelling reason to revise this projection. Last year's world energy growth rate was 3.6 percent, in good agreement with the projection. Also as projected, the growth in the developing countries and the socialist countries has been considerably higher than that in North America, Europe, and Japan.

The persistently sluggish economic development in many countries might indicate that the near term demands would be lower than projected. However,

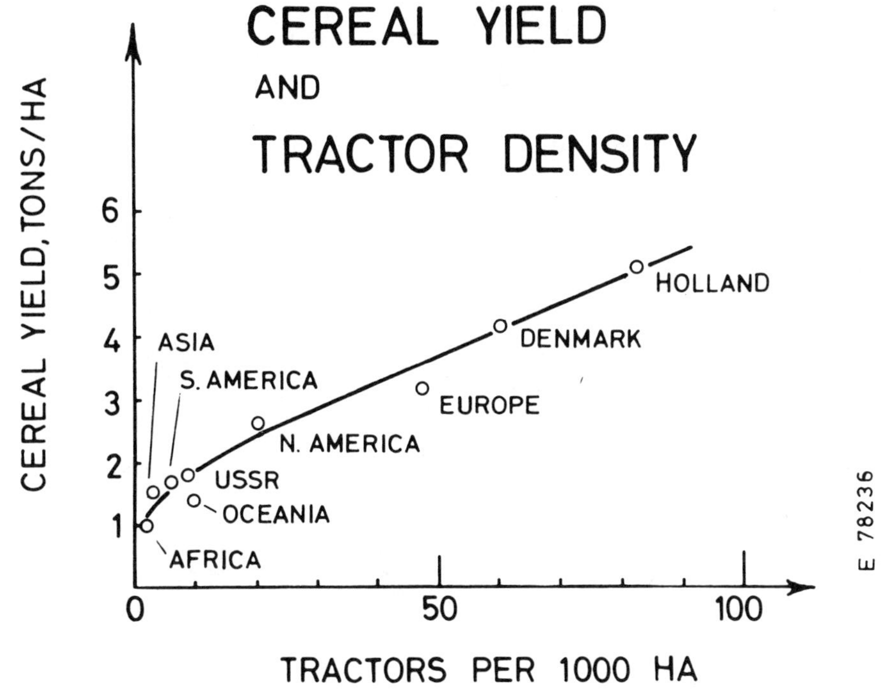

Source: *FAO Production Yearbook* (New York: 1974).

Figure 2-5. Yield of Cereals in Different Regions as a Function of Tractor Density.

if this turns out to be the case, it is matched by a slower introduction of new supply. The energy gap implied by the projection toward the end of the century would therefore still develop. As already said, this could cause us to enter the energy transition phase in a state of unpreparedness.

Energy projection to the year 2000 can be made without introducing too many ad hoc assumptions, and the results of most world projections today do not differ markedly. Even if a "soft energy path" to the future of the developed world were chosen, the demand growth in the rest of the world would dominate. Therefore, the problems would at most be postponed a decade. If, however, we try to prolong even moderate projections through what I have called the transition period, one easily gets exponential headaches.

Most energy projections are coupled to the growth of the GNP. Recent projections have all payed considerable attention to the coupling between GNP and energy and have examined how the coupling coefficients might change with time and affluence. However, few have postulated a complete decoupling of the two descriptors. In the distant future the world must approach a state of energy equilibrium. There are several arguments based on resources, climate, and social

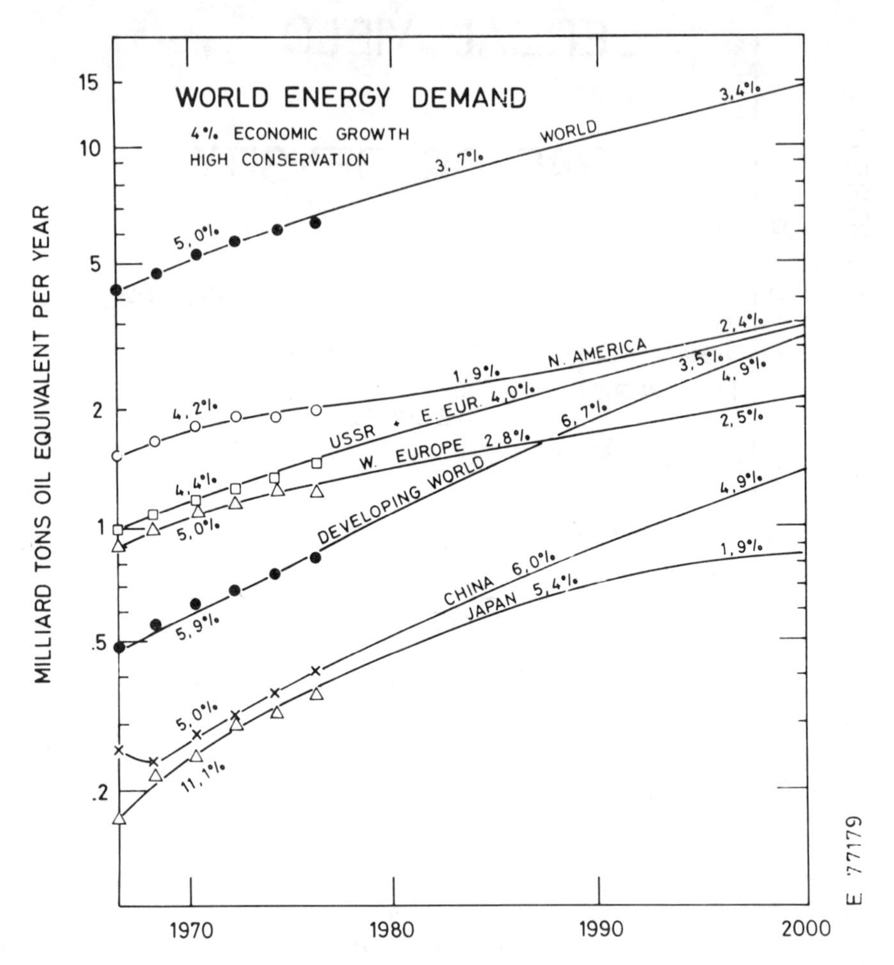

Source: Bent Elbeck, "Worldwide Energy Demand," in O.K. Kadiroglu, A. Perlmutter, and L. Scott, *Nuclear Energy and Alternatives* (Cambridge, Mass.: Ballinger, 1978), ch. 3.

Figure 2-6. Energy Demand to Year 2000 (high growth case).

needs, making it plausible that energy equilibrium could or must be reached sometime in the next century. If this is so, the transition period I have been speaking about would also be the time where the exponential curves do bend over.

As already discussed, the composition of the GNP is related to the state of economic development and so is the product mix and the energy intensity of the major economic sectors. These facts are important in the transition period for a lowering of the energy intensities of developed economies. However, they do not seem to lead to energy equilibrium as long as the economy continues to grow. Of course one can always beat an exponentially growing economy by an exponentially decreasing energy intensity, but physics and the data we have on past performances do not make this outcome likely.

Projecting energy demand into the distant future is therefore a task for which the tools are mostly lacking. One normative approach, indicated earlier in this chapter, is to say that the world ought to be able to live contentedly, say, on West German material standards. This might be rejected as impossible, in some countries for political reasons. Another approach is to prolong the curves in Figure 2-6, but with declining growth rates (see Figure 2-7). Finally, one can say that world energy demand has been growing by 2.5 percent per annum through the ages and that it will continue to do so. For the time span one would dare to consider, for example until the year 2050, it perhaps does not matter so much what approach one would choose. They all lead, in a reasonably surprise-free world, to a world energy demand in the distant year 2050 (the pensioners of that year are being born now) that is four to five times that of today, but with considerable differences among regions. In some sense, the year 2050 is not far enough away to really force us to make up our minds as to whether energy demand growth can or must be halted. But of course demand growth could be brought to an end because of lack of supplies.

DEMAND LIMITED BY SUPPLY?

Several studies have indicated that in the near period, world energy demand can be satisfied by the traditional sources, including a substantial nuclear contribution, under the assumption of somewhat higher energy prices and no major political restrictions on international energy trade. However, in the transitional period, the situation is less clear. Around the year 2000, oil supplies in all probability will be declining and major new energy sources must be introduced. In view of the present uncertainty about which roads to follow, it is highly probable that new sources of sufficient magnitude will not be available.

In order to show the size of the problem, Figure 2-8 illustrates a case with low energy growth in the period until 2050 and a suggestion as to how the energy might be supplied. Declining oil production could partially be replaced by coal and nonconventional oil, but in total the fossil fuel supply is likely to fall. Although fossil fuels remain important, the gap between demand and supply is therefore rapidly increasing. Nuclear energy from thermal reactors could at the same time be severely limited by uranium supply, which would necessitate a rapid introduction of some type of breeder. Fusion might, some time early in the next century, make a contribution, but the timing and magnitude is speculative. Most estimates of solar energy in its various forms give a very limited contribution at the turn of the century. In order to have a sizable impact in the next century, solar energy must grow at an unprecedented rate for a long period. On the solar energy prospects I would like to point to the significantly different situation in regions with low population density, such as the

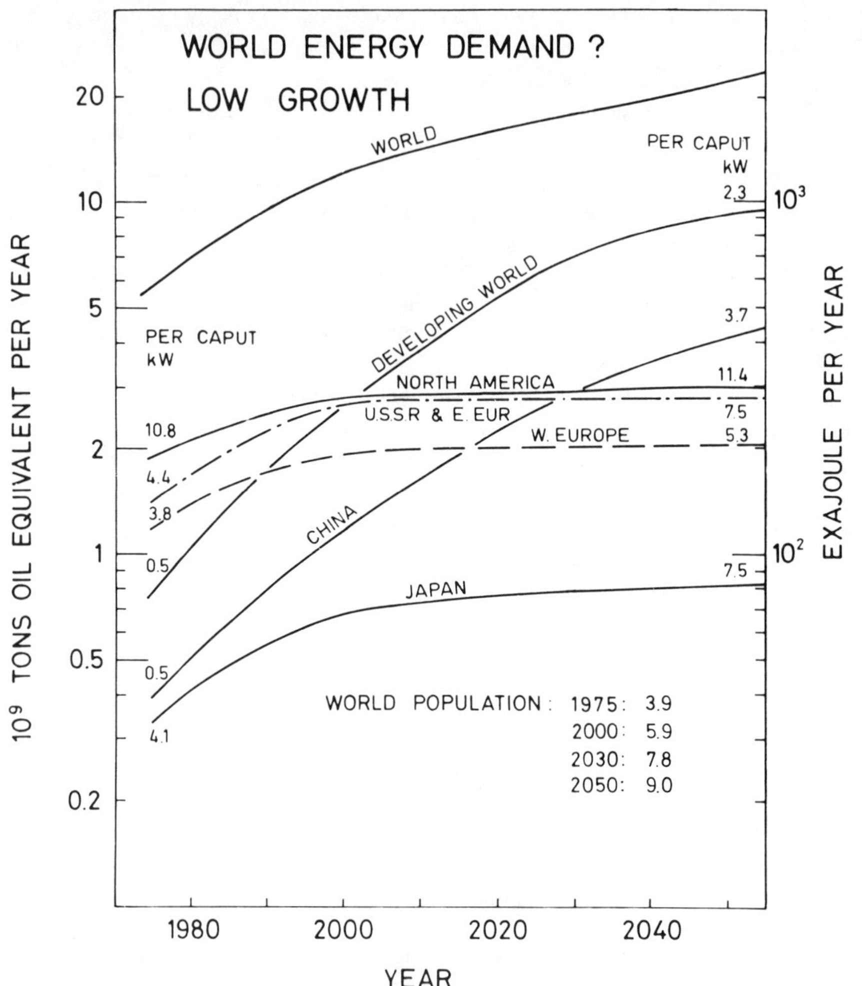

Figure 2-7. Energy Demand until Year 2050. Low and declining growth in accordance with considerations concerning energy intensities are given in text.

North American continent, and regions with high population density such as Europe, Japan, and many developing countries. In Europe an asymptotic energy demand of about 7 kW per capita seems conservative. With a population density of 250 km^{-2} as typical in central Europe, a 50 percent solar contribution, a 100 Wm^{-2} average insolation, and an overall 10 percent collector efficiency, we find a collector surface of about 9 percent of the land area. Solar energy projections should not be uncritically transferred from one area to another.

It is my personal conviction that energy demand will be severely supply limited in the next century. The scale of the technical solutions, even if we

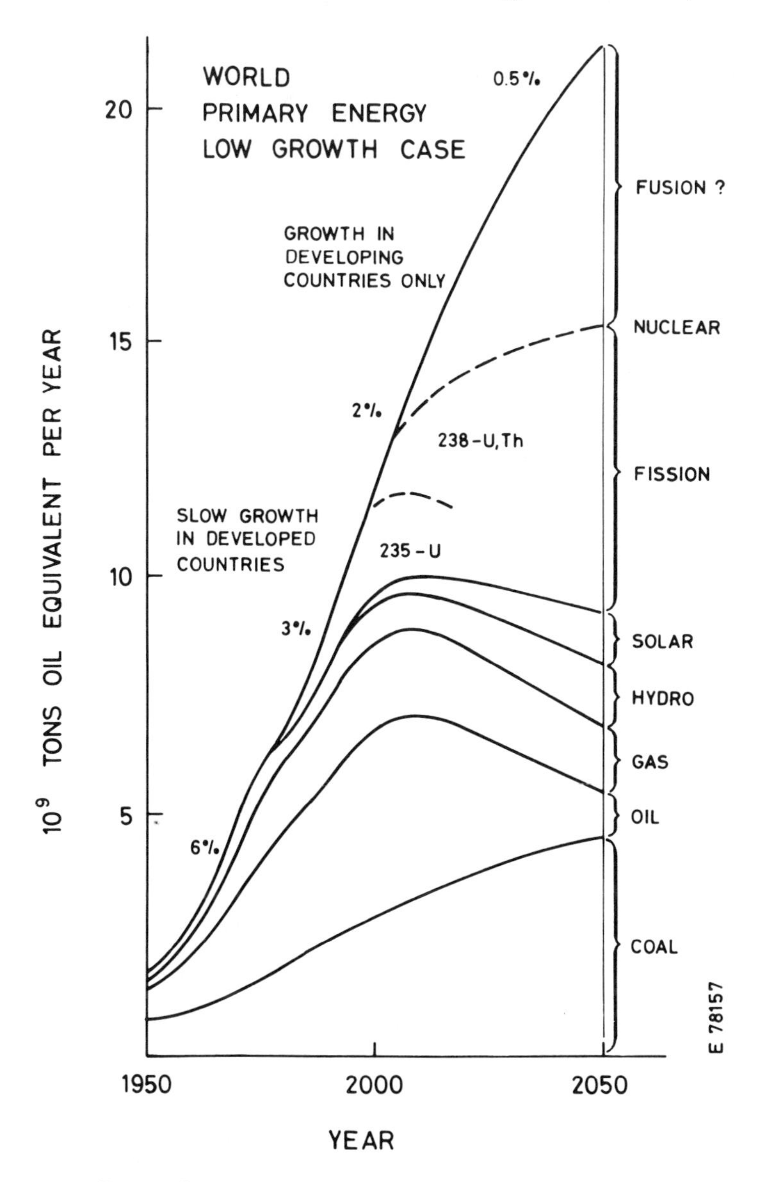

Figure 2-8. Energy Supply-Demand Matching for Low Growth Case until Year 2050. Until year 2000 in general agreement with *Energy: Global Prospects 1985-2000,* Report of WAES (New York: McGraw-Hill, 1977). After year 2000, the decline of oil and gas are based on total resources of 300×10^9 t of oil and 250×10^{12} m^3 of gas. Coal expansion by trend extrapolation. Solar is, because of land requirements, limited to regions with low population density and high insolation. Nuclear seen as balancing energy supply, but fusion contribution is purely speculative.

knew what they were, will be enormous, and the time available for development and deployment is limited. The transitional phase I have spoken about might therefore not be a transition to a period of eternal bliss, but a transition to a troubled world where aspirations are curtailed by insufficient energy supplies. The best we can hope for today is that we can buy a little time by making prudent use of the supply we have available, that the populations of the rich countries will understand that much is at stake for the less favored parts of the world, and that easy solutions simply do not exist.

REFERENCES

A Future Energy Scenario, Wolf Häfele and Wolfgang Sassin, Tenth World Energy Conference, Istanbul 1977.

World Energy Problems, Bent Elbek, Third International Summer College on Physics and Contemporary Needs, Nathiagali, Pakistan, 1978 (to be published by Plenum Press).

Chapter 3

A Global Energy Balance to Year 2000

Jay B. Kopelman
Electric Power Research Institute, Palo Alto

INTRODUCTION

Much of the material included in this chapter is extracted from a recently completed study of world energy supply and demand with which I was associated at SRI, International. The objective of that study was to try to assess what factors would most strongly affect the energy industry from the present to the year 2000 and beyond, and it involved, of course, some forecasts of what we think are likely developments.

Futurizing is an old business. It has been called the world's second oldest profession. Until quite recently in human history, the people who made these predictions were called seers, soothsayers, or oracles, and the techniques they used were crystal balls, star gazing, or examining the entrails of goats. More recently they are called economists, long-range planners, futurists, systems analysts, or weather forecasters, and they examine the entrails of computers. This is a grouping—I said it was the world's second oldest profession—whose collective reputation may not be much better than the reputation of the world's oldest profession. But it is improving—just as weather forecasting is getting better through the use of somewhat more sophisticated analytical techniques. So even though I have been bad-mouthing the forecasting business, that will not deter me from sharing with you what I believe to be some of the more robust conclusions of our work on global issues in energy supply and demand.

What are the issues in energy supply and demand today? The energy issue is a comprehensive and often confusing topic. It has precipitated discussions not only on energy, but on themes as ambitious as the future of humanity in general. I believe that is why it has attracted some of the world's foremost scientists such as many of those who are gathered at this forum.

Although it is widely recognized that our supplies of hydrocarbons are finite and that they will become increasingly scarce over the long term, it is important in making intelligent decisions for this transition to try to assess accurately how much time and urgency there is. It is no good crying "wolf!" too soon, just as it most assuredly is disastrous to wait too long. Since the 1973–1974 shortages during the Arab oil embargo, many cries of "wolf" have been raised. An important question in terms of credibility is, How desperate is the situation?

The problems with credibility are exacerbated, of course, by press reports that condense 500 page reports into one column summaries with flashy headlines using words like "energy crisis" on the one hand or "oil glut" on the other. However, there remain very real differences of opinion among people doing research about the gravity of the situation and the measures to be taken.

It is relatively well known that in the area of energy resource management, more uncertainties have developed in the last five years than in the previous twenty. In Figure 3–1, the posted price of world crude oil is shown from 1960 (preembargo price) through 1978. After a long period of relative stability, this price rose from $2.59 per barrel to $11.65 per barrel in one year. A strong cartel (OPEC) developed that now asserts control over a large fraction of petroleum traded in international markets. This development has substantially affected the price outlook for other fuels as well. International energy markets and other segments of world economies have been changing rapidly in the past five years in a still emerging response to the formation of the cartel.

Nevertheless, five years have been enough time for a body of conventional wisdom to be developed in this field. A common perception of the near future is that there will be increasing disparity between growing demand and dwindling non-OPEC supplies and ever-rising energy prices. In this view, the United States and other major industrial nations are confronted with the likelihood of a growing need for oil imports and greater dependence on foreign supplies. The resulting political and economic problems impel government planners to consider sometimes desperate responses.

Government policies to decrease long-term energy dependence can be encompassed by programs to either decrease consumption, increase indigenous production of conventional supplies, subsidize alternative technologies, or penalize imports. These programs include various combinations of end use taxes, import tariffs or quotas, price controls, research and development support, tax credits and subsidies, and the like. All of these programs represent some costs to taxpayers and consumers; in many cases these costs may become quite large.

In making choices of this kind, it is necessary, of course, to consider the most pessimistic scenarios and to try to plan adequately for them. However, prudent planning also requires an examination of the possibilities and impacts of other developments. In this discussion I would like to present the case for a world energy scenario somewhat different from conventional wisdom. Although it is

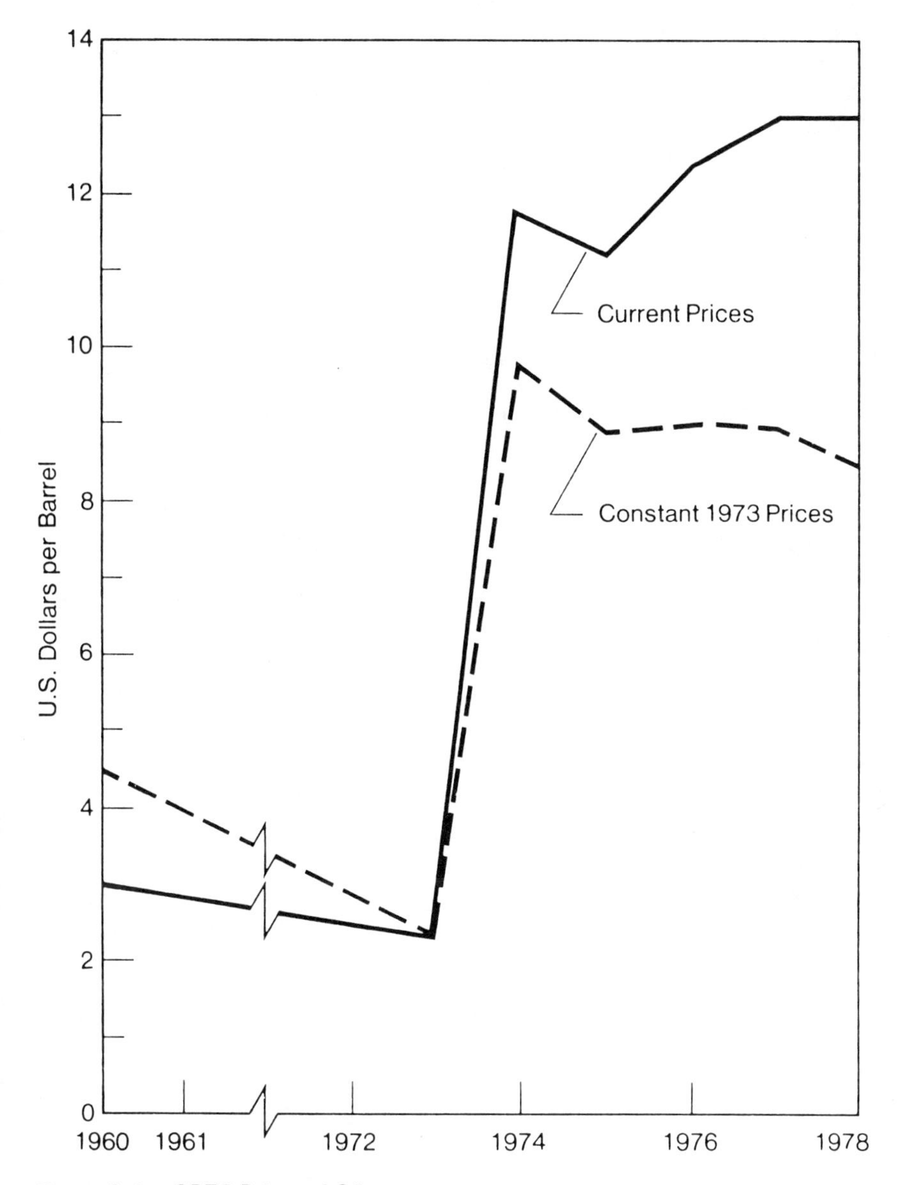

Figure 3-1. OPEC Prices of Oil.

not possible to give a discussion of our full analysis here, I would refer you to "Energy Modeling as a Tool for Planning," in the February 1978 issue of *Long Range Planning,* for a more complete discussion of methodology, as well as for references to other documentation of the model.

AN UNCONVENTIONAL VIEW OF ENERGY SUPPLIES

The principal conclusion of our analysis of world energy supply and demand is that there are adequate opportunities for increasing conventional supplies of hydrocarbon resources on a worldwide basis, diversifying the sources of supply, and substituting among fuels to allow an orderly development of alternatives through the remainder of this century and for some time into the next without enormous price increases. This conclusion is based on the following series of observations and estimates:

- The conventional techniques used to make estimates of resource availability in terms of a single number representing reserves are not considered adequate information to determine future supply-demand balances. Resource availability should be discussed in terms of recovery costs and market prices.
- There is no long-term condition of imbalance in supply and demand where "gaps" occur. In the absence of artificial price controlling regulations, prices respond well in advance of this impending situation to prevent just such a possibility.
- International energy supply and demand price elasticities have been broadly underestimated.
- Current estimates of proven oil reserves do not accurately reflect the long-range impact of higher petroleum prices. Higher oil prices will encourage exploration and the application of advanced recovery techniques so that future additions to proven reserves will be greater than they would have been at lower oil prices.
- The opportunities for fuel substitution are greater than has generally been recognized.
- There will be an expansion of the class of major oil exporting nations to include some new members in the next few years (Mexico, the United Kingdom, Norway, China, etc.). The disparity of national interests will cause highly varied responses to changing supply-demand situations in the world markets. Those who join OPEC will increase the diversity of cartel membership and the complexity of production allocation and pricing decisions to be settled.

As is well known, this view of world energy availability has not been the consensus, and as mentioned earlier, many other analysts anticipate serious supply-demand imbalances in the 1980s. I believe the differences arise because it has been customary for energy analysts to hypothesize a "scenario" price or set of prices for some reference energy source, usually Middle East oil, and then to try to determine how the demand for oil will grow and what supplies will be produced at this price. In this approach, demand is growing exponentially, and supplies are relatively static. It is not surprising then that at some point demand exceeds the supply, creating an "energy gap."

The procedure outlined above simplifies the analysis by ignoring the feedback between energy supply, demand, and prices and the effects of prices on interfuel substitution. This approach is misleading because it ignores the dynamics of energy pricing through time.

Energy prices respond not only to the instantaneous supply-demand balance but also to producers' and consumers' perceptions of the future. If a producer perceives that his resource is growing scarce because of depletion of the reserves, he will demand higher prices now for incremental production. How much he can raise his price depends upon a variety of time dependent variables including:

- The rate at which the resource is being depleted now and what rate might be expected in the future;
- That producer's potential loss of market share to other producers of the same resource or to competing substitutes or to declining demand;
- The impact that higher prices would have on the stimulation of new technology to replace future demand for the product;
- The individual producer's preference for present income versus deferred, perhaps higher, income at some future time.

Therefore, at any given time in energy markets, a variety of individual decisions are being made by producers and consumers that rebalance the supply-demand equations.

The approach used in our study incorporates a large, regionalized interfuel competition model that utilizes demand estimates for nineteen different residential-commercial, industrial, and transportation categories in each of eleven different regions. Of each of the nineteen end use categories where interfuel competition plays a role, different fuels compete for market shares on the basis of economic and behavioral characteristics including price ratios, historic market share, estimates of consumer inertia, and secondary materials availability. This was done separately for each of the global regions of the analysis.

Price competition at any point in the energy network is obtained by adding the costs of transportation, distribution, and conversion to the primary resource recovery costs that are embodied in marginal cost resource curves. These curves estimate the cost of producing the marginal barrel of crude oil, or the marginal ton of coal, or the marginal cubic meter of gas as a function of the total cumulative production for each of the world's major oil, gas, and coal resource basins, as well as some estimates for the cost and availability of nuclear fuel supplies. An iterative solution is obtained for generalized equilibrium in which markets clear at a specific price for each fuel, in each region, in each time period.

AN OUTLINE OF THE SRI WORLD ENERGY MODEL

It is not possible in this discussion to go into a full description of the SRI world energy model. I know some are already familiar with the generalized equilibrium approach used in the original SRI-Gulf Oil Corporation National Energy Model for the United States, while others may not be interested at all. The primary difference between the two models is the addition of a cartel submodel that determines crude oil prices for each of the OPEC regions on the basis of an optimization routine whereby the present value of discounted future OPEC revenues is maximized. This is done in a system where future demand for OPEC oil is responsive to price changes. A quick but useful look at the world aggregate energy picture, showing global aggregated resource curves for each fuel, is given in Figure 3-2.

The cost of recovery of a marginal barrel of oil or a cubic meter of gas is shown as a function of the cumulative production of these resources from some date forward. Also shown in this figure is the current price of international petroleum as determined largely by OPEC's assessment of the market situation and some of the current estimates of the average cost of producing clean synthetic fuels. A careful look at the figure gives a very clear picture of much of what is happening in fossil fuel supplies in the world. The curves for gas and oil turn sharply upward after about 2.5 trillion barrels of oil recovery and about 1.5 trillion barrels of oil equivalent gas recovery, so that the marginal costs of retrieving additional energy from these sources would be well above the current market price of oil or gas. (The addition of shale oil, heavy oils, and gas in tight formations would tend to flatten the upper portion of these curves so that the rise in cost would not be quite so steep).

However, the coal resource base is enormous. There is more energy in the proven coal reserves of any one of the world's six largest coal producers (United States, USSR, China, Germany, United Kingdom, and Australia) than in the richest oil-producing nations. The ultimate probable coal resources of any one of these six countries far exceed the entire world's reserves of oil on an energy equivalent basis. Beyond the coal resources, there are also the very substantial possibilities of exploitation of gas and oil from unconventional resources such as tar sands and oil shales (estimated to perhaps quadruple the ultimate potential global recovery of liquids and gases) or the enormous potential in nuclear fission and fusion or solar energy.

A close look at the situation reveals that the principal problem is not the availability of supplies, but the approaching depletion of low cost supplies. Oil could be obtained from shale, but we would have to pay at least 50 percent more for it than the present market price of crude oil, and unforseen environmental problems with its recovery or disposal could raise that cost even higher. Gas could be obtained from coal, but only at about $25 per barrel or twice the present market price for gas.

Coal appears to be plentiful and low cost, but what is not shown here are the problems associated with its recovery, transporation, and consumption that add enormously to the cost of its use:

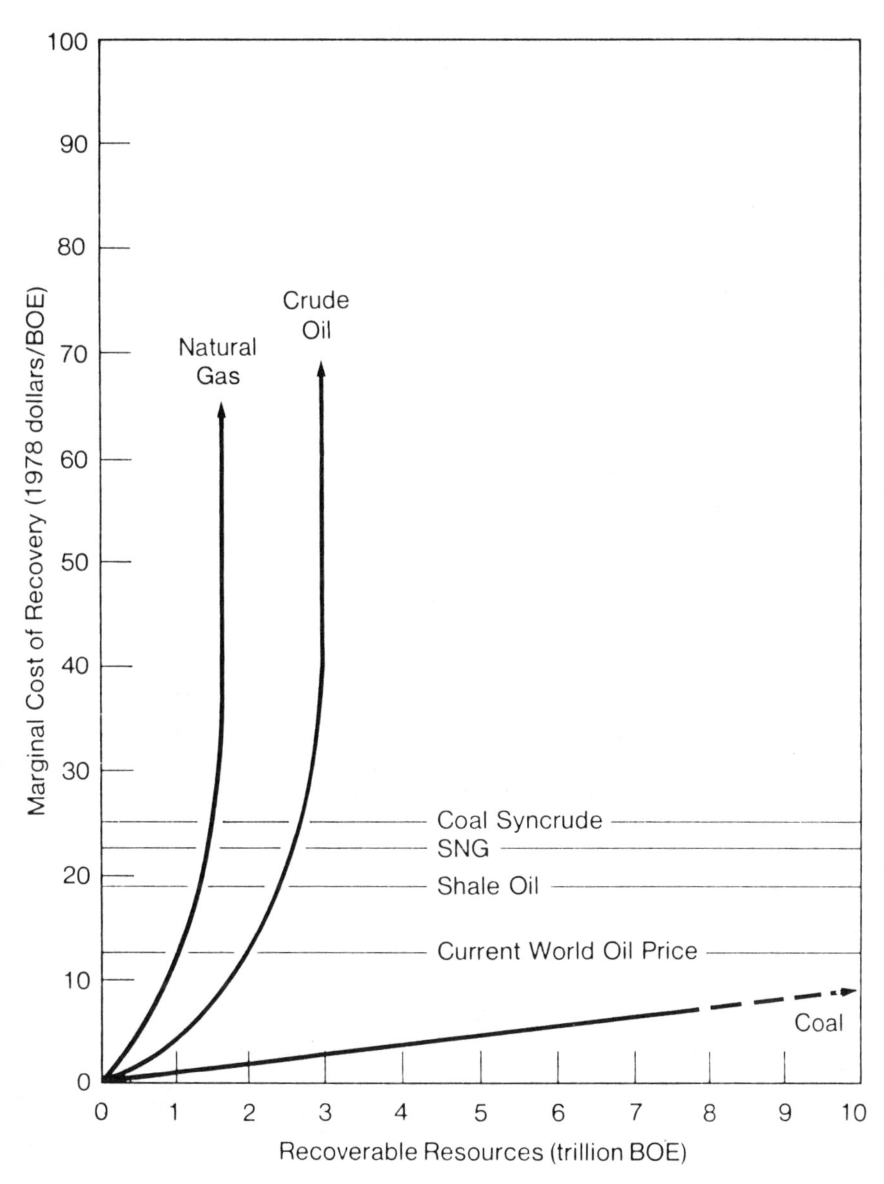

Figure 3-2. World Hydrocarbon Resources in 1975.

- Coal cannot be used effectively in automobiles, trucks, or airplanes.
- If we try to burn it in homes or businesses, it's cumbersome, dirty, and adds substantially to the costs of pollution control.

- Even burning it in industrial or utility furnaces creates problems of particulate emissions, sulfur oxides, nitrogen oxides, and carbon dioxide that we are only now beginning to appreciate. These large installations can afford much more effective emission control systems at lower costs than homes or businesses, but even this by no means eliminates the problem. If the true costs of these problems were reflected in the price of coal shown, the curve would be substantially higher.
- We can get some idea of what these additional costs might be by looking at the current estimates of the cost of converting coal into clean liquids or gases—the figure shows that the minimum cost for synthetic natural gas or syncrude from coal is well above the current market price for conventional oil or gas. Unless we are willing to pay as much as 50 to 100 percent more for our energy supplies than we are at present, it is just not economically feasible to produce these synthetics at current market prices. No producer is going to invest the enormous funds necessary to convert coal into gases or liquids until he is assured of a substantial market for his product at a reasonable price.

The same is true for oil from shale. Although the resource base is large, development of this resource begins at a minimum cost well above current market prices for conventional fuels. Although they are not shown here at present, the same would be true for almost all of the technologies for use of renewable resources in competition with conventional oil or gas. Through technological developments, the price of synthetics may decline or the market price of conventional supplies rise high enough to warrant the production of synthetics, but they cannot be justified economically in today's market.

A FORECAST OF THE GLOBAL ENERGY BALANCE

Now let us return to the forecast global energy balance. In doing a global energy balance, the results of this analysis (as with all the others) are critically dependent upon some basic assumptions concerning world population and economic growth rates. For the analysis presented here, these are:

- Population growth rates continue to decline in most regions. World average population growth will decline from about 1.8 percent at present to about 1.4 percent by the end of the century. In North America and Eastern and Western Europe and the Soviet Union, these rates will reach about 0.7 percent; in Japan they are even lower; elsewhere they are substantially higher.
- Gross domestic product (GDP) growth rates show substantial regional variations, but the world average declines from about 4.5 to about 3.7 percent by the year 2000. A large part of this decline is due to a slowing of the OECD nations' GDP growth by about 1 percent.

- The efficiency of energy consumption in most markets improves with the introduction of energy-saving conversion equipment. This trend is encouraged by the higher prices of recent years and by government policies mandating improvements in transportation markets.
- OPEC continues to operate to set international oil prices in accordance with its perception of its best interests. However, management of pricing and attempts at production allocation among the members of OPEC will become increasingly difficult.
- Indigenous resource production and pricing is determined in a "free market" system with generally consistent tax structures.
- Environmental goals are implemented through "best available technology" requirements, and the economic costs are absorbed into required revenues for each fuel.
- No political actions restrict the orderly development of economically competitive resources or technology; this implies no wars disruptive of energy trade, embargoes, or nuclear moratoria.

Although we have tested the conclusions and found that they are "robust" in that they are not sensitive to small changes in any one of the exogenous parameters, substantial variations from one or more of the foregoing assumptions would require a renewed look.

Long-range projections of world energy demand have been changing substantially in recent years. Our estimate for total primary energy demand for the year 2000 is more than 25 percent lower than earlier estimates that were made before the embargo. This is similar to Exxon's recent estimate of world energy demand for the non-Communist world in 1985 of 125 million barrels of oil equivalent per day—down 24 percent from estimates made in 1973 of 165 million barrels of oil equivalent per day. These differences are related to changing views of the future growth of regional macroeconomic indicators such as GDP and population, as well as to changes in the relationship of energy consumption to these variables and to price. The quadrupling of world oil prices since 1973 (shown in Figure 3–1) has had an important impact on all of these estimates. Even if demand often appears to be somewhat inelastic to price changes in the short term, long-term effects of the increases are making big changes in year 2000 projections.

In projecting the regional variation of total energy demand growth (as shown in Table 3–1) for the United States, Western Europe, Japan, the centrally planned economies, and the remainder of the world, we find that although OECD nations still account for nearly half of world energy consumption in the year 2000, other areas are experiencing greater growth and close the gap to some degree.

The changes that are foreseen as taking place in energy demand are perhaps best illustrated by examining the United States. Changes in prices, the enactment of a conservation ethic, and changes in the energy marketplace (saturation

Table 3-1. Primary Energy Demand (millions of barrels of oil equivalent per day).

	1960	*1970*	*1980*	*2000*
United States	20.7	31.7	38.9	58.6
Canada	1.8	3.0	4.5	7.3
Mexico/South America	2.2	4.2	7.5	20.9
Western Europe	12.5	22.0	27.3	46.3
Africa	1.0	1.8	3.4	7.6
Middle East	0.7	1.4	3.3	10.5
Japan	1.8	5.6	8.8	17.9
Centrally Planned Economies	18.2	27.1	42.9	87.5
Remainder	2.2	4.6	7.6	19.1
Total	61.1	101.4	144.2	275.7

Source: Stanford Research Institute International.

and increased energy use efficiencies) have reduced our forecasts for energy demand in the United States from estimates made in 1973 by about 37 percent. Total U.S. demand for primary energy, then, is expected to be about 58.6 million barrels of oil equivalent per day by the end of the century. In 1970, a barrel of oil equivalent energy was used to generate about $140 in gross domestic product (in constant 1978 dollars). By the year 2000, it is expected that the same amount of energy will generate about $191 in GDP or a 37 percent increase in economic efficiency in energy use. These efficiency improvements are all the more striking because in all regions electricity consumption is growing faster than total energy, and average electric power efficiencies are lower than for direct consumption of fuels. Such changes are occurring worldwide, although not so dramatically in nations that have historically not had access to the same abundance of relatively low cost energy as the United States.

The results of the interfuel competition analysis are shown in Table 3-2. Some of the principal conclusions from this work are:

- World demand for all forms of primary energy will continue to grow to the end of the century but at considerably lower rates than for the most recent twenty-five-year historical period. The annualized average historical growth for 1950 to 1975 was about 5 percent, whereas for the forecast period, the analysis indicates world average demand growth of 3.4 percent.
- The regional variation in anticipated primary energy consumption is substantial. In general, developing nations will show higher growth rates of energy consumption than developed nations because they are starting from much lower absolute values, have higher population growth rates, have less room for improvements in the efficiency of consumption, and are expected to make considerable effort to "narrow the gap" in GDP per capita.

Table 3-2. Primary Energy Supply (MBOE/Day).

	1975	*2000*
Oil	55	100
Gas	21	60
Coal	35	75
Nuclear	2	30
Hydro	7	15
Total	120	280

Source: Stanford Research Institute International

- Oil, coal, natural gas, and nuclear energy are all expected to play major roles in supplying the increased energy consumption, while hydroelectric energy supplies will become increasingly important in some of the less developed areas.
- Worldwide nuclear power generation growth rates, although still considerable, will be substantially lower than most projections of the recent past because of previously unanticipated political, social, and economic problems.
- Gas consumption, on the other hand, will probably grow more rapidly than expected.
- There will be considerable pressure on the price of crude oil traded on international markets in the near to intermediate term, and constant dollar prices are likely to decline somewhat (perhaps 20 percent) in the period from 1975 through the mid-1980s. Beyond that time, market conditions will enable producers to resume gradual price increases at the rate of about 1.5 percent per year (in constant dollars).
- Despite the near term international crude oil price decline, domestic consumers of refined petroleum products and other energy forms are not likely to see any price declines because of the mix in different regions of international and domestic supplies, consumer and producer taxes, protection of domestic supplies, and so forth. Nevertheless, it is not likely that consumer energy prices will make any dramatic increases in the years immediately ahead, except in those cases (primarily the United States and Canada) where government regulations have kept fuel prices artificially low with respect to world market prices.
- In this economic environment, because of price competition from conventional supplies, it is not likely that any unconventional supplies of energy will become commercially available on a global scale before the end of the study period. However, in specific locations, shale oil, heavy oils, and synthetic fuels from coal may contribute a measurable fraction of regional energy supplies by the end of this period. Our most recent estimate of the

cost of syncrude from oil shale is in the $20 per barrel range (1978 U.S. dollars), and shale is not developed on a commercial scale until oil prices reach this level in the latter years of the study period. By the year 2000, oil from shale contributes about 0.5 MBOE/day of petroleum supplies in the United States.

- Coal synthetics (high and low Btu gasification and liquefaction) do not become generally commercially available before the price of conventional supplies reaches the oil equivalent price of more than $20 per barrel (1978 dollars). By 2000, synthetics contribute less than 0.25 MBOE/day to oil and gas supplies. This same is true for direct use of solar energy and other renewable resources except hydro. Each may make some contribution in specific regions, but not enough to significantly affect the global supply-demand balances to the year 2000 that are the primary focus of this analysis.
- Although base case assumptions include the continued functioning of OPEC as a cartel, results of the analysis indicate that by the early part of the next decade, substantial pressures will be brought to bear on the cartel as increasing income is required to continue the many ambitious national development programs of OPEC members while total constant dollar cartel revenues do not increase correspondingly. This disparity will require extraordinary cooperation among cartel members to agree upon and to enforce pricing policies and production allocation schemes.

This is not meant to suggest a complacent attitude toward energy policy as much as to avoid an hysterical one. We are depleting the world's ultimately recoverable reserves of oil and gas and must be making plans for a future where these valuable commodities are consumed in only the most preferred uses, such as for petrochemical feedstocks and in other nonsubstitutable categories. However, our results indicate that there may be more time to make this transition than has been generally conceded in recent public forums on this vital question. It would be just as unwise to precipitate overregulation of energy markets or vast appropriations of scarce public and private funds to premature technologies of highly questionable value because of inaccurate perceptions of the availability of new supplies as it would be to ignore the problem and hope that it will go away. All of these alternatives would create more problems than are solved.

Public policy in this area should avoid creating artificial barriers to a smooth transition from dependence upon the more rapidly depleting resources to other more plentiful fossil fuel supplies where substitution is possible and ultimately to advanced nuclear designs and renewable resources. Furthermore, improvements in the efficiency of energy consumption ought to be encouraged, especially in those markets where demand is highly inelastic to price but where significant efficiency improvements may be achieved at relatively small cost (automobiles). It may not require so much the "wisdom of Solomon" as the "patience of Job," however, to avoid hasty decisions that disrupt markets and cause counterproductive results because of faulty perceptions of the urgency of taking some action—any action.

Part II

Living Through the Transition Period

Transition Topography

John C. Fisher
General Electric Company, Schenectady

A DESCRIPTION OF A TWO-COMPONENT SUBSTITUTION MODEL

We are interested in what the future holds but we have only the record of the past to guide us. I would like to describe a simple, and yet I believe a powerful, way of utilizing the record of the past to give us some insight as to what the future may hold.

Some years ago, R.H. Pry and I[1] made use of what may be the world's simplest model of technological change, and I would like to describe it to you. The model is based on three assumptions. The first is that many technological advances can be considered as competitive substitutions—namely, one method is substituting for another. An example would be the substitution of synthetic fibers for natural fibers in making various fabrics. The second assumption is that if a substitution has progressed as far as a few percent, it will go to a hundred percent. A few percent market penetration is necessary because enthusiasts can subsidize a new technology in its early stages until they run out of money. The important issue is, After they have run out of money and are no longer subsidizing it, how competitive is it? If it is still around and competitive, then the model says that it will go to completion. And the third assumption is simply that the rate of substitution follows a logistic curve (upper curve in Figure 4-1). The new technology rises exponentially at first, then slows down, and the old technology finally dies away exponentially at the same rate. The curve of substitution is a perfectly simple symmetric type of curve where the new technology eats up the old.

Now the basis for the model is that if the new technology can make inroads in market share when it is small, then surely it can do better as it grows. It will

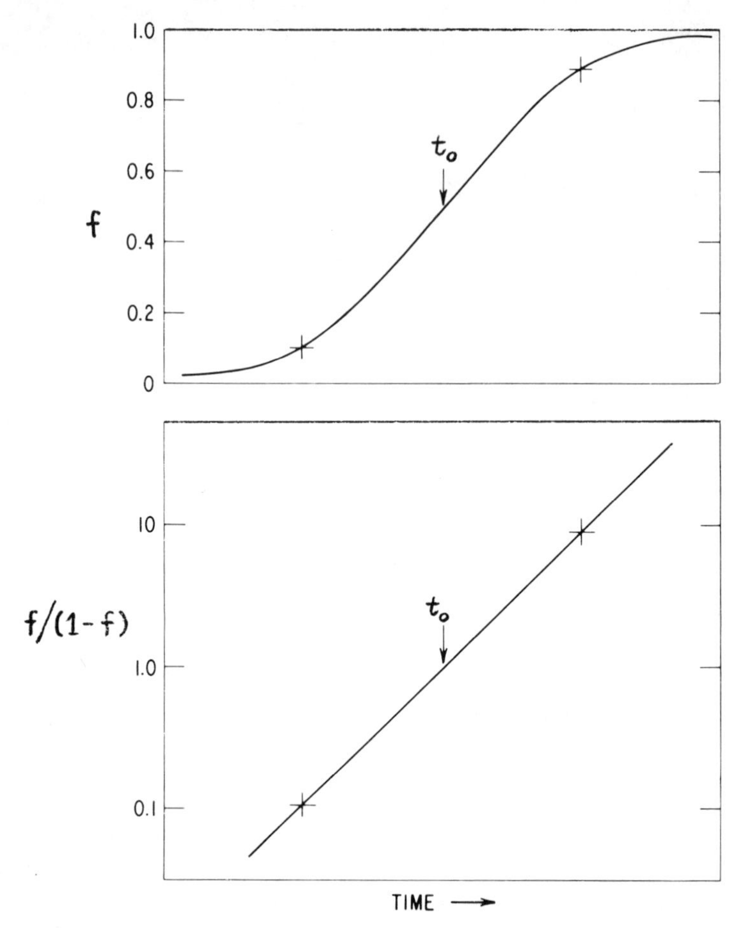

Figure 4-1. General Form of the Substitution Model Function.

increase its competitiveness, it will have advantages of scale economy as it gets bigger, it will have advantages of experience and learning so that its costs will go down as it gets bigger. And so we expect it to go all the way.

The model has a few advantages. There is little opportunity for the analysts to inject his own hopes, fears, or bias. You are stuck with this. It's so simple that there is nothing you can do, regardless of what you feel or hope. In other words, there is no opportunity for scenarios. This model has one scenario, the one it predicts, no others. And it is a clear-cut procedure with a definite projection. The model also has a lot of disadvantages. One I have described already is that although the new technology actually starts somewhere, the model says that it starts growing on an exponential curve that begins at minus infinity. And you have the problem of enthusiasts. So you have to wait until there has been a few percent market penetration before you can be sure that the new method or technology is going to take off and at what pace. The model says nothing about the price of either of the alternatives. It says nothing about the total quantity, only the mix.

Next, we assume the relation:

$$f = \frac{1}{2}\,[1 + \tanh \alpha\,(t - t_o)] \tag{4.1}$$

where f is the fraction of the new. This curve can be simplified by plotting $f/(1-f)$ on a logarithmic scale. Then you get straight lines, as in the lower half of Figure 4-1.

APPLICATIONS OF THE MODEL TO THE REAL WORLD

Figure 4-2 shows a plot for a definite substitution, $f/(1-f)$ on a log scale, versus years for the substitution of synthetic fiber for natural fiber. And there

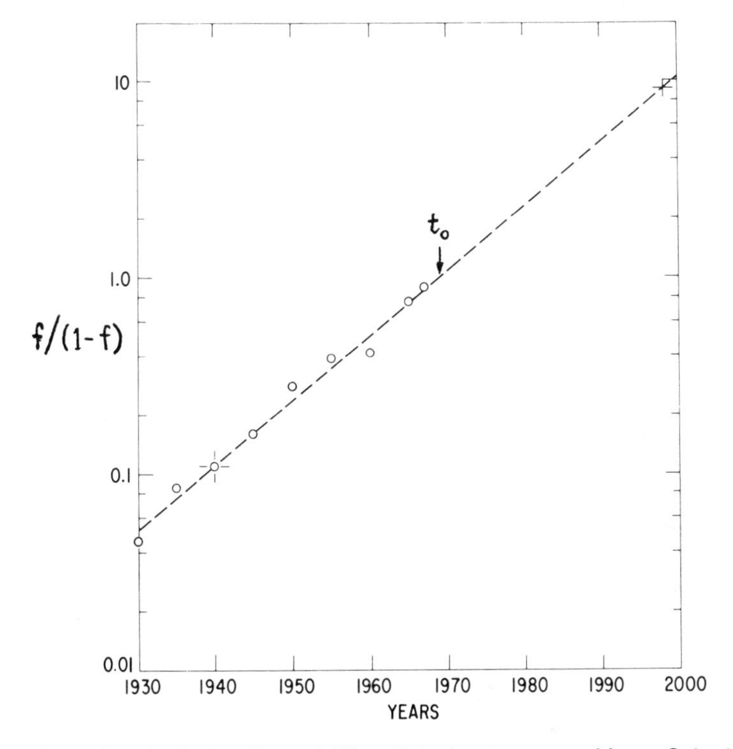

Figure 4-2. Synthetic for Natural Fiber Substitution versus Years. Substitution model fit to the data.

are a couple of interesting parameters indicated there. One is the time t_o when the substitution was half complete, which was in the late 1960s. That's when I began to worry whether I could still get cotton shirts if I lived so long. And the other is the takeover time, from about 10 percent to about a projected 90 percent market penetration, of about sixty years. One might live so long. As a matter of fact, it's hard to get a 100 percent cotton shirt now. They are mixing it with other stuff. So here, inexorably marching along, is the substitution of plastic shirts for field-grown shirts.

Now what about the absolute amounts? To get the absolute amounts we need to know the per capita consumption of fiber and the population growth, which are two different kinds of projections. Figure 4-3 shows one of them. Here is a history of the pounds of fiber consumed per capita in the United States from 1930 through about 1970. It goes up and down, and Pry and I approximated it by a straight line because we were interested in seeing what that would do. We took somebody's estimate of the population projection (I forget whose), and Figure 4-4 is the result.

If we can assume, as we did, that the fractional substitution of synthetic for natural is going to continue and that the per capita consumption of fiber in pounds per person goes up linearly the way we estimated and that the population grows modestly, then we find that synthetic fiber consumption goes from very small amounts in 1930 to quite a bit in 2000 and that natural fiber consumption peaks in the 1960s and goes down. That sort of projection is of academic interest to us, but if anybody in the business of growing cotton or owning cotton land saw it and believed it, it might have been of practical consequences to them. Thus, there is some useful information to be had from such projections of technological substitutions.

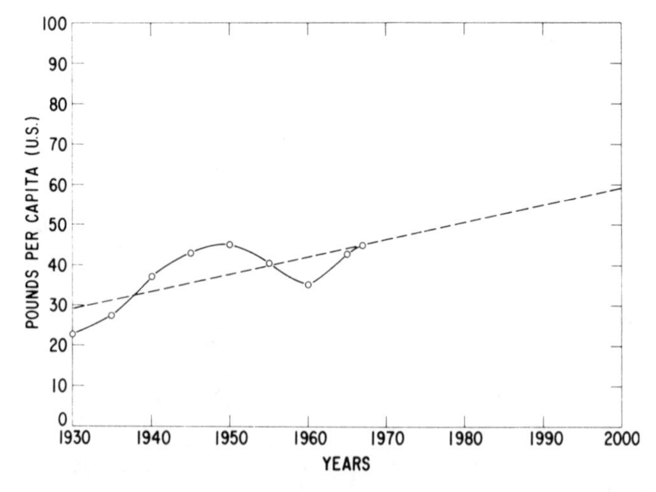

Figure 4-3. United States per Capita Fiber Consumption versus Years. Data and straight line projection.

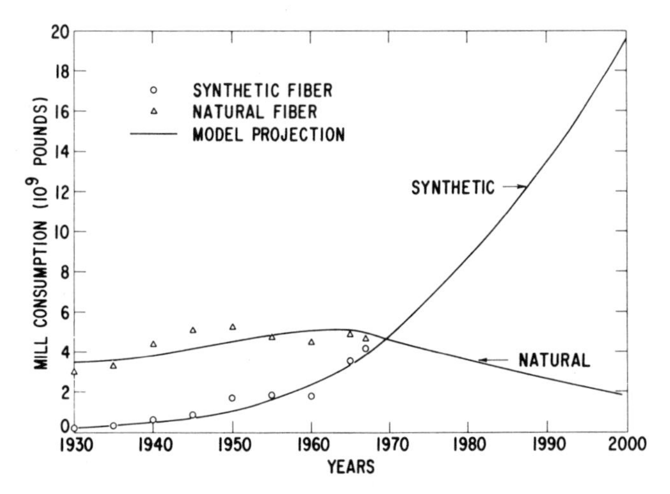

Figure 4-4. United States Fiber Consumption versus Years. Data and projection using the substitution model.

Figure 4-5 shows some other examples, plotted as $f(1-f)$ for the new technology on a logarithmic scale. The bottom curve shows how people accept margarine as a substitute for butter. It was half done in about 1960 with a take-over time of about sixty years. If you don't live too long, you'll still be able to get butter. The middle one is the one I have already showed you for fiber, and this funny-looking one on top is the substitution of synthetic rubber for natural rubber. This shows the extremes to which enthusiasm can go. During the war the United States felt that we really had to have rubber, so we heavily subsidized synthetic rubber and got a tremendous burst of synthetics. Then the war was over, the subsidy was removed, and things rolled back. But we did not fall back to zero; we fell back to a curve that is marching right along similarly to the other agricultural substitutions. Note, however, that synthetic rubber substitution is proceeding a lot earlier in history than it would have proceeded without that massive subsidy. The subsidy advanced the date of the substitution by a couple of decades, and the substitution continued at its characteristic pace after the subsidy was ended.

Figure 4-6 shows open hearth steel, Bessemer steel, water-based paints, oil-based paints, and various other items. A lot of things follow this kind of substitution. However, you notice that these are all substitutions of B for A, just two competitors.

EXTENSION TO MORE THAN TWO TECHNOLOGIES

A major advance in the model was made by C. Marchetti and his associates[2] at IIASA, the International Institute for Applied Systems Analysis in Austria. Marchetti found out how to apply the substitution model to more than two

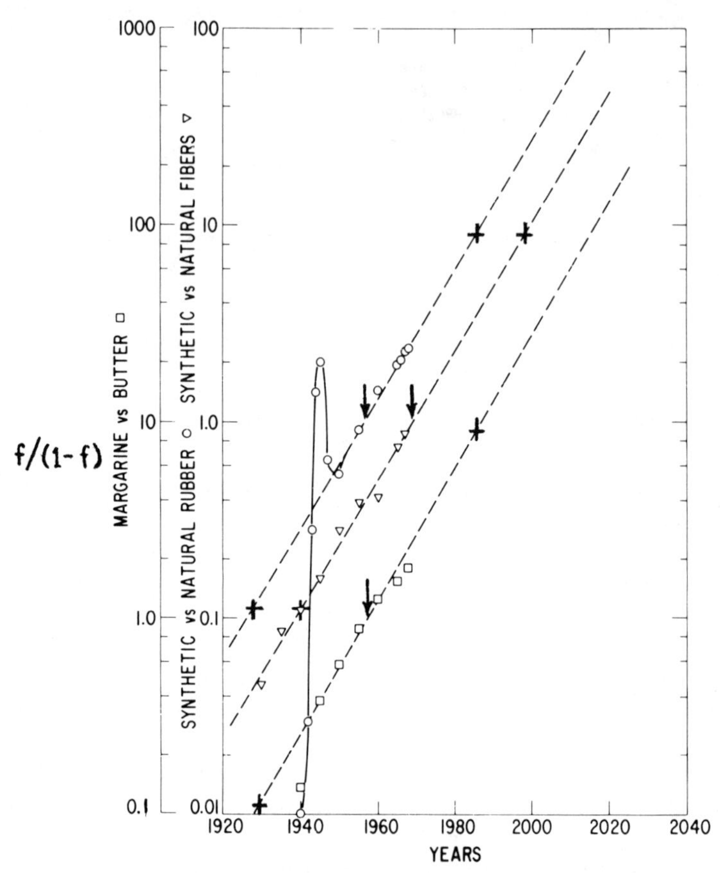

Figure 4-5. United States Consumption versus Years and Fit to Substitution Model for Three Agriculture-based Products. (See text)

technologies, and in Figure 4-7 we have a sketch of his method. Here we have an old technology *A* disappearing as a new one, *B*, displaces it. When plotting $f/(1 - f)$ on a logarithmic scale, *B* has a positive slope and *A* has a negative slope of the same magnitude. We are in an era where *B* is dominant. Suppose now something else starts up, an even newer technology, *C*. Now what are we going to do? As the total market fraction always must add up to one, something must be done to make everything add up to one. Marchetti's assumption is as follows: The new technology *C* increases market share on a logistic curve at the rate we observe; the old technology *A* loses market share on its same old logistic curve; and *B* gets what is left over (Figure 4-8). So as *C* goes up, the amount that is left over gets whittled away. Technology *B*, instead of continuing on up to 100 percent, begins to saturate, bends over, starts down, and approaches the same

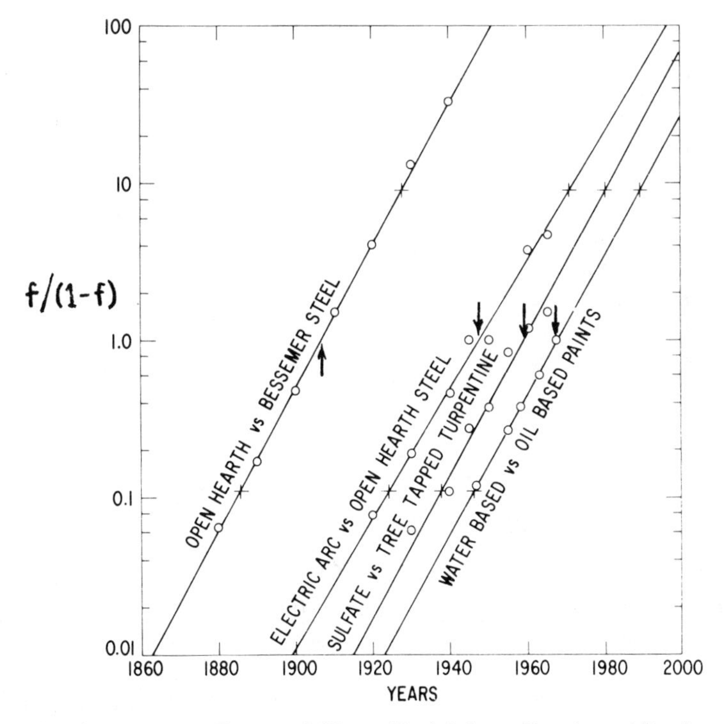

Figure 4-6. Substitution Data and Fit to Model for a Number of Products and Processes. All data United States except detergents for soap as noted.

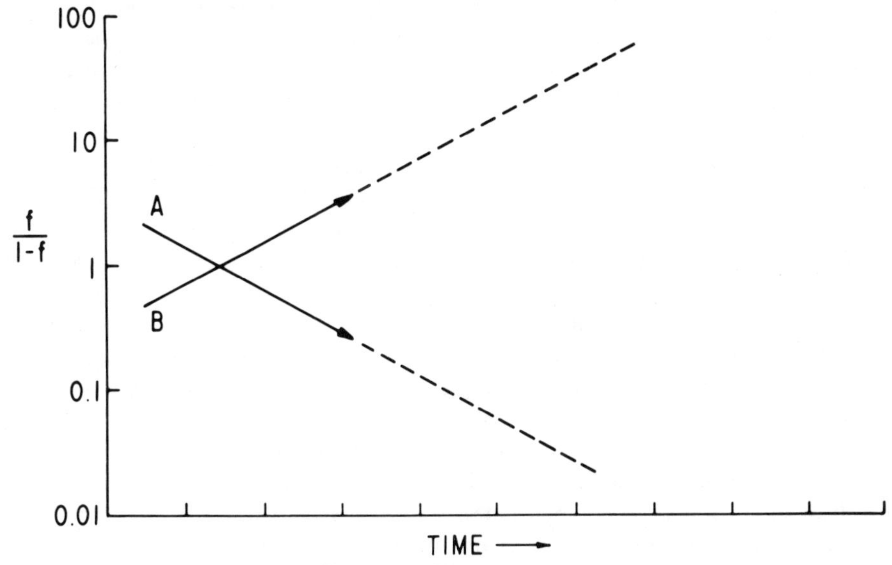

Figure 4-7. Technology *B* is Displacing Technology *A*.

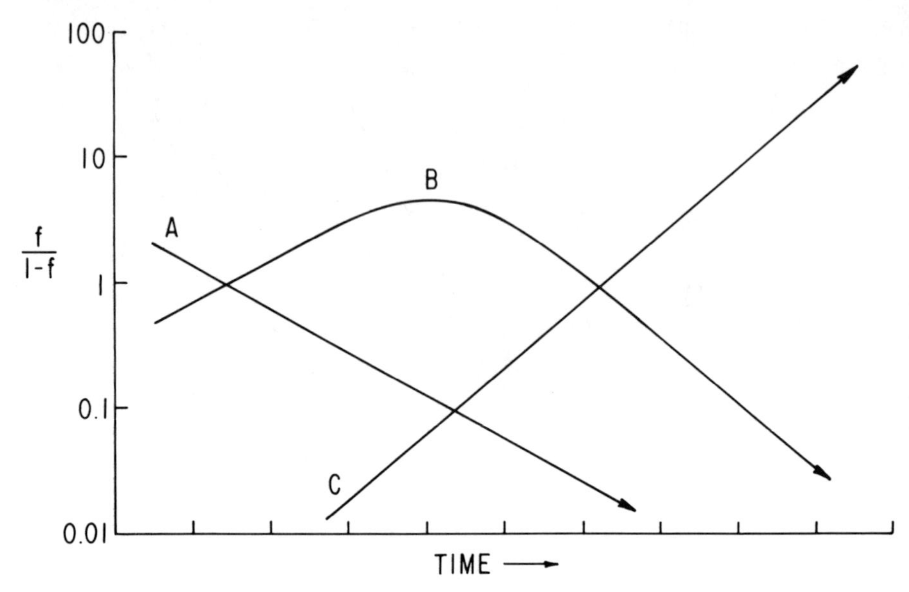

Figure 4-8. New Technology *C* Does Not Affect Declining Technology *A*, but Forces Technology *B* to Saturate and Then Decline.

logistic decline as *C*'s logistic rise. Then technology *C* becomes vulnerable to a still newer technology, *D*; and if *D* shows up, then *C* begins to saturate in its turn. The assumption is that there is only one technology at a time in the saturation phase. The others are either growing on logistic curves as we observe them to do or decaying on logistic curves to which they have been forced by younger and stronger competition.[3] The saturating technology has to make do with the residual market share, which it does until, in its turn, it is forced into logistic decline.

Figure 4-9 shows how Marchetti has applied his extension of the model to world fuels. By rolling the clock back and pretending that it is now 1920, we can see how this method will work. We have about twenty years of good data for world fuel consumption—that is, twenty years of data between 1900 and 1920 for the various fuels that are being consumed. Coal is the dominant fuel, with more than 50 percent of the market. Wood is on its way down, oil is on its way up, and way down here at about 2 percent in 1920 is natural gas just starting up. Now the model is very simple (Figure 4-10): it says that oil and gas, which are on their growth trajectories, will continue to grow on logistic curves, and wood, which is in decline, will continue on down its logistic curve. In order for that to happen, coal must be in the saturation phase, taking whatever market share is left over in its competition with other fuels. In due course coal's market share must start back down, as oil and gas grow ever larger. When coal returns to a logistic curve, then oil enters the saturation phase, and in its turn oil has got to turn around and come down. If we extend the projections ahead for a

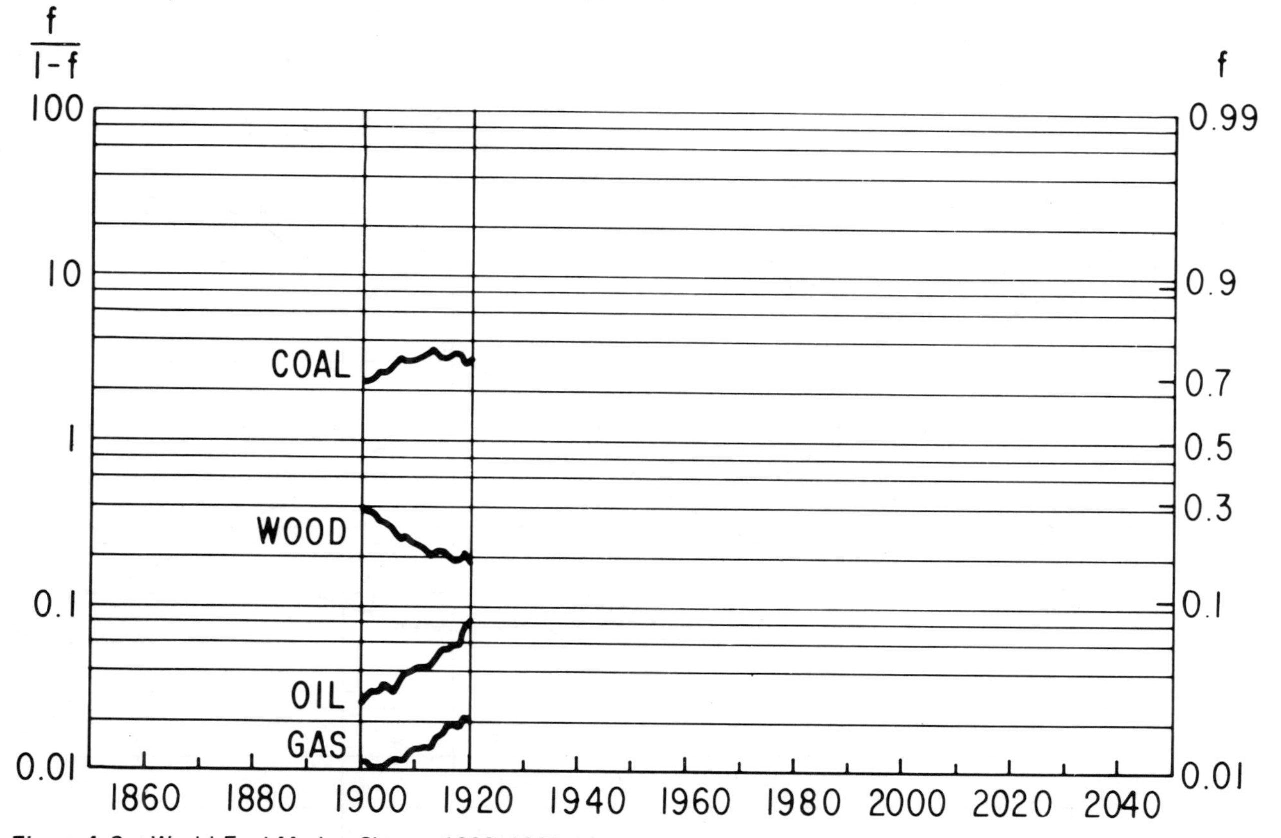

Figure 4-9. World Fuel Market Shares, 1900–1920.

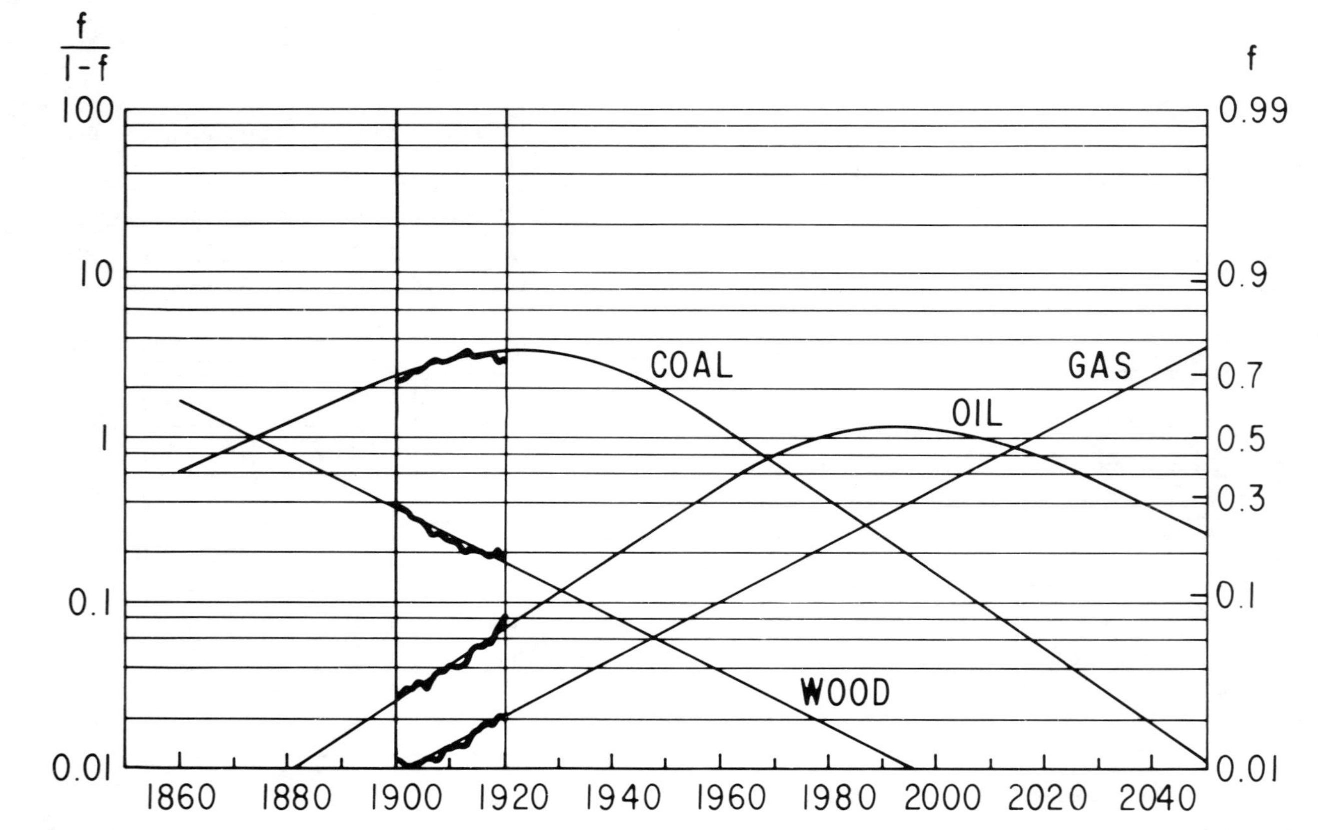

Source: C. Marchetti and N. Nakicenovic, "The Dynamics of Energy Systems and the Logistic Substitution Model. Volume 1: Phenomenlogical Part," Administrative Report AR-78-1B (Laxenburg, Austria: IIASA, July 1978).

Figure 4–10. Substitution Model Projection, as of 1920.

hundred years, we find a picture out here in the early part of the next century where natural gas has become the dominant fuel of the world and oil is on its way down and coal is farther on its way down and wood is practically out of the picture.

If we look at what actually happened, we can see whether our 1920 projection was any good. Figure 4-11 shows what actually happened. The model worked pretty well for coal up to 1974 and pretty well for oil. Coal did start on its downward trend; oil did continue to go up. The projection missed natural gas somewhat and completely missed nuclear. These are the two major mistakes in the projection. First, it didn't forecast that nuclear power would come in; but the model cannot say anything about a new technology that has not been introduced yet. Second, it missed the growth rate of natural gas, partly because natural gas had not achieved much market penetration in 1920. It is difficult to get the exponential rate right with so little market penetration. But all in all, the projection is not bad out to the present time.

REVISIONS IN THE MODEL FOR THE WORLD FUEL SUPPLY

Now, how do we revise this in light of what we now know after another fifty-four years of data? The main thing is that we ought to increase our rate of growth for natural gas a little bit. That will cause oil to bend down on the world basis a little sooner, so that the age of oil will peak near the end of this century and the first part of the next century will be the age of natural gas (Figure 4-12).

Now, how about nuclear? Clearly nuclear has had its enthusiasts and its subsidies, and we do not know if it is out of their hands yet. We do not know what its exponential growth rate will be when they are no longer able to afford to promote it. So the substitution model does not yet say anything about the growth of nuclear power.

Although the model cannot say anything about nuclear yet, I will depart from the model and share with you my own tentative estimate. My own feeling is that nuclear power is established as a technology and will survive on its own. In my view its timetable was clearly accelerated by the subsidy. Its growth rate will ultimately have to be determined by its competitiveness. The competitive penetration rate, I would guess, would be something like the other major energy sources. So taking off from where we are now, if nuclear should rise parallel with oil and gas, then it will grow to a significant fraction of the world's energy by 2050. Thus, my current feeling is that it is probable that the first half of the next century will be the age of natural gas and the last half of the next century will see the age of nuclear power on an overall average global scale.

Finally, I have three comments: You notice that we turned away from coal before we used it up. People stopped using coal not because the coal was gone,

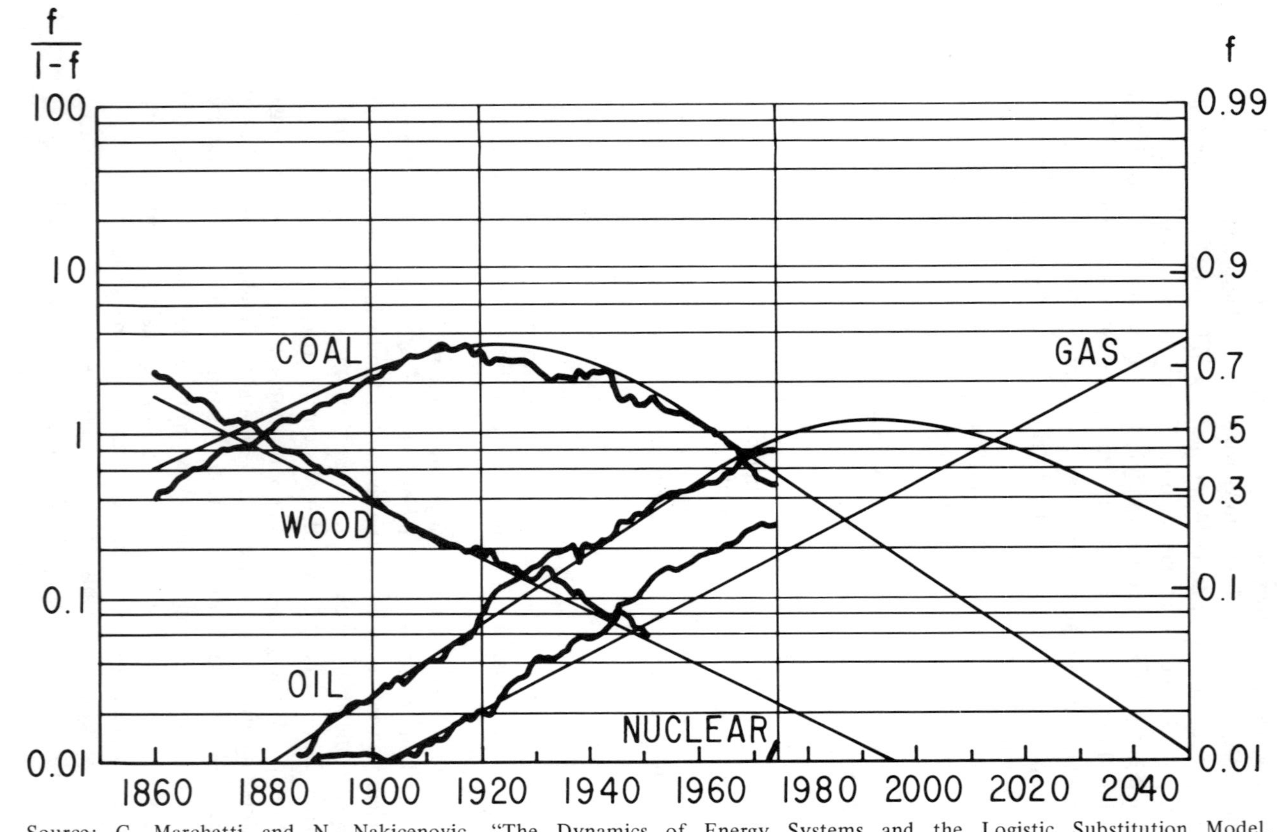

Source: C. Marchetti and N. Nakicenovic, "The Dynamics of Energy Systems and the Logistic Substitution Model. Volume 1: Phenomenological Part," Administrative Report AR-78-1B (Laxenburg, Austria: IIASA, July 1978).

Figure 4-11. Comparison of 1920 Model Projection with Historic Data, 1860–1974.

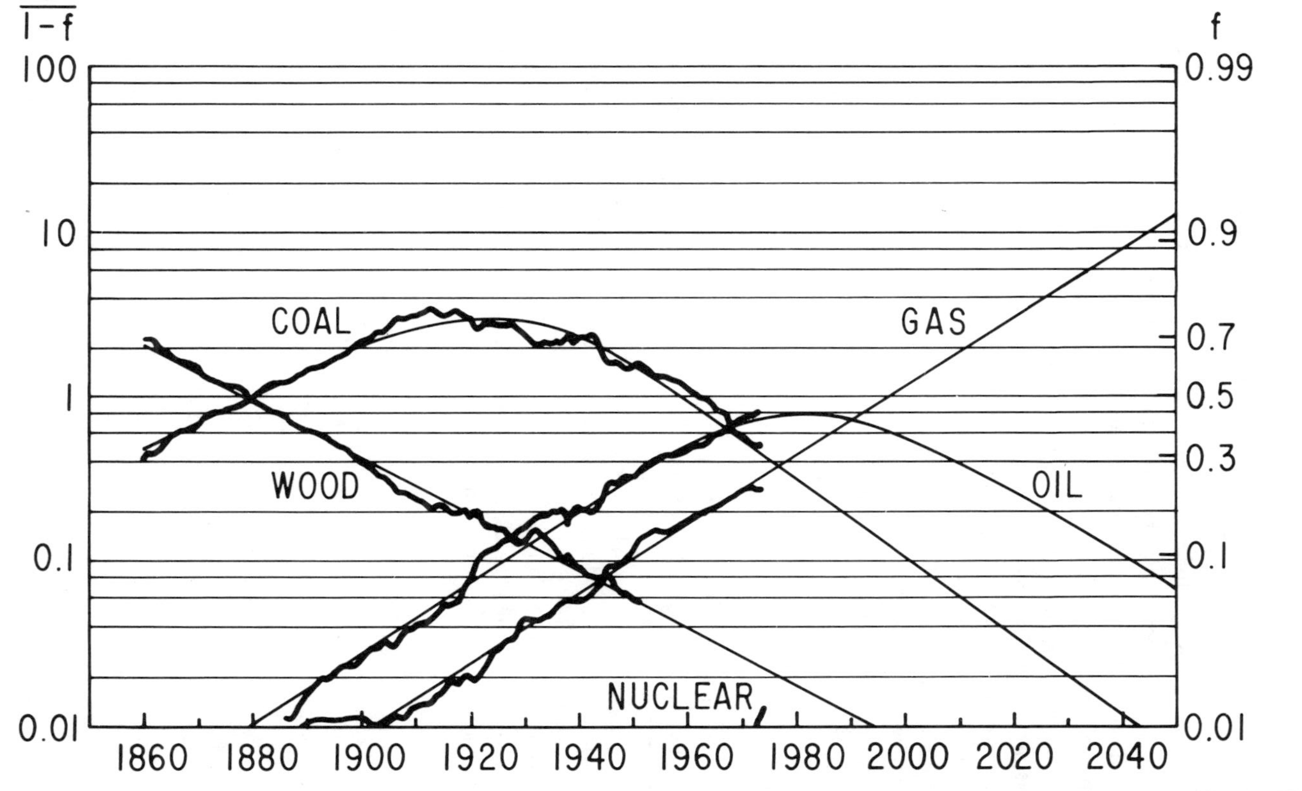

Source: C. Marchetti and N. Nakicenovic, "The Dynamics of Energy Systems and the Logistic Substitution Model. Volume 1: Phenomenological Part," Administrative Report AR-78-1B (Laxenburg, Austria: IIASA, July 1978).

Figure 4-12. Revised Projection based on Full Data Base to 1974, Omitting Nuclear. The future of nuclear power is discussed in the text.

but because they liked oil better. Now we are in the process of turning away from oil, not because the oil is gone, but because we like natural gas better. Certainly it is possible that we will turn away from natural gas before it is gone because overall we like nuclear power better.

Second, Marchetti and his associates have analyzed many nations independently. Not all nations are the same. I have shown you the aggregate of the world, but it is very interesting to look at how different nations are behaving within themsleves. I recommend that those of you who are interested read the material that Marchetti and his associates are publishing at IIASA.

Third, although this type of projection seems fatalistic, I do not want to suggest that anybody can relax. On the contrary, I view these market substitution curves as measuring the results of the efforts of millions of people in lifetimes of hard work directed toward various incompatible goals–social, economic, political. This competition is so fierce and so intense and so constant and incessant that hardly anything alters the progress of substituion it brings about. You can hardly notice the Great Depression or Second World War in those curves. What matters is the inherent strength and acceptability of each technology in the maelstrom of competition.

NOTES

1. J.C. Fisher and R.H. Pry, "A Simple Substitution Model of Technological Change," Report No. 70-C-215 (Schenectady, N.Y.: General Electric Research and Development Center, June 1970).

2. C. Marchetti and N. Nakicenovic, "The Dynamics of Energy Systems and the Logistic Substitution Model. Volume 1: Phenomenological Part," Administrative Report AR-78-1B (Laxenburg, Austria: International Institute for Applied Systems Analysis, July 1978).

3. See ibid. for the general rule governing the end of the saturation phase when there are more than three competitors.

Chapter 5

Some Alternative Demand Paths Through the Transition to a Long-term, Sustainable Energy Future

John H. Gibbons
University of Tennessee

The obscurest epoch is today.

R.L. Stevenson

INTRODUCTION

Mankind's rapidly rising use of energy, especially over the past 200 years, has facilitated an incredible growth in material wealth and human comfort. At the same time, constant efforts have been required to find additional energy supplies. Until the beginning of this decade, advancing technological sophistication enabled the industrialized world to provide ever-cheaper and more convenient energy as we moved from solar (wood, water) to fossil resources (coal, then oil and gas). Most recently we have added nuclear fission to the list.

While most fossil and nuclear fuels remain to be used, we face rising costs of supply at the margin. It is not surprising that in the United States, principal focus is fixed upon ways to increase supplies. Expansionism is a historical mind-set of Western civilization. This propensity is bolstered by the historical fact, albeit now obsolete, that incremental energy supply was obtained at lower than average cost.

The current consensus is that energy cost, properly accounted, is going to trend upward for a time as cheap resources are depleted and replaced with more expensive sources. Such a thesis implies that we should carefully examine patterns, dynamics, and technologies of energy demand to insure that future demand is economically tuned to the era of rising supply cost.

Over the past decade, especially since 1974, significant attention has finally been given to energy demand studies.[1] It is becoming clear that energy need not be a fixed factor in providing goods and services (the notion of a lock-step relationship between economic growth and energy consumption), but that *over time* energy is remarkably substitutable and has long-run sensitivity to price. Hence, in planning our energy future there are strong interactions between supply strategy and demand strategy. The interactions include price, timing, income growth, and market choices. If we assume higher energy (real) prices for the future, these will provide economic incentives for slower demand growth.

There are other reasons to seek moderation of demand growth. These include the problem of massive capital requirements to maintain supply growth; escalating demand from developing countries for the relatively inflexible supplies of oil traded on the world market; rapidly worsening domestic environmental problems associated with energy production; and vexing worldwide problems associated with energy (including influences of energy production on climate, nuclear proliferation, and waste management).

Given these imperatives, there are several distinguishable strategies that can be employed to reduce demand growth rates (Figure 5-1). Clearly there is a variety of options to substitute ingenuity for energy resource consumption. That is the main thesis of this chapter. Whether or not these options will or should be taken depends upon what is assumed about future total costs of energy. While consumer behavior patterns can and undoubtedly will be very influential, it is the direct substitution of technological sophistication for energy consumption that is emphasized below. Most of the results were developed by a major recent study[2] of energy supply and demand out to the year 2010.

FACTORS AFFECTING DEMAND FOR ENERGY

In the simplest sense, demand can be disaggregated to the product of the number of people (Figure 5-2) times the energy consumed by each person (at home, at work, and at play), as shown in Figure 5-3. Clearly for the United States, assuming we get the present inordinate amount of illegal immigration back in hand, population growth is rapidly slowing, and our labor force growth rate will soon drop sharply as the war time baby bulge is absorbed into the labor market.

The amount of energy consumed by each person depends upon disposable income and energy price and availability. It is in this area that one encounters the greatest range of plausible futures. It seems appropriate to think about future demand using energy price and personal income as parameters. Thus, one *projects* various demands that might follow from different explicit assumptions rather than *forecasting* or predicting a most likely demand.[3] Such an analysis has recently been completed,[4] using a combination of engineering and microeconomic techniques.

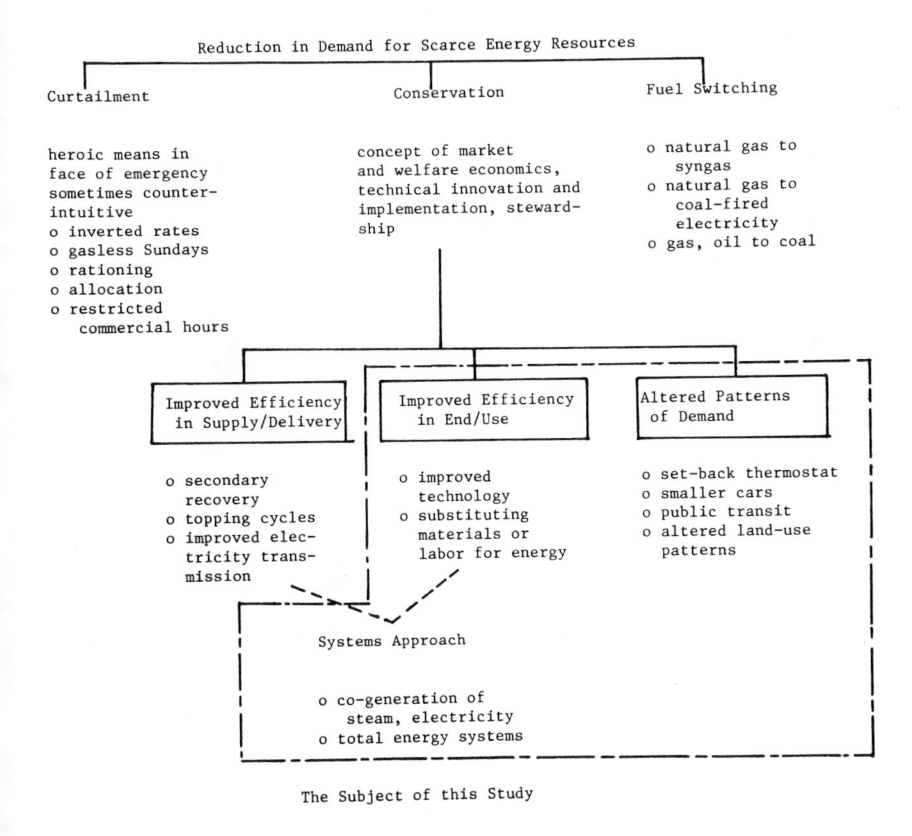

Figure 5-1. Classification of Strategies for Reducing Energy Demand.

DEMAND PATHS FOR CONSUMING SECTORS

Residential and Commercial Buildings

The energy required to provide heated, cooled, and lighted interior space depends sensitively upon the design and construction. For example, a small change in capital investment in a single family residence's thermal integrity (Figure 5-4) makes an impressive decrease in heating requirements. As energy prices move upward both new and existing structures merit considerable improvement. These improvements in the thermal envelope can be considerably augmented with more efficient heating, cooling, and lighting systems (such as heat pumps). Of course behavior changes can also make an impressive difference. A modest thermostat shift of, say, two or three degrees F (and/or night setback) can reduce the heating load by 10 percent or more. Recent evidence indicates that a combination of behavior change and insulation retrofit has

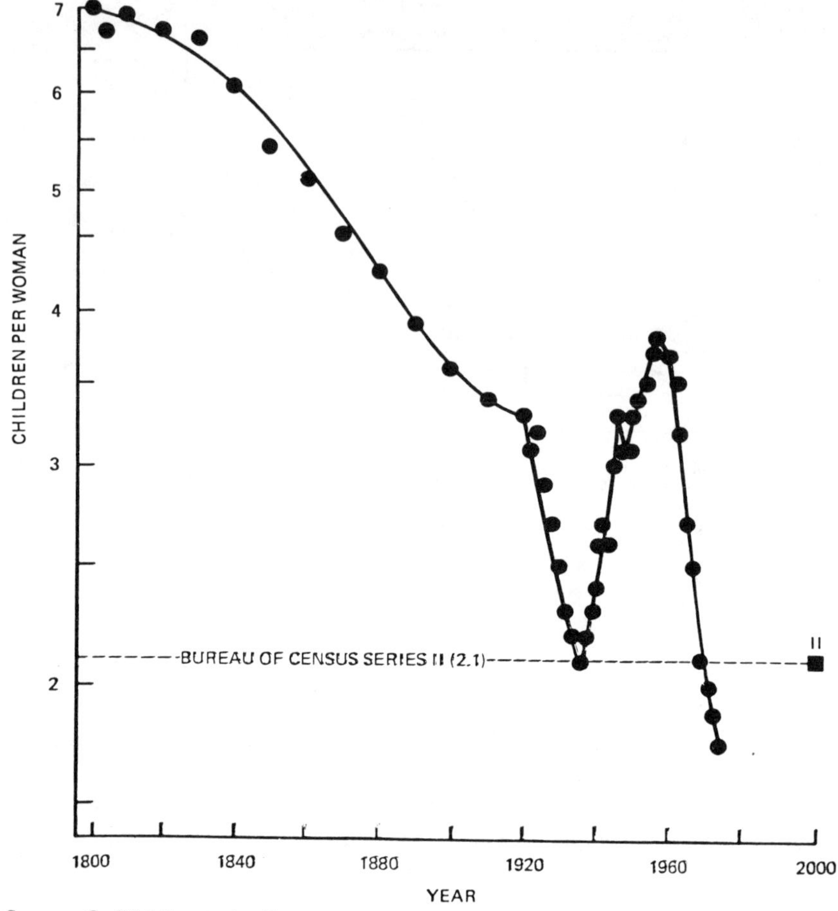

Source: C. Whittle et al., *Economic and Environmental Implications of a U.S. Nuclear Moratorium, 1985-2000.* (Oak Ridge, Tennessee: Institute for Energy Analysis Report ORAU/IEA, 76-4, Oak Ridge Associated Universities, 1976).

Figure 5-2. Birth Rate in the United States.

reduced gas heating loads per residence by more than 15 percent since the 1973-1974 crisis (Figure 5-5).

An analogy can be made between incremental cost curves for conservation investments and supply investments (Figure 5-6)—that is, the various actions one can take range from very cost effective to marginally effective. In fact, just as in the case of supply, one can make poor investments in saving energy! However, it appears that a great deal more investment can be made on the conservation side before reaching the equivalent incremental costs currently required on the supply side.[5]

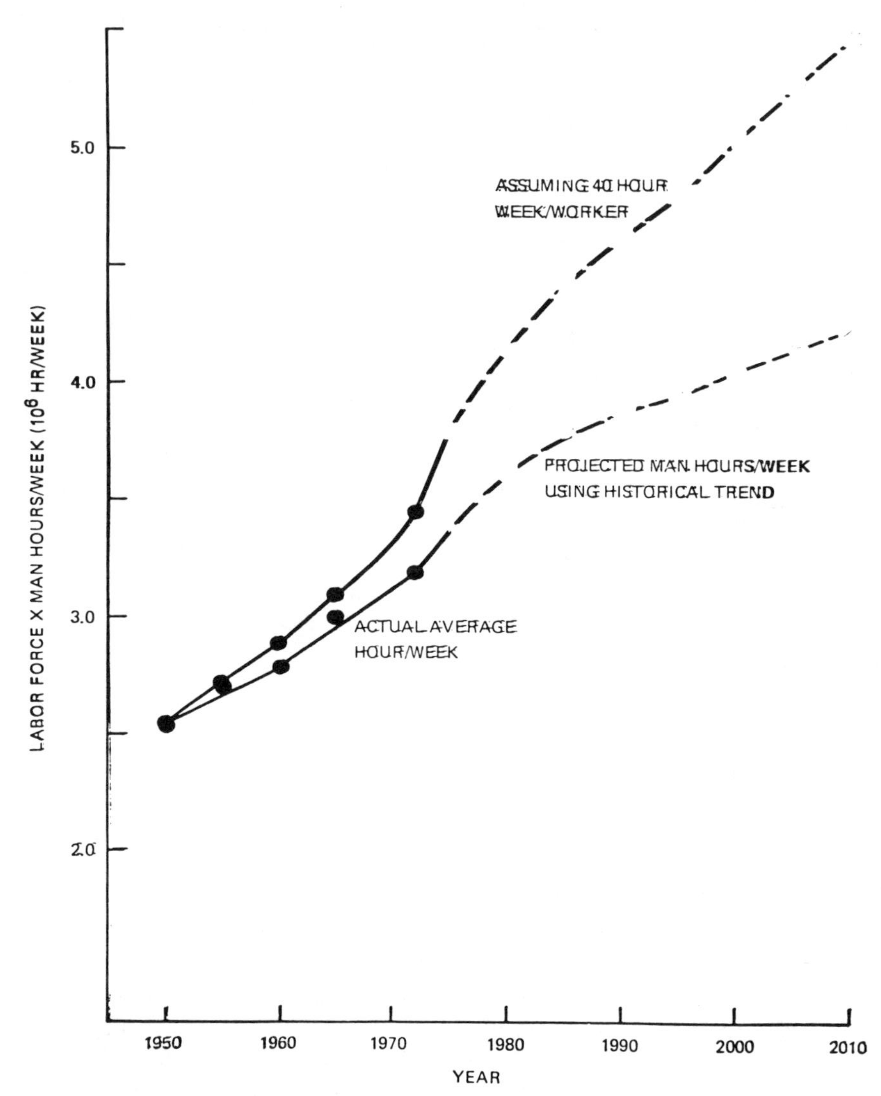

Figure 5-3. U.S. Labor Force.

Another example of the high substitutibility of capital for energy in new stock is the refrigerator (Figure 5-7). The data show that even at current prices, it pays to make a refrigerator about twice as enery efficient as the current U.S. average. Other countries facing high energy costs, especially developing nations that import U.S. goods, might wish to look more closely at the energy performance of imported products.

In the United States, unlike LDCs, another phenomenon is becoming

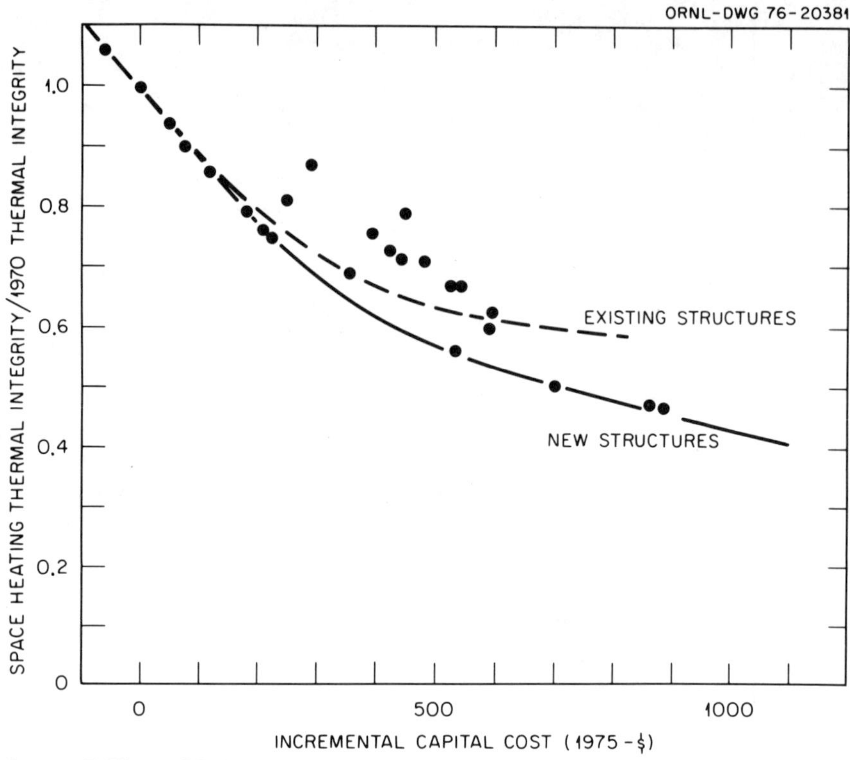

Source: E. Hirst and J. Jackson, *Energy* 2 (1977): 131.

Figure 5-4. Space-heating Thermal Integrity for Single Family Units versus Increased Capital Costs.

increasingly important in projecting future demand: market saturation (Table 5-1). In the U.S. many energy-intensive consumer items (e.g., space heating and cooling, refrigeration, washing) have already generated the "new" market and are rapidly shifting to a "replacement" market. This effect, which tends to reduce future demand growth, might be offset by the introduction of new *energy-intensive* products and services. However, it is difficult to identify serious candidates outside the transportation sector (notably airline travel).

In summary, for the buildings sector, higher energy prices combined with public policies developed since 1975 (e.g., building codes, incentives to insulate, appliance efficiency standards) and a slowdown in new family formation imply a considerably lower energy demand growth over the next two decades or more.[6]

Transportation

Energy intensiveness and use in this sector has changed markedly over the past twenty-five years and will likely continue to do so in the future

Table 5-1. Growth of Saturation and Electricity Consumption for Household Uses of Electricity, 1950-1970.

	Saturations (percent)			*Average Annual Use (kwhr use)*		
	1950	*1960*	*1970*	*1950*	*1960*	*1970*
Refrigerators	83	98	100	340	780	1,300
Air Conditioning	1	13	38	1,650	1,940	2,420
Lighting	100	100	100	450	500	600
Space Heating	1	2	8	10,000	12,900	14,600
Water Heating	11	21	25	3,670	4,000	4,500
Clothes Drying	1	12	29	520	930	990
Ranges	16	32	40	1,200	1,200	1,200
Television	13	90	95	290	330	420
Food Freezers	6	20	28	620	890	1,380
Clothes Washers	74	76	70	45	60	100

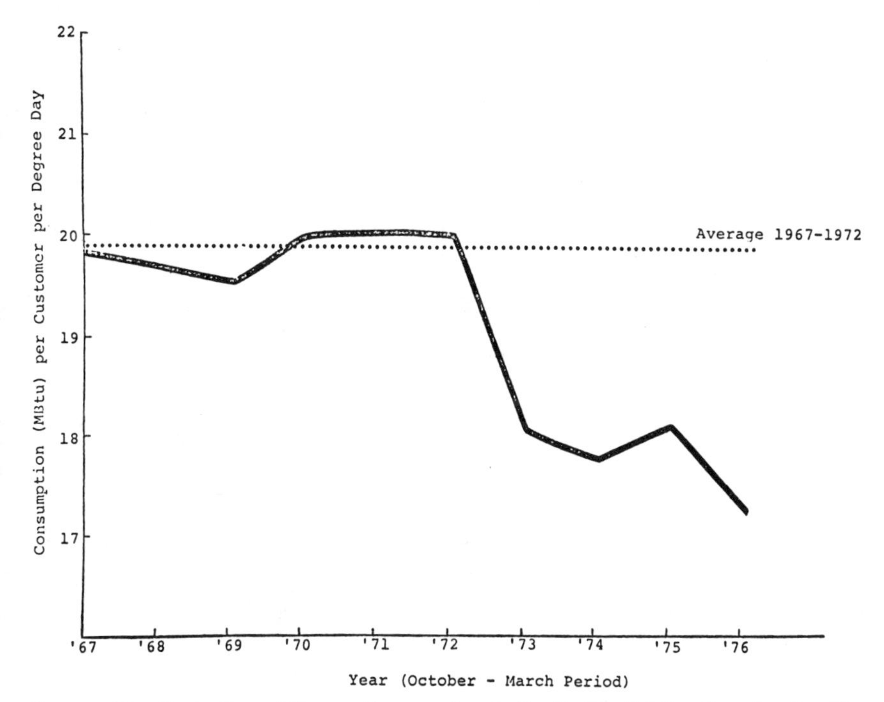

Source: "An Evaluation of Energy Conservation in the Residential Gas Spaceheating Market", American Gas Association Energy Analysis public 1978-1 (2/1/78).

Figure 5-5. Total U.S. Residential Gas Space-heating Consumption per Customer per Degree Day.

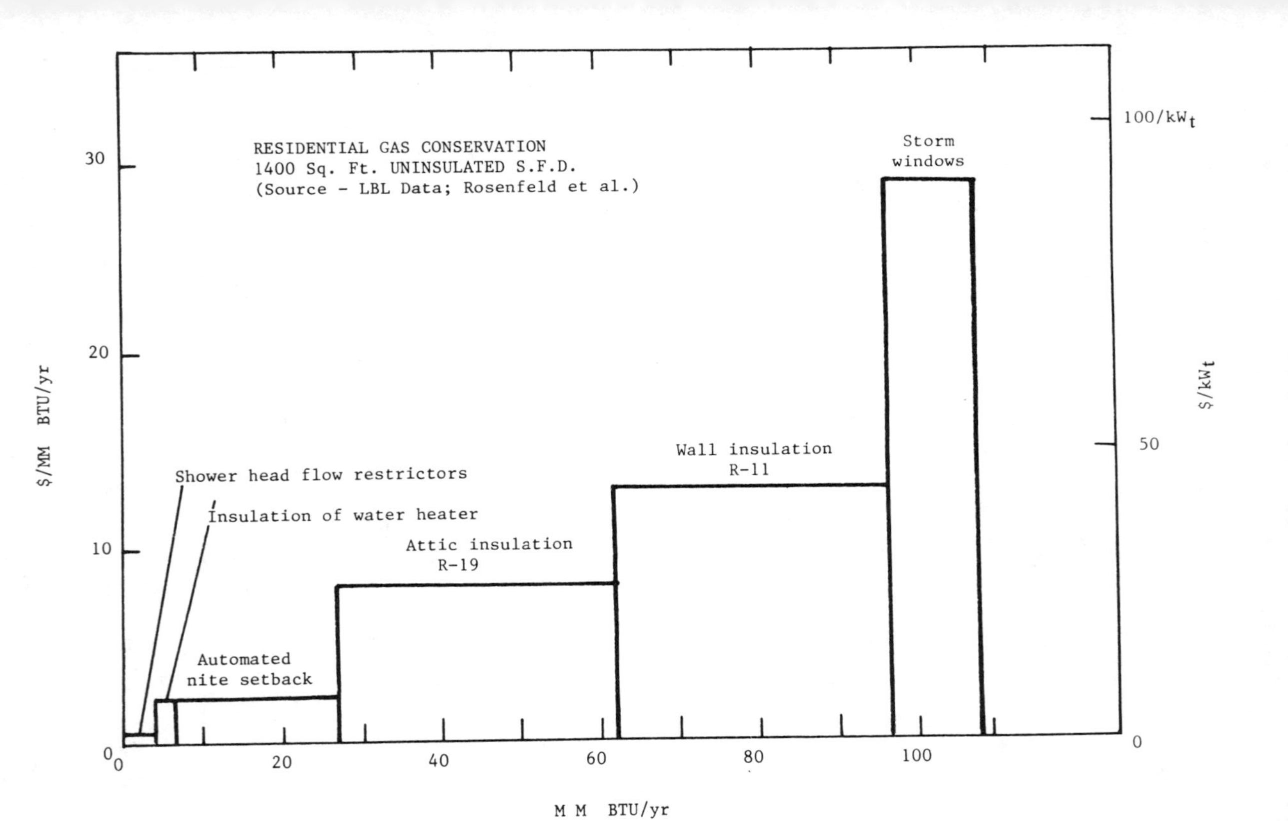

Source: A. Rosenfield et al., "Some Potentials for Energy and Peak Power Conservation in California," Lawrence Berkeley Laboratory Report LBL-5926 (1977).

Figure 5-6. Incremental Costs of Conservation Options and Corresponding Equivalent Cost of Supply Otherwise Required. Even for storm window investments (the least cost-effective shown) the option saves capacity otherwise required at a cost equivalent of only $100/kw.

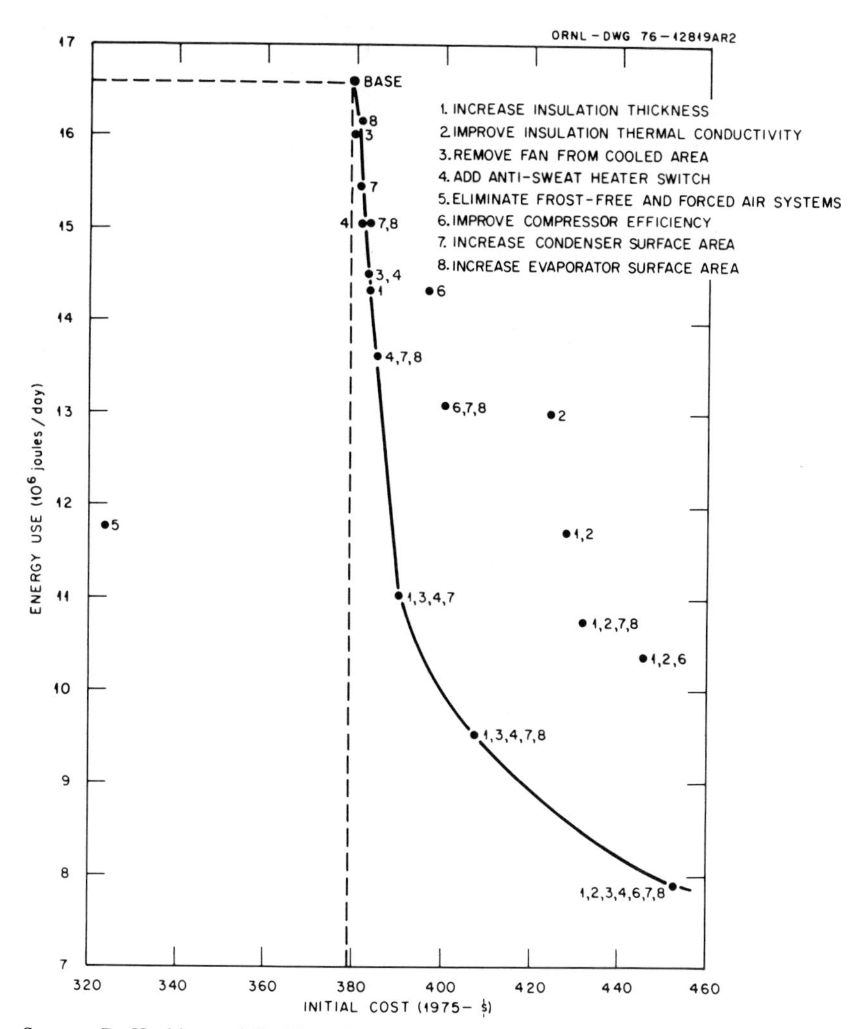

Source: R. Hoskins and E. Hirst, *Energy and Cost Analysis of Residential Refrigerators.* Report ORNL/CON-6 (Oak Ridge, Tennessee: Oak Ridge National Laboratory, 1977).

Figure 5-7. Electricity Use Versus Retail Price for a Typical Refrigerator.

(Figure 5-8). Jets have replaced commercial piston aircraft; all air travel has grown rapidly. The conversion to jets increased energy consumption per mile, but the convenience (and recent price deregulation) has led to rapid energy demand growth in this sector. At the same time the higher consumer demand accelerates the rate at which a new, much more energy-efficient fleet will be purchased by the airlines.

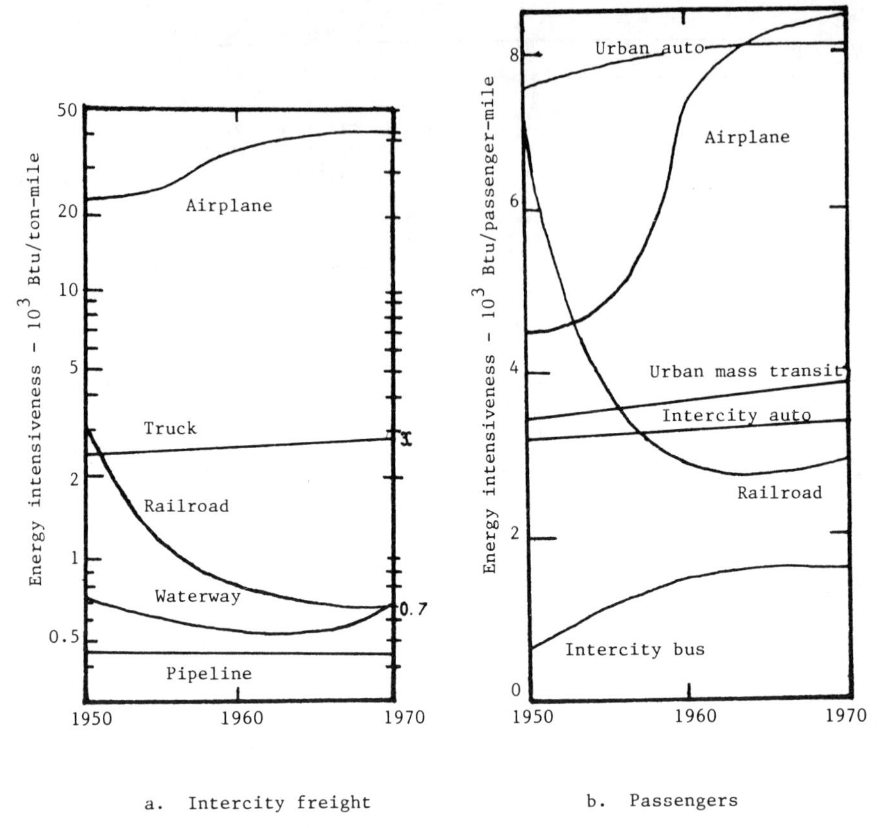

Source: E. Hirst, *Energy Intensiveness of Passenger and Freight Transport Modes:* 1950–1970 (Oak Ridge, Tennessee: Oak Ridge National Laboratory Report ORNL-NSF-EP-44, 1973).

Figure 5-8. Historical Variation in Energy Intensiveness of Various Transportation Modes.

The private automobile is an especially important actor because it accounts for over half of transportation energy. Fortunately, given time for a production changeover, the efficiency of a new car can be markedly improved for less than 10 percent increase in manufacturing cost (Figure 5-9). Consumers, even if they are responsive to total cost (to both own and operate) have little incentive to choose a highly efficient, more costly car instead of one that is less costly but uses more fuel. This insufficiency of the market to signal the performance that reflects national need for efficiency has led to public policies (Energy Policy and Conservation Act of 1975, National Energy Act of 1978) that require much more efficient cars, with intermediate and 1985 new fleet standards as well as a "gas guzzler" excise tax for poor performers.

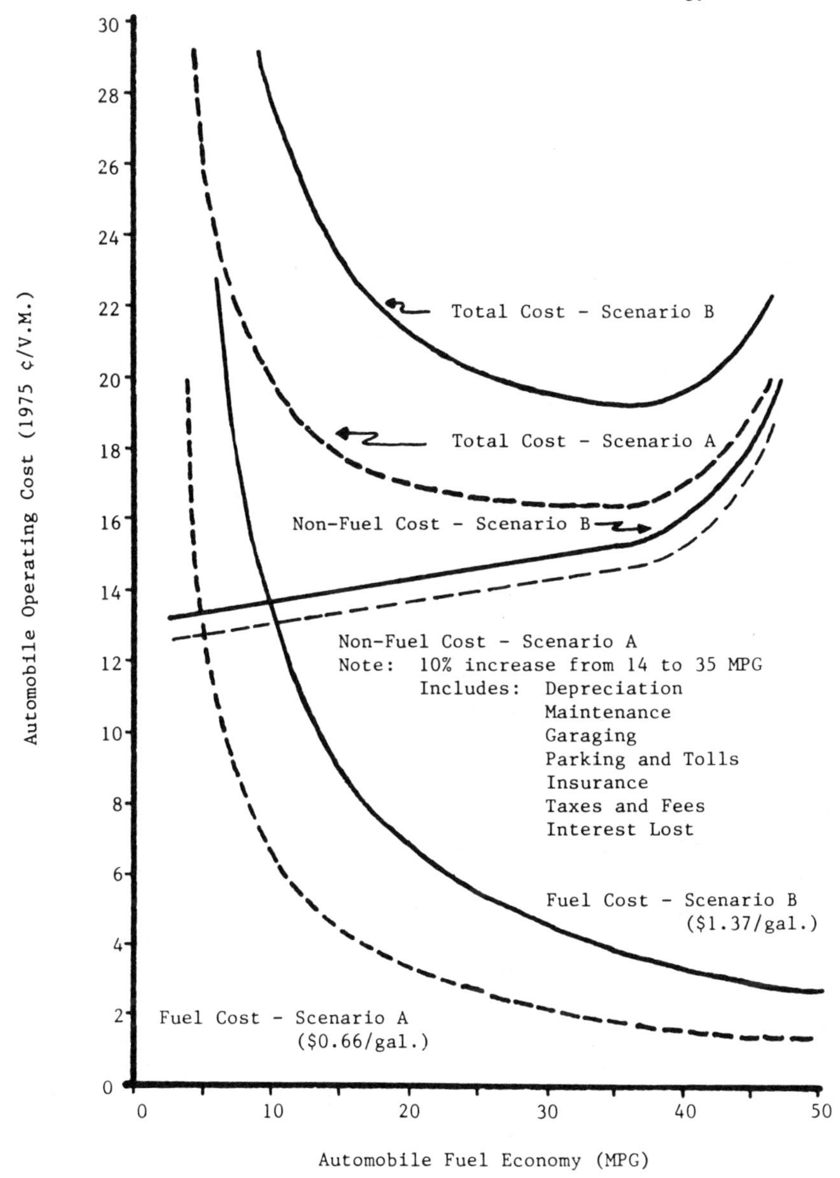

Source: J. Gibbons et al., "U.S. Energy Demand: Some Low Energy Futures", *Science* 200 (April 14, 1978): 142.

Figure 5-9. Automobile Costs to Own and Operate versus Fuel Economy. Two cases (A and B) are presented for different assumed gasoline price. The nonfuel cost is for automobiles produced after appropriate time for production change-over. Identical annual miles driven and nearly identical features of interior space and safety are presumed in all cases, but "performance" is sacrificed for the most efficient cars. The total cost to the consumer (to own and operate) is rather insensitive to choice of efficiency over a wide range of efficiency.

It is important to note that even in this relatively rapidly changing scene, with a turnover rate of about a decade, several decades will be required before the changes take full effect. This fact underscores the point that the most important actions of substitution in lowering demand growth can only occur over a protracted period of time.

Industry

Generally more cognizant of total cost than other sectors, industry has for many years been steadily reducing (at about 1 percent per year) the amount of energy required to produce a unit of output. Of course some of this reduction resulted from shifts from coal to gas and oil. Nevertheless, spurred by higher prices, it appears that the historical rate of improvement in energy productivity in this sector can continue for several decades or more. It is interesting to note that 60 percent of energy used by industry is for process steam and heat; furthermore the industries that account for 70 percent of the industrial energy budget are responsible for only 16 percent of industrial employment.

SOME OVERALL DEMAND RESULTS

One set of future demand paths (see Figure 5-10) was derived from four different sets of assumptions about future energy prices.[7] These ranged from constant real price (Scenario IV) to a quadrupling of real, delivered price by 2010 combined with vigorous public policies to cut demand growth (Scenario I). A second common assumption was that real GNP "only" doubles between 1975 and 2010. Depending upon these assumptions, total U.S. energy consumption in 2010 ranges from slightly less than current levels to about 140 quads, twice the 1972 level. It is also instructive to derive the ratio of energy to GNP, since this is roughly independent of growth rate assumptions. These results (Figure 5-11) indicate that even the heretically low demand scenarios seem to be within the long-term historical trend of the U.S. economy. Finally, the projected energy per capita (Figure 5-12), ignoring the dip during the Great Depression and the bump of World War II, can be seen as possibly continuing its post-WW II rise or, depending upon price, leveling out much as it did between 1910 and 1950. At a sufficiently high price, per capita consumption could fall as economic substitutes are introduced.

CONCLUSIONS

There exist a host of opportunities for more efficient use of energy, many of which are cost effective at today's average energy price. Many more options are cost effective when measured (as they should be) against present replacement (incremental) costs of energy supplies. Still more are effective if external costs (e.g., national security, environment, and health) and probable future cost increases are taken into account. However, the process of actually capturing

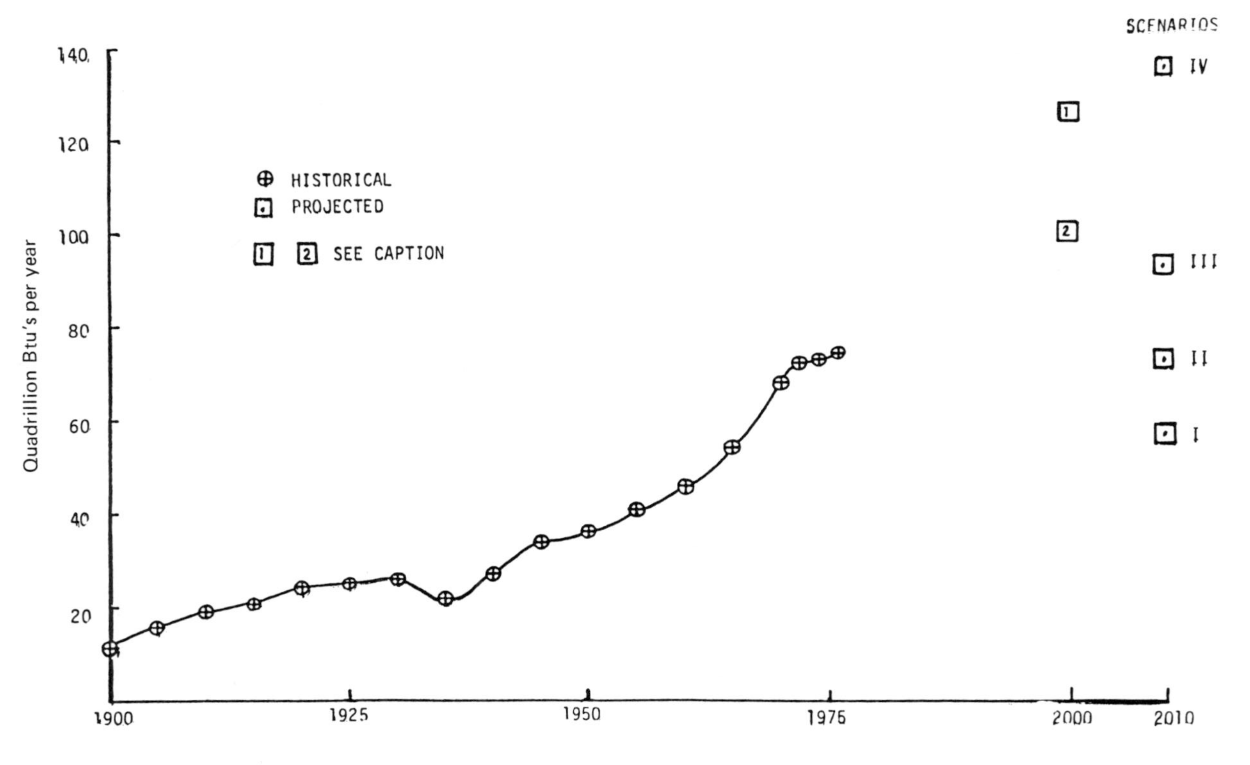

Source: J. Gibbons et al., "U.S. Energy Demand: Some Low Energy Futures," *Science* 200 (April 14, 1978): 142.

Figure 5-10. Summary of Total U.S. Energy Demand Scenarios. The results shown for comparison are [1] The Energy Policy Project "tech fix" case and the Institute for Energy Analysis "high" case and [2] The EPP "zero energy growth case" and IEA "low case".

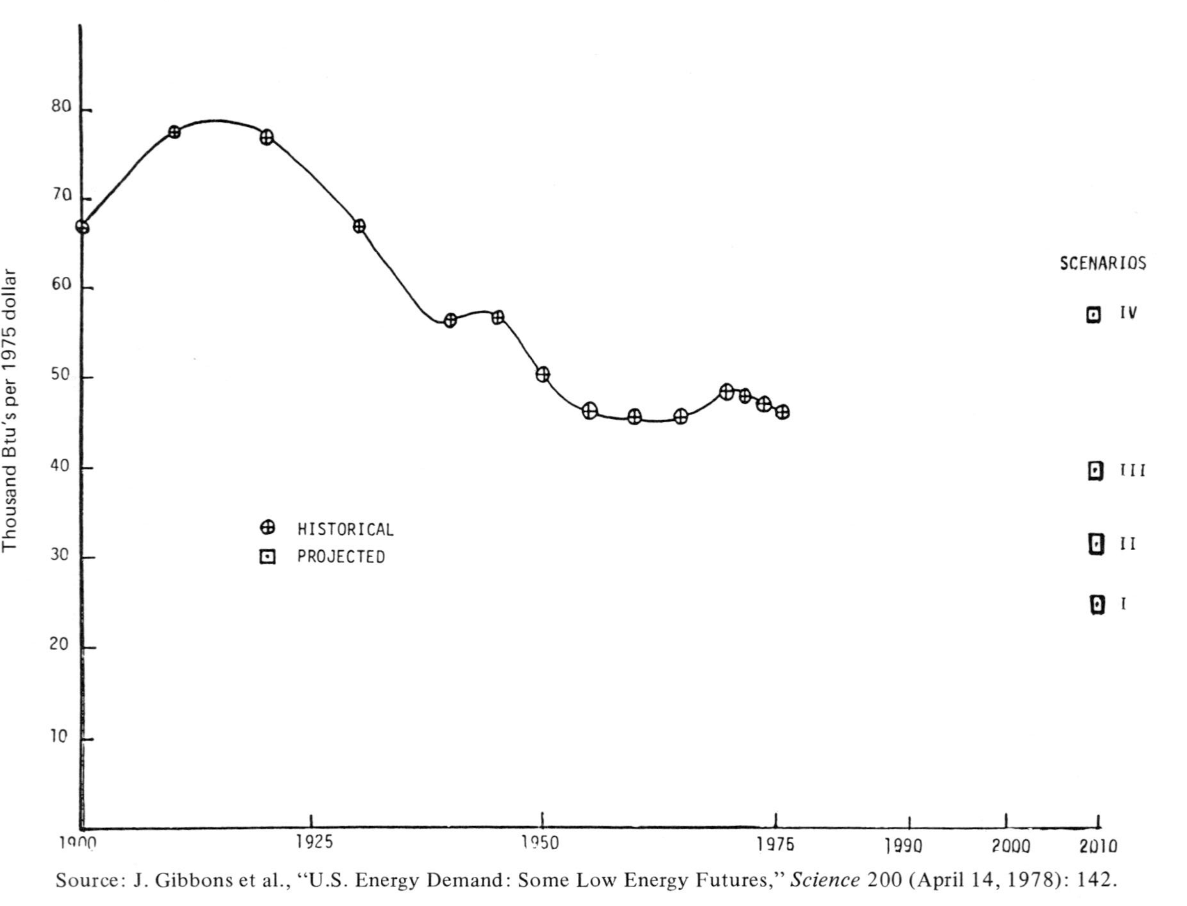

Source: J. Gibbons et al., "U.S. Energy Demand: Some Low Energy Futures," *Science* 200 (April 14, 1978): 142.

Figure 5-11. Time Dependence of the Energy Per GNP Ratio.

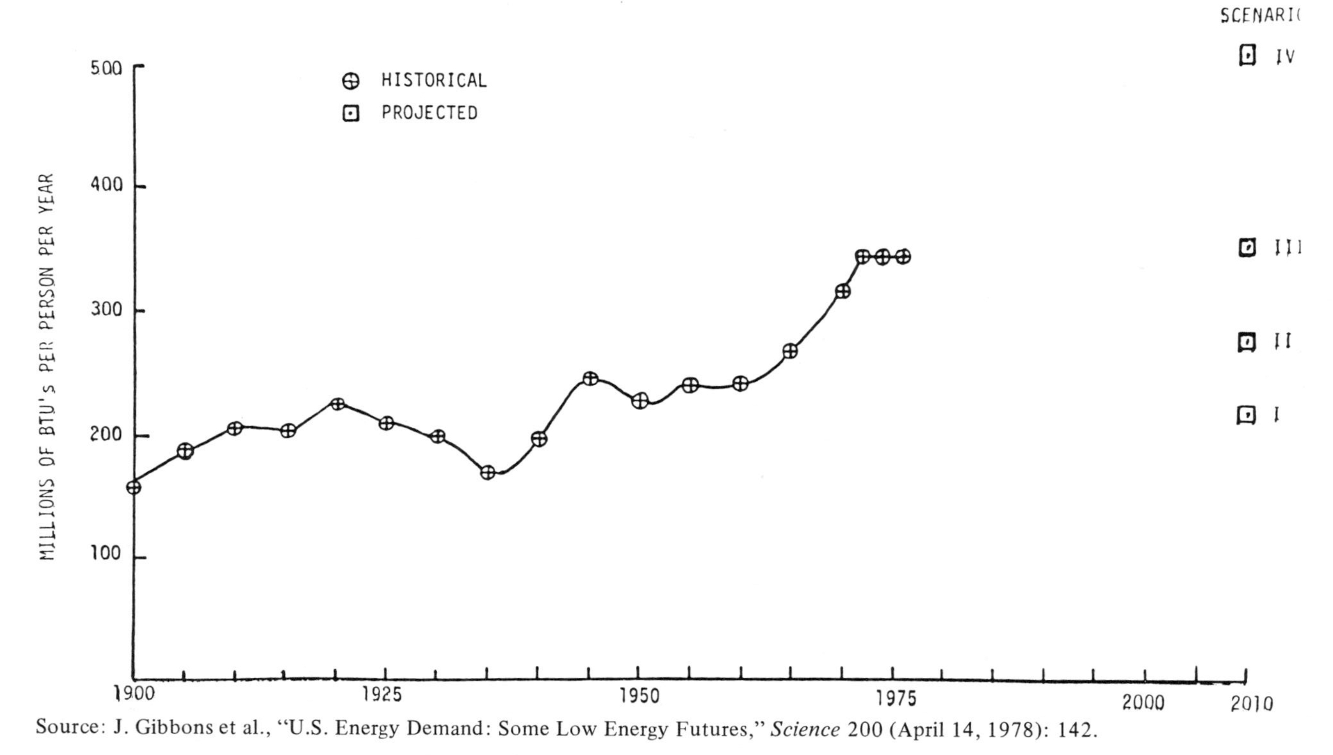

Source: J. Gibbons et al., "U.S. Energy Demand: Some Low Energy Futures," *Science* 200 (April 14, 1978): 142.

Figure 5-12. Per Capita Energy Demand History. Population growth rate was assumed to be that given by Census Bureau series II.

these opportunities is complex (although perhaps not as complex as some proffered new supply opportunities).

In supply various goods and services, energy is substitutible to a surprising degree, but such substitution derives from human ingenuity (in design, technology, and institutional innovation). Therefore, with their widely dispersed technical infrastructure, the industrial nations should be expected to lead the rest of the world in becoming more energy efficient and in reducing demand growth.

At a given time, energy demand reflects the characteristics of the existing energy-consuming capital stock and its rate of utilization. The long-term consumer choice of new stock depends upon energy price and other factors that affect consumer preference. Under "normal" conditions demand changes slowly, mostly with capital stock turnover corresponding to changes in population, income, energy price, and public policy.

Because major changes in energy demand have an inherently slow rate of change, there is a great imperative to plan and pace the process hopefully over decades, not years.

Lower demand growth has important implications for supply strategies during the "transition" over the next twenty to forty years. For example, (1) existing, lower cost resources will last longer, slowing the rate of price escalation and providing more time for transitions in new energy "sourcery". There are very important implications here for the timing and urgency of nuclear breeders.) (2) A given strategic reserve will last longer. (3) Some relief will be given to the international pressures on oil supplies.

NOTES

1. See, for example, Ford Foundation Energy Policy Project staff, *A Time To Choose: America's Energy Future* (Cambridge, Mass.: Ballinger, 1974); J. Gibbons et al., "U.S. Energy Demand: Some Low Energy Futures," *Science* 200 (April 14, 1978): 142. This is a summary of a more detailed report of the Panel on Demand and Conservation for the Committee on Nuclear and Alternative Energy Systems (CONAES), to be published Winter 1978–1979.
2. Gibbons et al.
3. Nils Bohr once remarked that "it is very difficult to make a prediction, especially about the future. . . ."
4. Gibbons et al.
5. See, for example, G.N. Hotsopoulos, E.P. Gyftopoulos, R.W. Sant, and T.F. Widmer, "Capital Investment to Save Energy," *Harvard Business Review,* March-April 1978, (pp. 111–123).
6. Eric Hirst and Janet Carney, "The Effects of Federal Residential Energy Conservation Programs", *Science,* Vol. 199, February 24, 1978, (pp. 845–851).
7. Gibbons et al.

Chapter 6

Realities of the Transition to New Energy Sources— A Manufacturer's View

Thomas H. Lee
General Electric Company, Fairfield

Changing energy sources is not a new experience for mankind. During the past 120 years or so, we have changed from wood to coal and from coal to oil, and it is most likely that we will be changing from oil to other sources in the future.

Although we are fortunate to have a hugh coal reserve in the United States, not every country is blessed with this good fortune. It is not surprising, therefore, that transition to renewable energy sources is receiving a great deal of interest and attention all over the world. But to bring about such a transition, there is a number of realities that we must face.

REALITIES OF THE ENERGY CRISIS

The first reality is an obvious one–the situation does not lend itself to quick and simple answers. The energy problem is an extremely complex mix of economic, technological, political, and sociological factors. We cannot provoke a solution merely by allocating enormous sums of money to develop new technology, and a vertically focused "Appollo Project" effort will not accomplish the task. Much broader aspects must be considered; for example, conflicting economic interests must be balanced. Foreign policy objectives, environmental concerns, and other evolving social values must be accommodated. Most important, public understanding and consensus will aid in the execution of difficult but sound political decisions. We must accomplish all of these tasks while preserving, to the maximum extent, our democratic principles and personal freedoms by avoiding mandated solutions.

Second, and contrary to public opinion, the Arab oil embargo and subsequent quadrupling of oil prices by OPEC did not cause our energy problem, any more than the cold winter of 1977 *caused* our natural gas shortage. These events have merely accelerated the need for coming to grips with some fundamental facts of life that were predicted by knowledgeable people a long time ago. In fact, in a letter to Congress in 1939, President Franklin Delano Roosevelt said, "Our energy resources are not inexhaustible . . ."; "It is difficult in the long run to envisage a national coal policy, or a national petroleum policy, or a national water-power policy without also in time a national policy directed toward all of these energy producers—that is, a national energy resources policy."

We are now depleting our own supplies of oil and gas at a prodigious rate; increasing dependence on others' supplies—more plentiful but still finite to be sure—while worldwide competition to obtain these supplies increases. If we stay on the present trajectory, the U.S. dependence on foreign oil will increase steadily, and in 1985 we may well be importing oil at double the current rate. Even with foreign supply and without taking foreign policy concerns into account, the reality is that *petroleum products cannot possibly continue to meet our needs for more than a few decades.*

Third, our energy needs will continue to increase over the foreseeable future. Some would interpret this as a value judgment of the growth-oriented business community. I intend no such value judgment, but submit that unless we recognize increasing consumption as a reality, at least for a while, we may be considered irresponsible planners for the future. Like it or not, economic development with a minimum of mandated allocations in order to maximize personal liberty has been our choice as the means for improving the human lot. That is, rather than ordering changes in the relative sizes of the pieces of the pie, we have chosen to increase the size of the whole pie. We also know that there has been a direct relationship between economic growth and the use of energy. While the connection between GNP and energy consumption may not be a basic law provable for all time, we do know that we cannot change it overnight without very disruptive consequences. Given that our population will continue to grow over the next century—though at a declining rate—increasing use of energy must be understood as a basic reality.

The fourth reality is that the United States, more than any other industrialized nation, has plentiful energy resources and alternative choices. Coal could supply our requirements for centuries. And nuclear fuel, with breeder reactors, would extend the energy cycle even longer. We have the potential, in time, to become self-sufficient to the degree warranted by political and economic considerations.

Within these bounds, it is clear that we must find the way to bring about a major shift in U.S. energy sources, regardless of any changes in lifestyle or personal values that may take place. The goal of transition in energy sources must be to responsibly make the energy available in order to maximize our country's options and opportunities. The price we will pay for being wrong or on the low side is very high and probably irreversible.

REALITIES OF THE TRANSITION TO NEW ENERGY SOURCES

Let us examine the realities of bringing this transition about. *First, the achievement of meaningful technological change takes a long time.* The conceptualization of an idea is only a short first step and is, unfortunately, where many who propose solutions stop. The technical feasibility must be proven in the laboratory, and pilot and demonstration stages must take place to prove practical feasibility and economic factors. These steps are required before commercialization and demand time-consuming investments. Additionally, other important questions must be answered, such as:

- Can it be developed into a viable source for a substantial percentage of our requirements?
- How long will it take?
- How much will it cost?
- What will the environmental consequences be?

It must be recognized that funding alone is a poor measure of actual emphasis and priority for a new technology when the technical life cycle is considered. Further, the development process cannot be arbitrarily accelerated simply by the allocation of funds. For most technologies, there are four stages, with an ascending order of funding needs:

1. *Conceptual stage.* This requires very little investment because brain power is relatively inexpensive, although hard to come by.
2. *Research stage.* To prove technical feasibility requires sophisticated experiments but not necessarily large investments.
3. *Developmental stage.* Building a prototype to investigate and study the problems that must be solved before a product can be made available for commercial application requires much larger investments.
4. *Design and manufacturing.* This stage requires facilities, tooling, process development, and support systems and usually requires more funds than all three previous stages combined.

For any technology, it is important to identify the stage it is in before deciding whether the program is adequately funded. As an example, $5 million a year to prove whether gallium arsenide is better than silicon for solar energy may be overfunded, while $300 million for nuclear breeders may be underfunded. And clearly, the premature heavy funding of technology efforts—for example, moving to the commercial facilities stage before technical and economic feasibility have been *demonstrated*—will not shorten the time needed to develop a commercially viable system. Misjudgments like this would be a blatant waste of our national resources.

Second, the magnitude of our energy requirements, particularly from oil, is enormous. In 1976, the United States used more than six billion barrels of oil, about 600 million tons of coal, and over two trillion cubic feet of natural gas. Thus, to have meaningful impact, any shift in energy technology must be able to contribute on a gigantic scale.

Much of the confusion and lack of realism in the energy debate results from failure to comprehend the requirements of timing and size on the management of technology transition. The development of nuclear energy provides a useful yardstick for reference. The underlying demonstration of scientific feasibility occurred in the late 1930s. The first small-scale engineering demonstration, the Staff field experiment, occurred in 1942. The initial 150 to 200 MW power reactor projects—Dresden and Shippingport—were completed in the late 1950s, and the first large reactors, capable of producing electricity at competitive costs, began to operate in the early 1970s.

Yet after over two decades and tens of billions of dollars, only 10 percent of the electricity in the United States is generated with nuclear power. Moreover, a huge job is still ahead (1) to make nuclear systems perform to the standards of excellence that we like to see, (2) to develop a fuel-reprocessing industry, and (3) to deal with long-term nuclear waste disposal. It is probably safe to say that the issue of public understanding and acceptance of nuclear power is a long way from being resolved.

Does this mean that we should abandon nuclear power or any other technology that similarly challenges us just because we are still in the process of commercialization? I would rather think not.

To bring about the transition from the present, largely petroleum-based, world economy to whatever future energy sources our children and grandchildren will depend on, there are many other realities that we must face. Some relate to how to manage a project with a sensible expectation, while others relate to the forecasting of the economics of new energy sources, and I will discuss them in this order.

In allocating our resources to bring about a transition of energy sources, we must strive for a balance between research and development. This may sound like a simple question of semantics. On the contrary, in my opinion the clear distinction between research and development is of vital importance when we face the task of allocating resources.

Rightly or wrongly, I have felt for some time that the successful launching of Sputnik by the Russians had a major impact on the American public, particularly on the industrial R&D effort. Our reactions to that event were strong, but unfortunately, in my opinion, not always logical. For example, Sputnik was basically an engineering success. The scientific laws governing the flight had been known for more than 200 years. To implement it, we needed advances in chemical engineering, in mechanical engineering, in guidance systems and instrumentation. Yet the reaction in the country as a whole was that we had better step up scientific activities. Many engineering schools practically stopped teaching engineering. Instead they produced pseudoscientists, and consequently, many Ph.Ds from electrical and mechanical engineering departments simply refuse to do engineering work. What they really wanted to do was physics, frequently in a very narrow aspect. Thus, a great engineering success in the Soviet Union caused a tremendous setback in our engineering education.

In industrial circles, the rush for R&D was on too. Most corporate executives felt that it would not be right if they did not have a research laboratory. Just what they expected from the researchers was another matter.

The built-in image for research projects was that they must be long range, glamorous, and something that no one had done before. Projects like these usually carry with them many scientific and technical problems that indeed need a great deal of research effort to solve. However, as businessmen, they wanted to see hardware–demonstration installations instead of research reports and curves. Therefore, they put pressure on our research people to build demonstration installations, and therein lies the dilemma of industrial research. Once one understood the dilemma, it was easy to trace the rise and fall of industrial research laboratories in this country. Even worse, our researchers, under the pressure of needing financial support, were willing to sacrifice some of their professional integrity in the process of preparing proposals. Instead of calling a spade a spade, instead of describing research in terms of studying one or more phenomena, research proposals were written with the hope of demonstrating the operation of prototypes. The latter type of work is really development.

Recognizing the differences between research and development is a management function. The output from research is usually information, while the output from development is some kind of prototype. Research projects usually require less money than a major development project. On the other hand, for any specific application, we must allow researchers to explore many possibilities, and we must be prepared to accept negative, disappointing answers as part of life. For major developments, we cannot possibly afford going down too many avenues, and the probability of success at the initiation of a project should be very much higher than that for a research project.

Figure 6-1A is a pictorial representation illustrating this point. If the magnitude of each arrow represents the dollars that we should spend on any single idea in research, the distribution of financial resources should look like the picture of the left–namely, a large number of modest exploratory projects in as many directions as possible. For development, it should look like the picture on the right. The real fact of life is either as shown in Figure 6-1B in business organizations or as shown in Figure 6-1C where money is no object.

Two other realities that are very important to the subject of this forum are the "learning curve" concept and economy of scale.

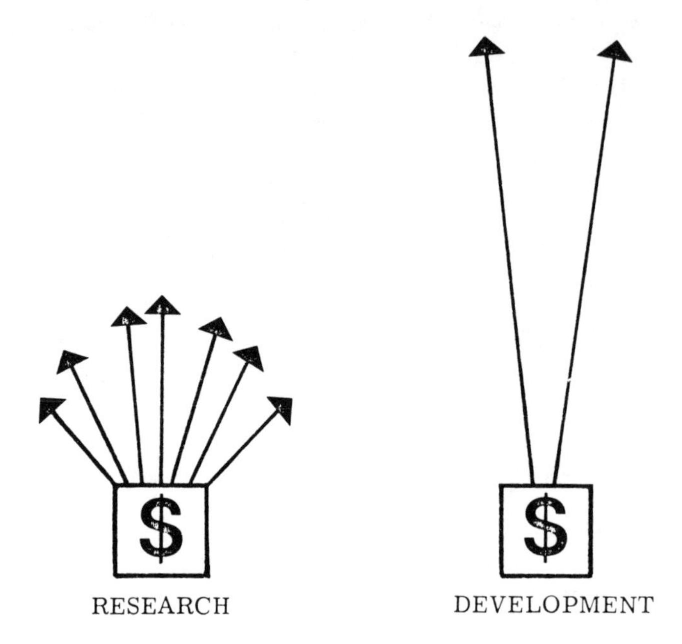

Figure 6-1A. Ideal Allocations of Research and Development Dollars

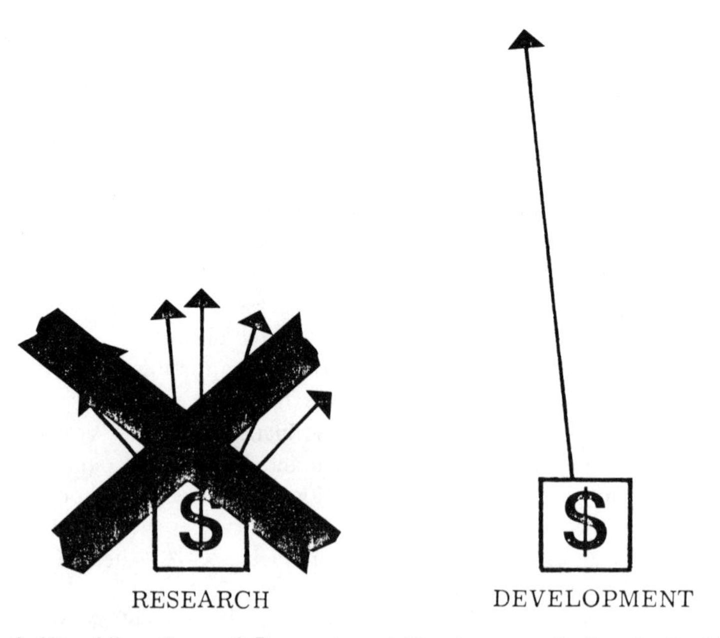

Figure 6-1B. Allocations of Research and Development Dollars in Real Business Organizations

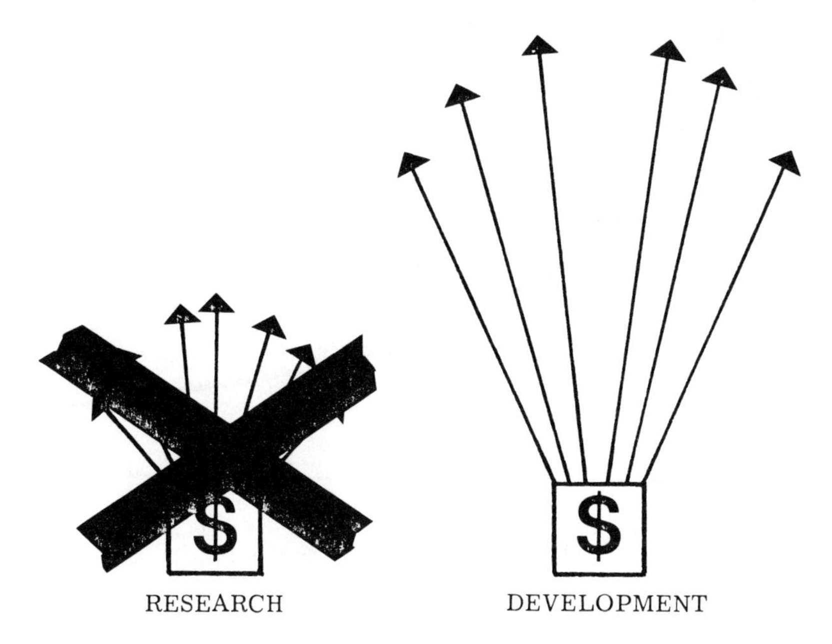

Figure 6-1C. Allocations of Research and Development Dollars Where Money is No Object

REALITY OF THE LEARNING CURVE

Instead of writing a dissertation on learning curves and the related theory of market share, I will quote from a publication of the Boston Consulting Group: "Experience Curves as a Planning Tool."

> The experience concept is related to the well-known *learning-curve* effect. The aircraft and electronics industries often use it to guide both their cost control decisions for assembly operations and their pricing policies. The Department of Defense procurement officers use the learning-curve effect in setting cost targets in cost-based contracts.
>
> The well-known learning curve relates the direct-labor hours required to perform a task to the number of times the task has been performed. For a wide variety of activities, this relation has been found to be of the form shown in [Figure 6-2], in which cost to perform decreases by a constant percentage whenever the number of trials is doubled. Plotted on log-log scales, this relation becomes a straight line with a slope characteristic of the rate of "learning," such as that shown in [Figure 6-3].
>
> A 20 percent reduction in hours for each doubling of performances—or what is called an 80 percent curve—is typical of a very wide variety of tasks.

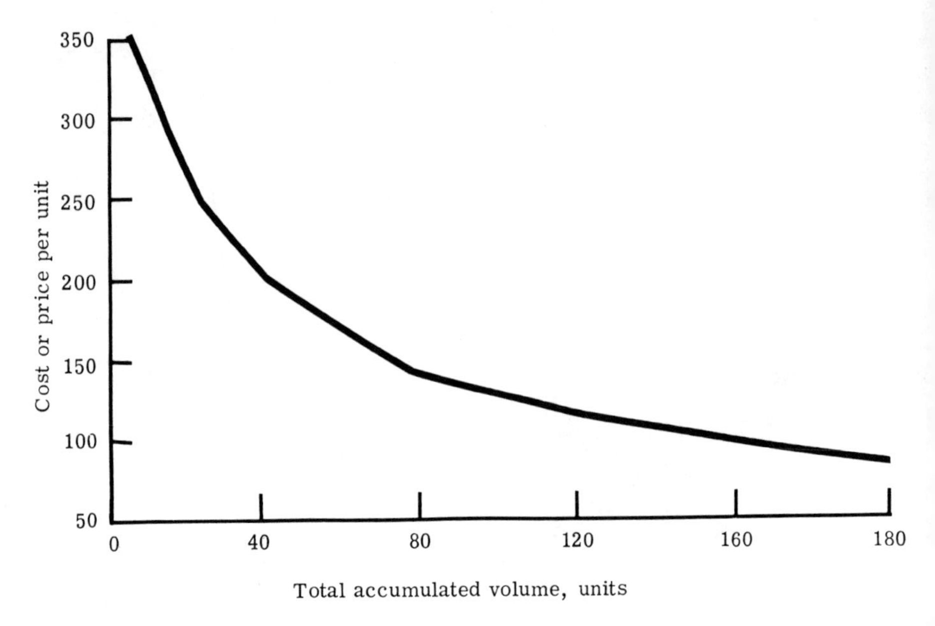

Figure 6-2. Learning Curve for Labor Costs

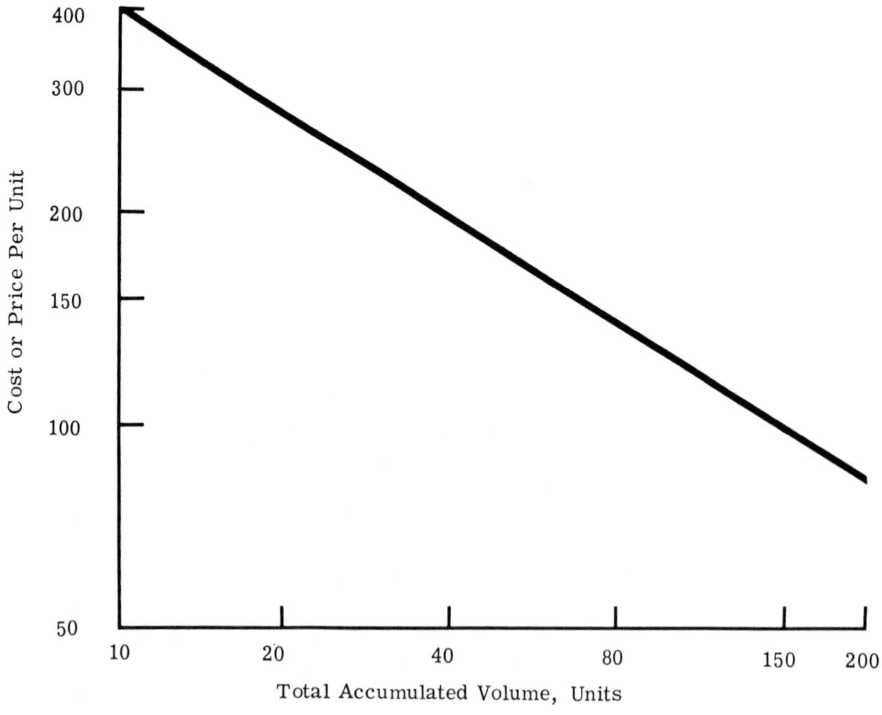

Figure 6-3. Learning Curve in a Log-Log Plot

> In spite of long-standing awareness of the learning-curve phenomenon and its effect on costs, . . . its obvious strategic implication seems until now to have been overlooked. If cost declines predictably with units produced, the competitor who has produced the most units will probably have the lowest cost. Since the products of all competitors have sensibly the same market price, the competitor with the most unit experience should enjoy the greatest profit. Furthermore, it should be clear that very substantial differences in cost and profit can exist between competitors having widely different unit experiences.
>
> Over a period of time when market positions are relatively stable, experience can be equated with market share. Thus, market share and profitability are closely related.

The paper then went on to show many examples that prove that the theory really works in many cases, from semiconductors to Japanese beer (see Figures 6-4 and 6-5). Therefore, one conclusion that can be drawn is that market share is a key strategic parameter. One should fight for higher shares. My purpose in this discussion is to prove not that the theory of learning curve is wrong, but that to apply this theory without understanding the business dynamics can be very dangerous.

In the case of nuclear industry, the application of market share strategy has led to serious problems for the entire industry. There are many reasons behind this:

- In the last fifteen years, the size of commercial reactors offered was increasing at a rate of approximately 50 MW per year (Figure 6-6)—that is, the largest reactor offered fifteen years ago was about 300 MW, and today the largest one is about 1200 MW. The total cycle time—from time of receipt of order to commercial operation has increased from five years to about 10 years. Yet in a short period of twenty years, industry has gone through six different designs. In other words, when the second design was being worked on, there was no feedback from the experience of the first design at all. The entire foundation of the learning curve was shattered.
- The nuclear industry is subject to regulatory restrictions. Regulations change from year to year. The design that was licensable fifteen years ago is not licensable today. Again, this makes the learning curve theory invalid.

In spite of all these problems, market share was the dominating strategy for nuclear suppliers. The consequence is: No one in the industry is profitable. Examples like this lead one to the very important conclusion that we must understand the dynamics of the industry in estimating the economics of a new energy technology. Blind application of the learning curve concept can lead to erroneous conclusions.

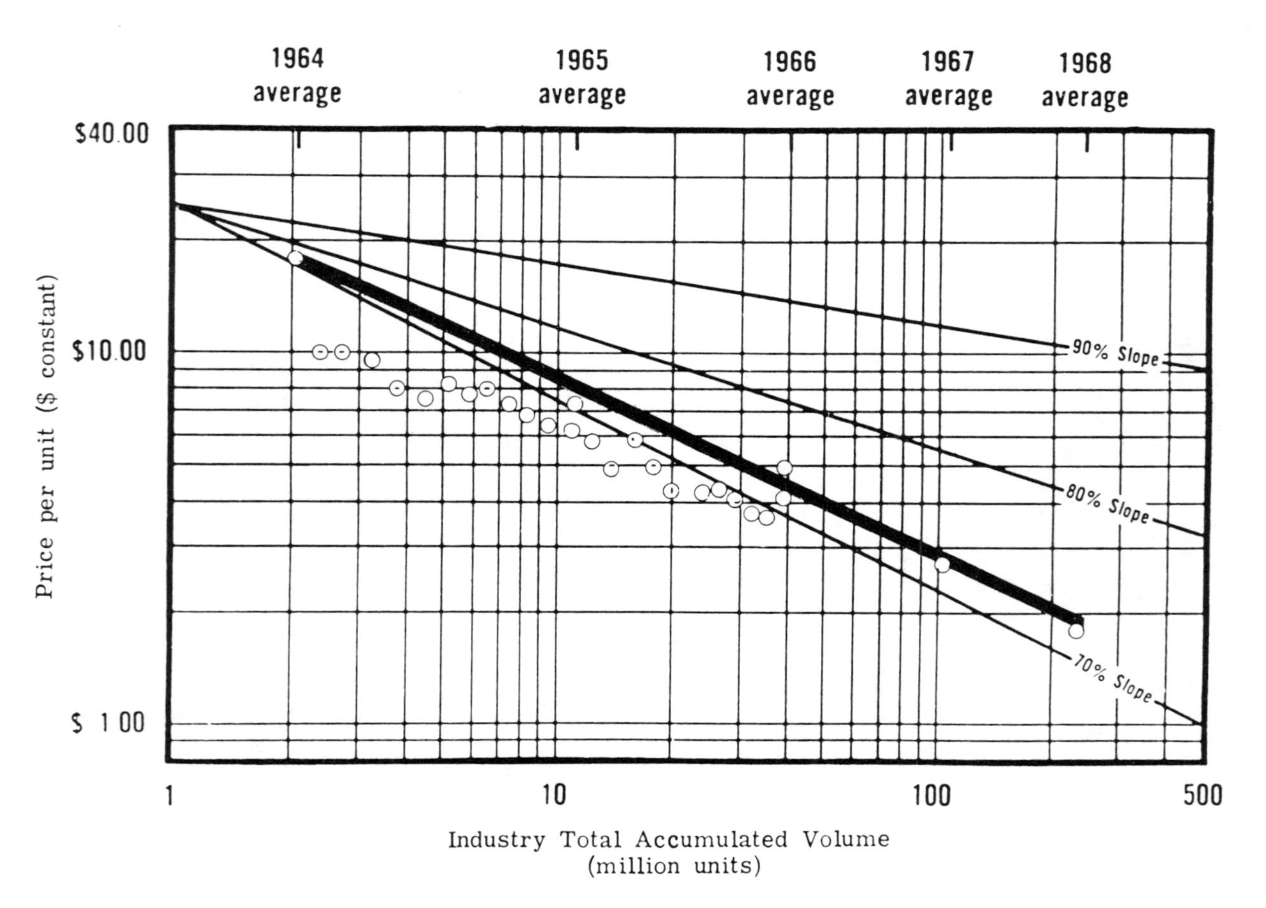

Figure 6-4. Learning Curve for Integrated Circuits

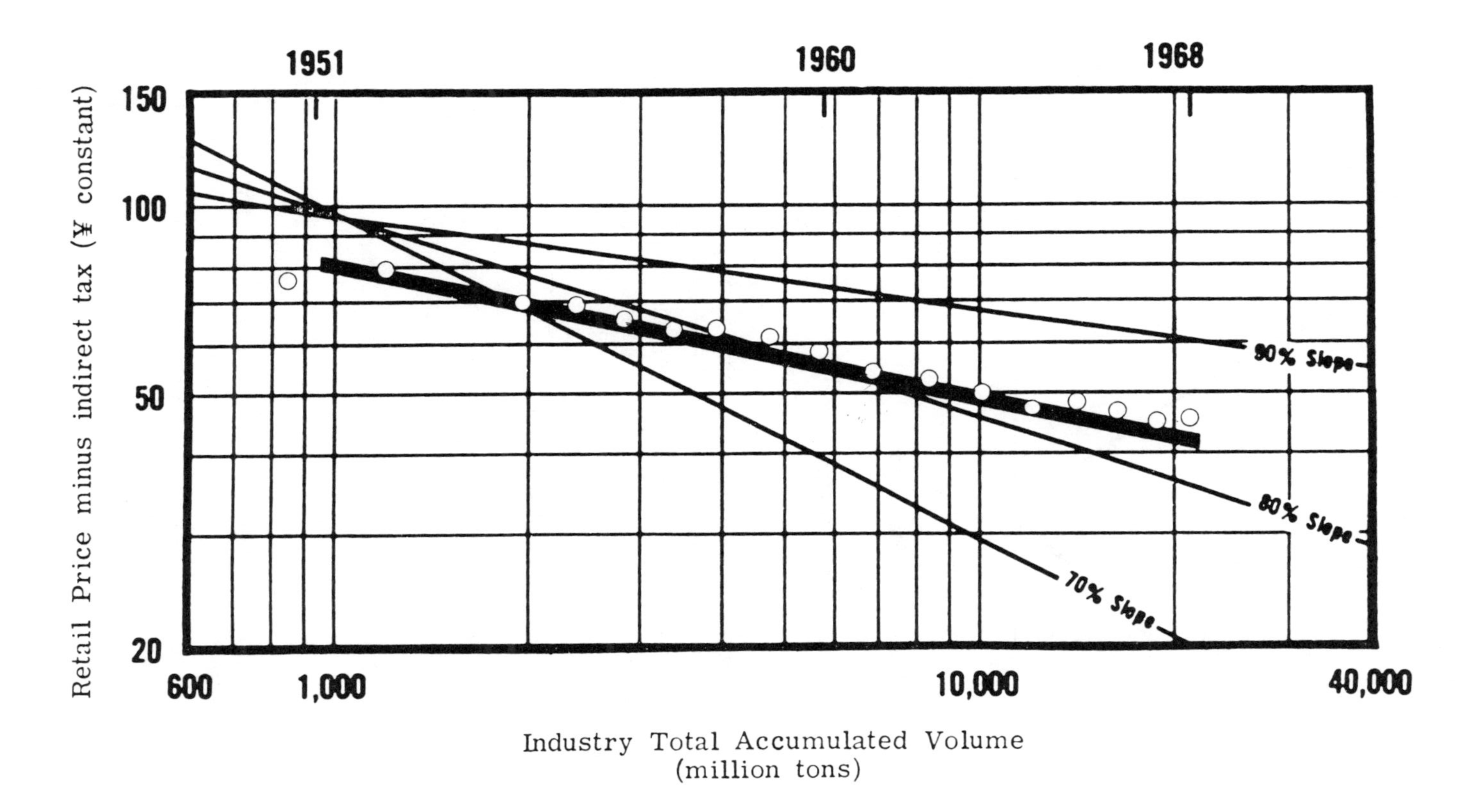

Figure 6-5. Learning Curve for Japanese Beer

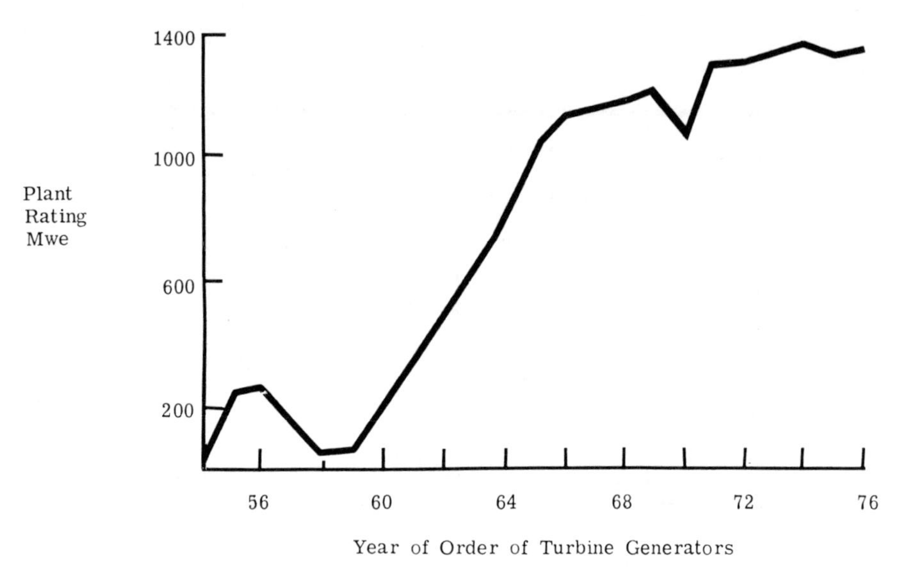

Figure 6-6. U.S. Nuclear Plant Orders

REALITY OF ECONOMY OF SCALE

The question of size, or economy of scale, is a concept that is even older and more established than the concept of learning curves. Figures 6-7 to 6-10 show the direct cost of four different kinds of equipment–aircraft engines, steam turbines, power transformers, and power electronic equipment. It is obvious that for each type of equipment, there is a significant economy of scale–that is, the $/kW or $/HP figure drops significantly as the size of the equipment is increased. This concept has been applied to the construction of electric power plants for many years. However, the initial capital cost is only one of the elements in the cost of electricity. Another very important element in the equation for cost of electricity is the capacity factor, which is the ratio of the actual kW hour output to the theoretical output. The effective capital cost of a plant is really given by:

$$\frac{\text{Plant Cost (\$)}}{\text{Plant Rating (MWE)} \times \text{Capacitor Factor}}$$

The performance of existing power plants indicates that there is a relationship between the capacitor factor and the size of the plant. For fossil plants, it is shown in Figure 6-11. The statistics for nuclear plants are much less but a similar trend exists (Figure 6-12). Whenever curves like these are shown, proponents of large size would argue on the ground of (1) inadequate statistics, and (2) learning effect–in due time the trend will be changed. And the believers of the statistics argue another way:

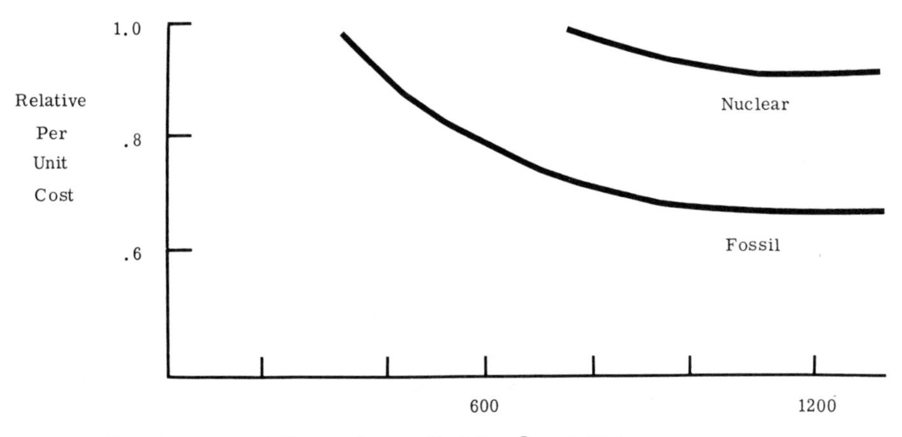

Figure 6-7. Economy of Scale—Steam Turbine Generators

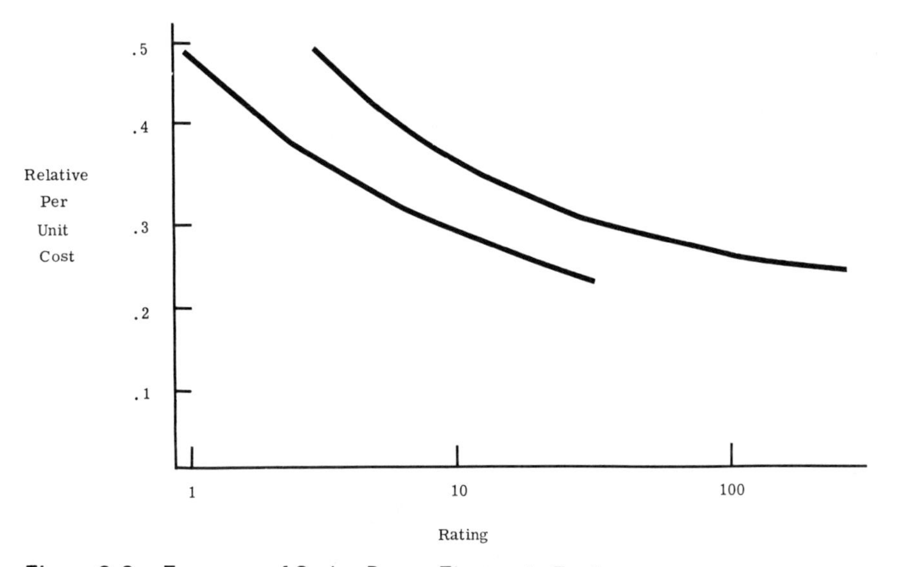

Figure 6-8. Economy of Scale—Power Electronic Equipment

1. For a standard design, a scaled up design usually contains more identical parts. The probability of failure for a given failure mode is therefore directly proportional to the size.
2. The downtime is a function of size. It is so much easier to fix a bicycle tire than the tire of a ten ton truck.

There are other technical reasons, such as that the design stress for a large machine may be much closer to the theoretical stress limit than a smaller

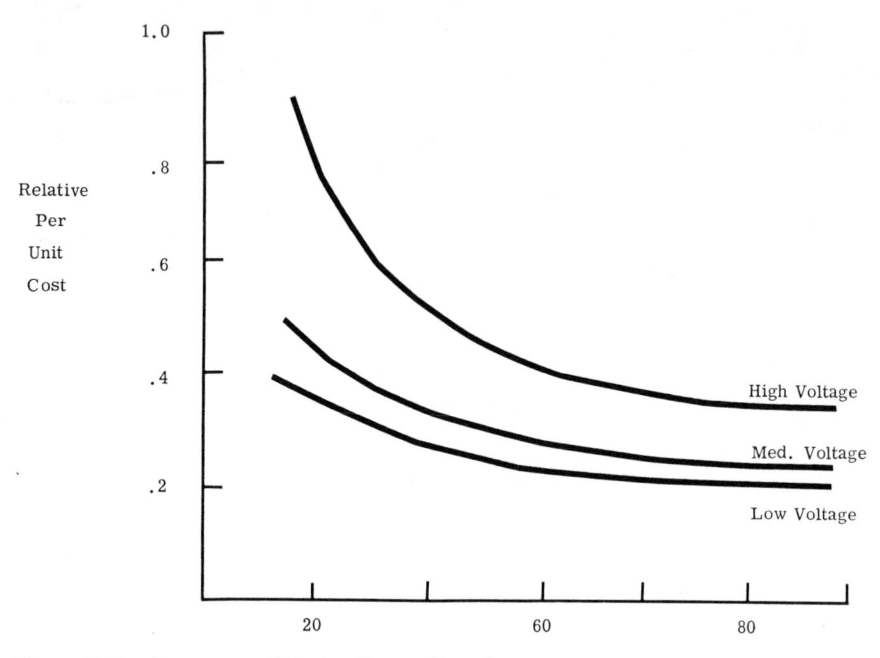

Figure 6-9. Economy of Scale—Power Transformers

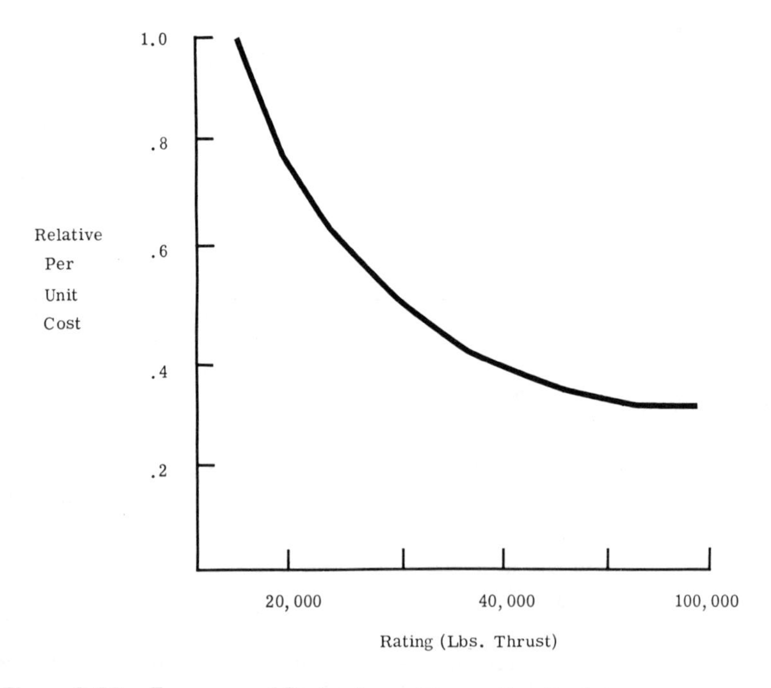

Figure 6-10. Economy of Scale—Large Bypass Fan Engine

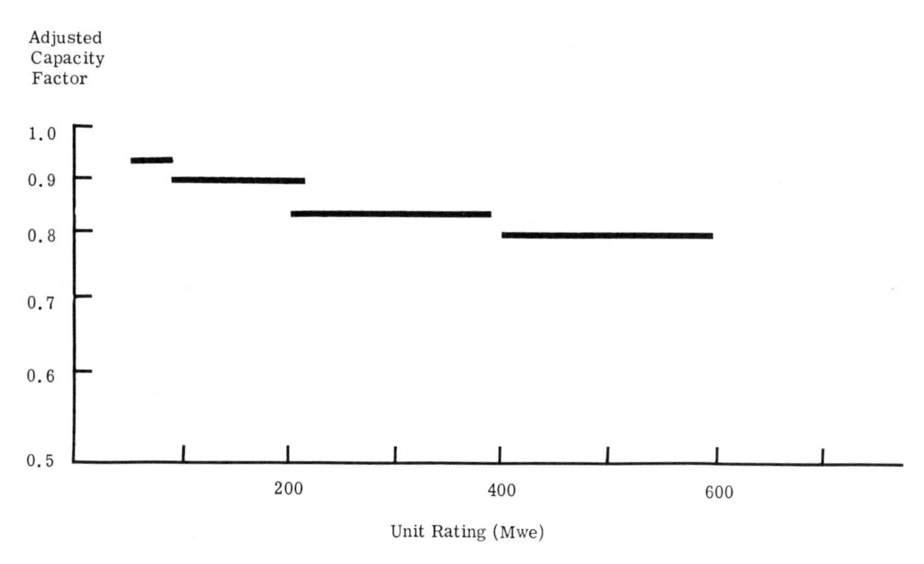

Figure 6-11. Performance of Mature Subcritical Fossil Steam Units

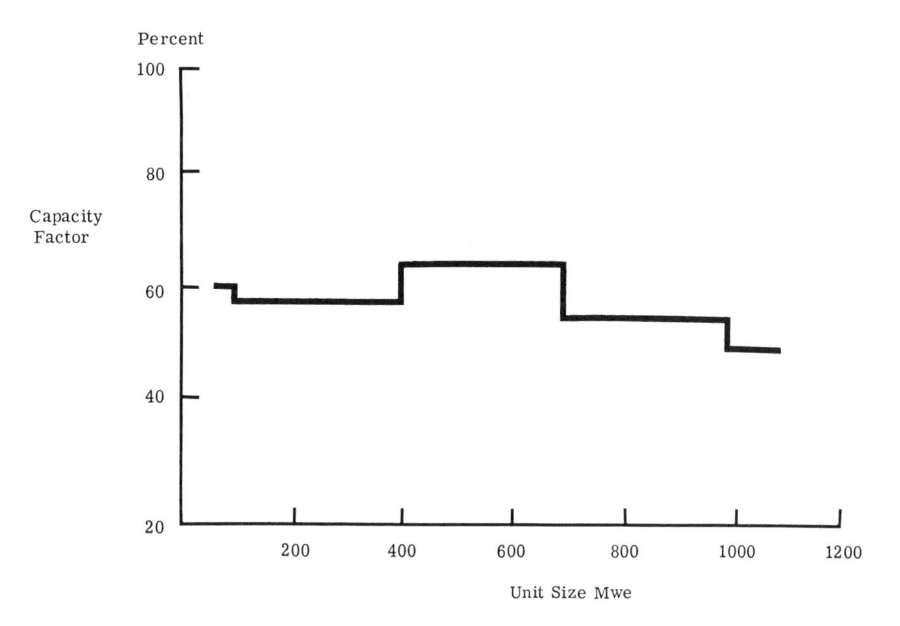

Figure 6-12. Annual Nuclear Unit Capacity Factor vs Size

machine, and yet, the random probability of failure of materials is determined not by the theoretical limit but by the probability of imperfections that could be related to size.

Both arguments are valid to some degree. The truth probably lies somewhere in between. Let us not worry about who is right, but rather, let us make a sensitive analysis. To do this, we will define two parameters:

1. A, which describes the economy of scale—that is, cost in \$/kW $\sim R^{-A}$, where A may vary between 0.3 and 0.5 (as shown by Figures 6-12 to 6-13C.)
2. C, which describes the fractional drop in capacity factor for each increase in plant rating of 1000 MW. If C is equal to 0.2, it means that the capacity factor drops by 20 percent for each increase in plant rating of 1000 MW.

If the cost and capacity factors indeed follow these relationships, it can be shown that there is always an optimum size. But merely knowing the optimum size is not enough. We must know the implications when we deviate from the optimum. This can be done by examining the relative costs for a range of As and Cs. For example, Figures 6-13A and 6-13B show the relative costs for C = 0.2 and C = 0.4. It is seen that for sizes larger than the optimum, the difference

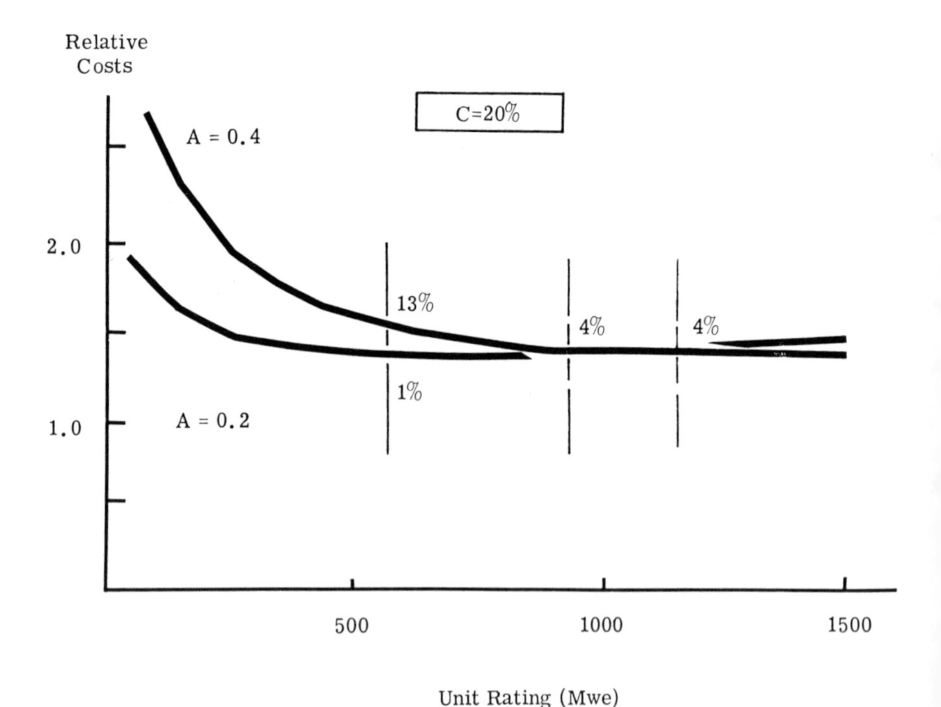

Figure 6-13A. Relative Costs for $C = 20\%$

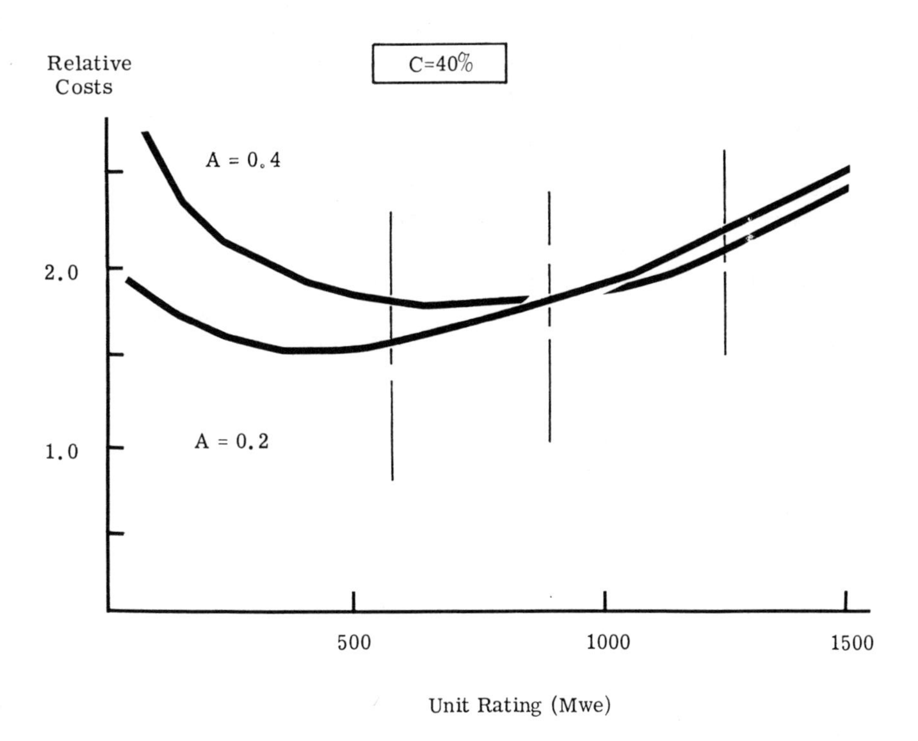

Figure 6-13B. Relative Costs for $C = 40\%$.

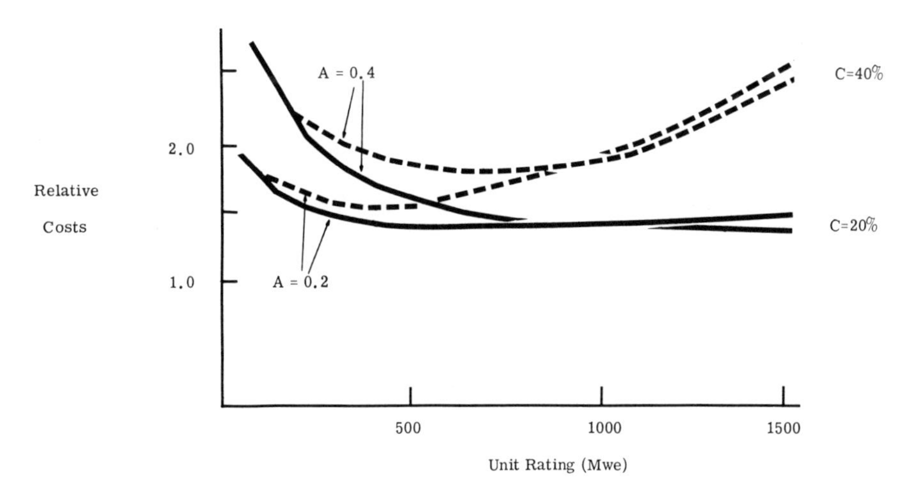

Figure 6-13C. Relative Costs for both $C = 20\%$ and $C = 40\%$

between the relative costs for the cases $A = 0.2$ to 0.4 is small—that is, economy of scale is relatively unimportant, but the difference is significant for sizes smaller than the optimum. Figure 6-13C is a combination of the two previous figures. In performing the following sensitivity analysis, basically, we do not know either A or C. But let us assume that we know A is 0.4. If we had guessed that $C = 40$ percent, we would have chosen a smaller plant, say about 600 MW. Suppose that guess was wrong and that C is actually closer to 20 percent. We should have chosen a plant closer to 1,200 MW. The interesting fact is that on the curve for $A = 0.4$ and $C = 0.2$, the difference between the relative costs of 600 MW and 1,200 MW is rather insignificant. Now suppose that we had guessed $C = 0.20$ and chosen the plant size of 1,200 MW and it turned out that C is actually equal to 0.4. The costs for a plant of 1,200 MW with $C = 0.4$ is very much higher than for a plant with $C = 0.2$. One can try other sensitivity analyses, such as assuming that we know C but not A, and so on. These analyses will all lead to the conclusion that it is safer to err on the smaller side.

CONCLUSIONS

Transition of energy sources from the present petroleum-based economy to something else, renewable or not, appears to be inevitable. But a successful and smooth transition for this time will be far more difficult because of social, political, and environmental reasons. Understanding of the realities is a necessary condition for the successful management of the forthcoming transition. In this chapter we have suggested a direction that more comprehensive analyses might take.

Session B

Energy Models for Exploring Alternative Evolving Balances

Part I

Impacts of Energy Strategies on the U.S. Economy

Chapter 7

Alternative Futures With or Without Constraints on the Energy Technology Mix

Tjalling C. Koopmans
Yale University

My name is Tjalling Koopmans. I started professional life as a student of theoretical physics under the guidance of Professor Hans Kramers. In 1934 I moved away and by stages became an economist. I am now at Yale University. My memories of physics concern the elements, but not all the astounding developments that have taken place since then.

I will make a few introductory remarks to this session. After that, I will turn around and temporarily assume the role of Dissertator.

Energy modeling has taken its stride in the last six or eight years. Its intellectual roots lie in earlier work at the Electricité de France and in more general econometric model building in the Netherlands, in the United States, and thereafter in many other countries. Particular recent flowering of energy-economic modeling has occurred in the United States and at the International Institute for Applied Systems Analysis in Austria. Similar work has also been done in India, Mexico, and several European countries.

The aim of energy modeling is to visualize alternative energy futures in a quantitative way. The information inputs to the models are drawn from various fields of knowledge and of informed speculation. The technological information is drawn from the sciences and engineering. The behavioral information is drawn from economics in regard to consumers' and producers' behavior and, where appropriate, from the other social sciences. Resource availability and cost data are drawn from geology and from mining engineering economics. This is not a complete list, but those are perhaps the principal founts of information and data.

The speculative element is inevitable in any modeling activity that looks forward into an extended future. But the words to avoid are "prediction" or "forecast." What the new modeling profession produces is neither predictions nor forecasts. I prefer to call them *projections.* They are statements of the form: If such and such, then so and so follows. The findings are not limited to stating the "thens." They take the form of a collection of "if . . . then . . . " statements.

The modeler's expertise and responsibility concern the statements so formed. The modelers should also be ready to defend their choices of alternative ifs just looked at by themselves and to explain why, where, and how they obtained them. But the ultimate responsibility for assessing the "ifs" is with the fields of knowledge that are drawn upon in their formulation.

The first chapter of this session puts one and the same question to three different models and compares the answers obtained. This question concerns the effects of given constraints that for whatever reason may be imposed on the supply or the use of energy. The principal effect explored is that on the total output of the economy.

The second chapter has a conceptual intent. It looks at various ways in which that question can be formulated and examined.

The third chapter was written independently of the second and conversely. It combines a so-called process model of the energy sector with an econometric model of the rest of the economy, in order to obtain or refine empirically based answers to the same question. All three chapters have one important set of assumptions in common; they presuppose the underlying model of a competitive market economy, and they study the impacts of specific hypothetical interventions on the part of the government in such an economy. In other respects and in other assumptions, these three formulations will differ from each other, to give us the benefit of a variety of conceptual and empirical approaches.

This chapter is concerned with the impacts of various constraints on energy supply as seen by different models. It is mostly a report on one component of a recent energy modeling study that was undertaken for the Committee for Nuclear and Alternative Energy Systems (CONAES). This in turn is a committee of the National Research Council engaged in a study of long-run energy problems for the two national academies (of sciences and of engineering). As part of the work of its "synthesis panel," a "modeling resource group" (MRG) was set up. Several members of that group, for which I served as chairman, are in the program of the present Forum. I mention Kenneth Hoffman, and William Hogan, Alan Manne, Robert Litan. My remarks here consist mostly of a summary of the work of the group as a whole. If my own opinions enter in, I may be unaware of it.

Figure 7-1 gives a compressed listing of the ifs used whenever possible in all models compared. I will not give the numerical values in all cases. The *ifs* fall into three categories. The first category is called *realization variables.* It consists of assumptions about variables that, in first approximation, are generated by processes independent of the energy policies under consideration. First comes an assumption about the GNP growth rate from 1975 to 2010. While this is assumed to vary somewhat over time, it averages 3.2 percent per annum over

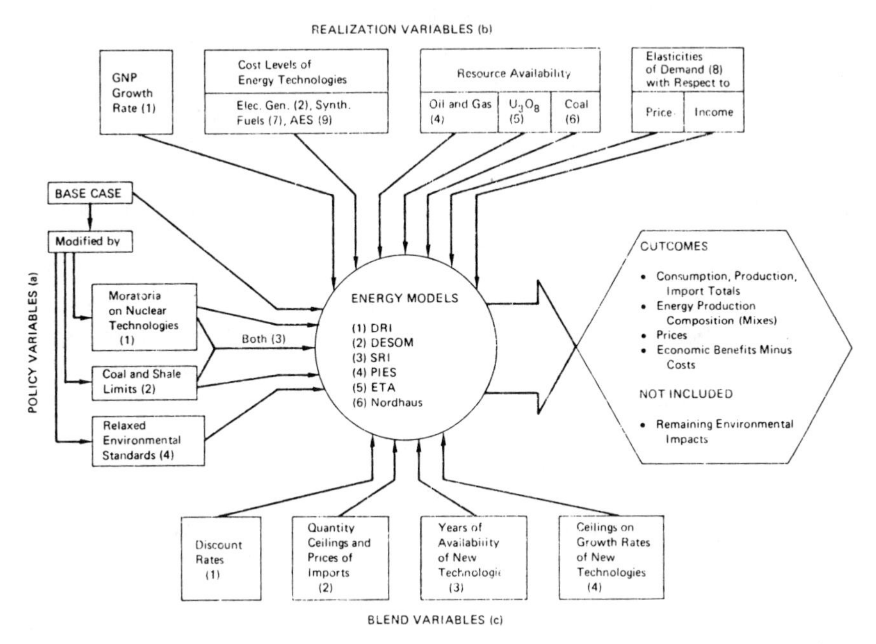

Figure 7-1. Compressed View of Driving Variables, Energy Models, and Their Outcomes.

that period. Second are the cost levels of energy technologies, with particular emphasis on the capital cost of electricity generation. All costs are expressed in real dollars as of 1975: coal-fired stations at $520 per kilowatt-electric of capacity, light water reactors at $650; advanced converter reactors at $715; liquid metal fast breeder reactors at $810; and solar central station at $1,730, more than twice the highest cost of the other technologies. The next category of realization variables estimates resource availability with regard to oil, gas, uranium, and coal. An important additional realization variable is the price elasticity of demand for energy. This parameter is defined in absolute value as the percentage decrease in demand divided by the percentage increase in price that calls forth that decrease in demand. It is represented by a negative number. While it can be affected by energy policies, this is not assumed to be the case for the policies here considered.

On the left is a listing of what I shall call *policy variables.* First of all there is the base case. This is defined as an energy policy regime not different from what was in force in 1975. Three variants of this case are considered, which I will spell out later on. The motivation for looking at these cases is to provide estimates of what would happen if alternative policies that might be considered are in fact put in place. It is not claimed, assumed, or denied that any of these policies is desirable. To answer such questions is not a modeler's job. Of the alternatives considered, the first set specifies nuclear moratoria—that is, decisions whereby for certain classes of reactors no new construction will be started after the date from which the moratorium is initiated. The second set, coal and shale limits, is considered in recognition of concerns about the environmental impacts of coal mining, such as mining accidents, acid rain, or increases in the concentration of CO_2 in the atmosphere. The latter, a consequence of increased use of any fossil fuels, is still uncerain in its effects. However, enough worries have been expressed to make it of interest also to consider a possible limitation on this account. The last one is a relaxation of environmental standards.

The third category consists of *blend variables*—that is, variables that have both a realization aspect and a policy aspect. As an example, discount rates reflect both the actual rates of return earned in the capital market and policy decisions as to how to relate the discount rates used for evaluation of benefits and costs to the readings of the market. As a matter of, if you like, innovation in methods, the MRG has used two different discount rates in different parts of its calculations. One rate, the pre-corporation-tax rate estimated at 13 percent, was assumed to apply to actual future business decisions on investment and pricing. The other, a 6 percent posttax discount rate, was assumed to determine the balance of contemporaneous future consumption flows that the consumer has open to him when he makes his savings decisions. This rate was used to reduce future benefits and costs to the reference point of 1975 in estimating the net social benefit from alternative energy policies.

Other blend variables included import ceilings and import prices, particularly on petroleum and liquefied natural gas, as well as the assumed year of commercial availability of the various technologies, put at a uniform year 2000 for the nuclear breeder reactor; the advanced converters; and the solar electricity-generating station.

The questions of the effect of the policies listed on the left were then asked of a number of models. To this end the study projected a number of items in our list of outcomes: energy consumption, production, imports, prices. The emphasis in this chapter is on the aggregate net economic effects from technology choices or constraints.

I have already mentioned one further assumption that is not recorded in this diagram—that of a competitive market system, which in modeling technique is simulated by an optimization. This goes back to Adam Smith, who spoke of the invisible hand acting through the price incentives to bring about what we now call an efficient use of resources. This idea is here extended into the future. It is implicitly assumed—I do not know to what extent Adam Smith speculated along these lines—that there is a great deal of correct foresight on the part of decisionmakers. For instance, if an energy R&D policy is announced for the long run, the model assumes that all investors in future technologies correctly estimate the totals that will be invested by the energy industry and fit their own plans into that estimate.

Three of the six models listed in the center of Figure 7-1 were used. All of these were both long-range and optimization models. The first of these was DESOM (evolved at Brookhaven Laboratory from BESOM, originally developed by Kenneth Hoffman; since then, William Marcuse has been in charge of the application and further development of that model). DESOM distinguishes itself from the other models in that it postulates an exogenously given time path of the vector of energy end uses. There is, of course, in reality a dependence of end use on price, but for reasons connected with the original purpose of the model, this assumption was made. The second was ETA (Energy Technolgoy Assessment), modeled by Alan Manne of Stanford University; and the third, a model developed by William Nordhaus, then of Yale University, but at present on the Council of Economic Advisors to the President.

I have already mentioned that the elasticities of demand for energy, particularly the price elasticity of demand, turned out to be quite important to the results (see Table 7-1). For the DESOM model we do not have a numerical price elasticity of demand. Its authors may have estimated one, but for our purposes it is sufficient to note that it is quite small, because the only price response of demand for primary energy that can occur in that model is the application of more capital or of ingenuity in order to save energy by greater efficiency in the various conversions and other steps from primary energy extraction to energy end use. The ETA and Nordhus estimates (shown in the second and fourth columns in the table) are rather different numerically. That fact, though puzzling in itself, was of quite some help in tracing the effect of the price elasticities. In any case, these elasticities came with the models, and therefore are inherent in all the projections made with these two models, except that Dr. Manne also made available some alternative projections based on a price elasticity indicated in the third column by −0.5. And that, again, was very useful and informative.

Table 7-2 gives a list of policies in more detail. Two separate definitions of moratoria are considered. The first is on all advanced converter reactors and fast breeder reactors (and this is, again, a moratorium only on starting new construction, not in regard to construction already underway). The second policy is a moratorium on all nuclear technolgoies, including construction of new light

Table 7-1. Estimated Price and Income Elasticities of Demand for Aggregate Energy in Three Models.

	DESOM	*ETA*	*ETA (−0.5)*	*Nordhaus*
Elasticity of demand for aggregate energy with respect to:				
Price	small	−0.25	−0.50	−0.40
Income	0.75	1	1	0.90

For further explanations see Table III.22 of the MRG report.

Table 7-2. Estimated Differences in Net Economic Benefits from Six Technology Mixes, Also Equal to the Net Economic Costs of Five Alternative Policies to Reduce Environmental Impacts[a] (unit: billions of 1975 dollars).

	DESOM (costs only)	*ETA*	*Nordhaus*
Shortfall below base case of benefits minus costs in six policy scenarios			
1. Base case	(0)	0	0
2. Moratorium on all ACRs and the FBR[b]	(43)	8	2
3. Moratorium on all nuclear technologies	(105)	46	136
4. Coal and shale limits	(914)	159	64
5. Moratorium on all ACRs and the FBR and coal and shale limits	(1,012)	181	72
6. Nuclear moratorium and coal and shale limits	(2,325)	358	457

[a]In all policy scenarios, total benefits and costs are the sums of year-by-year benefits and costs, discounted to 1975 at 6 percent per annum. DESOM computes only discounted costs, through 2025; ETA computes discounted benefits and costs through 2050; and Nordhaus through 2060. For each year, benefits estimate the value to the consumer of total amounts of energy consumed, on an incremental basis. For further explanations, see Section III.8 of the MRG report.

[b]For ETA this policy includes a moratorium on the LWR with plutonium recycle; for the other models it does not.

Note: Costs of R&D are excluded.

water reactors. The third policy imposes limits (that increase with time) on the production and processing of coal and oil shale. The last two policies are combinations of the preceding ones. They contain the coal and shale limits together with one or the other of these moratoria.

With regard to the numbers in Table 7-2, the economic benefit is estimated from the consumer's response to the price of energy. This is where the price elasticity of demand for energy comes in. If price is decreased just a little bit, then correspondingly demand, given enough time for the response, will stabilize at a level just a little bit higher. Take that price as per unit valuation of the small slice of demand added. If the price decrease is larger in extent, then a mathematical integration to evaluate an area under the demand curve is needed to estimate the economic benefits from a decrease in price, say, or the economic losses from a policy-induced increase in price. That is the method used in estimating the benefits. The costs were estimated from the cost data that I have already described.

Now let me explain the two titles of the table. The expression "differences in net economic benefits from six technolgy mixes" assumes that we speak of technologies that are not yet in operation. For instance, entry 2. in the table estimates what economic benefit can be obtained from availability of the reactors in question. (It is not prejudged whether one or more than one reactor type will actually be used in that case.) Alternatively you can interpret the entries in the table as net economic costs of denying yourself the use of these technologies—the economic cost then being due to a higher energy price and correspondingly less consumption. The same numbers in the table are subject to both interpretations. The DESOM figures are cost figures only because, as explained, DESOM does not have an end use demand response to price. The ETA and Nordhaus figures do reflect such a response. For various reasons explained in the report of the MRG, the ETA and Nordhaus estimates don't agree in the specific numbers too well. However, some patterns stand out in all three columns. One is a certain nonadditivity. In the ETA column the all-nuclear moratorium shows $46 billion as the economic cost figure, whereas the coal and shale limits by themselves show $159 billion as the economic cost. If you impose both sets of constraints, you obtain $358 billion, which is much more than the sum of the other two numbers. This superadditivity has to do with the mathematical fact that the production side is represented, in geometric terms, by a convex body in the output and input space, while the preference side is represented by a concave utility (or preference) function. Similar effects are found in the DESOM and the Nordhaus numbers.

Another important finding for the two models that recognize a response of demand to price again stands out in spite of the differences between corresponding specific entries for these two models. This is the smallness of the effects of the various policies relative to the GNP projection as a whole—even though in absolute terms the impacts are very large sums in dollars of 1975. This is shown in Figure 7-2 where, on the horizontal axis, we set off the ratio of energy use projected for 2010 under the policy in question divided by energy use projected for the base case (primary energy equivalents in both cases). On the vertical axis we set off a similar ratio of projected GNP for 2010 in the case of the policy in question, divided by that in the base case.* Thus the base case itself is represented by the vertex point *B* of the unit square. All but one of the other ETA and Nordhaus points hug the upper side of the unit square to such an extent that the differences between points are hard to see. In order to show the details at all, we give in Figure 7-3(a) a variant of Figure 7-2 in which the unit of the vertical scale is stretched by a factor of 4.

The one exception in Figure 7-2 arises from a policy not included in the list in Table 7-2, but recorded in Section A of Table 7-3. Both the ETA and Nordhaus models were also used for experiments with another fictitious policy, that of placing various upper bounds on the rate of growth of energy use ranging from 0 to 1.5 percent per annum, out to 2010. Such a policy could be implemented, for instance, by means of a conservation tax on primary energy use.

*The underlying calculations are recorded in Table 7-3(b).

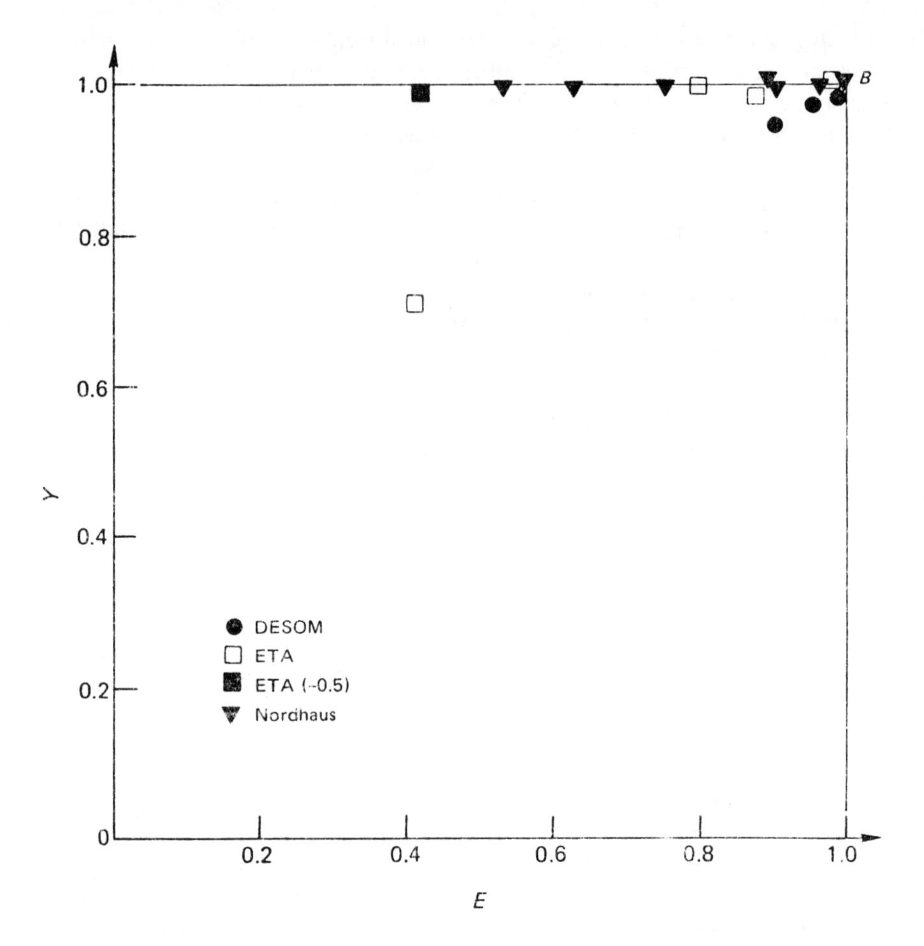

Key:

E on Horizontal Axis: Ratio of aggregate energy consumption in 2010 projected for each policy to that for the base case.

Y on Vertical Axis: Ratio of cumulative discounted GNP, 1975–2010, projected for each policy to that for the base case.

Symbols for Models are shown in the legend. Projections for DESOM extend to 2025, for ETA to 2050, for Nordhaus to 2010. ETA and ETA (–0.5) have price elasticities of –0.25 and –0.5, respectively.

Symbols for Policies: B = Base case for all models; for the lowest energy ETA and ETA (–0.5) points the policy constraints energy use to 70 quads throughout; for all other points see Figure 7–3(a).

Figure 7-2. Estimates of the Long-run Feedback from Aggregate Energy Consumption on Cumulative Discounted Real GNP.

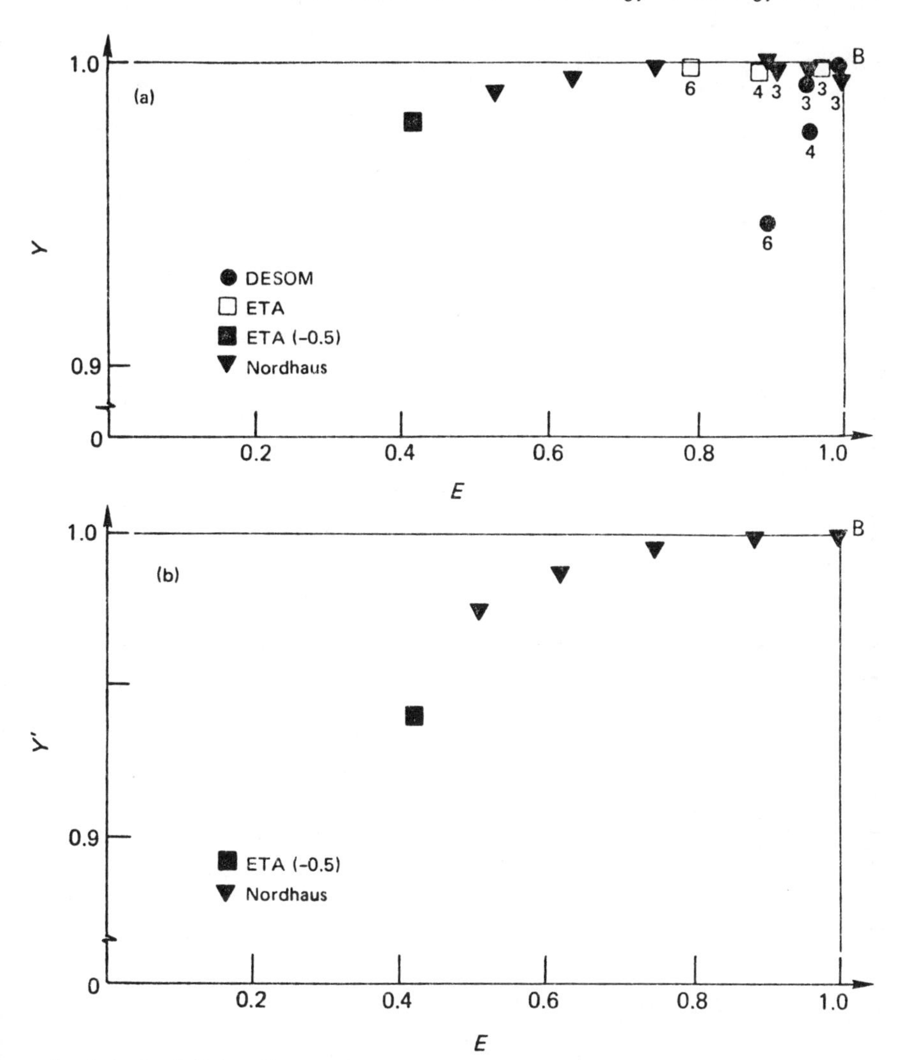

Symbols for Policies: Numbered points in (a) refer to policies 3, 4, and 6 listed in Table 7-2; the string of unnumbered Nordhaus points in both figures represents policies of successive curtailments of the growth rate of energy use, see Table 7-3; the lowest-energy ETA point has run off the scale.

Y' on Vertical Axis of (b): Ratio of undiscounted GNP in 2010 for each policy to that for the base case.

Other symbols as in Figure 7-2.

Figure 7-3. Estimates of the Long-run Feedback from Aggregate Energy Consumption on (a) Cumulative Discounted GNP, 1975-2010, and (b) Undiscounted GNP for 2010.

Table 7-3. Estimates of the Long-run Feedback from Aggregate Energy Consumption on Cumulative Discounted Real GNP.

Scenarios (policies)	*Implied Conservation Tax in 2010*	*Nordhaus* *A (35.0)* *B (50.0)*		*ETA (49.4)* *Price Elasticity* $= -0.25$		*ETA (49.4)* *Price Elasticity* $= -0.50$		*DESOM (43.0)*	
		$\bar{E}^a$	$\bar{Y}^a$	$\bar{E}^a$	$\bar{Y}^a$	$\bar{E}^a$	$\bar{Y}^a$	$\bar{E}^a$	$\bar{Y}^a$
A. Upper limits on growth rate of energy consumption, percent per annum	$ (75)/ 10^6 Btu								
None	0	1.000	1.000	1.000	1.000	1.000	1.000	--	--
1.5	0.34	0.893	0.999	--	--	--	--	--	--
1.0	1.02	0.751	0.997	--	--	--	--	--	--
0.5	1.98	0.631	0.995	--	--	--	--	--	--
0.0	3.19	0.531	0.991	--	--	--	--	--	--
Upper bound of 70 Quads	b	--	--	0.419	0.711	0.419	0.982	--	--
B. MRG scenario (see Table 7-2)									
1. Base case	--	1.000	1.000	1.000	1.000	--	--	1.000	1.000
3. Nuclear moratorium	--	0.968	0.997	0.980	0.999	--	--	0.984	0.998
4. Coal and shale limits	--	1.000	0.999	0.894	0.997	--	--	0.952	0.979
6. Nuclear moratorium and coal and shale limits	--	0.901	0.991	0.795	0.993	--	--	0.900	0.946

[a] $\bar{E}$ = ratio of aggregate energy consumption in 2010 in the indicated scenario to that of the base case;
$\bar{Y}$ = ratio of cumulative discounted GNP in the indicated scenario to that of the base case.

[b] Price elasticity, −0.50 only, tax of electricity, 126 mills/kWh; oil and gas, 8.9 $/$10^6$ Btu.

Note: Number in parentheses after name of model indicates total discounted GNP, in trillions of 1975 dollars, calculated over the horizon of the model.

Points representing the outcomes of these calculations were also plotted in Figures 7-2 and 7-3. In particular, Figure 7-2 shows for the ETA model two different estimates for the zero energy growth policy, with the price elasticities of demand of −0.25 and −0.50, respectively. The absolutely lower elasticity figure of −0.25 leads to a curtailment of GNP (out of the base case) larger by almost three-fifths than the figure −0.50 does–so much larger that in Figure 7-3 the corresponding point has run off the scale. Another indication of the effect of an (absolutely) small estimate of the price elasticity of demand is given by the DESOM points in Figure 7-3(a). These points already turn away from the base case GNP level for much smaller curtailments of energy use.

All these indications emphasize the importance of as accurate econometric estimates of the price elasticity of demand for energy as can be made.

Chapter 8

Energy and Economic Growth: A Conceptual Framework

James L. Sweeney*
Stanford University

Since the 1973 oil embargo there has been intense interest throughout the world in the economic impacts of changes in the price or availability of energy. This concern is reflected in U.S. energy policy debates, OPEC pricing deliberations, and international meetings such as the Conference on International Economic Cooperation.

The relationships between energy and the economy are complex and multifaceted. Sudden changes in energy price or availability can lead to structural and aggregate unemployment, to sharp increases in inflation, and to major changes in the distribution of wealth, both among nations and within a given country. In the longer run, when transitional impacts subside and after compensating monetary and fiscal policies are implemented, changes in the price or availability of energy may influence the long-run growth prospects of a nation.

The purpose of this chapter is to provide a simple conceptual framework within which the long-run impacts of energy on economic growth can be considered. Within this framework the question is asked: What will be the impact on aggregate output, consumption, and welfare of changes in the availability or price of energy?

In this long-run analysis it will be assumed that the economy always maintains full employment of all factors of production. Differences among the energy carriers—coal, oil, natural gas—will be ignored in order to facilitate the conceptual discussion, although the extension to consider different fuels is possible.

*Many of the ideas presented here were developed jointly with Tjalling Koopmans. I would like to thank Ernest Berndt, Dale Jorgenson, William Hogan, Alan Manne, John Weyant, and David Wood for their helpful comments throughout the research project. Partial financial support has been provided by the Office of Conservation Planning and Policy, U.S. Department of Energy. Of course, all errors remain the sole responsibility of the author.

Many economic or policy changes may influence the price or availability of energy. Examples include changes in the OPEC-administered world oil price, a moratorium on nuclear construction, a tariff or quota on imported energy, a domestic price control program, a tax on domestic energy production, or a successful energy supply R&D program. Each change can influence the use of energy in the growth of its economy, although the ratio of the two magnitudes may vary widely among programs.

In order to focus on the essential differences and similarities among the various economic and policy changes, each can be represented analytically as one (or more) of four types of changes: an import price change, a domestic cost function change, an import tariff, or a domestic tax. The first two analytical representations are associated with changes in the *resource cost* structure of energy in the economy, in the value of other resources that are used up or foregone within the domestic economy in order to gain access to a given quantity of energy. The latter two analytical representations are associated with changes in the *price* structure of energy but not with changes in the *resource cost* of energy. An import tariff or domestic tax will change the price of energy to consumers or producers but will not directly entail changes in the resources foregone to gain access to a given amount of energy. As will be shown in the subsequent analysis, policies that can be modeled by the first two analytical representations will lead to greater impacts on the economy than will policies that can be accurately modeled by the latter two analytical representations.

Policies can also be separated into those directly influencing imported energy and those directly influencing domestically produced energy. The analytical representations maintain this distinction. However, as will be shown, this difference is not fundamental to explaining differences in the aggregate economic impacts of alternative policies.

Table 8-1 lists several policy or economic changes influencing the energy system and their corresponding analytical representations. Each of the energy system changes listed on the left margin is equivalent in its impact on the aggregate economy to one or more of the analytical representations listed across the top margin. The elements in the table indicate which analytical representations are equivalent to a given energy system change and what are the correct directions of change. The pluses represent increases in the specific analytical representation, while the minus signs represent decreases. For example, a limit on the domestic use of coal will increase the cost to the United States of producing a given quantity of energy. Thus, coal limits can be represented as an increase in the domestic cost function; a plus sign appears in the appropriate box. An energy conservation tax would influence all energy, both domestic and imported. This tax would be equivalent to an import tariff plus a domestic tax; pluses appear in two boxes. A successful R&D program[1] to produce inexpensive solar collectors would reduce domestic cost; a negative sign appears. An import quota would limit the amount of energy imported but would not change the

Table 8-1. Analytical Representation of Energy System Changes.

	Analytical Representations			
Energy System Changes	*Import price*	*Domestic cost function*	*Import tariff*	*Domestic tax*
OPEC price increase	+			
Domestic price control with entitlements			−	+
Average cost pricing of electricity				−
Nuclear moratorium		+		
Coal limits		+		
Import tariff			+	
Import quotas			+	
Domestic resource depletion		+		
Conservation program – uneconomic restrictions			+	+
Energy conservation tax			+	+
Successful energy supply R&D		−		

resource costs borne by the economy to gain access to the imported energy. An import quota would have the same aggregate impacts as an import tariff. More complex is the crude oil price control program with entitlements. This program subsidizes imports and taxes domestic production. Thus, it can be represented as a combination of a negative import tariff (a subsidy) and a positive domestic tax.

Many energy system changes are equivalent to one or more of these four analytical representations. Therefore, subsequent discussions will be couched in terms of the impacts of changes in the four analytical representations.

EFFECTS ON NET NATIONAL PRODUCT

The Model

Although much of the current U.S. congressional energy policy debate is rooted in discussions of wealth redistribution, one of the most important summary variables for an economy is the gross national product (gross domestic product) or the net national product (net domestic product). This section will discuss the latter aggregate variable, although interpretations in terms of gross national product follow directly.

The basic model is that of a three sector economy, consisting of an energy-importing sector, a domestic energy-producing sector, and the rest of the economy (ROE). Energy is an intermediate good; output from the ROE may be either a final or an intermediate good.

The first sector imports some quantity of energy (E_I) at a fixed price (P_I), which includes the price paid to the exporting source plus any cost of the import activity. The net resource cost to acquire this energy is simply $P_I E_I$, which becomes one claim on the gross output from the rest of the economy.

The second sector uses output from the rest of the economy in order to produce domestic energy. The quantity of ROE outputs used to produce E_D units of energy will be represented as $H(E_D)$. This is simply the cost function for producing domestic energy. The total energy used in the economy is the sum of the domestically produced energy plus the imported energy, as represented in equation (8.2), below.

The rest of the economy uses as inputs a quantity of capital services (K), a quantity of labor services (L), and a quantity of energy (E) and produces as output an aggregate quantity of goods and services (which depends on K, L, and E). This aggregate quantity will be referred to as the gross output of the economy. The relationship between K, L, E, and the resulting gross output can be expressed in terms of an aggregate production function for the economy, $F(K, L, E)$. The net national product (Y) is the difference between the gross output and the costs of importing plus domestically producing energy. This is expressed in equation (8.1) below. Equation (8.1) follows directly since, by definition, net national product equals total value of final products produced in the economy or equals the sum over all sectors of the economy of value added.

In a competitive economy,[2] several marginal conditions must also hold.[3] The price of energy (P_E) will equal its marginal productivity, since competitive firms are assumed to be price-taking profit maximizers. And in a competitive economy, the price of energy facing the domestic supplier (P_D) will equal the marginal cost of producing energy. Supply price will equal the energy price plus any tax on domestic production (T_D). These relationships are expressed in equation (8.3). Similar relationships hold for imported energy: the energy price in competitive equilibrium will equal the import price (P_I) plus the tax on imported energy (T_I). This is represented by equation (8.4). Finally, there are necessary marginal conditions for market clearing in the capital and labor markets. The price of labor services (P_L) or of capital services (P_K) is simply equal to the appropriate marginal productivity. These conditions appear as equations (8.5) and (8.6).

Six equations thus represent the economy. Equation (8.1) defines net national product; equation (8.2) equates energy consumed to domestic plus imported energy; and equations (8.3) through (8.6) represent the marginal conditions describing a competitive economy in equilibrium.[4] Note that supply conditions for capital and labor have not been written. These will be discussed at a later point.

$$Y = F(K, L, E) - H(E_D) - P_I E_I. \qquad (8.1)$$

$$E = E_D + E_I. \tag{8.2}$$

$$P_E = \partial F/\partial E = dH/dE_D + T_D. \tag{8.3}$$

$$P_E = \partial F/\partial E = P_I + T_I. \tag{8.4}$$

$$P_K = \partial F/\partial K. \tag{8.5}$$

$$P_L = \partial F/\partial L. \tag{8.6}$$

Impacts of energy system changes on NNP can be evaluated by use of equations (8.2) through (8.6), since the impacts of each of the four analytical representations can be evaluated by means of these equations.

For small changes in any analytical representation, the impact on NNP can be calculated by taking the total differentials of equations (8.1) and (8.2) and by using equations (8.3) through (8.6). This allows one to derive equation (8.7), which expresses changes in net national product as a sum of weighted changes in capital, labor, domestic energy production, imported energy production, import price, and domestic cost. In equation (8.7) the change in the domestic cost function is expressed as a change in the average cost function (equal to δAC_D) times the quantity of domestic energy (E_D).

$$\Delta Y = -E_I \Delta P_I - E_D \delta AC_D + T_I \Delta E_I + T_D \Delta E_D + P_L \Delta L + P_K \Delta K. \tag{8.7}$$

Equation (8.7) will be fundamental to all subsequent analysis of the relationships between energy sector changes and net national product changes. Therefore, it is repeated in the first column of Table 8-2. The various terms on the right-hand side of equation (8.7) are interpreted as components of NNP change. These components are direct effects of cost changes, consisting of changes in the resource cost function; "welfare costs" of price changes stemming from divergences between prices and costs; and induced effects via labor quantity and capital quantity. The change in the net national product equals the sum of the components.

To use equation (8.7) (or equivalently, Table 8-2), it is necessary to specify a complete model that describes capital, labor, and energy quantity changes in response to changes in the underlying environment. This can be done by more completely specifying the production functions and cost functions described previously and modeling the supply functions for capital and labor, or else simpler models can be postulated in order to illustrate the fundamental interactions. The latter procedure will be followed in this chapter.

Table 8-2. Components of NNP and Welfare Changes in Response to Energy System Changes.

Effects / *Gauges*	*Changes in NNP*	*Monetary Equivalent Welfare Change*
Direct Effects of Cost Changes:		
Import price	$-E_I \Delta P_I$	$-\sum_t \alpha_t E_{It} \Delta P_{It}$
Shift of domestic cost function	$-E_D \delta A_{CD}$	$-\sum_t \alpha_t E_{Dt} \delta A C_{Dt}$
"Welfare Cost" of Price Changes:		
Import tax	$T_I \Delta E_I$	$\sum_t \alpha_t T_{It} \Delta E_{It}$
Domestic tax	$T_D \Delta E_D$	$\sum_t \alpha_t T_{Dt} \Delta E_{Dt}$
Effects via:		
Labor input		
Efficient allocation	$P_L \Delta L$	0
Two labor prices		$\sum_t \alpha_t T_{Lt} \Delta L_t$
Capital input		
Efficient allocation	$P_K \Delta K$	0
Two interest rates		$\sum_t \alpha_t T_{Kt} \Delta K_t$

$$\alpha_t = \sum_{\tau=1}^{t} \frac{1}{1+r_\tau}$$

Taxes versus Cost Increases

As a first illustration, equation (8.7) can be used to examine the differences in NNP impacts stemming from resource cost increases and those motivated by tax increases. It will be assumed that capital and labor are supplied perfectly inelastically and that there are no changes in the cost function for domestically produced energy. Therefore, the first two terms and the last term on the right-hand side of equation (8.7) will be identically zero.

If all taxes remain zero while imported energy prices change, then equation (8.7) is reduced to the simple differential equation:

$$\frac{dY}{dP_I} = -E_i \,. \tag{8.8}$$

If the import price remains constant while the tariff on all energy (imported plus domestic) changes, equation (8.7) is reduced to the differential equation:

$$\frac{dY}{dE} = T, \tag{8.9}$$

where T is the tax rate on all energy.

Note that in the case of import price changes, the impact on NNP of a price change is proportional to energy imports. However, in the case of a tariff on all energy, the impact is independent of the quantities imported, but depends on the change in the quantity of energy consumed.

Upper and lower bounds on the NNP impacts of tariffs and of changing import prices can be evaluated very simply by use of equations (8.8) and (8.9). If imported energy price were to increase from P_{IO} to P_{IF}, then the change in NNP (Y) could be calculated by integrating equation (8.8):

$$\Delta Y = \int_{P_{IO}}^{P_{IF}} (-E_I)\, dP_I.$$

Since E_I is decreasing in P_I and increasing in Y,, this allows simple limits to be set:

$$E_{IO}\,\Delta P_I > -\Delta Y > E_{IF}\Delta P_I, \qquad (8.10)$$

where E_{IO} and E_{IF} are the initial and final import levels, respectively, and ΔP_I is the change in energy price:

$$\Delta P_I = P_{IF} - P_{IO}.$$

Inequality (8.10) provides tight limits on the NNP reductions of an import price increase whenever the level of imports occurring with the higher prices is near to the level occurring with the lower prices. However, when the price change leads to large changes in imports, the bounds are very loose.

Inequality (8.10) allows an unambiguous comparison between the NNP reduction and the revenue gains accruing to the exporter of energy. Initial revenue is $P_{IO}E_{IO}$, while final revenue is $P_{IF}E_{IF}$. Then the revenue gain for the exporter (ΔR) is:

$$\Delta R = P_{IO}(E_{IF} - E_{IO}) + E_{IF}\Delta P_I. \qquad (8.11)$$

This equation provides an upper limit to the revenue gain as prices increase:

$$\Delta R < E_{IF}\Delta P_I.$$

Comparing this to inequality (10.10) shows that for price increases:

$$-\Delta Y > \Delta R. \qquad (8.12)$$

The revenue gain to the exporting country from increasing energy price must be strictly smaller than the resulting NNP loss borne by the importing country. Furthermore, if the exporter's marginal cost of energy production is no greater than the initial price, then the gain in revenues minus costs (net profits) must also be strictly smaller than the resulting loss in NNP borne by the importing country.

A similar analysis can be conducted for a tax rate increase. Integrating equation (8.9) provides the inequality:

$$T_O \, \Delta E > Y > T_F \, \Delta E, \tag{8.13}$$

where T_O and T_F are the initial and final tax rates on all energy, respectively.

Inequality (8.13) can be used for any change in tax. For example, assume that the initial tax is zero and that either a positive or a negative tax (a subsidy) is imposed. In either case, inequality (8.13) shows that ΔY must be negative: any tariff or subsidy will reduce NNP. The larger the tariff or subsidy and the larger the change in energy consumption, the greater will be the possible NNP reduction.

In order to use equations (8.8) and (8.9) to obtain more precise estimates of NNP change than are provided by the above inequalities, it is necessary to describe the model more fully. In particular, the aggregate production function must be specified more completely in order to relate energy demand or energy imports to energy price and NNP. For this discussion it will be assumed that all energy is imported, so that $E_I = E$. And it will be assumed that the aggregate production function is such that in a competitive equilibrium the quantity of energy demanded is the following constant elasticity function of NNP and energy prices: Equation (8.14) will be somewhat loosely referred to as a demand function for energy in this economy, and ϵ will be referred to as the price elasticity of demand for energy.

$$E = AYP_E^{-\epsilon}. \tag{8.14}$$

To calibrate equation (8.14) it will be assumed that in the year 2010, if the OPEC price increase had not occurred, the energy price would be $0.80/MMBtu, the energy demand would be 220 quads, and the NNP would be $4,400 billion. These assumptions correspond to those of Hogan and Manne,[5] except that those authors assume a constant elasticity of substitution *production* function rather than a constant elasticity *demand* function.

What then are the impacts on NNP of an import price change? Equation (8.14) can be used to eliminate energy consumption in equation (8.8), giving a simple differential equation relating Y and P_I:

$$\frac{dY}{dP_I} = -AYP_I^{-\epsilon}.$$

This differential equation can be solved to give the following equation relating NNP to imported energy price:

$$\frac{Y}{Y_O} = \exp\left\{\frac{S_O}{1-\epsilon}\left[1 - \frac{P_i}{P_{IO}}^{\;1-\epsilon}\right]\right\}, \tag{8.15}$$

where

$$S_O = \frac{P_{IO}E_O}{Y_O}. \tag{8.16}$$

S_O is the expenditure on energy in the base case as a fraction of base case NNP and will be referred to as the *value share* of energy in the economy. The symbols E_O, Y_O, and P_{IO} are the base case values of E, Y, and P_I, as indicated previously.

Equation (8.15) shows that the impact of increasing energy import prices on NNP is increasing in S_O: The greater the base case expenditure on energy (as a fraction of NNP), the greater will be the economic impact of an import price change. Also, the greater the elasticity of demand, the smaller the impact of a given import price increase on NNP.

Equation (8.15) has been solved under the base case assumptions indicated earlier. These assumptions give a value of S_O of 4 percent, which corresponds to the preembargo experience. The results appear in Figures 8-1 and 8-2 for several different values of demand elasticity (ϵ).

Figure 8-1 shows net national product as a function of imported energy price for increases or decreases in energy price from the 1972 levels. Note that increasing imported energy price always reduces NNP and that decreasing price always increases net national product. The impacts depend significantly upon the elasticity of energy demand. For low elasticities, an increase in energy price has a larger impact on the economy than would occur with a high elasticity. But even with relatively high elasticities, increases in energy price can lead to significant reductions in net national product. For example, an increase in the energy price from \$0.80/MMBtu to \$3.00/MMBtu reduces net national product by almost 10 percent if the elasticity of demand is 0.1 and by 6 percent if the elasticity of demand is 0.7.

Figure 8-2 shows the values of net national product and energy use as the energy prices change. The greater the elasticity of demand, the smaller will be the reduction in net national product for a given reduction in energy use. A 30 percent reduction in energy use (from 220 quads to 154 quads) reduces net national product by \$97 billion (2.2 percent) if the elasticity of demand is 0.7, but reduces net national product by \$70 billion (16.3 percent) if elasticity of demand is 0.1. Note, however, that the former reduction is motivated by an import price increase of roughly 33 percent, while the latter is motivated by a 370 percent increase in imported energy price.

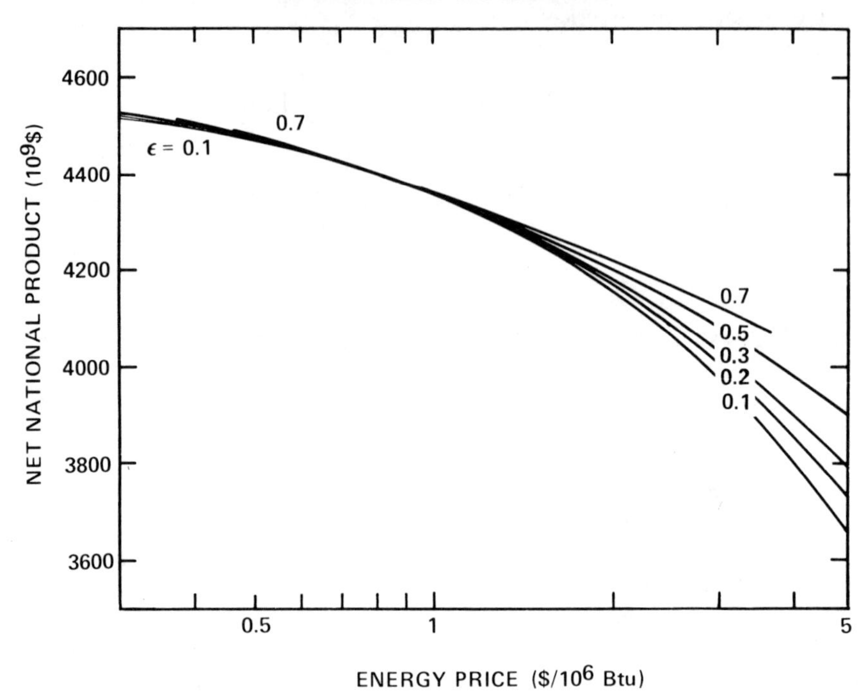

Figure 8-1. Net National Product as Function of Imported Energy Price

This analysis has focused upon increases in imported energy price when there is no domestic production. If some energy is domestically produced, then the impact of import price changes on NNP is reduced correspondingly, as shown by equation (8.8).

Increases in the domestic cost function would have impacts on NNP analogous to those stemming from increasing import price, as can be shown by equation (8.7). In fact, an analysis could be conducted for all energy domestically produced with an infinitely elastic domestic supply curve. In this case, exogenous shifts in the supply price would have NNP effects identical to those analyzed here for an import price increase.

The key issues for evaluating NNP impacts of cost increases are (1) the magnitude of the cost increase, (2) the expenditure on energy as a fraction of NNP in the base case, and (3) the elasticity of demand for energy. Whether the cost increase stems from domestic cost function increases or import price increases is irrelevant for evaluating impacts on NNP.

What are the impacts on NNP of a tax change? Under an assumption that a tax T is imposed on all energy, whether imported or domestically produced, equation (8.9) is the fundamental differential equation. It will be assumed that the domestic supply price of energy is determined by the fixed import price

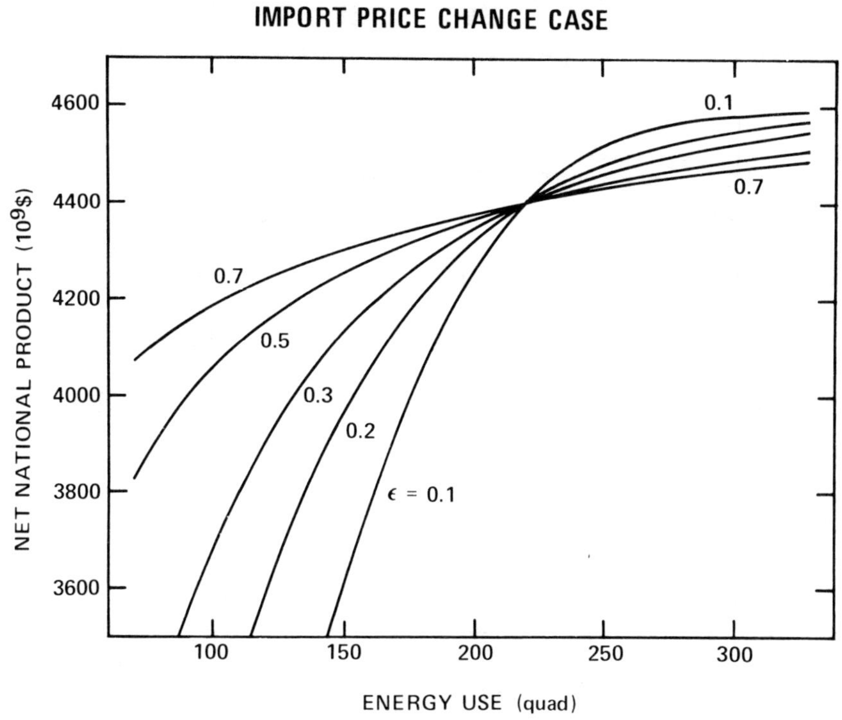

IMPORT PRICE CHANGE CASE

Figure 8-2. Net National Product as Function of Energy Use

of P_I. Then, from equation (8.14), demand for energy can be expressed as follows:

$$E = A\ Y(P_I + T)^{-\epsilon}. \tag{8.17}$$

Equations (8.9) and (8.17) are sufficient to determine NNP and energy consumption as functions of the tariff. These equations are independent of the fraction of energy imported, and thus the impact of a tariff on NNP will be independent as well.

No closed form solutions to equations (8.9) and (8.17) have been found. However, solutions can be obtained numerically. These results are provided in Figures 8-3 and 8-4.

Figure 8-3 shows NNP as a function of energy price (P_E) as the tariff changes. When energy price is \$0.80/MMBtu, the tariff is zero. NNP will decline for increases or decreases in energy price away from the import price. New national product is maximized at zero tariff for energy price equal to import price. Small changes in tax around zero lead to virtually no loss in NNP, while equivalent tariff changes beginning at nonzero levels lead to greater NNP impacts. These results all occur because, as shown in equation (8.9), the change in NNP associated with a given change in energy use is proportional to the

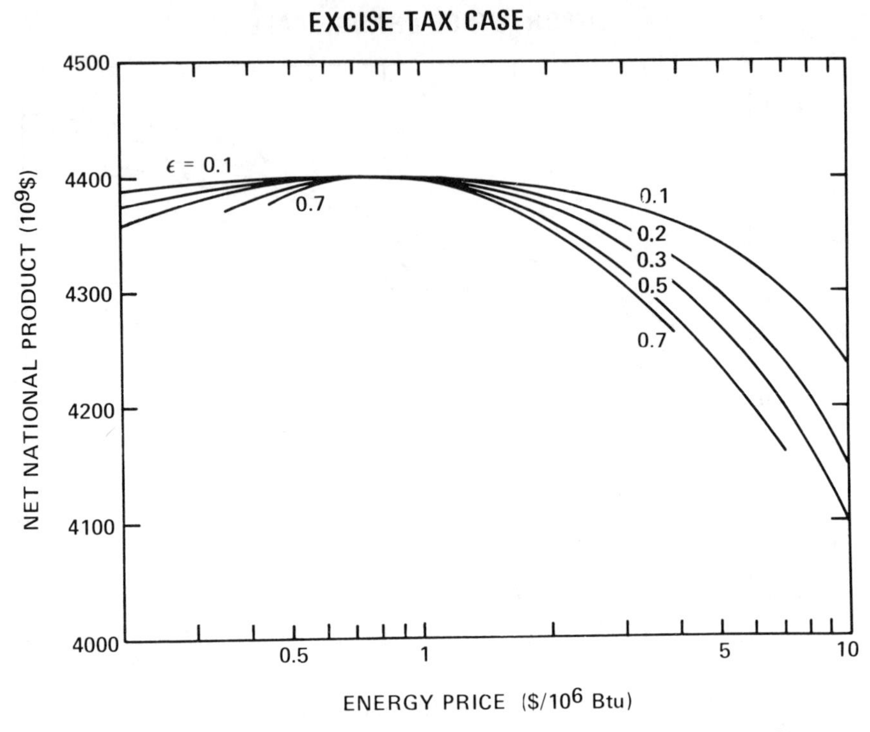

Figure 8-3. Net National Product as Function of Energy Price as Tariff Changes

tariff, which in turn is the difference between the marginal value of energy (P_E) and the marginal cost (P_I) of attaining energy.

Figure 8-3 also casts light on the role of the demand elasticity. For a given tariff, the higher the elasticity, the greater the impact on NNP. For example, an increase in energy price to $3/MMBtu reduces net national product by $23 billion if the elasticity of demand is 0.1 and by $95 billion if the elasticity of demand is 0.7. This is precisely opposite from the result obtained for a change in import price. The explanation is that the higher the elasticity, the greater the change in energy consumption for a given tariff and thus the greater the NNP change (see equation 8.9). The consumption of energy is reduced by 28 quads (13 percent) in the former case and by 135 quads (61 percent) in the latter.

Figure 8-4 plots changes in net national product against changes in energy consumption motivated by changing excise taxes. For relatively high elasticities, large percentage reductions of energy use occur with relatively small percentage reductions in net national product. For example, a 30 percent reduction in energy consumption leads to a $14 billion reduction (0.3 percent) in net national product if elasticity is 0.7, but a $270 billion reduction (6 percent) if elasticity is 0.1.

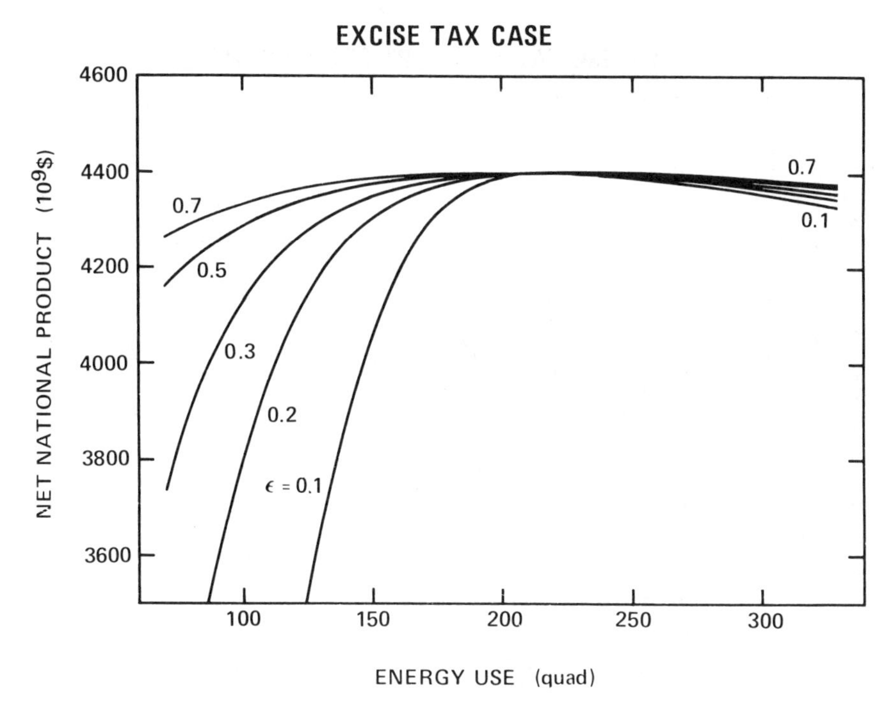

Figure 8-4. Net National Product as Function of Energy Use as Excise Taxes Change

Note that an increase in energy use motivated by a negative excise tax (a subsidy) will also reduce net national product. Therefore, attempts to control energy prices below marginal costs or to otherwise subsidize energy consumption will reduce NNP. However, the result is not symmetrical. A 50 percent increase in energy use motivated by a subsidy leads to a much smaller reduction in net national product than does the same proportionate reduction in energy use motivated by a tax.

Figure 8-5 compares the NNP impacts of import price increases to those of excise tax increases for two different demand elasticities (0.2 and 0.5). The numbers under the curves are the demand elasticities, while the letters I or R denote import price change or excise tax change, respectively. A given energy consumption reduction will be associated with a greater impact on NNP if motivated by an import price increase than if motivated by an excise tax increase.

The effects of simultaneous shifts of import price and tariff are illustrated in Figure 8-6. The downward sloping curve, labeled "Tariff=*O*, P_I Changing," shows NNP as a function of import price for zero tariff. Each of the three curves intersecting this curve illustrates NNP for import price constant but tariffs

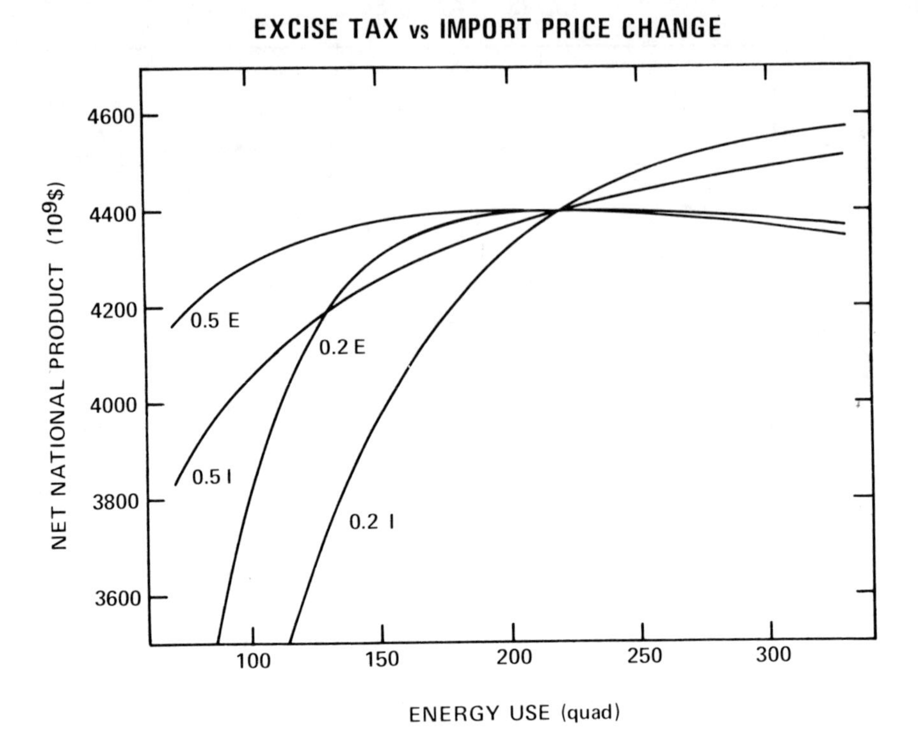

Figure 8-5. Impact on Net National Product of Excise Tax versus Import Price Change for Demand Elasticities 0.2 and 0.5.

changing. At the points of intersection the tariff equals zero. As shown, a given energy price may be associated with many possible values of NNP. Higher values of NNP occur when the import price is low but the tariff is high.

The differences between the NNP impacts of excise taxes and of import price changes cast some light on the potential value of a tariff on energy, if the tariff influences import prices. Suppose that importing countries were to impose a tax on oil. If the import price of energy were to remain unchanged, this tariff would reduce the net national product of each importing country, as is illustrated by the move from point A to point B in Figure 8-6. However, since the tariff would reduce demand for OPEC oil, it might induce the OPEC nations to reduce the oil price from what it would have been otherwise. This price reduction would increase net national product, corresponding to a move from point B to point C. Whether the consuming nations would be made better off or worse off by the combination of changes would depend upon which of these effects were to dominate. For a relatively small excise tax, however, such a strategy would increase net national product even if the reduction in import price were significantly smaller than the excise tax.

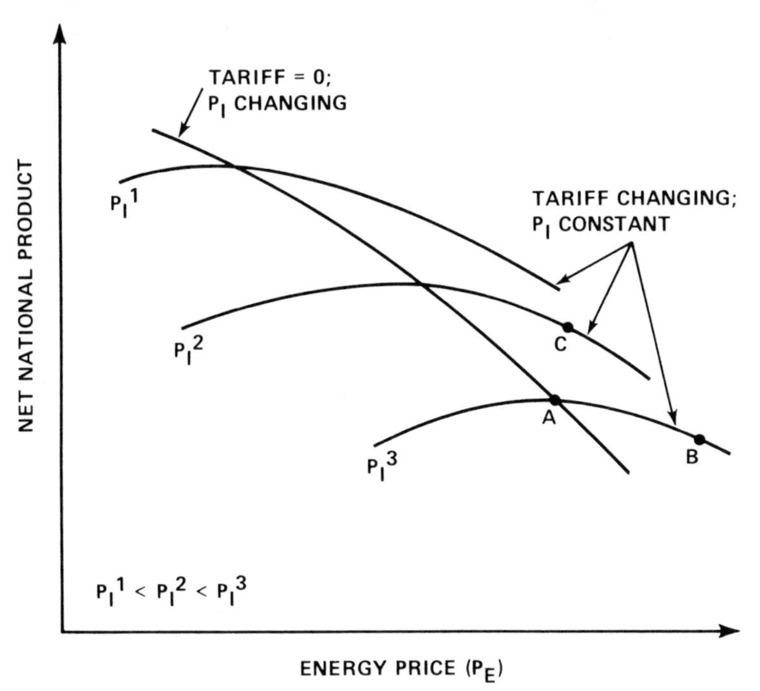

Figure 8-6. Effects on Net National Product of Simultaneous Changes in Tariff and Import Price.

The value of such a strategy depends upon the fraction of energy imported. Imposing the excise tax would have the same proportional negative impact on NNP for a nation virtually self-sufficient in energy as it would for one importing all energy. The gains from the import price reduction, however, would vary greatly between the two, with the virtually self-sufficient nation gaining little, while the nation totally dependent upon imports would gain much. The latter may see a net gain in NNP and, therefore, may be motivated to support the plan, while the former may see a net loss and may not be so motivated.

The example may become more relevant when it is remembered that some mandatory energy conservation programs may be analytically represented as an excise tax on energy. Thus, nations that are heavily dependent upon imports would be more willing to engage in domestic actions to force a world price decrease than would nations that were virtually self-sufficient.

NNP Effects Via Capital and Labor

The preceding section incorporates an assumption that changes in energy availability will not influence the quantities of other input factors. However, equation (8.7) is more general. Any induced increase in capital or labor services will increase NNP by an amount equal to the factor price times the quantity increase. Conversely, energy sector changes that reduce the capital formation or the labor supply would reduce NNP. The magnitude and direction of labor or capital service changes depend upon characteristics of the aggregate production function and of the supply functions for these factors.

Characteristics of the production function are important because changes in the energy price, with other prices constant, may either increase or decrease the demand for capital or labor. If an increase in the price of energy leads to an increase in the demand for labor, then labor and energy are said to be *supplementary* commodities. Conversely, if labor demand decreases, then energy and labor are said to be *complementary* commodities.[6] Whether energy and capital or energy and labor are complementary or supplementary depends upon properties of the production function. Empirical research is necessary to determine which relationship is correct.

Empirical studies have indicated that energy and labor tend to be supplementary products: an increase in the price of energy tends to increase the demand for labor. This occurs fundamentally because increases in energy price cause firms and consumers to use more labor-intensive and less energy-intensive processes and products as energy prices increase.

While the result is still far from conclusive, empirical evidence tends to support the hypothesis that capital and energy are complementary. Berndt and Wood,[7] in their excellent review and reconciliation of the evidence, conclude that capital and energy are complementary products: increases in energy price probably reduce the demand for capital. This conclusion, however, is not accepted by all researchers in the field.

Supply functions for capital and labor are also critical in determining changes in capital and labor quantities. If these supplies are independent of prices, then the issue of complementarity versus supplementarity is irrelevant, for changes in energy or capital demand functions will not influence the equilibrium capital or labor quantities, but only their prices. If, on the other hand, capital or labor supply is very responsive to price, then the change in equilibrium quantity roughly equals the constant price demand change.

Labor supply depends both upon labor price and upon NNP, with either increasing wage rates or decreasing NNP leading toward increases in the labor supply. While the current empirical evidence is inconclusive, it suggests that the elasticity of supply of labor is small. Therefore, it can be expected that increases in energy price will only slightly change NNP via changes in labor quantity. Rather, increases in energy prices will tend to increase wage rates.

The capital stock of the economy increases slowly over time whenever net investment (the net rate of capital formation) is positive and decreases when net investment is negative. Thus, any factor that increases the rate of capital formation will, in the long run, increase the capital stock, although the effects would not be felt rapidly.

The rate of capital formation depends upon both NNP and upon the interest rate, which is the relevant price of capital services. While empirical evidence is inconclusive, the rate of capital formation seems to be a significantly increasing function of the interest rate. Under the assumption that capital and energy are complementary, increases in energy price will decrease capital demand. In the short run, the supply of capital is relatively inelastic, so this will simply tend to depress interest rates. However, in the long run, reductions in capital demand would induce significant reductions in the equilibrium capital stock and thus would reduce NNP as energy prices increase.

For fixed interest rates, capital formation is roughly proportional to NNP. Therefore, reductions in NNP reduce the rate of capital formation and, after some time, reduce the capital stock. This further reduces NNP. The net result of this feedback loop is a multiplied impact on NNP.

The various effects via capital and labor are summarized in Figure 8-7 under an assumption of capital-energy complementarity and labor-energy supplementarity. This illustrates the positive feedback loop through capital and NNP and the negative feedback loop through labor and NNP.

In summary, in response to increases in the price of energy, several effects tend to reduce NNP. First are the direct effects of cost changes or "welfare costs" of price changes, as described in the previous section. Second, if capital and energy are complementary, then price changes induce a reduction in the demand for capital and the equilibrium quantity of capital, which in turn reduces NNP. Finally, the first two effects, by reducing NNP, reduce the supply of capital, further reducing its equilibrium quantity, and thereby lead to further NNP reductions.

EFFECTS ON WELFARE

The previous section has been focused on net national product. However, there are several difficulties with the use of NNP as a measure of well-being, two of which are particularly relevant here. First, increases in labor use expand output but decrease leisure time. The increase in output is included in NNP but the value of the foregone leisure time is not deducted. Second, increases in the rate of capital formation (for fixed NNP) imply decreases in the current output available for consumption. Thus, current well-being is not uniquely determined by NNP whenever the investment rate changes. Finally, since people make trade-offs over time, even if NNP were a good static measure it would be necessary to find some method of aggregating NNP changes over time to obtain a welfare criterion. This section focuses attention on a better measure of welfare than NNP.

It will be assumed that both individuals and the economy as a whole can preferentially rank consumption trajectories and alternative consumption-leisure time (or labor time) combinations and that the individual and the societal

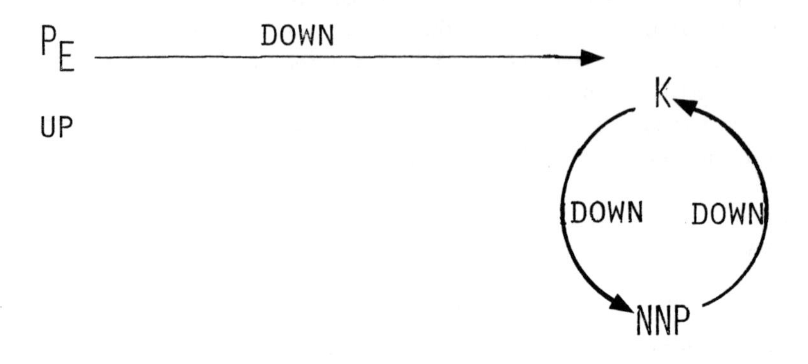

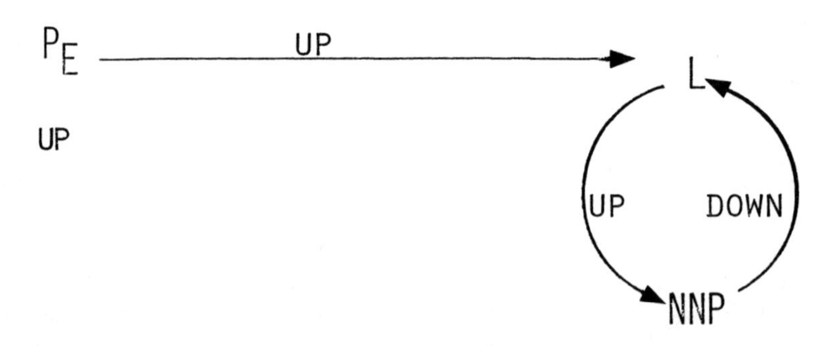

Figure 8-7. Effects via Capital and Labor of an Increase in Energy Price.

rankings are identical. Under this assumption, one can define, as a measure of performance, some welfare functions, W:

$$W = \sum_{t=0}^{\infty} U(C_t, L_t) e^{-\rho t} , \tag{8.18}$$

where C_t is consumption at time t, L_t is labor used at time t, $U(\cdot)$ is a utility function (which may itself depend on time), and $e^{-\rho t}$ (for $\rho \geqslant 0$) is the discount factor on future utilities.[8] Welfare, under this well-known utilitarian formulation, is represented by a weighted sum of individual utilities of consumption and labor. This formulation incorporates the notions that well-being depends on the goods and services available to be consumed, rather than simply upon the net output of the economy, and that well-being also depends upon leisure time available. Additionally, it allows for intertemporal tradeoffs.

Equation (8.18) provides the welfare criterion to be used in the subsequent analysis. However, it could easily be generalized if necessary. In order to evaluate how welfare changes in response to changes in the energy sector, it is useful to have a more intuitive measure than that provided by the welfare function itself. For small changes in welfare, define the monetary equivalent welfare change (ΔMW) as the ratio

$$\Delta MW = \frac{\Delta W}{\partial U/\partial C_O} \,. \tag{8.19}$$

The monetary equivalent welfare change equals the change in welfare divided by the welfare change that would occur with a unit increase of consumption at time O. Thus, the monetary equivalent welfare change can be interpreted as the number of dollars' worth of consumption in time period O that would provide the same change in welfare as that stemming from the arbitrary change in the consumption and labor trajectories. Thus, the monetary equivalent welfare change is a conceptually simple, workable measure of welfare changes.

The Model

All elements of the model of NNP determination from the previous section will remain valid in this section. However, in addition, it is necessary to specify the relationships between NNP, consumption, and capital formation. NNP can be allocated between two purposes–consumption or capital investment. The capital change from one period to the next is just equal to the net investment in the economy. Therefore, the change in capital stock plus consumption must equal NNP at each time:

$$C_t + [K_{t+1} - K_t] = Y_t, \tag{8.20}$$

where K_t and Y_t are capital stock and NNP at time t, respectively.

Marginal conditions of market equilibrium, in addition to those specified previously for the energy sector, must be specified here. On the demand side of labor markets, labor price (P_{Lt}) must equal the marginal productivity of labor:

$$\frac{\partial F}{\partial L_t} = P_{Lt}. \tag{8.21}$$

On the supply side, consumers in the economy face a labor-leisure choice. If consumers maximize well-being under a budget constraint, each must satisfy a marginal condition relating labor price (net of taxes) to a ratio of marginal utilities of labor and consumption. If the utility function $U(\cdot)$ in the welfare function adequately represents individual preferences, then the marginal conditions describing labor supply can be written as follows:

$$-\frac{\partial U/\partial L_t}{\partial U/\partial C_t} = P_{Lt} - T_{Lt}, \tag{8.22}$$

where T_{Lt} is the tax on labor.

Similar relationships hold for the capital stock. On the supply side, consumers choose the allocation of their income between consumption and capital formation. A unit reduction in consumption at period of time t allows the consumer to increase consumption at the next period of time by an amount equal to $(1 + r_t)$, where r_t is the interest rate facing the consumer (the consumption interest rate). An optimizing consumer, choosing consumption and rate of capital formation, will select that consumption rate for which the ratio of discounted marginal utilities of consumption between the two time periods is equal to $1/(1 + r_t)$:

$$\frac{(\partial U/\partial C_{t+1})}{(\partial U/\partial C_t)} e^{-\rho} = \frac{1}{1 + r_t}. \tag{8.23}$$

On the demand side of capital markets, firms choose the capital to use based upon its price and productivity. The price facing the user of capital (P_{Kt}) is the supply price (the interest rate) plus the tax on capital (or on profits earned using capital), denoted by T_{Kt}. This gives:

$$\frac{\partial F}{\partial K_t} = P_{Kt} = r_t + T_{Kt}. \tag{8.24}$$

The entire market over time is now represented by equations (8.1), (8.2), and (8.20), which relate physical flows, and by equations (8.3) through (8.6) and (8.21) through (8.24), which express necessary conditions for a competitive equilibrium. The measure of welfare is provided by equations (8.18) and (8.19).

In order to evaluate the effects of energy system changes on welfare, equations (8.1), (8.2) and (8.20) can be differentiated totally, the marginal conditions (8.3) through (8.6) and (8.21) through (8.24) can be inserted, and the results can be combined with equation (8.19). This process provides the following equation for monetary equivalent welfare changes:

$$\Delta MW = \sum_{t=0}^{\infty} \alpha_t \left\{ -E_{It}\ \Delta P_{It} - E_{Dt}\ \delta AC_{Dt} + T_{It}\ \Delta E_{It} + T_{Dt}\ \Delta E_{Dt} + T_{Lt}\ \Delta L_t + T_{Kt}\ \Delta K_t \right\}, \tag{8.25}$$

where

$$\alpha_t = \prod_{\tau=1}^{t} \left(\frac{1}{1 + r_\tau} \right). \tag{8.26}$$

α_t is simply the discount factor on monetary flows obtained using the consumption interest rate. This factor is quite different from $e^{-\rho t}$, the discount factor on utilities.

Equation (8.25) allows the monetary equivalent welfare change to be obtained as simply a discounted sum of individual components. These individual components are presented in the second column of Table 8-2.

The components of the monetary equivalent welfare change are, for the most part, simply discounted sums of the components of NNP change as shown in this table. The direct effects of import price changes or shifts in the domestic cost function and the "welfare costs" of an import tax or a domestic tax correspond precisely. However, the effects via the labor input and the input of capital are not discounted sums of changes in the corresponding NNP components.

A comparison of the two columns of Table 8-2 shows that if a given policy leads to no changes in capital or labor quantities, then the monetary equivalent welfare change associated with that policy equals the discounted sum of NNP changes. If a given energy system change were to induce capital or labor to increase, however, then the monetary equivalent welfare change would be strictly less than the discounted sum of NNP changes.

The monetary equivalent welfare change via capital or labor always has the same sign, but a smaller magnitude[9] than the discounted sum of the corresponding component of NNP change. Therefore, discounted sums of NNP changes will not correctly measure changes in monetary equivalent welfare whenever there are induced changes in capital or labor.

In evaluating the welfare changes via labor and capital input, it is necessary to differentiate between economies that are perfectly efficient and those that are not. In an efficient allocation, the price of labor facing a consumer and that facing a firm are identical; the tax on labor is zero. Similarly, the tax on capital is zero.

In an efficient world, induced changes in labor or capital input do not change welfare even though they do change NNP. In the case of labor markets, this occurs because the consumption benefits obtained through an increase in labor input are just equal to the losses due to the decreased leisure time. In the case of capital markets, an increase in capital formation at one time period leads to a stream of additional consumption at later times. In an efficient allocation, the present value of this stream of future consumption changes is just equal to the reduction in current consumption required to increase the capital stock. Thus, if there is an efficient allocation of capital, changes in the capital stock have benefits just equal to their cost and, therefore, lead to a zero change in welfare.

Whether energy is complementary or supplementary with capital or labor is totally irrelevant for evaluating welfare in an efficient allocation. Only in the case of taxes on labor or capital will these relationships be significant for welfare evaluations.

The situation is different if the allocation is not efficient or if there are taxes on labor or capital. In this case, the monetary equivalent welfare changes via capital or labor inputs can be important. There will be a welfare gain from increasing labor supply whenever there is a tax on labor because the marginal productivity of labor will be greater than the marginal value of leisure foregone, the difference (per unit) being equal to the tax. Similarly, there will be a welfare gain from increasing the rate of capital formation whenever there is a tax on capital because the discounted sum of future consumption gains will be greater than the requisite current consumption reduction.[10]

An example is helpful in showing that with two different interest rates, a change in capital formation may have significant impacts on welfare. Assume that the consumer interest rate is r for all time, the tax is r and, therefore, that the price of capital facing a firm is $2r$.[11] Suppose that in year O, investment is increased by \$1 and that, therefore, consumption is decreased by \$1. The capital stock at all subsequent times will be increased by \$1 as long as the additional output produced each year is simply consumed. This assumption will be made.

Under these assumptions, the additional consumption in years 1 and beyond will be $2r$ dollars, the marginal productivity of capital. This sum of $2r$ dollars per year, discounted at an annual interest rate of r is \$2. Thus, the discounted sum of NNP increases stemming from the \$1 increase in capital formation is \$2. The discounted sum of consumption changes is \$2 minus the \$1 reduction in year O. Thus, the monetary equivalent welfare change of investing \$1 more is \$1, if the interest rate facing firms is twice the consumption interest rate! The same result could be obtained directly from the corresponding component in Table 8-2: the tax of r, discounted at a rate r, gives a monetary equivalent welfare change of \$1. Induced changes in capital stock can have proportionately large welfare impacts.

The divergence between the consumption interest rate and that facing firms may have implications for energy policy. A simple example will illustrate. To the assumptions of the preceding example, add the assumptions that all energy is imported and, further, that there is no concern for international vulnerability (a clearly difficult assumption!). Assume that capital and energy are complementary. The question then is, What tariff on imported energy would maximize welfare?

To illustrate the point, particularly simple mathematical forms will be specified, as indicated

$$\Delta K_t = -\gamma \, \Delta P_E = -\gamma T,$$

$$\Delta E_t = -\theta \, \Delta P_E = -\theta T.$$

What is the monetary equivalent welfare change associated with the imposition of a tariff? In an efficient world, the change in monetary welfare is equal to simply:

$$\frac{-\theta T^2}{2r} .$$

In this situation, any positive or negative tariff reduces welfare: the optimal tariff is zero. In this case of no externalities, no import vulnerability costs, and an efficient world, the optimal tariff on energy is zero.

The answer is different, however, in the situation of two interest rates. In the case assumed, the monetary equivalent welfare change is:

$$-\gamma T - \frac{\theta T^2}{2r} .$$

The optimal tariff is no longer zero. All positive tariffs reduce welfare by reducing capital formation and by reducing energy consumption below its optimal level. However, a small negative tariff (a subsidy on energy consumption) would increase welfare by increasing capital formation. The welfare gain from the increase in capital formation would dominate the welfare loss from excessive energy consumption. A negative tariff having a value of $-r\gamma/\theta$ would minimize welfare. Thus, in the situation of no externalities or vulnerability costs, capital-energy complementarity, and too little capital formation, welfare can be increased by subsidizing the use of energy.

This example, though unrealistic, does suggest that the existence of taxes on labor and capital that lead to nonoptimal rates of capital formation or labor supply may imply different policy prescriptions than those appropriate in a fully efficient world. The magnitude of these effects is unknown. And even the sign of the optimal tariff in the simple model is open to question if there is uncertainty whether capital and energy are complementary. However, the example does suggest that the interactions between energy and capital need to be considered before simple policy prescriptions are adopted.

SUMMARY AND CONCLUSIONS

This chapter has presented a conceptual framework and a few simple examples of the long-run relationships between energy sector changes and economic growth. Both a simple descriptive measure–NNP–and a more complex normative measure–welfare–have been considered.

It has been argued that many energy system changes could be represented by combinations of four analytical representations–tariffs, taxes, import price changes, and domestic cost function changes. Thus, the conceptual framework was developed considering impacts of only the four analytical representations, rather than separately examining each of the many possible energy system changes. However, the conclusions developed here can be applied to the many basic changes in the energy system.

Changes in welfare and in NNP can be decomposed into components representing the direct effects of import price or cost increases, the "welfare" effects of tariff or tax increases, and the effects via induced changes in the quantities of other factors of production, particularly capital and labor. The first two classes of components influence both net national product and welfare, with the components of monetary equivalent welfare change being equal to a discounted sum of the corresponding components of NNP change. However, the welfare change components via induced shifts in capital and labor are not discounted sums of the corresponding NNP components. In an efficient world, changes in capital and labor will influence net national product but will have no impact on welfare. In a world of two different interest rates, changes that induce an increased rate of capital formation will increase welfare. The monetary equivalent welfare change via capital or labor is of the same sign but of smaller magnitude than the corresponding discounted sum of NNP change components.

Policies that change costs of energy (import price changes, domestic cost function changes) have far larger impacts on the economy than policies that simply change energy prices (tariffs and taxes). The relative differences are particularly pronounced for small changes in energy price or cost. For equivalent cost increases, whether the cost increase is imposed on domestic energy or on imported energy is not relevant to the NNP or welfare impact.

The impact of a tariff or a cost increase depends upon the elasticity of demand for energy, a measure of the ability of the energy-using sectors to adjust to changing prices. For a given import price increase, the greater the demand elasticity, the smaller the economic impact. For a given tariff, the greater the demand elasticity, the larger the economic impact (and the energy impact). However, for a fixed reduction in energy consumption, the greater the demand elasticity, the smaller the economic impact of either an import price change or a tariff.

For increases in the world price of energy, the NNP loss borne by importing countries will always be greater than the revenue gain by the exporting country. And if the world price exceeds the exporting country's marginal cost of energy supply, then the NNP loss will also exceed the net revenue gain (net of costs).

Simple examples were produced that suggested conflicting policy advice. A tariff on all energy consumption (or an energy conservation program) would increase NNP and welfare if the tariff, by reducing energy demand, motivated a decrease in imported energy price. On the other hand, if capital and energy were complementary, if there were less than optimal capital formation, and if the import price were fixed, then an energy subsidy would increase both NNP and welfare. In the face of the conflicting advice, only more carefully articulated quantitative analysis (to say nothing about a consideration of other interactions) could be relied upon to develop the appropriate recommendation.

This chapter has presented only a simple conceptual structure for thinking through the relationships between energy and economic growth. More complete analysis depends upon the use of carefully articulated models of the energy system and at least of capital and labor markets. Such studies have been initiated through the CONAES project[12] and the Energy Modeling Forum,[13] but many issues are still far from resolution.

NOTES

1. Note that although the cost of conducting the R&D program does not appear in this table, it must be included in an overall evaluation of the economic impact of such a program.

2. The results are also applicable to an efficient centrally planned economy.

3. The implicit assumption here is either that there are no market failures in the economy other than those explicitly discussed or that the specific market failures in the economy remain unchanged as energy system changes occur. In particular, this assumption may be violated if an energy conservation program that eliminates market failures is implemented. In that case, the function $F(K, L, E)$ or the marginal conditions in equations (8.3) and (8.4) may change as a result of the program.

4. While this chapter discusses E as a scalar variable under the implicit assumption that energy can be viewed as a single aggregate quantity, E can also be viewed as a vector, thereby allowing a disaggregation by energy type. The interpretations in each equation will be straightforward.

5. W. W. Hogan and A. S. Manne, "Energy-Economic Interactions: The Fable of the Elephant and the Rabbit?" in *Energy and the Economy,* vol. 2, EMF Report 1, Standord University, (Stanford: September 1977). Also in C. Hitch, ed., *Modeling Energy-Economy Interactions: Five Approaches* (Washington, D.C.: Resources for the Future, 1977).

6. There are several different, but related, definitions of complementarity and supplementarity. A complete discussion is beyond the scope of this paper, but is discussed in W. W. Hogan, "Capital-Energy Complementarity in Aggregate Energy-Economic Analysis," in *Energy and the Economy,* vol. 2, EMF Report 1 Stanford: Stanford University, September 1977.

7. D. O. Wood and E. R. Berndt, "Engineering and Econometric Approaches to Industrial Energy Conservation and Capital Formation: A Reconciliation," M.I.T. Energy Laboratory Working Paper (77–040WP), November 1977.

8. Not to be confused with a discount factor on future consumption streams.

9. This occurs because the tax on capital or labor must always be smaller than the capital and labor price facing consumers (unless the supply prices are zero or negative).

10. A tax on labor would be associated with too little labor use. A tax on capital would be associated with too little capital formation.

11. This perhaps corresponds to the situation in the United States with a 50 percent corporate income tax coupled with a personal income tax on the earnings.

12. *Energy Modeling for an Uncertain Future,* T. C. Koopmans et al., Report of the CONAES Modeling Resources Group Washington, D.C. (National Research Council of the National Academy of Sciences, January 1978).

13. Energy Modeling Forum, *Energy and the Economy,* EMF Report 1 (Stanford: Stanford University, September 1977).

Chapter 9

The Economic Impact of Policies to Reduce U.S. Energy Growth

Edward A. Hudson
and
Dale W. Jorgenson
(Presented by Dale W. Jorgenson)
Harvard University

INTRODUCTION

The purpose of this chapter is to quantify the impact of alternative energy policies on future energy prices, energy utilization, and economic growth in the United States. Growth in energy consumption has become an important issue in U.S. economic policy as a result of the establishment of an effective international petroleum cartel by the Organization of Petroleum Exporting Countries (OPEC). The OPEC cartel has succeeded in raising world petroleum prices fourfold since 1973. This has resulted in rapid increases in the real price of delivered energy in the United States; the real price of delivered energy rose 7.0 percent annually between 1973 and 1977. To put these price increases in historical perspective, it can be noted that from 1950 to 1973 the real price of delivered energy in the United States declined at the rate of 1.8 percent per year.

Since 1954, the Federal Power Commission has maintained a system of wellhead price controls for natural gas entering interstate commerce. These controls have maintained prices below market-clearing levels and have necessitated the development of a system for the quantitative allocation of interstate natural gas. A similar situation has developed in the petroleum market. The U.S. government responded to the increase in world petroleum prices beginning in 1973 with a system of controls on domestic crude oil prices and on prices of petroleum products. These controls have been maintained and have been accompanied by an increasingly complex system for the allocation of petroleum.

The system of price controls on petroleum products within the United States has involved averaging the price of imported crude with that of domestic crude in the pricing of refined products. In effect, foreign producers receive a subsidy paid by means of a corresponding tax on domestic producers. By maintaining petroleum prices below world levels, U.S. domestic demand for petroleum has been allowed to rise more rapidly than it should have in the absence of price controls. The effective tax on domestic production has permitted domestic supply to fall more rapidly than it should have in the absence of price controls. The net impact of price controls has been to increase imports of petroleum products dramatically in the face of higher world petroleum prices. The Federal Power Commission approved a similar pricing system for domestic natural gas and natural gas imported in liquefied form (LNG), creating a system of effective subsidies for imported natural gas.

Higher world energy prices have not been passed on to domestic energy consumers in full, due to continued price controls on domestically produced petroleum and natural gas. Energy policy measures now in prospect will reduce or eliminate the effects of these controls and will expose large sectors of the U.S. economy to the full impact of higher world energy prices. The Carter Administration has proposed that the prices of petroleum and natural gas be set on the basis of "replacement cost" on the world petroleum market. In addition, the administration has proposed taxes on the utilization of petroleum and natural gas and measures to promote conversion to coal and other fuels that would have the effect of raising prices for petroleum and natural gas above world market levels. Higher energy prices will result in a reduction in the growth of energy consumption. However, higher energy prices can also have an important impact on future U.S. economic growth.

ECONOMETRIC AND PROCESS ANALYSIS MODELS

A satisfactory framework for analysis of the effect of alternative energy policies on energy prices, energy utilization, and economic growth requires an approach that encompasses both process analysis and econometrics. Process analysis provides for a detailed characterization of technology for energy conversion and energy utilization and permits the analysis of effects of introducing new energy technologies. Econometrics provides for the incorporation of behavioral and technical responses of patterns of production and consumption to alternative energy prices and permits an analysis of the impact of energy prices on the demand for energy, nonenergy intermediate goods, capital services, and labor services. In the process analysis approach, energy flows and energy conversion processes can be described in physical terms. In the econometric approach, flows of economic activity, including energy flows, are described in terms of economic accounts in current and constant prices.

To analyze the effect of alternative energy policies, we employ a dynamic general equilibrium model of the U.S. economy. For each of the commodities endogenous to the model—energy resources, energy conversion process, energy products, nonenergy products, capital services, and labor services—the model incorporates a balance between demand and supply that determines relative prices. In addition, the model includes a balance between saving and investment that determines the rate of return and the growth of capital stock. Economic growth is modeled as a sequence of one period equilibria determining demand and supply and relative prices for all commodities. Investment in each period determines the level of capital stock available in the following period. Dynamic adjustment to changes in energy policy is modeled by tracing through the impact on future levels of capital stock.

Since the output of the energy-producing industries is largely utilized by other industries rather than by final consumers, the matrix of interindustry transactions—representing flows of commodities, including energy, among industrial sectors—is a natural focal point for the study of the impact of energy policy. By representing energy sector transactions in physical terms, we can provide a link to process analysis models. By representing these transactions in economic terms, in current and constant prices, we can provide a link to econometric models. By using both forms for representing energy transactions, process analysis and econometric modeling can be combined within the same framework. This integration of process analysis and econometric approaches permits a detailed characterization of energy technology to be combined with a complete representation of the impact of energy prices on the economy as a whole.

The first component of our modeling framework is the "Long Term Interindustry Transactions Model" (LITM) developed by Hudson and Jorgenson.[1] This model is based on a system of interindustry accounts for the private domestic sector of the U.S. economy, divided among ten producing sectors.[2] Six sectors cover energy conversion and extraction—coal mining, crude petroleum, crude natural gas, petroleum refining, electric utilities, and gas utilities—the remaining four sectors—agriculture, manufacturing, transportation, and services—cover the production of nonenergy products. Final demand is divided among four categories—personal consumption expenditures, gross private domestic investment, government purchases of goods and services, and exports. Primary input is divided among capital services, labor services, and imports.

In the LITM framework the technology of each producing sector is represented by an econometric model based on the price possibility frontier, giving the supply price of output corresponding to given prices of primary and intermediate inputs and a given level of productivity.[3] For any given set of prices, technical coefficients giving primary and intermediate inputs per unit of output of the sector can be derived from the price possibility frontier. Given the level of output of the sector, the technical coefficients determine demand for energy, nonenergy intermediate goods, and primary factors of production.

The preferences of the household sector are represented by an econometric model determining demand for consumption goods, supply of labor, and supply of saving.[4] This model allocates personal consumption expenditures among the outputs of the ten producing sectors, services of housing and consumers durables, and domestic labor services.

The second component of our modeling framework is the "Time-Phased Energy Systems Optimization Model" (TESOM) developed at the Brookhaven National Laboratory by Hoffman and associates.[5] This model is based on the Reference Energy System, which provides a complete physical representation of technologies, energy flows, and conversion efficiencies from extraction of primary energy sources; through refining and various stages of conversion from one energy form to another; and to transportation, distribution, and storage of energy sources.[6] In the Reference Energy System, energy supplies such as nuclear fuels, fossil fuels, and hydropower are allocated to energy demands defined on a functional basis—such as space heating, industrial process heat, and automotive transportation. Energy supplies and demand are linked by energy conversion processes such as steam generation of electricity from coal.

Energy flows in TESOM are based on the cost-minimizing pattern of allocation of energy supplies to satisfy energy demands. The minimization of cost can be formulated as a linear programming model of the transportation type. Given levels of demand for energy services, available supplies of energy resources and conversion capacities, conversion efficiency, and capital and operating cost of utilizing technologies, the energy sector optimization model determines a set of energy conversion levels that minimize cost; the dual to the linear programming problem determines shadow prices for energy demands, energy resources, and energy conversion capacities. In the time-phased version of this model the energy conversion capacities are the result of previous investments.

The combined LITM-TESOM framework models the U.S. interindustry transactions—flows of energy resources, energy conversion activities, energy products, nonenergy products, capital services, and labor services—as a result of a dynamic general equilibrium of the U.S. economy.[7] In each period the supply prices of all commodities are determined by price possibility frontiers for the nonenergy sectors and by processes selected for the energy sectors—given the prices of capital and labor services and exogenously given prices of imports. Given these prices and the supply prices for each product, technical coefficients for inputs of intermediate goods and primary factors of production are determined. Finally, given the technical coefficients, demands for the output of each sector of the economy are determined by final demands for all products. While the level of investment for the private domestic economy as a whole is endogenous to the model, the allocation of investment among producing sectors is given exogenously. Final demand for government purchases and exports is also determined exogenously.

We can complete the description of a dynamic general equilibrium analysis of the U.S. economy by describing the markets for capital and labor services. In each period, the supply of capital is fixed initially by past investments. Variations in demand for capital services by the producing sectors and household sector affect the price but not the quantity of capital services. Similarly, in each period the available labor time is fixed by past demographic developments. Variations in demand for labor time by the producing sectors and by the household sector for consumption in the form of leisure affects the price and the allocation of labor time between market and nonmarket activity. Finally, the supply of saving by the household sector must be balanced by final demand for investment by the producing sectors. Investment generates the level of capital stock available at the beginning of the following period and creates the conditions for a new equilibrium of product and factor markets, given the time endowment available in that period.[8]

ALTERNATIVE ENERGY POLICIES

The starting point for our analysis of the impact of alternative energy policies is a base case projection of future energy and economic growth with no change in energy policy. We assume that any quantity of petroleum imports is available at the world price, where the world price of petroleum rises at a rate of 1 percent per annum relative to the rate of growth of the U.S. GNP price deflator until 1990 and 2.5 percent annually, in real terms, thereafter. The annual rate of growth or real GNP is projected to average 3.2 percent from 1977 to 2000. This growth is considerably slower than the average annual growth rate of 3.8 percent between 1950 and 1973. The decline is partly due to a reduction in population growth and partly due to a reduction in productivity growth resulting from higher energy prices. Primary energy input is projected to rise from 76 quadrillion Btu in 1977 to 139 in 2000, an average annual growth rate of 2.6 percent. This growth is also slower than that experienced in the past—the average rate of growth of primary energy input between 1950 and 1973 was 3.2 percent per annum. Part of the reduction is due to decreased economic growth and part is caused by the continuing rise in real energy prices.

The increase in the relative price of petroleum over time leads to a steady decline in the relative importance of petroleum in total energy use. Natural gas is also projected to decline in relative importance due to supply constraints and price increases. Coal and nuclear sources sustain much of the growth in energy use; both direct use of coal and the use of electricity grow relatively rapidly. Imported petroleum accounts for approximately half of all petroleum use in 1977; in the base case projection, this share rises to almost 60 percent by 2000. The base case projections therefore imply continued reliance on imports for a substantial fraction of energy supply.

The reduction of import dependence, in order to reduce the associated economic and political risks, is an important objective underlying U.S. energy policy proposals. To analyze such policies, and their energy and economic effects, we consider the following set of policy packages:

Policy 1: Taxes are imposed on U.S. petroleum production to bring domestic petroleum prices to world levels; natural gas prices are increased but price controls are retained; energy conservation is stimulated by taxes on use of oil and gas in industry, restriction of oil and gas use by electric utilities, subsidies for insulation of structures, and mandatory performance standards for energy-using appliances.

Policy 2: The measures included in Policy 1 are combined with tariffs on imported oil rising to $4.50/bbl in 1985 and to $7.00/bbl in 2000 and with corresponding taxes on natural gas.

Policy 3: Policy 2 is combined with exicse taxes on delivered energy sufficient to reduce total primary energy input in 2000 to 90 quadrillion Btu.

Policy 4: Policy 3 is combined with excise taxes on delivered energy sufficient to reduce total primary input in 2000 to 70 quadrillion Btu.

OVERVIEW OF ECONOMIC EFFECTS

The processes by which the economy adjusts to higher energy prices and reduced availability of energy involves many simultaneous changes and takes time to accomplish.[9] For expository purposes these processes can be represented as in Figure 9-1. The changes in the prices and availability of energy, and induced changes affecting other inputs, give rise to changes in the pattern of prices of inputs into production. This leads to a changed pattern of output prices. This in turn induces a redirection of the pattern of final demand spending–in particular consumption, but also investment, government, and foreign purchases–away from the now more expensive energy-intensive goods and services. Simultaneously, producers react to the changed pattern of prices that they must pay for inputs by altering their operating procedures, processes, and products to economize on expensive energy input. Both final demand and production patterns are restructured. Therefore, the pattern of sectoral output and the pattern of use of productive inputs are redirected away from energy and energy-intensive products. Finally, these changes are reflected in changes in summary measures of economic performance such as energy input and real GNP.

The adjustment processes involve shifts in the composition of final spending and of production input patterns away from energy. While energy output will show the most noticeable changes, use of capital, labor, and intermediate materials can also be significantly affected. These changes occur both in each year and over time. Within each year, substitutions between products and processes will lead to a different mix of inputs being used. This restructuring will permit part of the energy reduction to be absorbed without loss of overall economic output. However, as these different inputs are only imperfect substitutes and as these

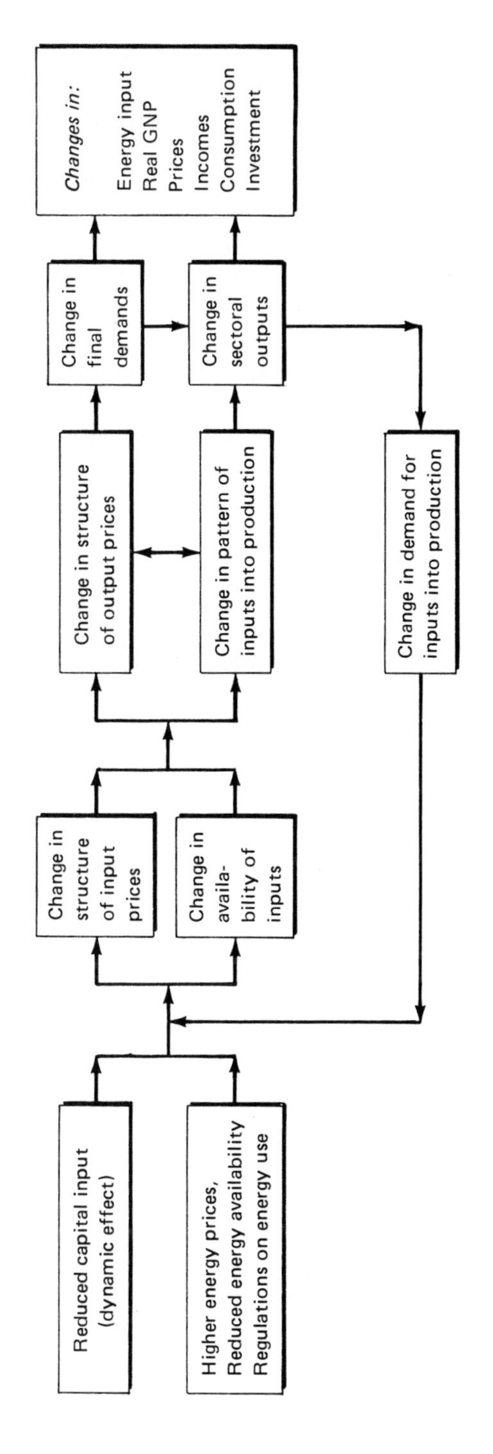

Figure 9–1. Overview of Adjustment to Reduced Energy Input.

other inputs must be diverted from other uses, some output will be lost through this restructuring. In addition there is a dynamic adjustment that operates through investment and capital. If saving and investment are reduced as a result of energy restrictions due to lower income or lower rates of return, the rate of accumulation of capital is slowed. This in turn slows the rate of growth of productive capacity and of output and incomes, providing an additional source of economic restructuring and economic cost.

The impact of energy changes on GNP reflects both the substitution and the dynamic effects. Table 9-1 indicates the aggregate nature of these effects for Policies 1 and 2 in the year 2000. Real GNP in Policy 2 is 3.2 percent less than in the base case. This decline is partly due to the dynamic effect operating through the capital stock—capital input is 3 percent lower than in the base case. This decline in capital input accounts for approximately a one percentage point reduction in real GNP or one-third of the total GNP reduction. At the same time, there are substitutions away from energy as an input intro production. Of the 16 percent reduction in energy input, 3.2 percent is accounted for by reduced total economic output while approximately 13 percent is sustained by substitutions toward other inputs and toward non-energy-intensive products. Both capital and labor are partially substituted for energy; the energy-capital

Table 9-1. Capital, Labor, and Energy Inputs in 2000.

	Base Case	*Policy 1*	*Policy 2*
Quantities of Input			
Capital Services [a]	831.5	821.9	806.3
Labor Services [a]	1281.3	1281.2	1281.1
Energy [b]	138.5	126.6	116.3
Real GNP	2721.7	2679.8	2634.9
Input Quantities, Percent Change from Base Case			
Capital		−1.2	−3.0
Labor		−0.0	−0.0
Energy		−8.7	−16.0
Real GNP		−1.5	−3.2
Gross Input Productivities [c]			
Gross Capital Productivity	1.0	0.9961	0.9984
Gross Labor Productivity	1.0	0.9846	0.9682
Gross Energy Productivity	1.0	1.0771	1.1529
Ratios between Input Quantities [d]			
Energy:Capital	1.0	0.9240	0.8606
Energy:Labor	1.0	0.9134	0.8406
Capital:Labor	1.0	0.9885	0.9698

[a]Measured in billions of 1972 dollars.

[b]Quadrillion Btu of primary energy input.

[c]Gross input productivity is real GNP per unit of the input in the policy case divided by this ratio in the base case.

[d]Ratios are divided by their base case values.

ratio is reduced by 14 percent, the energy-labor ratio by 16 percent. Labor and energy turn out to be substitutes in all production sectors. This leads to an increase in labor input relative to GNP. The level of gross labor productivity, or real GNP per unit of labor input, is therefore reduced; it declines by more than 3 percent from the base case level. This corresponds to lower real per capita incomes and is the counterpart of GNP reductions.

The relationship between capital and energy is more complex. In some sectors, such as services, capital and energy are substitutes, so that increased investment in capital such as insulation is the pattern of response to higher energy prices and to requirements forcing energy reductions. In other sectors, such as manufacturing, capital and energy frequently move in a complementary fashion so that the response to higher energy prices is to reduce the rate of investment. In addition, the shift of spending patterns between sectors, each sector with a different capital intensity, results in a change in the relation between capital input and total ouput. The net result of these capital changes is that capital input rises relative to energy input, that capital input declines relative to labor input, and that capital input per dollar of real GNP rises, with a corresponding decline in capital productivity.

ADJUSTMENTS IN THE PRICE STRUCTURE

The immediate point of impact of energy policy measures is in the structure of relative prices. Energy becomes more expensive relative to others goods and services. In the attempt to allocate purchases so as to minimize production costs, producers reduce energy use, moving toward less energy-intensive inputs and processes. Similarly, in the attempt to derive maximum value from their consumption budgets, households redirect their expenditure patterns to economize on the now expensive energy and energy-intensive goods and services. Thus, higher prices for energy, and so far energy-intensive products, lead to reductions in energy use and to slower growth in energy consumption. In addition, the non-price energy conservation measures contained in the Policy 1 package add considerable further pressure to the shift toward less intensive energy use.

The price adjustments commence with the relative prices of inputs. Table 9-2 shows these changes for the simulations for 2000. The changes in capital, labor, materials, and energy input prices feed through the production structure to alter the whole pattern of relative prices of produced goods and services. The price of capital services is taken as the numeraire in the model system, so the observed variation in prices is for other prices to adjust relative to the prices of capital services. The variation involves an increase in the price of energy, an increase in the price of intermediate materials, and a decrease in the price of labor services.

The rise in energy prices directly reflects the taxes and other policy measures. Two measures of the energy price increase are given. The first, referring to actual energy prices, relates to dollar prices paid for the purchase of energy.

Table 9-2. Adjustments in Input and Output Prices in 2000.

	Policy 1	*Policy 2*	*Policy 3*	*Policy 4*
Input Prices, Percent Change from Base Case				
Capital	—	—	—	—
Labor	–0.68	–2.04	–5.43	–9.90
Energy	15.73	29.74	108.24	217.01
Intermediate Materials	0.24	0.80	1.12	2.61
Output Prices, Percent Change from Base Case				
Agriculture, Nonfuel Mining, Construction	1.07	1.31	2.29	5.54
Manufacturing	0.08	–0.23	1.14	5.91
Commercial Transportation	3.11	4.57	7.83	13.38
Services, Trade, Communications	–0.26	–0.87	0.56	2.24
Energy, Actual prices [a]	0.19	10.75	87.71	187.33
Energy, Effective costs [b]	15.73	29.74	108.24	217.01

[a] Actual energy prices refer to the average dollar cost of delivered energy in terms of dollars per Btu.

[b] Effective energy costs refer to the average cost of energy services to energy purchasers, allowing for both price and nonprice conservation measures and calculated using a fixed weight quantity index.

These increase, relative to the base case, by 0.2 percent for Policy 1 and by 187 percent for Policy 4. However, the effective cost of energy rises by more than this due to the effect of nonprice policy measures. Nonprice regulations that require the use of more capital equipment or of different production processes increase average costs of production. The estimate of these additional costs, together with the direct price increases, is given by the second measure, effective energy cost. For Policy 1, this overall measure of energy cost is 16 percent above the base case, for Policy 4 it is 217 percent higher. Thus, both price and nonprice measures contribute significantly to the pressure to reduce consumption of energy.

Labor prices show a relative decline. This means that labor prices under the policy simulations show a less rapid growth over the forecast period than under base case conditions. The reason for this slower growth lies in the employment constraint imposed on the simulations—the rate of unemployment is constrained to be the same as in the base case. As the energy measures lead to a reduction in the level of economic activity, the demand for labor is reduced. This reduction is more than enough to offset the additional labor use due to energy-labor substitution. A relative reduction in labor prices is required in order to stimulate labor use. Thus, labor prices in Policy 2 are 2 percent less than in the base case and are 10 percent less in Policy 4. Finally, the prices of intermediate materials show a small increase as a result of the policy measures. This is due to the effect of higher energy costs raising production costs for these materials to an extent greater than the cost reduction following the decline in labor prices.

These changes in input prices lead to adjustments in the level and pattern of output prices. The overall level of output prices is increased. However, since energy represents, compared to capital, labor, and materials inputs, only a small component of total production costs in most industries, the higher energy prices lead to relatively small increases in average output prices. The patter of relative prices is more substantially altered. Energy prices rise significantly–in Policy 3, for example, by 88 percent in terms of dollars per Btu and by 108 percent in terms of effective costs to purchasers. Other output prices change by smaller proportions–in Policy 3, transportation prices rise by 8 percent; agriculture, nonfuel mining, and construction prices rise by 2.3 percent; manufacturing prices rise by 1.1 percent; and services prices are 0.6 percent higher.

The pattern of price changes is closely related to the mix of inputs used in each sector. Energy output, of course, is highly energy intensive, and it shows a substantial price increase. Commercial transportation is the next most energy-intensive sector, and its price also rises significantly. Agriculture, nonfuel mining, construction, and manufacturing are less energy intensive and show smaller price increases. Services, trade, and communications are the least energy-intensive sector and are also relatively labor intensive, so that these prices either decline or show only a small increase. The pattern of changes in prices and effective costs alters to make energy-intensive goods and services relatively expensive. Together with the associated nonprice measures, this is the force motivating producers and consumers to redirect their expenditure patterns and to use less energy in production and consumption activities.

CHANGES IN THE PATTERN OF FINAL DEMAND

Final demand expenditure is a critical determinant of the overall level and composition of activity in the economy. Final demand–personal consumption expenditures, private investment, government purchases and exports–dictate what is produced in the economy. Changes in final demand are therefore important in securing reductions in the energy intensity of economic activity. Reduction in final demand purchases of energy directly result in energy saving. Also, reductions in final demand purchases of energy-intensive goods permit less energy to be used in production activities and so accentuate the overall saving in energy. Both these types of expenditure adjustments–less energy purchases and less use of energy-intensive products–are induced by the restructuring of output prices.

Table 9-3 summarizes the changes in final demand spending resulting from the energy policy measures. The broad pattern of spending is similar under all policies, with services absorbing a little more than half and manufacturing

Table 9-3. Final Demand Patterns in 2000.

	Base Case	*Policy 1*	*Policy 2*	*Policy 3*	*Policy 4*
Composition of Real Final Demand (percent)					
Agriculture, Nonfuel Mining, Construction	8.38	8.30	8.22	8.13	8.06
Manufacturing	30.48	30.48	30.45	30.30	30.02
Commercial Transportation	3.64	3.48	3.36	3.30	3.21
Services, Trade, Communications	54.08	54.70	55.15	55.91	56.94
Energy	3.43	3.04	2.82	2.36	1.77
Total	100.00	100.00	100.00	100.00	100.00
Real Final Demand, Percent Change from Base Case					
Agriculture, Nonfuel Mining, Construction		−3.1	−5.6	−10.6	−15.9
Manufacturing		−1.9	−3.9	−8.4	−13.9
Commercial Transportation		−6.2	−11.0	−16.6	−22.8
Services, Trade, Communications		−0.8	−2.1	−4.7	−8.0
Energy		−13.1	−20.9	−36.5	−54.9
Total		−2.0	−3.9	−7.9	−12.6

about one-third of total real final demand. However, there are some significant shifts in final spending induced by the price and nonprice energy policy measures.

The changes in the price structure and the other energy conservation measures lead to a substantial adjustment in real final demand. The overall level of real final demand is reduced; this in itself results in a significant reduction in energy use. For example, final demand for energy in Policy I is 13 percent below the base case. This energy saving is compounded by a redirection of spending patterns away from energy and energy-intensive goods and services. Non-energy-intensive output, particularly services, becomes relatively more important within the pattern of final spending.

CHANGES IN PRODUCTION PATTERNS

The patterns of inputs into each production sector also change in response to the energy policy measures. Higher energy prices create an incentive for producers to alter input patterns away from energy and thereby to reduce production costs. The nonprice direct regulations concerning energy use provide additional pressure to reduce energy purchases. Also, the changes in labor prices and in prices of other intermediate inputs provide incentives for further adjustments in the mix of inputs and processes. The net result of these pressures is to induce, or to force, producers to adjust their purchase patterns,

Table 9-4. Input Patterns in Nonenergy Production 2000 (input-output coefficients for aggregate input categories).

	Base Case	*Policy 1*	*Policy 2*	*Policy 3*	*Policy 4*
Agriculture					
Capital	0.1946	0.1938	0.1932	0.1921	0.1900
Labor	0.2542	0.2575	0.2598	0.2661	0.2722
Energy	0.0242	0.0232	0.0225	0.0200	0.0176
Materials	0.5271	0.5255	0.5245	0.5218	0.5193
Manufacturing					
Capital	0.1194	0.1176	0.1161	0.1143	0.1131
Labor	0.2815	0.2845	0.2881	0.2963	0.3046
Energy	0.0235	0.0231	0.0226	0.0193	0.0179
Materials	0.5756	0.5748	0.5732	0.5701	0.5644
Transportation					
Capital	0.1971	0.1956	0.1939	0.1920	0.1888
Labor	0.4016	0.4031	0.4049	0.4076	0.4100
Energy	0.0384	0.0373	0.0361	0.0327	0.0298
Materials	0.3629	0.3640	0.3651	0.3677	0.3714
Services					
Capital	0.3389	0.3405	0.3418	0.3456	0.3493
Labor	0.3526	0.3585	0.3627	0.3738	0.3864
Energy	0.0186	0.0180	0.0175	0.0151	0.0131
Materials	0.2899	0.2830	0.2780	0.2655	0.2512

economizing on energy use by changed production practices and processes and to place greater reliance on nonenergy inputs. Thus, not only is energy input reduced but the entire pattern of inputs into each sector is changed. The estimated changes in input patterns are shown in Table 9-4, which gives the aggregate input-output coefficients for each of the major non-energy-producing sectors—agriculture, nonfuel mining, and construction; manufacturing; commercial transportation; and services, trade, and communications. These coefficients measure the proportion of the total real input into the sector that is of the specified form, whether capital services, labor services, energy, or intermediate materials.

The unit input requirement of energy into agriculture, nonfuel mining, and construction is reduced by the policy measures by 4 percent in Policy 1 and by up to 27 percent in Policy 4. Energy and capital show a complementary relationship in this sector; when energy prices increase, the use of capital input is reduced. Therefore capital input is reduced as part of the adjustment process, although this change is not large. The reduction in energy and capital inputs must be compensated for by an increase in other inputs. Labor is the key input that provides this compensating increase. The input-output coefficient of labor rises from 0.2542 in the base case to 0.2722 in Policy 4, a sufficient rise to offset the move away from energy, capital, and materials. Finally, the input of intermediate materials is reduced slightly as part of the adjustment process.

The manufacturing sector follows a similar pattern of adjustments between inputs. The use of energy per unit of output is significantly reduced. Also, energy and capital show a strong complementary relationship. This means that a reduction in capital input is associated with the higher energy costs and with reduced energy input. In contrast, labor is a substitute for both capital and energy. Consequently, the more expensive energy and capital input mix is partially replaced by the now relatively less expensive input of labor services. Input of intermediate materials declines slightly. Overall, energy saving in manufacturing is achieved by a reduction in energy and capital use accompanied by an increase in labor input.

The nature of the energy-capital link can be complex. Some types of capital have a direct complementary relationship with energy in that the more capital that is used, the more energy input is required. This is true for many types of motive power uses of energy and capital and for some types of process uses. In other instances, however, there is a substitution relationship between capital and energy; energy can be saved by the use of more capital equipment. For example, energy required per unit of output can often be reduced by the use of more sophisticated and more expensive capital. However, even this energy-capital substitution relationship is consistent with an overall appearance of complementarity between energy and capital. The reason for this operates through a separate input, labor, which is typically a substitute for both capital and energy. A rise in the price of energy rise to the following adjustments: energy use is reduced; capital input tends to be increased (energy-capital substitutability); labor input is increased (energy-labor substitutability); capital input tends to be decreased (labor-capital substitutability). There are, then, pressures to reduce and to increase capital use. In manufacturing, the net result is that a reduction in capital use accompanies the reduction in energy input.

Transportation shows a substantial decline in energy intensity as a result of the policy measures—for example, the input coefficient for energy falls from 0.038 to 0.036, an 11 percent decline, between the base case and Policy 2. Complementarity between energy and capital leads to a reduction in the output of capital services; the capital input coefficient rises by 2 percent for Policy 2. Both labor and intermediate materials can substitute for energy and capital in transportation, and the coefficients for these inputs in Policy 2 rise by 0.8 percent and 0.6 percent respectively.

The services, trade, and communications sector responds somewhat differently from the other non-energy-producing sectors. The extent of the energy reduction, 6 percent in Policy 2, is comparable to that achieved in the other sectors, but the manner in which this reduction is achieved is different. The principal difference is in the role of capital. In services the relationship between energy and capital is one of substitutability; higher energy costs and reduced energy input are associated with greater use of capital services. The reduction in energy input is, in part, secured by an increase in capital input. A central reason for the different energy-capital relationship in this sector lies in the type of use made of energy. In services, a great deal of energy is used for space heating and for air conditioning. Reduction in this use of energy can be achieved through improved design and insulation of structures and more sophisticated heating and cooling equipment. Each of these changes uses additional capital. In Policy 2, for example, the reduction in the energy input coefficient is associated with a 0.9 percent increase in the capital input coefficient. Labor can also be substituted for energy; in fact, the degree of substitution is greater than that between energy and capital, and the labor input coefficient is increased by 2.9 percent. The final set of inputs, nonenergy intermediate goods, are reduced substantially, the input coefficient in Policy 2 being 4 percent less than in the base case. In sum, the input restructuring in services is to move away from energy and other produced inputs toward capital and labor inputs.

The economywide pattern of capital, labor, energy, and materials inputs is determined jointly by the pattern of these inputs in each sector and by the relative size of each sector in the economy. Higher energy prices and nonprice restrictions on energy use result in a restructuring of input patterns away from energy, away from capital (except in servcies), and toward labor. In addition, the sectoral composition of output shifts away from energy, transportation, and agriculture and toward services. The overall effect of these changes is to substantially reduce the overall energy intensity of production; the average labor intensity increases, and the average capital intensity shows a very small increase.

The interrelationships between inputs can be formalized by means of the Allen partial elasticity of substitution. This elasticity indicates the changes in the relative quantities in which two inputs are used, caused by a change in their relative prices. The elasticities, for each pair of inputs into each sector and evaluated at the 1971 data point, are shown in Table 9-5. It should be noted that the model system does not use these elasticities explicity; rather, it incorporates models of producers' behavior, some of whose characteristics can be summarized in numerical terms by means of these elasticities. The numerical values of these elasticities can be interpreted as follows: a value of zero means that the two inputs are used in fixed proportions; a negative value means that there is a complementary relationship between the inputs—a rise in the price of one input is associated with reduced use of the second input; a positive value implies a substitution relationship—a rise in the price of one input leads to increased use of the other. Also, the greater the absolute value of the elasticity, the stronger the relationship or interaction between the inputs. The own elasticity of substitution will be negative; this simply implies that when the price of this input rises, demand for the input will decline.

In agriculture, nonfuel mining, and construction there is a reasonably strong substitution relationship between energy and labor and a substantial response of energy demand to energy price. In manufacturing, there are three strong interactions: energy and labor and capital and labor are substitutes while energy and capital are complements. The commercial transportation sector exhibits complementarity between energy and capital and a significant own price elasticity of demand for energy. In the services, trade, and communications sector,

Table 9-5. Interrelationships between Inputs: Allen Partial Elasticities of Substitution.

	INPUT			
	Capital	*Labor*	*Energy*	*Intermediate Materials*
Agriculture, Nonfuel Mining Construction				
Capital	–1.7673			
Labor	0.3553	–2.5018		
Energy	–0.0591	1.4148	–29.6499	
Intermediate materials	0.6134	1.0442	0.5987	–0.8289
Manufacturing				
Capital	–3.1820			
Labor	1.1004	–1.6181		
Energy	–1.4156	1.8900	–4.8410	
Intermediate materials	0.0963	0.5072	–0.4732	–0.2435
Commercial Transport				
Capital	–1.4036			
Labor	0.1755	–1.0920		
Energy	–0.8577	–0.0574	–11.5998	
Intermediate materials	0.5747	1.1309	1.7739	–1.7267
Services, Trade, Communications				
Capital	1.6979			
Labor	1.0903	–0.8795		
Energy	1.2110	2.3065	–49.3616	
Intermediate materials	0.0660	0.0440	–1.8201	–0.0245

there are four strong interdependecies: energy and capital, energy and labor, and capital and labor are substitutes, while energy and intermediate materials are complements. In addition, there is a high own price elasticity of demand for energy.

COMPOSITION OF TOTAL OUTPUT

Total output from each sector depends both on final demands and on purchases as inputs into other producing sectors. Final demand expenditure and the pattern of input purchases are each adjusted away from energy-intensive products as a result of the higher energy prices and associated nonprice conservation measures. Therefore, total demand for the output of each sector changes, with output in general declining and with the output of energy-intensive products declining most substantially. The patterns of changes in real gross outputs are shown in Table 9-6. Gross output in the economy declines as a result of the energy policy packages. In 2000, total real gross output in Policy 1 is 1.9 percent less than in the base case, while in Policy 4 the reduction is 12.7 percent. The enery sector is most affected, with its output, in constant dollar terms, declining by 6.2 percent in Policy 1 and by 42.2 percent under Policy 4. The

Table 9-6. Sectoral Real Gross Output in 2000.

	Base Case	*Policy 1*	*Policy 2*	*Policy 3*	*Policy 4*
Composition of Total Gross Output (percent)					
Agriculture, Nonfuel Mining, Construction	8.59	8.53	8.46	8.39	8.37
Manufacturing	35.93	35.88	35.84	36.05	35.98
Transportation	5.04	4.93	4.90	4.77	4.74
Services	45.67	46.09	46.48	47.43	47.76
Energy	4.77	4.56	4.31	3.35	3.16
Total	100.00	100.00	100.00	100.00	100.00
Total Gross Output, Percent Change from Base Case					
Agriculture, Nonfuel Mining, Construction		−2.9	−5.0	−9.7	−15.0
Manufacturing		−2.0	−3.8	−7.3	−12.6
Transportation		−4.0	−6.3	−12.6	−17.9
Services		−1.0	−1.8	−4.1	−8.7
Energy		−6.2	−12.8	−35.1	−42.2
Total		−1.9	−3.6	−7.6	−12.7

next largest decline is in output from the transportation industry, the decline being 4 percent and 17.9 percent in Policies 1 and 4, respectively. Output from the agriculture, nonfuel mining, and construction sector is also reduced by 2.6 percent for Policy 1 and 15 percent for Policy 4. The manufacturing sector maintains its relative size in the economy; the reduction in the output of manufactured goods is almost identical to the overall reduction in economic output. Output from service, trade, and communications activity is reduced by less than average. The reduction in this output is only 1 percent in Policy 1 and 8.7 percent in Policy 4. Service types of activities therefore assume a greater importance within the overall productive structure of the economy.

These sectoral changes are closely related to the energy intensity or production. The largest reductions in output occur in those industries that are most energy intensive. The smallest reductions occur in sectors, particularly services, that use relatively little energy per unit of output. The result is a shift of production toward those industries whose output is relatively nonenergy intensive. The average energy content of production is therefore reduced; the energy required per dollar of output from the economy is reduced under these new economic structures relative to the energy requirement in base case conditions.

DYNAMIC ADJUSTMENTS THROUGH INVESTMENT

The above analysis has focused on adjustments to spending and input patterns that were essentially substitution responses to changes in relative prices. These are the principal means by which the economy adjusts to a less energy-intensive structure. However, another very significant, and related, set of effects is through investment and the capital stock. Higher energy prices lead to a reduction in capital income and to a reduced rate of return on capital. Part of this reduction in the rate of return is related to the energy-capital complementarity observed in some sectors. Additional investment in energy-conserving capital, particularly where it is forced by direct regulation and mandatory performance standards, can have low total productivity and can accentuate the decline in yield on capital. Lower rates of return lead directly to reduced saving and investment in the private economy. In addition, the income reductions due to the substitution adjustments considered above lead to further cutbacks in the volume of private saving and investment. Thus, private investment is reduced below base case levels. This directly results in a slowing of the rate of growth of capital stock. In fact, the growth paths of capital stock under the energy policy measures are projected to be permanently below the base case growth path.

The significance of these reductions in investment and the capital stock is that one of the principal inputs into production, capital services, is reduced. Reduced investment and capital means that the productive capacity of the economy is reduced throughout the forecast period. Further, this lowering of productive potential, compared to base case level, becomes progressively greater over time. This capital reduction or dynamic effects is a fundamental mechanism through which the shift to a less energy-intensive configuration of spending and production can lead to slower growth and can impose output and income loss on the economy.

The magnitudes of the investment and capital effects are shown in Table 9-7. Investment rises over time under the energy policies as well as in the base case, but the rate of growth of investment is less under the policy measures—that is, the level of real investment in each year is less than in the base case. The relative reduction in investment levels is not large—by 2000 it is 1.6 percent for Policy 1 and 4.9 percent for Policy 2—but the cumulative impact on the level

Table 9-7. Investment and Capital Stock.

	1985	*1990*	*2000*
Investment: Percent Change from Base Case			
Policy 1	–0.4	–1.0	–1.6
Policy 2	–1.4	–3.0	–4.9
Policy 3	–1.8	–5.2	–10.5
Policy 4	–2.4	–6.7	–16.3
Capital Stock: Percent Change from Base Case			
Policy 1	–0.17	–0.47	–1.17
Policy 2	–0.49	–1.27	–3.03
Policy 3	–0.63	–1.85	–7.00
Policy 4	–0.81	–2.40	–10.75

of capital stock is significant. Under Policy 2 conditions, the level of capital stock in 1990 is 1.3 percent below the base case and by 2000 it is 3 percent below. This slowing in the rate of growth of capital directly implies that the productive potential of the economy grows less rapidly than in the base case.

AGGREGATE ECONOMIC COST OF ENERGY REDUCTIONS

Two types of adjustments of the economy to changes in energy conditions have been analyzed. The first involves the restructuring of production and spending patterns away from energy and from energy-intensive inputs and production. The second adjustment operates through changes in savings and investment and results in a slower growth of capital and aggregate productive capacity. Both of threse adjustment mechanisms impose costs on the economy.[10] From an aggregate point of view, these costs take the form of a reduction in the volume of final output that can be obtained from the economy, compared to that possible under base case conditions.

The substitution of labor, capital, and nonenergy goods and services for energy input into production is not perfect; some output is lost as a result of the restructuring. In other words, additional labor and other inputs can help to compensate for less energy input, but some reduction in net output is still probable. Also, additional labor and other inputs used to replace energy must be obtained from other uses, thus reducing the total volume of potential output. These same adjustments can be viewed in terms of factor productivities. The process of reducing intensity of energy use involves increasing the intensity of labor use and, in some cases, capital use. The input-output coefficients for labor and, in an aggregate sense, for capital increase. Thus, more labor and capital are used per unit of output. This is equivalent to saying that the productivities of labor and of capital are reduced as a result of the energy adjustments. At any time these inputs are limited in supply, so reduction in their productivities translates directly into a reduction in the potential output of the economy. Real GNP declines, or its growth rate slows, as a result of the substitution away from energy input into production.

The dynamic costs of energy reduction follow in part from these substitution costs and in part from separate mechanisms. The reduction in output and income as a result of the substitution processes leads directly to a reduction in the aggregate level of saving and investment. In addition, the rate of return on capital can fall as a result of the higher energy prices and the greater input of capital per unit of output in the economy as a whole. A decline in rates of return leads to further reductions in saving and investment. Therefore, capital growth under the energy policies is lower than in the base case. This corresponds directly to a slowing of the growth of the productive capacity. At any given time, this involves a lower real GNP than under base conditions, compounding the economic cost caused by the substitution process.

The magnitudes of the economic costs of restrictive energy policies are shown in Table 9-8 in terms of real GNP. For each of the four policy packages, real GNP is less than in the base case. Further, the reduction is larger, the more restrictive the policy measures. Also, the reduction under each policy becomes larger over time in both absolute and relative magnitude. For Policy 1, the economic cost in 1985 is 0.6 percent of real GNP while in 2000 the cost rises to 1.5 percent of GNP. Under Policy 4 the cost is 2.7 percent of GNP in 1985, rising to 11.9 percent in 2000. Economic growth is reduced under each policy, but positive growth continues in all cases. For example, Policy 1 reduces the annual rate of economic growth by about 0.1 percentage points.

These economic costs are substantial. One way to calculate the overall cost is to find the present value of the real GNP loss over the entire 1977 to 2000 period. These present values (as at 1977 using a 5 percent discount rate) are $148 (1972) billion for Policy 1, $350(1972) billion for Policy 2, $615(1972) billion for Policy 3, and $919(1972) billion for Policy 4. To place these in perspective it can be noted that U.S. real GNP in 1977 was about $1330 (1972) billion. Thus, although energy reductions can be achieved, they do involve a substantial real cost in loss of potential income and output.

The GNP loss from energy policies can be separated into a part resulting from the substitution cost and a part resulting from the dynamic or capital cost. This separation is only approximate, since both costs arise from interdependent adjustment processes, but it does indicate the relative magnitudes

Table 9-8. GNP Effects of Energy Policies.

	1985	*1990*	*2000*
Change in Real GNP from Base Case, $(1972) bn			
Policy 1	−10.6	−24.0	−41.9
Policy 2	−29.9	−53.5	−86.8
Policy 3	−36.9	−79.6	−197.1
Policy 4	−48.4	−107.3	−324.7
Change in Real GNP from Base Case, percent			
Policy 1	−0.60	−1.18	−1.54
Policy 2	−1.69	−2.64	−3.19
Policy 3	−2.08	−3.91	−7.24
Policy 4	−2.73	−5.27	−11.93
Growth in Real GNP (average percent per annum)			
Base Case	3.65	2.82	2.94
Policy 1	3.58	2.70	2.90
Policy 2	3.43	2.63	2.88
Policy 3	3.38	2.44	2.58
Policy 4	3.30	2.28	2.19

of these two cost components. The GNP loss resulting from reduced capital input is calculated by using a result from the macroeconomic theory of growth that states that if factor inputs are paid at rates equal to the value of their marginal products, a 1 percent change in capital input leads to an S_k percent change in real GNP, where S_k is the share of capital income in national income. This income share in the projections is approximately 0.35. The relative GNP loss arising from the dynamic adjustment mechanism is indicated by 0.35, multiplied by the percentage reduction in capital stock relative to the base case. The remaining GNP loss is attributed to the substitution effects of moving toward less energy-intensive input patterns. Table 9-9 shows the separation of real GNP loss into substitution and dynamic effects. The greater part of the loss arises from the substitution effect. Thus, in Policy 2 in 1985, the 1.69 percent real GNP reduction comprises a 1.54 percent decline due to substitution effects and a 0.15 percent decline due to dynamic effects; over 90 percent of the reduction is due to substitution effects. Over time, however, the dynamic effect increases in relative importance. By 2000, the 3.19 percent GNP reduction in Policy 2 comprises a 2.28 percent substitution cost and a 0.91 percent dynamic cost; 2.28 is about 70 percent of 3.19, smaller than the previous 90 percent;

Table 9-9. Substitution and Dynamic Effects in GNP Reduction.

	1985	*1990*	*2000*
Change in Real GNP (percent change from base case)			
Policy 1	−0.60	−1.18	−1.54
Policy 2	−1.69	−2.64	−3.19
Policy 3	−2.08	−3.19	−7.24
Policy 4	−2.73	−5.27	−11.93
Dynamic Effect: Change in Real GNP due to Capital Reduction (percentage points)			
Policy 1	−0.05	−0.14	−0.35
Policy 2	−0.15	−0.38	−0.91
Policy 3	−0.19	−0.56	−2.10
Policy 4	−0.24	−0.72	−3.23
Change in Real GNP due to Substitution Effects (percentage points)			
Policy 1	−0.55	−1.04	−1.19
Policy 2	−1.54	−2.26	−2.28
Policy 3	−1.89	−3.35	−5.14
Policy 4	−2.49	−4.55	−8.70
Proportion of Real GNP Change due to Substitution Effects (percent)			
Policy 1	92	88	77
Policy 2	91	86	71
Policy 3	91	86	71
Policy 4	91	86	73

hence *only* 70 percent of the loss is now due to substitution effects. The cumulative and durable nature of capital means that the relative importance of the dynamic changes further increase in the more distant future.

SUMMARY AND CONCLUSION

Analysis of each of the four energy packages suggests that substantial reductions, relative to the base case, can be achieved in the volume of energy use. These reductions occur as a result of adjustments in the pattern of energy use and in the structure of economic activity. However, a consequence of these adjustments is a reduction in the level of real GNP relative to the base case. For example, real GNP in 2000 for Policy 2 is predicted to be 3.2 percent or $87 (1972) billion below the base case, while the loss in real GNP in Policy 4 is 11.9 percent or $325(1972) billion. Alternatively, the effects may be viewed in terms of growth rates—the growth of energy use can be slowed, but at the cost of some decrease in the rate of aggregate economic growth. For example, the policies can reduce average annual real GNP growth rates by up to 0.7 percentage points. Thus, a significant economic effect and economic cost can be predicted as a result of energy conservation policies.

However, a significant result of the analysis is that the economic impact as measured by the loss in real GNP is relatively less than the reduction in energy use. Adjustments in the pattern of energy use and of economic activity permit the energy intensity of spending and production to be reduced. This reduces the average energy content of each dollar of economic activity. Conversely, it means that the decline in real GNP caused by energy policy is less than the proportionate decline in energy use. Table 9-10 summarizes the relationship between the decline in real GNP and the reduction in energy input. On average, each percentage point reduction in energy input leads to only a 0.2 percentage point reduction in real GNP. Thus, in Policy 2 in 2000, for example, the 16 percent reduction in energy input is associated with a substantial improvement in the aggregate economic efficiency of energy use and requires only a 3 percent reduction in the total output of the economy. The relative economic cost of energy reduction becomes greater as the strength of the policy measures increases, since energy saving becomes progressively more difficult and costly.

Table 9-10. Aggregate Relationship between Energy Input and Real GNP (ratio of percentage in real GNP to percentage change in primary energy input, changes relative to the base case).

	1985	*1990*	*2000*
Policy 1	0.18	0.18	0.18
Policy 2	0.19	0.19	0.20
Policy 3	0.19	0.20	0.21
Policy 4	0.20	0.21	0.24

Also, the relative cost increases over time, due to the investment and capital reductions caused by restrictive energy policies.

This chapter has focused on the economic adjustment mechanisms that provide the flexibility in energy use underlying the result that energy reductions can be achieved with less than proportionate reductions in the level or growth of economic output. In particular, two features of this energy-economy relationship have been analyzed. The first covers the nature of the adjustment mechanisms and the reasons for partial rather than total flexibility in the relationship between energy input and economic output. The second covers the reasons for the variation in this relationship, particularly the increasing economic impact over time of energy reductions.

Our empirical finding that there is a reasonable degree of flexibility in the energy-economy relationship is highly significant. From a policy point of view it suggests that it is possible to implement energy policies designed to restrict energy growth without having to suffer comparably large economic costs in terms of reduced GNP and slower economic growth. This makes it possible to contemplate restrictive energy policies designed to reduce petroleum imports. At the same time there is an economic cost associated with such policies. Only if policymakers judge this cost to be less than the benefits obtained from promoting energy-related objectives are the measures justified. From a forecasting point of view, the finding provides important information for assessing the likely economic impacts of increases in energy prices, whether due to government policy, to increasing relative scarcity of resources, or to external influences.

NOTES

1. The original form of this model was presented in a report to the Energy Policy Project of the Ford Foundation, see E.A. Hudson and D.W. Jorgenson "Interindustry Transactions," in D.W. Jorgenson et al., *Energy Resources and Economic Growth,* Final report to the Energy Policy Project (Washington, D.C.: 1973), ch. 5; E.A. Hudson and D.W. Jorgenson, "U.S. Energy Policy and Econonomic Growth, 1975–2000," *Bell Journal of Economics and Management Science* 5, no. 2 (Autumn 1974). The model has subsequently been revised and extended. A comprehensive description of the current version of the model is given in E.A. Hudson and D.W. Jorgenson, *The Long Term Interindustry Transactions Model: A Simulation Model for Energy and Economic Analysis,* Final Report to the Applied Economics Division, Federal Preparedness Agency (Washington, D.C.: 1977).

2. Annual interindustry accounts for the United States for the period 1947 to 1971 have been prepared by Jack Faucett Associates, *Data Development for the Input-Output Energy Model,* Final Report to the Energy Policy Project (Washington, D.C.: 1973). These accounts, on the same sectoral basis as the LITM system, give transactions in both current dollar and constant dollar terms.

3. The model of production is also described in E.R. Berndt and D.W. Jorgenson, "Production Structure," in D.W. Jorgenson et al., *Energy Resources and Economic Growth,* Final Report to the Energy Policy Project (Washington, D.C.: 1973), ch. 3, and in L.R. Christensen, D.W. Jorgenson, and L.J. Lau, "Transcendental Logarithmic Production Frontiers," *Review of Economics and Statistics,* 55, no. 1 (February 1973). A related application of the production model for the manufacturing sector is given in E.R. Brendt and D.O. Wood, "Technology Prices, and the Derived Demand for Energy," *Review of Economics and Statistics* 57 (August 1975).

4. The econometric model of consumption is described in D.W. Jorgenson, "Consumer Demand for Energy," in W.D. Norhaus, ed., *International Studies of the Demand for Energy* (Amsterdam: North-Holland, 1977). The theory of this model is also developed in L.R. Christensen, D.W. Jorgenson and L.J. Lau, "Transcendental Logarithmic Utility Functions," *American Economic Review* 65, no. 3 (June 1975) and in D.W. Jorgenson and L.J. Lau, "The Structure of Consumer Preferences," *Annals of Social and Economic Measurement* 4, no. 1 (January 1975).

5. The Brookhaven optimization models are described by K.C. Hoffman, "A Unified Framework for Energy System Planning," in M.F. Searl, ed., *Energy Modeling* (Washington, D.C.: Resources for the Future, 1973); K.E. Hoffman and E.A. Cherniavshy, *Interfuel Substitution and Technological Change* (Upton, New York: Brookhaven National Laboratory, 1974P; and E.A. Cherniavsky, *Linear Programming and Technology Assessment* (Upton, New York: Brookhaven National Laboratory, 1975).

6. The Reference Energy System is described in M. Beller, ed., *Sourcebook for Energy Assessment* (Upton, New York: Brookhaven National Laboratory, 1975).

7. The integration of the LITM system with the TESOM framework is discussed in detail by K.C. Hoffman and D.W. Jorgenson, "Economic and Technological Models for Evaluation of Energy Policy," *Bell Journal of Economics* 8 (Autumn 1977). An application of the integrated framework is given in D.J. Behling, R.C. Dullien, and E.A. Hudson, *The Relationship of Energy Growth to Economic Growth Under Alternative Energy Policies* (Upton, New York: Brookhaven National Laboratory, 1976).

8. A theoretical treatment of the dynamic adjustment process, in the context of a macroeconomic growth model, is given by W.W. Hogan, "Capital Energy Complementarity in Aggregate Energy-Economics Analysis," (Energy Modeling Forum, Institute of Energy Studies, Stanford University, Stanford, California, 1977).

9. The impact of energy policy on future U.S. economic growth, and the types of economic adjustments resulting from energy policy, are considered in E.A. Hudson and D.W. Jorgenson, "Energy Policy and U.S. Economic Growth," *American Economic Review* 68, no. 2 (May 1978).

10. A detailed analysis of the reduction in productivity, capital, and real GNP resulting from higher energy prices (in this case the 1973 and subsequent rises in world oil prices) is given in E.A. Hudson and D.W. Jorgenson, "Energy Prices and the U.S. Economy, 1972–1976," *Natural Resources Journal* (forthcoming).

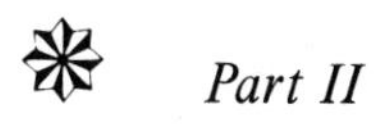

Part II

World Models

Chapter 10

A Long-run Model of World Energy Demands, Supplies, and Prices*

Hendrik S. Houthakker
Harvard University
and
Michael Kennedy
The Rand Corporation
(Presented by Hendrik S. Houthakker)

The World Energy Model, which this chapter describes, is a disaggregated dynamic partial equilibrium model designed for projections under alternative assumptions of the world's energy markets at ten-year intervals (1985, 1995, and so on). It is an *equilibrium* model in that supply and demand in each market are equilibrated by a set of prices calculated within the model; there are no "gaps" between supply and demand. However, it is a *partial* equilibrium model because developments outside the energy markets affect, but are not affected by, energy developments. The model is *dynamic* not merely in the trivial sense of allowing for growth, but also because it recognizes time patterns in the discovery and depletion of natural resources and in the durable nature of electric power plants. Finally, it is *disaggregated* in that the energy markets are divided three ways: (1) by fuel (oil, natural gas, coal, uranium, and hydro); (2) by area (United States, Canada, Latin America, Europe, the Middle East and Africa, Asia and the Pacific—the Communist countries are not included in the model); and (3) by technology (only in the case of electricity).

Disregarding the dynamic aspects for the moment, the model is an application of nonlinear mathematical programming. The objective function, subject to certain qualifications mentioned below, is the sum of producers' and consumers' surpluses in the energy markets. The maximization is constrained by four types of relations:

*Recent work on this project was done under a research contract with the Hoover Institution, Stanford, California.

- Supply functions describing the response of production of primary resources to changes in prices (for some fuels these functions are further distinguished into long-run and short-run supply functions);
- Demand functions describing the response of final consumption to changes in prices and income;
- Technological conditions describing the transformation of primary energy resources into electricity; and
- Transportation vectors describing the movement of energy resources between regions.

These four constraints will be discussed in more detail below, though not in that order.[1] While some of the supply and all of the demand functions are assumed to be nonlinear, the model uses a locally linear approximation which permits solution by means of quadratic programming. The basic algorithm is a variant of the capacity method of quadratic programming;[2] it produces a solution in a finite number of steps whose accuracy is verified by the Kuhn-Tucker conditions. The algorithm was taken over with minor modifications from the World Oil Model;[3] it will not be restated here except for an example of its operation. Despite this common feature, the World Oil Model, apart from being restricted to oil, is not dynamic in the sense that the World Energy Model is.

DEMAND FUNCTIONS

The demand functions are double logarithmic in all prices, income, and population.

$$Q_i = \alpha_i \left(\frac{Y}{N}\right)^{\beta_i} N \prod_{j=1}^{5} P_j^{\gamma_{ij}}, \tag{10.1}$$

where

Q_i is consumption of the ith commodity,

Y is an index of economic activity (called "income"),

N is population, and

P_j is the price of the jth commodity.

Regional indexes have been suppressed for simplicity. The parameters α_i, β_i, and γ_{ij} can all be made specific to the region and commodity.

However, as will be seen below, our solution algorithm demands a linear approximation of the demand curve. Any demand curve linear in prices can be written

$$Q_i = a_i + \Sigma\, b_{ij} P_j. \tag{10.2}$$

We derive a linear approximation of form (10.2) to our basic demand curve of form (10.1) in the following manner: We first guess a set of prices, $P_1^*, \ldots, P_5^*$. These prices, together with assumed income and population levels, generate a guess of quantity demanded, Q^*. (The regional subscript has been suppressed here for convenience.) The parameters a_i and b_{ij} are then generated through the formulae

$$b_{ij} = \gamma_{ij} \cdot Q_i^* / P_j^*,$$

$$a_i = Q_i^* - \Sigma\, b_{ij} P_j^*. \tag{10.3}$$

Income itself is projected from assumptions concerning the growth rate of GNP per capita and the growth rate of the population; the default assumptions may be found in Houthakker and Kennedy.[4] The price effects involve not only the own price of an energy source, but also the prices of competing energy sources; thus the cross-price elasticity between oil and natural gas is assumed to be 0.1. The price and income elasticities are stated in Table 10–1.

TRANSPORTATION

Every fuel except electricity can be transported to each region from any other region and will be transported if the price difference between two regions (taking export and import dutires, if any, into account) is sufficient to cover the cost of transporting the fuel. The transportation cost is divided into a fixed cost specific to the fuel and a variable cost depending on the distance. In the case of oil, the variable costs per barrel in 1972 dollars (as everywhere in this chapter) are given in Table 10–2. The variable costs for other primary fuels are derived from those of oil by a multiplier, which is 0.25 for natural gas, 12.5 for coal, and 1 for uranium. Fixed transportation costs are used only for natural gas (60 cents per thousand cubic feet, to allow for liquefaction and gasification) and

Table 10–1. Default Values of Income and Price Elasticities of Final Demand.

		Price of				
Fuel	*Income*	*Oil*	*Gas*	*Coal*	*Uranium*	*Electricity*
Oil	0.8	–0.45	0.1	0.1	0	0.15
Gas	0.7	0.1	–0.45	0.1	0	0.1
Coal	0.6	0.1	0.1	–0.8	0	0.15
Uranium	0.1	0	0	0	–0.01	0
Electricity	1.4	0.15	0.1	0.15	0	–1.0

Table 10-2. Default Values of Interregional Oil Transport Costs.

(1972$ per barrel)

	United States	*Canada*	*Latin America*	*Europe*	*Middle East and Asia*	*Asia and Pacific*
United States	0	0.25	0.35	0.65	1.50	1.20
Canada	0.25	0	0.55	0.80	1.65	1.20
Latin America	0.35	0.55	0	0.45	1.20	1.25
Europe	0.65	0.80	0.45	0	1.00	1.50
Middle East and Africa	1.50	1.65	1.20	1.00	0	0.50
Asia and Pacific	1.20	1.20	1.25	1.50	0.50	0

for uranium ($2 per pound). However, for gas transport from Canada and Latin America (i.e., Mexico) to the United States, liquefaction is presumably unnecessary, and the total transportation cost for these gas flows, if any, is put at 30 cents per thousand cubic feet by default.

SUPPLY FUNCTIONS: COAL AND ELECTRICITY

Because of their essentially dynamic character, the exposition of the supply functions for oil, gas, and uranium will be postponed till the next section. The supply functions for coal and hydroelectric power are simply constant elasticity functions of the domestic price of coal and electricity, respectively. The base to which these elasticities are applied is production in the previous period (initially 1972).

In the case of coal, the exhaustion of reserves is probably a less important factor than the availability of labor and the development of productivity. Consequently, no explicit accounting of coal reserves is included in the model; however coal production is subject to a decline rate that moves the supply curve to the left over time. The annual decline rate is 1 percent for the United States, Canada, and the Middle East and Africa; 2 percent for Latin America and the Asia-Pacific region; and 3 percent for Europe.[5] The supply elasticity of coal has a default value of 1.2 in all regions.

The supply of hydroelectricity (which includes solar and geothermal power) is also governed by a supply elasticity, with a default value of 1.5. However, no decline rate is used for this energy source; on the contrary, it is assumed that hydroelectric facilities, once built, remain in existence for the remainder of the projection. In addition, the program takes account not only of hydro capacity in existence in 1972, the initial year, but also of capacity under construction and expected to be operative by the middle 1980s.

Secondary electricity—that is, electricity derived from fossil fuels and from uranium–is subject to a linear programming technology, which means an infinite elasticity of supply. These forms of electric capacity are built whenever their total cost is covered by the regional price of electric power excluding a "handling charge," which represents the cost of transmission and distribution. The capital cost per kilowatt hour of each type of secondary electricity is calculated by a standard formula from the following three determinants:

1. The cost per kilowatt for each type of electric power (see Table 10–1);
2. The load factor, put at 500 hours per year by default; and
3. The rate of return prevailing in the region concerned.

The annual rate of return is set by default at 15 percent in Latin America, 13 percent in Asia and the Pacific, and 10 percent in the other four regions.

As in the case of hydro, the initial capacity in 1972 and the capacity under construction during 1972-1985 are considered given. Capacity built in one period is again transferred to the next period, but this time was a depreciation rate of 3 percent per year. Thus, even if no new oil-fired electric power stations are built, those existing and under construction in 1972 will continue to produce electricity and to consume oil, albeit at a declining rate.

Table 10–3 lists the cost per kilowatt hour of each type of secondary electricity and also the input of primary fuel per kilowatt hour. For instance, the average oil-fired station is assumed to require 0.0019 barrels of oil per kilowatt hour. The uranium requirements for light water reactors and for breeders reflect both initial charges and refueling; in this version of the model, all uranium is supposed to be both enriched and reprocessed (see also the appendix to this chapter).

SUPPLY FUNCTIONS: OIL AND NATURAL GAS

For these two fuels, production is modeled as a two stage process. The first stage is the *development* of reserves out of the resource base, and the second stage is

Table 10–3. Default Values of Electric Capacity Cost and of Input Coefficients.

Fuel and Process	*Capacity Cost per kW (1972$)*	*Units of Input per kWh*	*Fuel Unit*
Oil	380	0.0019	barrel
Gas	330	0.011	1000 cu. ft.
Coal	450	0.00385	ton (2000 lbs.)
LWR	660	0.00011	lb. (U_3O_8)
Breeder	900	0.000004	lb. (U_3O_8)

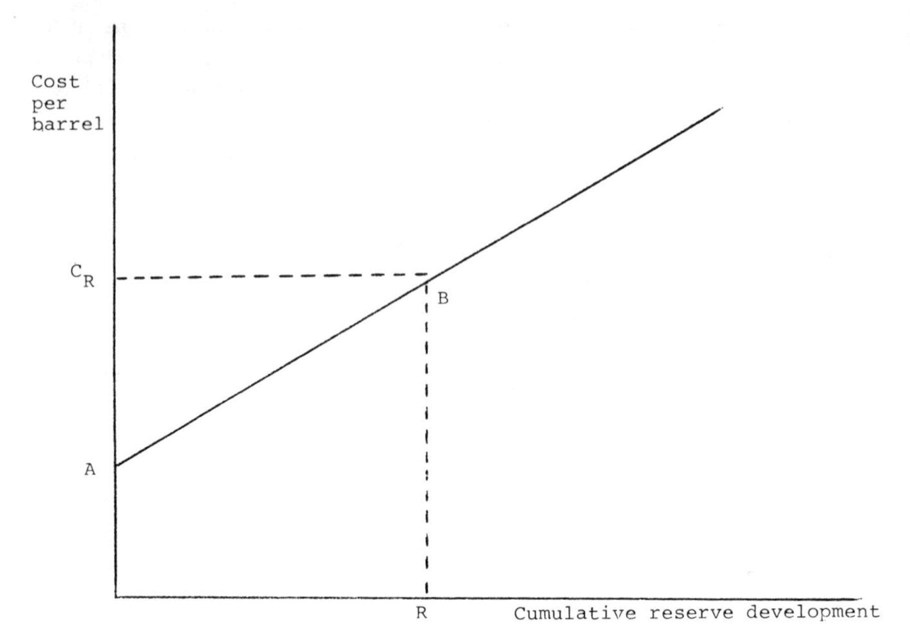

Figure 10-1. Assumed Relation Between Cost per Barrel of Reserve Development and Cumulative Reserve Development. (See text for definition of symbols)

the *production* of fuel out of reserves according to the geological characteristics of the reserve accumulation.

We assume that reserves can be developed at an increasing cost per barrel and that this cost is determined by cumulative reserve development. Figure 10-1 shows the assumed relation between cost per barrel of reserve development and cumulative reserve development. This curve has no time dimension, being simply a cost-depletion relation. Since we assume the curve is linear, it can be set by the model users (and most conveniently summarized) by specifying two points: A, the initial cost of reserve development; and B the point representing the amount of reserves available (say R) at a cost up to some preset reference cost, C_R.

What quantity of reserves will competitive producers develop in any given time period? Ignoring the impact of future prices on current decisions, it could be assumed that they will develop reserves up to the point where the price of the resource equals the marginal cost of development. However, it is well known that the temporal competitive price path of an increasing cost resource of the kind depicted in Figure 10-1 will be rising and that competitive producers will, therefore, develop fewer reserves than those indicated by the point where price equals marginal cost. The model is not fully intertemporal in that future prices do not influence current decisions, and we must therefore make some adjustment in market behavior to reflect the influence of future higher costs and prices on current reserve development. The solution adopted is *ad hoc*: in each time period, we add to the cost as a function of cumulative reserve development curve a second linear cost curve, only a function of current reserve development. This additional marginal cost is a proxy for the user cost of reserve development that would be automatically included in profit calculations in a fully intertemporal model with perfect foresight. This additional linear marginal cost curve, in the absence of any other compelling formulation, is assumed to have the same slope as the first curve. Thus, the total (resource and user) marginal cost curve has twice the slope of the resource curve above. The situation is displayed graphically in Figure 10-2. We assume that R_0 reserves have been developed up to the current time period. Marginal resource cost of new development is given by line *CE*, while marginal resource plus user cost is given by line *CD*, whose slope is twice that of *CE*. By competitive assumptions then, if price is *P* in this period, new reserve development will be $(R_1 - R_0)$.

Total marginal cost in the next period is shown in Figure 10-3. The marginal resource cost of research development is still shown by line *AFE*, the same line that appeared in Figure 10-2. We now, however, are starting from a base in which R_1 cumulative reserves have been developed up to the current time period, and *FG* represents the new total marginal cost. *FG* is simply the translation of line segment *CD* up line *AFE* until its endpoint rests over point R_1 instead of R_0.

Thus, we see the recursive nature of the model; after resources have been developed in any given period, the resource cost curve is moved up until it starts again from the currently unexploited fuel base. Although the adjustment for

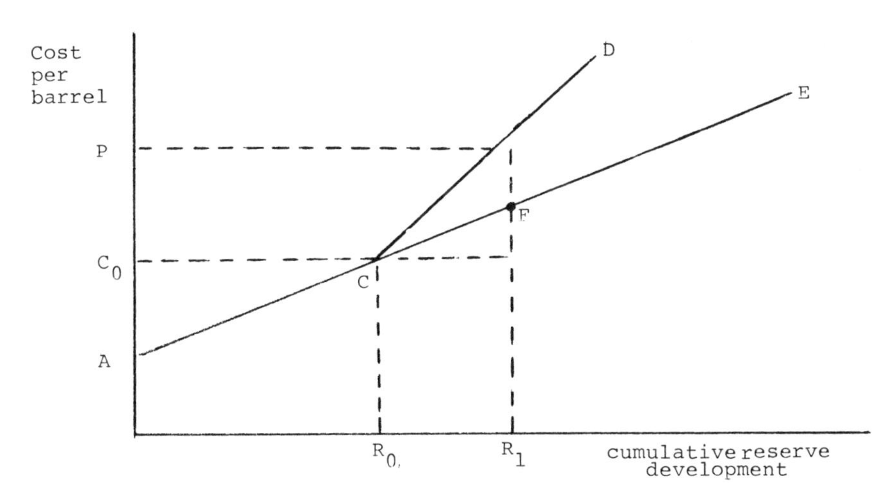

Figure 10-2. The Effect of Future Higher Costs and Prices on Current Reserve Development for the Model of Figure 10-1. (See text for definition of symbols)

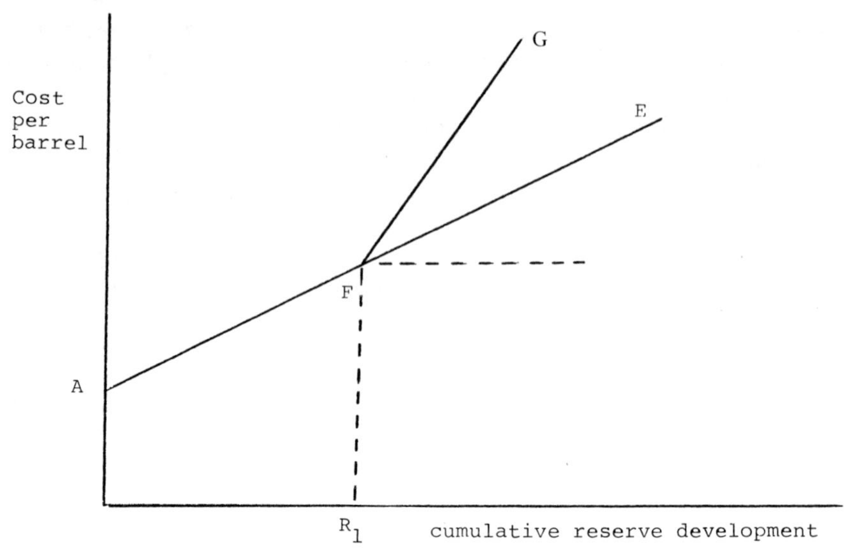

Figure 10-3. The Effect of Marginal Cost for New Reserve Development

future price increases that we have just outlined is an *ad hoc* approximation to the full intertemporal production optimization process that would be appropriate if future prices were known, it would seem to be reasonable given producer uncertainty in the real world. We also point out that the resource cost curve implies that new reserves are developed in strict order from lowest to highest cost. Thus, "surprises" of new cost resource deposits are ruled out.

This concludes the discussion of oil and gas reserve development.[6] However, reserve development is not synonymous with production. Instead, oil and gas are only produced from reserves over a period of time; the assumption is made that output flows from a given resource deposit according to a constant decline rate. That is, if d is the decline rate, and some amount R of reserves is developed, production is assumed to be dR in the first period and then to fall by a factor of d each year into the future.

Production in any given period is thus a sum of two components—production from reserves newly developed in that period, plus continuing production from reserves developed in past periods. This can be conveniently expressed arithmetically. If ΔR_τ is new reserve development in period τ, Q_t is production in year t, and d is (again) the decline rate, then

$$Q_t = d\Delta R_t + (1-d) \sum_{\tau=1}^{\infty} d^{\tau} \Delta R_{(t-\tau)}. \tag{10.4}$$

The first part of this expression, as seen above, is a linear function of price in the tth period, while the second part is predetermined, and thus a constant in the tth period. Oil and gas production in any period are, therefore, linear functions of price in that period.[7]

In the model there is one important exception to this formulation, which occurs in the case of oil production in the Middle East and Africa. This region is used to represent OPEC behavior. It must make the cartel price of effective and is therefore assumed to produce virtually unlimited amounts of oil at a price equal to marginal resource cost plus a fixed tax per barrel. This tax, representing monopoly profit per barrel, can be varied by the model user from period to period to investigate the consequences of various kinds of OPEC behavior.[8] However, the model at present treats the OPEC tax as exogenous and has no built-in theory of cartel policy.

URANIUM SUPPLY

Virgin uranium is also a depletable resource and can be characterized by an upward sloping marginal cost curve that is a function of cumulative reserve development, as in Figure 10-1. We treat it the same as we treat oil and gas, except that reserves are assumed to be completely producible in the period of development. Since our periods are ten years in length, this seems to be a reasonable hypothesis. Thus, formula (r) with $d = 1$ holds in the case of U_3O_8, and production in any period is simply a linear function of price in that period. We again add a second marginal cost curve, of equal slope, to the marginal resource cost curve in each time period to approximate the influence of future higher costs and prices in today's market.

PROGRAMMING REPRESENTATION OF MODEL

As seen above, supply and demand for each primary energy source are approximated locally by price. Thus, excess demand (demand minus supply) is locally also a linear function of price. If e is a vector of excess demands, and p a vector of prices, this relation can be written

$$p = a + Ae. \tag{10.5}$$

Next, electricity generation and product transportation are considered constant returns to scale processes and can be represented as column vectors in an activity matrix T. The material balance constraint of the system (letting t be a vector of activity levels and h a vector of outputs from inherited electric production capacity) is then

$$e = Tt + h. \tag{10.6}$$

Finally, the economic constraints on transport and electricty activities (letting k be a vector of capital costs associated with these activities) can be represented as

$$-T'p \leqslant k, \quad (10.7)$$

$$(k + T'p)'t = 0. \quad (10.8)$$

Conditions (10.5) through (10.8) are sufficient to determine the equilibrium in the energy market at any point in time. As shown in Kennedy,[9] they can be regarded as the first order conditions of a quadratic programming problem.

Since all parameters except one (the OPEC tax) have default values, only minimal user input is necessary to start the program. However, the following sets of parameters can be set by the user to override the default values:

1. The supply elasticities by region (only for coal and hydro);
2. Own and cross-price elasticities of demand;
3. The income elasticities of demand for fuels and regions;
4. Fixed supplies–namely, supplies discovered between 1972 and the present and likely to be in production by 1985 (e.g., Mexican oil);
5. Capacity costs per kW of electric power by fuel;
6. Rates of return by region;
7. Freight multipliers (used to calculate variable transportation costs on fuels other than oil);
8. Initial uranium supplies by region;
9. The load factor in electric power;
10. Import and export duties by fuel and region;
11. Excise tax rates over and above those existing in 1972, by fuel and region;
12. The rate of a tax on Btus;[10]
13. The population growth rates for each region;
14. Electric capacity under construction by fuel and region; and
15. Growth rates of GNP per capita.

As the model moves from one period to the next, many internal parameters are adjusted by the program. Population growth, income growth, and resource depletion affect a and A, while electric capacity construction changes h. After these adjustments are made, the model proceeds to solve for prices and quantities in the next period.

In order to give the interested reader some insight into the efficiency of the algorithm, Table 10-4 presents the successive steps in attaining a one period equilibrium. The solution involves the successive introduction of vectors from the matrix T into the "basis" (or the elimination of previously included vectors from the basis). A vector is introduced if its operation is just profitable at the current value of the driving variable (here called "scale") of the capacity method

Table 10-4. Example of Iterations Required to Produce a One Period Solution.

Iteration Number	*Number of Vectors in Basis*	*Key*	*Type of Vector*	*Identifiers*			*Scale*
0	0	2	2	5	6	1	44.978
1	1	2	2	3	4	1	40.946
2	2	2	2	2	1	1	32.864
3	3	2	2	5	4	1	12.122
4	4	2	2	3	1	1	9.918
5	5	2	2	3	4	4	8.551
6	6	2	2	2	4	4	8.364
7	7	1	2	3	4	1	7.703
8	6	2	2	5	4	4	6.955
9	7	2	2	1	3	3	5.597
10	8	2	2	2	1	2	5.034
11	9	2	2	5	1	1	4.305
12	10	2	2	3	1	2	2.201
13	11	2	2	6	4	4	1.918
14	12	2	1	1	3	0	1.814
15	13	2	1	5	2	0	1.439
16	14	2	1	6	2	0	1.414
17	15	1	2	2	1	1	1.340
18	14	2	1	6	4	0	1.297
19	15	1	2	6	4	4	1.264
20	14	1	1	6	2	0	1.129
21	13	2	1	4	4	0	1.117
22	14	2	1	2	3	0	1.025
23	15	2	2	5	2	1	1.019
24	16	2	1	5	4	0	1.013

of quadratic programming.[11] This value is steadily lowered until it reaches unity, at which point the one period problem is solved.

The first column in Table 10-4 shows the number of the iteration; and the second, the number of vectors that were in the basis prior to this iteration. The third column labeled "Key" contains a 2 if a vector was introduced in this iteration and a 1 if a vector was dropped. The next column indicates the type of vector that was added to or dropped from the basis; it is 1 for new electric power production (i.e., production from stations that were not inherited from the previous period) and 2 for transportation.

The meaning of the next three columns, labeled "Identifiers," depends on whether the vector represents transportation or electricity production. For vector type 2 (transportation) the first identifier refers to the region of origin (1 for the United States, 2 for Canada, etc.), the second to the region of destination, and the third to the fuel transported (1 for oil, 2 for gas, 3 for coal, 4 for uranium). For an electricity vector (type 1) the first identifier is the region and the second is the fuel used,[12] the third identifier is not used.

The rightmost column shows the value of "scale" corresponding to the current iteration. Thus when scale equals 44.978 (the lowest value at which no new

Table 10-5. Projections for 1985 with OPEC Tax at $10 per Barrel and All Parameters at Default Values.†

	DEMAND	*SUPPLY*	*INPUT*	*PERCENT ELECTRICITY*	*IMPORTS*	*PRICE*	*CUMULATIVE PRODUCTION*
USA							
OIL	−6.08	3.26	−0.42	6.71	3.24	11.755	42.09
NGAS	−16.32	15.68	−2.78	7.65	3.41	1.673	245.07
COAL	−0.15	0.87	−0.71	55.68	−0.02	16.655	9.21
URAN	−0.00	0.06	−0.06	17.31	0.00	30.525	0.58
ELEC	−3299.82	417.53	2882.29	12.65	0.00	0.020	0.00
CAN							
OIL	−0.66	0.67	−0.01	1.04	0.00	11.905	8.50
NGAS	−1.56	4.13	−0.11	2.23	−2.46	1.373	43.26
COAL	−0.01	0.03	−0.02	11.57	0.00	18.463	0.31
URAN	−0.00	0.02	−0.00	7.78	−0.02	29.620	0.22
ELEC	−452.10	349.78	102.31	77.37	0.00	0.021	0.00
L AM							
OIL	−1.60	2.23	−0.30	30.46	−0.33	11.405	26.70
NGAS	−2.67	3.62	−0.00	0.00	−0.95	1.373	32.23
COAL	−0.02	0.02	−0.02	8.39	0.02	21.030	0.21
URAN	−0.00	0.00	−0.00	5.03	−0.00	29.970	0.03
ELEC	−511.76	287.19	224.57	56.12	0.00	0.024	0.00
EUR							
OIL	−4.91	1.45	−0.57	13.79	4.03	11.255	10.35
NGAS	−4.53	5.21	−0.68	2.87	0.00	1.665	63.74
COAL	−0.21	0.39	−0.18	21.65	0.00	23.336	4.73
URAN	−0.00	0.02	−0.07	31.19	0.06	32.420	0.13
ELEC	−2160.87	620.91	1539.96	28.73	0.00	0.022	0.00
BRDR				1.77			

ME&A							
OIL	−1.33	11.82	−0.23	29.10	−10.25	10.255	133.68
NGAS	−2.26	3.33	−1.07	23.10	0.00	0.988	31.30
COAL	−0.08	0.10	−0.02	10.16	0.00	22.831	1.10
URAN	−0.00	0.04	−0.01	16.24	−0.04	29.420	0.36
ELEC	−422.78	90.51	332.27	21.41	0.00	0.022	0.00
AS&P							
OIL	−3.49	0.84	−0.65	23.09	3.30	10,755	9.96
NGAS	−1.18	1.34	−0.16	0.99	0.00	1.228	13.37
COAL	−0.20	0.29	−0.09	15.09	0.00	27.149	3.08
URAN	−0.00	0.06	−0.06	36.12	0.00	31.717	0.39
ELEC	−1491.38	368.51	1122.86	24.71	0.00	0.025	0.00
WORLD							
OIL	−18.08	20.26	−2.18	13.76	−0.00		
NGAS	−28.51	33.32	−4.81	5.24	−0.00		
COAL	−0.67	1.70	−1.03	32.00	0.00		
URAN	−0.00	0.21	−0.21	22.95	0.00		
ELEC	−8338.70	2134.44	6204.26	25.60	0.00		
BRDR				0.46			

	UNIT	*USA*	*CAN*	*L AM*	*EUR*	*ME&A*	*AS&P*	*WORLD*
ENERGY TRADE BAL.	BLN$	−43.5	3.9	4.7	−47.3	106.2	−35.5	
PRIMARY EN. PROD.	QUAD	64.1	13.7	18.8	30.1	72.9	22.2	221.8
PRIMARY EN. CONS.	QUAD	83.8	9.7	16.6	57.0	15.4	39.6	222.1
NEW ELEC. CAPAC.*	GWE	374.3	20.9	46.0	205.4	59.1	162.2	867.8
NEW ELEC. INVEST†	BLN$	193.6	7.1	16.1	123.0	25.2	93.8	458.8

*Excludes Hydro. Geothermal & Solar

†Elapsed Time for One Period Run = 3.08.1 min

technology is used), it becomes profitable to transport oil (1) from the Middle East (5) to the Asia-Pacific region (6). At a scale value of 40.946, a second vector enters the basis, representing transportation of oil from Latin America to Europe.

Naturally the value of "Key" is 2 for the first few iterations, since vectors cannot be dropped before they are introduced.[13] The first time a vector is dropped is in iteration 7, where oil transportation from Latin America to Europe become unprofitable at a scale value of 7.703. Three more vectors added to the basis for high scale values are all transportation vectors; new electricity production does not come in until iteration 14, where the United States starts to build additional coal-fired stations.

Finally, with iteration 25, a scale value of 1 is reached, and the solution obtained. In this iteration, the basis does not change, so it contains 17 vectors out of 150 in the matrix T.

The actual solution for this case (corresponding to 1985 with an OPEC tax of $10 per barrel and all parameters at their default values) is analyzed in Table 10-5. All quantities are per year and in billions of units (barrels of oil, 1000s of cubic feet of gas, tons of coal, pounds of U_3O_8, and kilowatt hours), and all prices and other money figures are in 1972 dollars. "Demand" means final consumption, "supply" primary (as distinct from intermediate) production, "input" is input of fuels into electric energy, "percent elect" is the percentage of electricity generated from each primary source (including hydro, shown in the supply column), "imports" are exports if the sign is negative, the "price" is the one charged to final consumers, and "cum prod" is cumulative production of the exhaustible energy sources starting in 1972. The signs in each column are chosen so as to make the algebraic sum of demand, supply, input and imports equal to zero. The only line item that may need explanation is "BRDR," which refers to the breeder, used by 1985 in Europe but only because some breeders are already operating or under construction there not because they are competitive. For discussion of the numbers we refer to Houthakker and Kennedy.[14]

Since the present chapter is concerned with methods rather than with results, attention may be called to the last line of Table 10-5, which shows the elapsed time ("ET") needed for this one period run. It is about three minutes, nearly all of which is spent printing out the results; for multiperiod runs, the elapsed time is roughly proportional to the number of periods. Computation in the strict sense is almost instantaneous, demonstrating the advantages of a compact model and an efficient algorithm.

APPENDIX: THE NUCLEAR FUEL CYCLE

In the version of the model described in the main text the nuclear fuel cycle is treated only implicitly; both enrichment and reprocessing are assumed in the input coefficients, but there is no convenient way of assessing the extent and impact of these activities explicitly. Since trade in enriched uranium and re-proecssing have become major items of public concern, the World Energy Model is now being revised to permit better analysis of the policy options with respect to these activities.

This is being done by introducing two new energy commodities in addition to the five already included. These commodities are "fissile materials" and "spent fuel." The first may be interpreted as the excess of U-235 over and above the quantity present in unenriched uranium.[15] Enrichment produces fissile material using virgin uranium and electricity as inputs. The input coefficients depend on the "tails assay" and on the technology (gaseous diffusion or centrifuge). Present plans are to have two enrichment activities–gaseous diffusion with high tails and centrifuge with low tails. The introduction of fissile materials also makes it possible to distinguish two kinds of nuclear generation–light water reactors, which use the additional fissile material; and heavy water, which do not. Breeders also use enriched uranium, but at a different rate. Both uranium and fissile material can be traded, depending on favorable price differences and other restrictions.

All nuclear reactors produce spent fuel, and reprocessing makes it possible to recycle some of this into fissile material. Whether this will actually be done, of course, depends on the economics (on which the model will yield information) and on institutional constraints, which can also be incorporated in the model. It will then be possible to estimate, for instance, what effect restrictions on reprocessing will have on energy prices, on uranium supplies, and on the trade balance.

NOTES

1. The version of the World Energy Model described here is essentially the same as the one for which results were presented in H.S. Houthakker and M. Kennedy, "Long-Range Energy Prospects," in R. & D.H. ElMallakh, eds., *Energy Options and Conservation* (Boulder, Colorado: International Research Center for Energy and Economic Development, 1978). Subsequent additions to the model, particularly with respect to the nuclear fuel cycle, are summarized in the appendix to this chapter.

2. H.S. Houthakker, "The Capacity of Quadratic Programming," *Econometrica* 28, no. 1 (January 1960).

3. M. Kennedy, "An Economic Model of the World Oil Market," *Bell Journal of Economics & Management Science* (Autumn 1974), H.S. Houthakker, *The World Price of Oil, A Medium-Term Analysis* Washington, D.C.: American Enterprise Institute, 1976).

4. Houthakker and Kennedy, Table 1 (under the Leading "bare case."

5. Any such numbers quoted in this chapter are "default values." The program is designed to make substitution of alternative parameters very easy. The default values should be considered merely the authors' best guesses and are not

an inherent part of the model. Even when a single default value is given for all regions, the user can generally specify different values for different regions.

6. Since the default values of the parameters shown in Figure 10-1 vary by region and by fuel, it would be too complicated to state these values here.

7. The earlier discussion of the resource cost of reserves was not precise about the definition of marginal cost per barrel. Now that the geology of production has been specified, we can give that marginal resource cost a clear interpretation. If V is the development cost per barrel (including capitalized operating costs), d the decline rate, and r the rate of return on capital, marginal resource cost per barrel is $(r + d)V$.

8. Houthakker and Kennedy, pp. 33-34.

9. Kennedy, "An Economic Model of the World Oil Market" and "A World Oil Model," in Dale Jorgenson, ed., *Econometric Studies of U.S. Energy Policy* (Amsterdam: North-Holland, 1976).

10. Houthakker and Kennedy, pp. 36-37.

11. This term recalls the use of "world scale," a measure of transportation cost, in early versions of the World Oil Model. In terms of the present model, "scale" can be interpreted as a multiplier applied to capital cost.

12. As mentioned earlier, hydroelectricity is covered by a supply function rather than by a technology vector.

13. The capacity method, unlike the simplex method, does not normally need slack vectors. The basis, in the former method, is consequently of varying size, greatly reducing both the size of the matrices that have to be inverted and the number of iterations.

14. Houthakker and Kennedy, pp. 18-23. Table 2 in that paper is not identical with Table 10-5 here because of a number of improvements in data and programming. These have particularly affected the supply projections for hydroelectricity and oil and to a lesser extent those for natural gas. The general pattern, however, is much the same.

15. We are indebted to Dr. Eugene Kroch, now at Columbia University, for suggesting this approach.

❋ *Chapter 11*

Energy Transition Strategies for the Industrialized Nations

Alan S. Manne*
Stanford University

ABSTRACT

On the basis of a "no surprise" scenario, the following projections appear plausible: Through the year 2000, OPEC will retain considerable flexibility in its pricing or production policy. OPEC's policies may be determined either by economic or by political objectives and will not be constrained by petroleum resource limits.

After the year 2000, it is likely that the world's conventional hydrocarbon resources will become rapidly exhausted. It will take time, and it will be expensive to make a transition to new energy supply technologies. From the year 2000 onward, it will be difficult to ensure a smooth transition unless international oil prices are raised by more than 2 percent per year (expressed in dollars of constant purchasing power) during the 1980s and 1990s. From the viewpoint of both OPEC and the consumer nations, it would be preferable to initiate gradual price increases during the near future (say, 4 percent per year from 1980)–rather than to continue OPEC's 1975-1978 policy of a constant (or decreasing) real price level for crude oil.

METHODOLOGY AND KEY ASSUMPTIONS

These projections have been performed separately for the United States, the other industrialized OECD nations, and the less developed countries. Together,

*Research supported by Control Analysis Corporation, Electric Power Research Institute, and Stanford University Institute for Energy Studies. The individual author is solely responsible for the views expressed here. Thanks go to Karen Epstein and Paul Preckel for their helpful assistance.

this group of nations is sometimes known as WOCA, the world areas outside the CPE (centrally planned economies).

Each of the WOCA nations has its individual characteristics, but all are linked to each other through the international price of oil at the Persian Gulf. Each consuming nation is viewed as price taker—making whatever adjustments are economical in response to an OPEC-determined price trajectory. Depending on the scenario, individual nations may impose upper bounds on imports or on domestic production. In these cases, there may be a differential between the domestic and the international price of energy.

This chapter explores the logical implications of several alternative rate of price increase, but does not attempt to determine formally which of these might be optimal from the viewpoint of OPEC or from that of the consuming nations. Our modeling approach (similar to that adopted by WAES, the Workshop on Alternative Energy Strategies) is not altogether satisfactory. It does, however, have three important advantages: (1) it avoids endless debates on the nature of OPEC's motivation but instead focuses on its capabilities; (2) it permits us to analyze individual nations or groups of nations in parallel rather than simultaneously; and (3) it avoids the need for precise estimates of the amount of petroleum resources available to OPEC and the other developing nations. Each price trajectory implies a different production pattern. It is only the final stage of this analysis that requires a comparison of the cumulative production requirements against alternative geological estimates of petroleum resources.

For the OECD countries,[1] these calculations are based on ETA-MACRO, a model that simulates a market economy over time (see Table 11-1). It is assumed that producers and consumers will be sufficiently farsighted to anticipate the scarcities of energy that are likely to develop during the twenty-first century. Supplies, demands, and prices are matched through a dynamic

Table 11-1. LDC Oil and Gas Production Requirements, OPEC and Non-OPEC Sources.

Case 1: 2% Year OPEC Price Rise (from 1980)

Unit: quads of Btu

Year	OPEC Export Price ($/MMBTU)	LDC Consumption OPEC	LDC Consumption Non-OPEC	NET EXPORTS TO: U.S.	NET EXPORTS TO: OECD, Non-U.S.	NET EXPORTS TO: CPE	Total Annual	Cumulative (from 1980)
1985	2.2	9.2	21.1	23.6	33.0	0.0	86.9	434
1990	2.4	12.8	27.0	33.0	33.3	0.0	106.1	965
1995	2.7	16.4	30.4	43.1	34.8	0.0	124.7	1588
2000	3.0	20.0	34.2	52.1	40.8	0.0	147.1	2324
2005	3.3	24.3	40.3	55.4	46.0	0.0	166.0	3154
2010	3.6	29.6	47.5	63.3	50.8	0.0	191.2	4110
2015	4.0	36.0	56.1	66.8	54.3	0.0	213.2	5176
2020	4.4	43.8	66.1	70.5	63.0	0.0	243.4	6393
2025	4.9	53.3	77.9	69.5	71.5	0.0	272.2	7754

Table 11-1. (continued)

Case 2: 4% Year OPEC Price Rise (from 1980)

Unit: quads of Btu

Year	*OPEC Export Price ($/MMBTU)*	*LDC Consumption OPEC*	*LDC Consumption Non-OPEC*	*NET EXPORTS TO: U.S.*	*NET EXPORTS TO: OECD, Non-U.S.*	*NET EXPORTS TO: CPE*	*Total Annual*	*Cumulative (from 1980)*
1985	2.4	9.2	20.3	21.4	31.3	0.0	82.2	411
1990	3.0	12.8	25.4	21.4	29.2	0.0	88.8	855
1995	3.6	16.4	27.8	29.5	27.8	0.0	101.5	1362
2000	4.4	20.0	30.5	31.6	29.8	0.0	111.9	1922
2005	5.3	24.3	34.9	26.8	28.7	0.0	114.7	2495
2010	6.5	29.6	39.9	27.2	27.3	0.0	124.0	3115
2015	7.9	36.0	45.7	23.0	23.3	0.0	128.0	3755
2020	9.6	43.8	52.3	15.4	15.7	0.0	127.2	4392
2025	11.7	53.3	59.8	7.4	8.4	0.0	128.9	5036

Case 3: 6% Year OPEC Price Rise (from 1980)

Unit: quads of Btu

Year	*OPEC Export Price ($/MMBTU)*	*LDC Consumption OPEC*	*LDC Consumption Non-OPEC*	*NET EXPORTS TO: U.S.*	*NET EXPORTS TO: OECD, Non-U.S.*	*NET EXPORTS TO: CPE*	*Total Annual*	*Cumulative (from 1980)*
1985	2.7	9.2	19.6	19.4	29.9	0.0	78.1	390
1990	3.6	12.8	24.1	16.7	26.0	0.0	79.6	788
1995	4.8	16.4	25.6	14.6	22.0	0.0	78.6	1181
2000	6.4	20.0	27.2	15.6	20.3	0.0	83.1	1597
2005	8.6	24.3	30.2	7.0	11.7	0.0	73.2	1963
2010	11.5	29.6	33.6	0.0	2.9	0.0	66.1	2293
2015	15.4	36.0	37.4	0.0	0.0	0.0	73.4	2660
2020	20.6	43.8	41.6	0.0	0.0	0.0	85.4	3087
2025	27.5	53.3	46.3	0.0	0.0	0.0	99.6	3585

Case 4: 4% Year OPEC Price Rise (from 1980)
No Additional Nuclear after 1985.

Unit: quads of Btu

Year	*OPEC Export Price ($/MMBTU)*	*LDC Consumption OPEC*	*LDC Consumption Non-OPEC*	*NET EXPORTS TO: U.S.*	*NET EXPORTS TO: OECD, Non-U.S.*	*NET EXPORTS TO: CPE*	*Total Annual*	*Cumulative (from 1980)*
1985	2.4	9.2	20.3	21.5	31.5	0.0	82.5	412
1990	3.0	12.8	25.4	21.3	31.9	0.0	91.4	869
1995	3.6	16.4	27.8	29.0	30.7	0.0	103.9	1389
2000	4.4	20.0	30.5	34.7	34.7	0.0	119.9	1988
2005	5.3	24.3	34.9	35.7	37.1	0.0	132.0	2648
2010	6.5	29.6	39.9	37.9	39.1	0.0	146.5	3381
2015	7.9	36.0	45.7	35.4	36.6	0.0	153.7	4149
2020	9.6	43.8	52.3	28.9	29.1	0.0	154.1	4920
2025	11.7	53.3	59.8	20.1	21.2	0.0	154.4	5692

Notes on Oil and Gas Demands Shown in Table 11-1

- *OPEC domestic consumption* is assumed to be independent of the export price. The price and income effects will work in opposite directions within OPEC nations. Numerical values are based upon WAES, p. 289:

(unit: MBDOE)

	1985 (cases C and D)	2000 (cases C-2 and D-8)
Oil	2.76 - 3.30	4.50 - 7.80
Natural Gas	1.15 - 1.54	2.76 - 3.97
Total	3.91 - 4.84	7.26 - 11.77
midpoint of range:	4.38	9.52

Percent/year growth rate, 1985-2000: 5.3

Values for 1990 and 1995 have been interpolated linearly between these midpoints. From 2000 onward, it is assumed that there will be a 4 percent annual growth rate in domestic consumption. Conversion factor from MBDOE to quads per year: 2.1.

- *Non-OPEC domestic consumption* is based upon Gordian Associates, "LDC Energy Supply/Demand Balances and Financiing Requirements", Final Report, February 27, 1978. Note that these values refer to a "base case," assuming "gradual reduction in GNP growth to 2020" (pp. S-5-6):

(unit: MBDOE)

	1975	1990	2000	2020
Oil	5.460	11.290	14.148	27.927
Natural Gas	.669	1.583	2.144	3.534
Total	6.129	12.873	16.292	31.461

The Gordian Associates report is the primary data source for "Energy Needs, Uses and Resources in Developing Countries," BNL 50784, March 1978. The Gordian methodology is described as follows in the BNL document (pp. 27-29):

> Commercial fuel demand was estimated based on assumed GNP growth rates and correlations between energy demand, GNP growth and energy prices. . . . The trends in income elasticity over time take into consideration the fact that income elasticity of oil and gas consumption generally tends to rise as per capita income levels increase from "low" to "middle," then typically decline as income rises to higher levels. . . . The assumption was then made that energy prices would increase in real terms by 50% by 2000 and 100% by 2020. A price elasticity of demand. . . of −0.3 was used to reduce total commercial energy demand in the future reference years. . . . In a number of ways these assumptions are conservative, leading to a low estimate of future LDC energy demand.

In effect, the BNL document describes a scenario in which OPEC prices are increasing at 2 percent per year from 1980. To adjust for a 4 or 6 percent price rise by 0.6 percent or 1.2 percent compounded annually (again using a demand elasticity of −0.3).

Compound growth interpolation has been used between each of the four benchmark years specified in the Gordian report and also to extrapolate from 2020 to 2025.

- *Net exports to OECD* are identical with import requirements generated by individual ETA-MACRO runs for the United States and other OECD nations.
- *Net exports to CPE* (centrally planned economies) are shown here as zero. It is assumed that exports from China will aproximately offset future imports by the Eastern European countries and the USSR.

nonlinear programming model. The higher that prices rise, the greater will be the volume of future supplies that are likely to become available and the greater will be the inducements for consumers to conserve energy. ETA-MACRO was built by combining two dynamic submodels:

- ETA (an acronym for energy technology assessment), a process analysis of the energy sector; and
- MACRO, a macroeconomic growth model providing for substitution between capital, labor, and electric and nonelectric energy.

Figure 11-1 provides an overview of the principal links between the sectoral and the macro submodels at a single time. Electric and nonelectric energy are supplied by the energy sector to the rest of the economy—employing a wide range of conventional and unconventional conversion technologies. The economy's gross output depends upon the inputs of energy, capital, and labor. In turn, output is allocated between current consumption, investment in building up the stock of capital and current payments for energy costs.

The model allows for price-induced conservation (e.g., using insulation to replace heating fuels in homes and other structures) and also for interfuel substitution (e.g., using electrically driven heat pumps in place of oil and gas burners). The model incorporates the effect of rising energy prices upon capital formation—hence slowing down economic growth. For the United States and the other OECD countries, the key assumptions are given in Table 11-2.

NUMERICAL RESULTS AND POLICY IMPLICATIONS

Of these alternative scenarios, perhaps the most plausible is the one associated with a 4 percent annual price increase. From Figure 11-2, note that the LDCs'

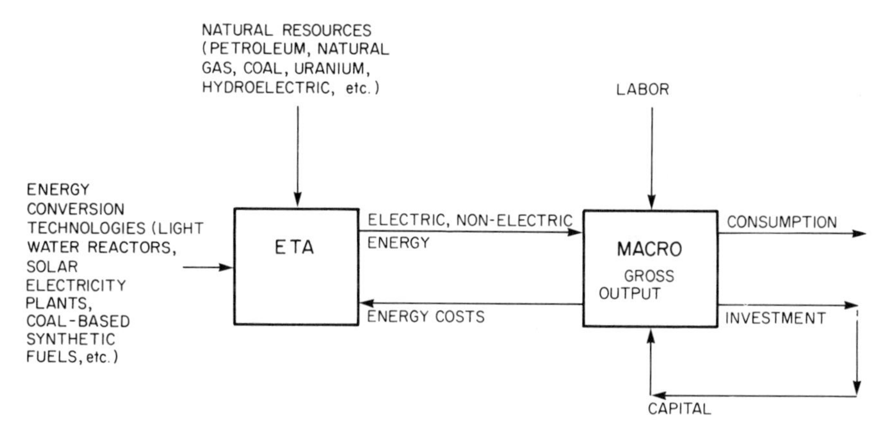

Figure 11-1. An Overview of ETA-MACRO.

Table 11-2. Key Assumptions for U.S. Energy Policy.

Energy Demands

- Potential GNP growth at constant energy prices, 1970–2000: 3.4%/year (realized growth rate ≈ 3.0%), After 2010, 2.0%/year.
- Elasticity (ease) of substitution between energy and capital-labor = .30.
- Price controls to be modified gradually so that domestic consumers pay incremental replacement costs of energy.

Domestic Nonnuclear Energy Supplies:

- Coal consumption limited to 40 quads in 2000 and to 60 quads asymptote (1970 and 1975 levels were 13 quads).
- Domestic oil and gas production to rise at 1% per year, 1980–90.
- Remaining conventional domestic oil and gas resources recoverable at prices of $2–$5/MMBTU: 1600 quads.
- Cost of inexhaustible nonelectric alternative energy systems (AES) = $6/MMBTU.

Imported Energy Supplies:

- Cautious and surprise-free OPEC pricing and production policies. Annual price increases of 2, 4, or 6% from 1980.

Principal Differences Between Other OECD Nations and the United States

- In 1975 (benchmark year for other OECD nations), they consumed 4% more primary energy, but produced 63% more GNP than the United States.
- Aggregate capital-output ratio of 3.0 in 1975 (versus 2.5 for U.S.). Hence a higher investment-GNP ratio.
- Outside the United States, there are limited prospects for increased coal consumption. Next most economical alternative to nuclear would be oil- and gas-fired electricity.
- Outside the United States, remaining oil and gas resources recoverable at prices of $2–$5/MMBTU: 2000 quads. This would permit production to increase until 1995. Thereafter a production ceiling of 35 quads (twice the current level).
- Fossil share of additional electric energy production: at least 25% (versus 40% for the United States.
- Hydroelectric, geothermal, and the like limited to 1% annual growth rate (versus 2.5% for the United States).
- Nuclear capacity limited to 400 GW in year 2000 (versus 300 GW for U.S.).
- Uranium resources available at prices up to $150/lb: 6 million short tons (versus 4 million for the United States).

domestic consumption requirements would be low during the 1980s, but rise rapidly thereafter. Exports to the OECD countries would increase moderately through the year 2000, but then would decline sharply. This picture appears reasonably consistent with today's conventional wisdom on hydrocarbon resources—hence it is said to constitute a "no surprise" scenario.

Figure 11-3 summarizes the implications of three alternative rates of OPEC oil price increase—2, 4, and 6 percent on the average from 1980 onward, measured in real terms, net of inflation. Given the oil and gas resources available

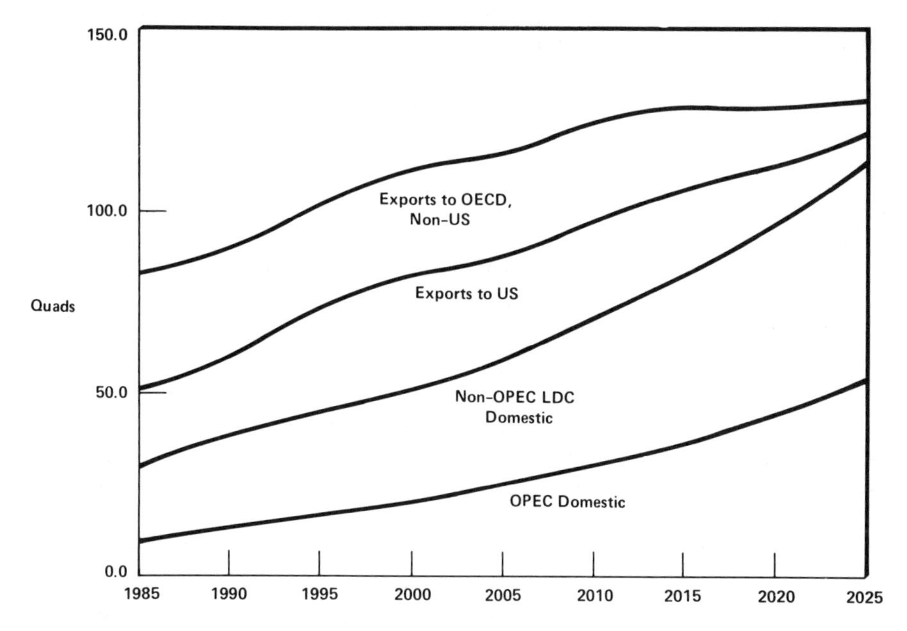

Figure 11-2. Demands for LDC Oil and Gas Production, OPEC and Non-OPEC Sources (4%/year OPEC price increase), Case. 2.

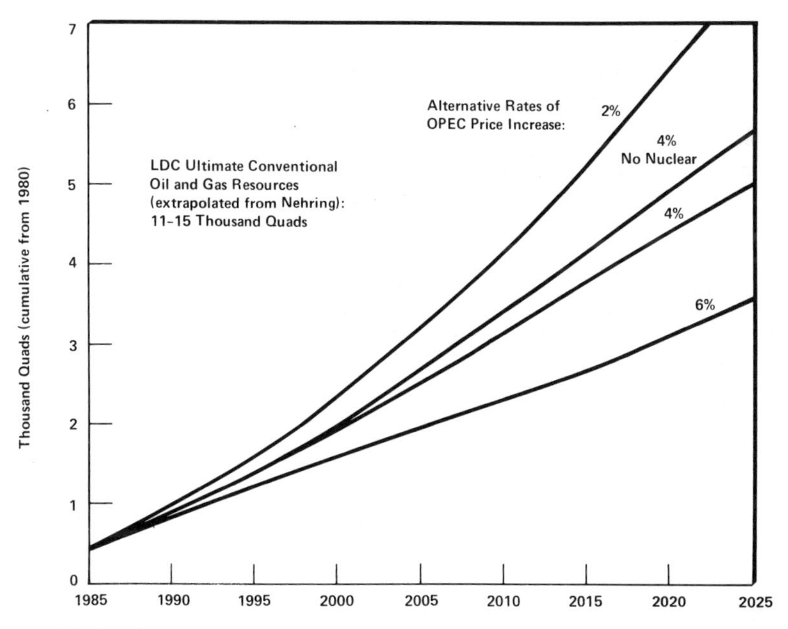

Figure 11-3. Cumulative LDC Oil and Gas Production, OPEC and Non-OPEC Sources.

within the OECD nations, this figure shows the production requirements that remain to be supplied by the OPEC and non-OPEC developing nations.

It appears reasonably clear that a 6 percent price increase policy would be nonoptimal from the viewpoint of OPEC. With this policy, the cartel's production requirements would stabilize (or possibly decrease). Given our assumptions on energy supply and conservation options available to the OECD nations, their imports from OPEC would virtually disappear after the year 2000.

Even with a 2 percent price increase policy, it should be feasible for the OPEC and non-OPEC LDCs to keep production moving up to satisfy domestic and export demands through the year 2000. After that date, however, the cumulative requirements would rapidly approach the limits on ultimate conventional hydrocarbon resources that have been estimated by Nehring and others. There is, of course, a wide margin for error in estimating the magnitude of these resources. One can anticipate that individual LDCs will be exceedingly cautious in relying upon "undiscovered recoverable" oil—until those resources have in fact been identified at specific locations.

Figures 11-4 through 11-7 indicate the rough dimensions of the effort that would be required to achieve a transition away from oil and gas within the OECD countries. This strategy would have to be many faceted. It would involve oil and gas conservation—plus electrification—plus a rapid rate of introduction of unconventional supply technologies. None of this would be easy or inexpensive.

From the year 2000 onward, it will be difficult to ensure a smooth transition unless international oil prices are raised by more than 2 percent per year during the 1980s and 1990s. From the viewpoint of both OPEC and the consumer nations, it would be preferable to initiate gradual price increases during the near future (say, 4 percent per year from 1980). This would be better than to continue OPEC's 1975-1978 policy of a constant (or decreasing) real price level for crude oil and then unexpectedly to be confronted with another oil crisis in which real prices double within a one or two year period.

These broad conclusions would hold—regardless of whether or not the industrialized nations continue to develop nuclear electric power. Specific dates could, however, change significantly if AES (unconventional nonelectric alternative energy systems) costs were to be much higher or lower than the assumed value of \$6/MMBtu. The role of nuclear energy would be more significant if the AES were based primarily on nuclear process heat or hydrogen. From Table 11-3, note that the cumulated economic losses from a "no nuclear" policy would be \$330 billions for the United States and an additional \$550 billions for the other OECD nations (discounting at 5 percent through 2030)—given a 4 percent annual OPEC price rise. Other numerical assumptions could of course lead to different estimates of the cost of such a policy. Except for the United States and Canada, there are only limited prospects for coal to replace nuclear energy within the other major OECD nations.

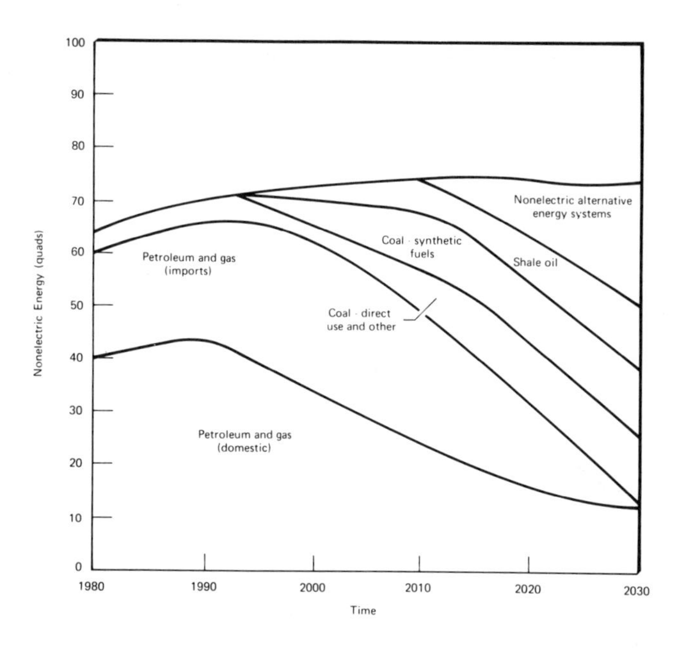

Figure 11-4. U.S. Projections, Alternative Annual Rates of OPEC Price Increase, Case 2: 4 Percent OPEC Price Increase. (Nonelectric energy)

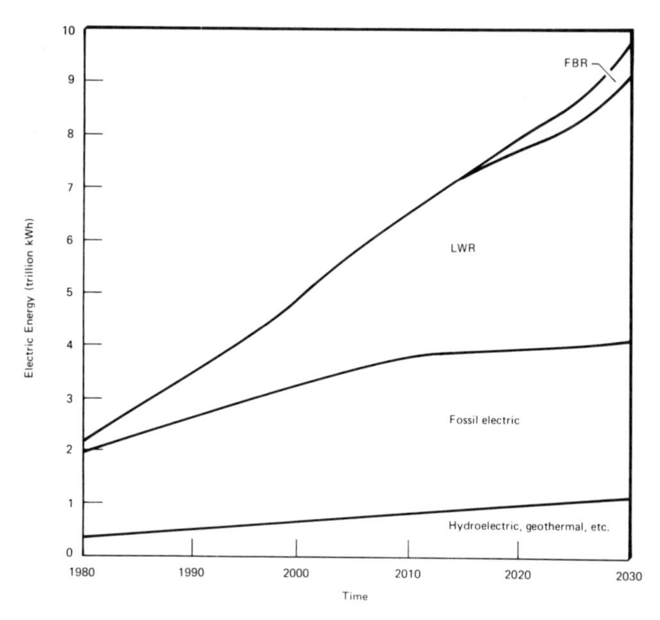

Figure 11-5. U.S. Projections, Alternative Annual Rates of OPEC Price Increase, Case 2: 4 Percent OPEC Price Increase. (Electric energy)

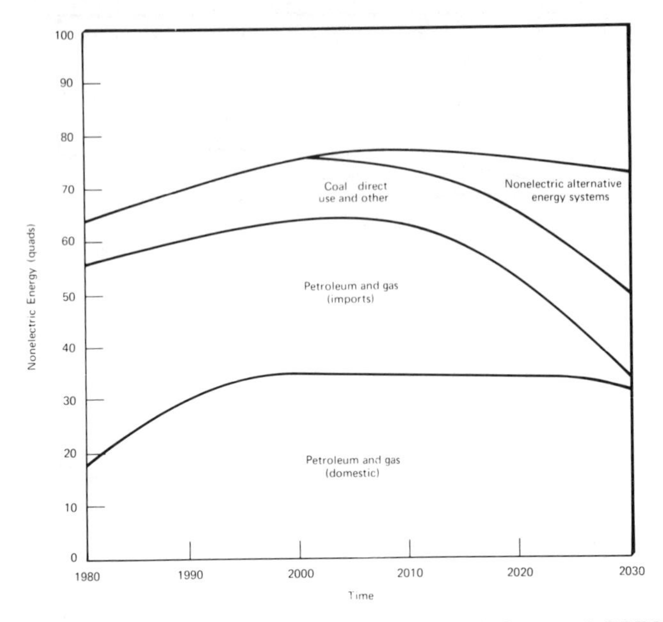

Figure 11-6. OECD, Non-U.S., Alternative Annual Rates of OPEC Price Increase, Case 2: 4 Percent OPEC Price Increase. (Nonelectric energy)

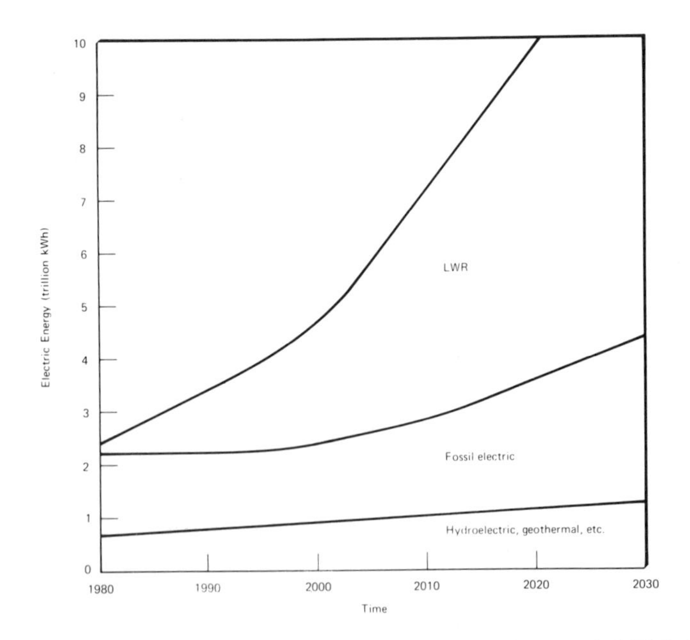

Figure 11-7. OECD, Non-U.S., Alternative Annual Rates of OPEC Price Increase, Case 2: 4 Percent OPEC Price Increase. (Electric energy)

Table 11-3. Summary of ETA-Macro Results.

Case Identification Number	*1*	*2*	*3*	*4*
Percent Annual OPEC price increase (from 1980)	*2 Percent*	*4 Percent*	*6 Percent*	*4 Percent*
Nuclear capacity additions after 1985	*with nuclear*	*with nuclear*	*with nuclear*	*no nuclear*
GNP ($ trillions):				
U.S. (2000	3.34	3.24	3.15	3.23
U.S. (2030	6.00	5.34	5.63	4.96
Other OECD (2000	4.99	4.91	4.80	4.87
Other OECD (2030	9.10	8.67	8.90	8.23
Energy consumption (quads):				
U.S. (2000	129.4	119.9	111.9	117.9
U.S. (2030	206.3	172.6	201.6	131.7
Other OECD (2000	122.7	113.9	105.3	107.8
Other OECD (2030	199.7	194.8	193.5	132.9
Electricity generation (trillion kWh):				
U.S. (2000	4.8	4.8	4.9	4.7
U.S. (2030	9.4	9.9	9.4	5.8
Other OECD (2000	4.4	4.6	4.6	3.5
Other OECD (2030	9.2	13.2	10.0	5.8
Nuclear capacity (TW at 65 percent capacity factor):				
U.S. (2000	0.3	0.3	0.3	0.1
U.S. (2030	0.8	1.0	0.9	0
Other OECD (2000	0.4	0.4	0.4	0.1
Other OECD (2030	1.0	1.5	1.1	0
Present value of consumption: ($ trillions, cumulative from 1975, 5 percent discount rate):				
U.S. (2000	31.56	31.41	31.33	31.41
U.S. (2030	47.70	46.70	46.20	46.37
Other OECD (2000	48.41	48.25	48.09	48.21
Other OECD (2030	72.67	71.86	71.41	71.31
Present value of total OECD import costs for oil and gas ($ trillions, cumulative 1985-2025):				
0 percent discount rate	16.40	11.26	4.43	16.51
5 percent discount rate	5.38	4.41	2.71	5.63
10 percent discount rate	2.61	2.38	1.89	2.74

NOTES

1. For oil and gas demand projections within the developing nations, a simpler methodology was employed. See footnotes to Table 13–1.

2. Further details are provided in two appendixes available upon request to the author. Also see "ETA-MACRO: A Model of Energy-Economy Interactions," EA-592, Electric Power Research Institute, December 1977.

Chapter 12

The Underestimated Potential of World Natural Gas

Henry Rowen
and
Jaffer Syed
Stanford University
(Presented by Henry Rowen)

In the past few years, several analyses have been published that show marked slowing of U.S. and world energy demand projections over the next several decades by comparison with the trend in the pre-oil-price-increase period.[1] Some analyses also give weight to increased supplies from important new sources of oil, principally Mexico.[2] The result of these analyses is to suggest that a further sharp increase in the oil price is not likely to occur in the 1980s as was suggested a year ago by several U.S. government agencies and outside commentators; the crisis might be deferred to the 1990s or even beyond.

One of the more remarkable features of the published analyses is the relative neglect of the impact of the future development of natural gas on all energy markets but especially on that for oil. It is not that the forecasters have ignored natural gas; projections of sizeable increases in its usage worldwide can be found in reports by PIRINC, OECD, WAES, and so forth. It is the apparent underestimation of the potential significance of natural gas development–together with accumulating evidence of probable low growth in energy demand–that is the central proposition of this chapter.

OUTLINE OF THE ARGUMENT

First, throughout we are discussing only conventional sources of natural gas. No geopressured methane or other exotic sources are considered.

Second, we concentrate on natural gas prospects outside of the United States.

Table 12-1. Approximate Oil and Gas Production in Quads

	1975	
	U.S.	*Non-U.S.*
Oil Production	18	95
Gas production (Marketed)	20	27

Sources: *Petroleum Encyclopedia* (1978), vol. 11; American Petroleum Institute, *World Oil* 187, no. 3 (August 15, 1978).

This is not to hold that the United States does not contain large undiscovered resources of conventional natural gas, perhaps larger than is widely credited. In any case, this is not part of our argument.

Third, there is a good reason to believe that there is a vast potential for natural gas development outside the United States and that changes in the relative price of fuels since 1973 make it economical to develop this potential. Consider the pattern of oil versus gas usage in 1975 as shown in Table 12-1.

This low production of gas relative to oil outside of the United States apparently does not have a geological basis. Estimates of original oil in place versus original gas in place do not suggest that the United States was endowed more generously with gas relative to oil by comparison with the rest of the world.

Instead we can find an explanation in the economics of energy. Consider the comparison of oil versus gas prices and costs shortly before and in the period since 1973 (all expressed in 1975 dollars per million BTU), as shown in Table 12-2.

These numbers, which simplify a complex reality, nonetheless tell an important story: before 1973 it did not pay to liquefy and transport LNG or to transport it by pipeline for long distances or under costly circumstances.[3] After 1973, the cost of moving LNG is comparable to that of the delivered cost of crude oil. It also clearly pays to move gas in more costly pipelines (perhaps from the Arctic or under the sea) or over much longer distances.

The pre-1973 pattern of usage reflected these economic factors. The high energy-consuming regions of the world were the United States and southern Canada, Western Europe, Japan, and Eastern Europe (including the western USSR). Natural gas was found in proximity in markets most notably in the United States and Canada, and the infrastructure for its transport and use were built there in the quarter century after World War II. Aside from the Netherlands and, later, the British sector of the North Sea, little nonassociated gas from the oil-producing area of the North Sea could be expected to be gathered and delivered ashore. Some gas was also being moved to Western Europe from the USSR during this period. As for Japan, no substantial amount of gas had been found in or near it. The USSR, the world's largest oil producer, had substantial amounts of gas relatively near major consuming areas in the western USSR.

Table 12-2. Oil and Gas Costs

	Pre-1973	*Post-1973*
	Costs in dollars/million Btu (1975 dollars)	
Oil		
Crude oil (delivered to users' ports)[a]	0.50	2.40
Natural Gas		
LNG (Liquefication, Transport, and Gasification only)Algeria to U.S. East Coast)[b]	2.00	2.00
Pipeline–wholesale transport only		
(1000 miles, nonextreme conditions)[c]	0.22	0.22
(1000 miles, extreme conditions)–e.g., e.g., sub-Arctic)[d]	0.40	0.40
(1000 miles, 400 feet sea depth off shore pipeline at 1,000 MMscfd)[e]	1.40	1.40

Sources:

[a]*International Crude Oil and Product Prices* (Middle East Petroleum and Economic Publications, April 1978); *Energy Prices 1960–73* (Washington, D.C.: Foster Associates).

[b]Nordine Ait Laoussine, "LNG Exports–A Contribution to World Energy Supplies in the Decades to Come," *OPEC Review* 1, no. 7 (October 1977): 21–35.

[c]*Oil and Gas Journal*, August 22, 1977, p. 73.

[d]Roughly computed from estimated transportation costs for Alaskan North Slope to lower forty-eight states, quoted in *Oil and Gas Journal,* January 16, 1978, p. 42.

[e]Roughly computed from the costs given in Magne Ostby and S.S. Marsden, "Gas-to-Methanol conversion proposed for offshore gas fields," *Oil and Gas Journal,* November 7, 1977, pp. 83–87.

However, it chose to develop oil much more intensively than gas (it produces 20 percent more oil and 50 percent less gas than the United States),[4] a choice that probably correctly reflected the cost of capital in that country.

In short, natural gas not found near major consuming areas was not likely to be worth much. As a consequence, it is not surprising that natural gas was flared for many years in the Middle East and in many other oil-producing regions. Indeed, this was widely done in the United States until the post–World War II construction of a vast gas collection and distribution system.

Now the economics have greatly altered, and these changes are being reflected in market behavior. It pays to look for natural gas and to develop it under circumstances when in the past it would have been left undeveloped or developed much more slowly. Not surprisingly, important gas discoveries have been made in the past few years in offshore Thailand, northwest and southeastern Australia, Argentina, Columbia, Bolivia, Malaysia, and New Zealand.

Second, pipelines are planned or under construction over much greater distances and through more costly environments: under the Mediterranean from North Africa to Italy, from eastern Siberia to Western Europe, from the North Slope of Alaska to the United States, and if further gas finds are made, from the remote Canadian arctic islands,[5] from Mexico to the United States (assuming an agreement on price can be reached), and so on.

Third, now that it pays, LNG projects are being developed around the world.[6] Currently there are in operation systems capable of moving .67 TCF per year.[7] Projects with a total capacity of around 4 Tcf are under construction for completion by the early 1980s. Additional planned (with varying degrees of firmness) projects have a total capacity of around 5 Tcf, for a possible mid- or late 1980s total of around 9 Tcf. The point is sometimes made that the OPEC countries will not export LNG if this threatens to undermine the price of oil. This is a complicated issue, some aspects of which we discuss later. We note, however, that several LNG projects are not in OPEC countries and that quite a few countries with significant reserves of natural gas are not members of OPEC, including the United States, Canada, Norway, the Netherlands, the United Kingdom, Malaysia, Pakistan, China, the USSR, Australia, and Mexico.

A fourth consequence of the rise in the real price of oil is the movement of heavy users of energy to the sources of the gas. Table 12-2 compared the cost of oil only with the wholesale transportation costs of natural gas; it did not include any value for the gas at the wellhead. The value of gas that has no alternative beneficial use is zero, and this is why so much gas has been flared over the years. But as the possessors of natural gas often observe, the net price to them after paying the high costs of LNG transportation is very much less per unit of energy than for oil. The owners therefore may choose to use the gas at home. A high value use now is in reinjection of gas to sustain oil production. In addition, producers may choose to create industries for whom energy, or hydrocarbons as a feedstock, is a large component of total factor costs. Examples of energy-intensive industries that one should expect to see developing in gas rich centers remote from major consuming areas include aluminum production, fertilizer production, petrochemicals, and carbon black. In fact, we can observe a large increase in such projects.[8]

However, even for such industries, energy is not the only major factor of production. Although many of these countries may also be rich in capital, many are scarce in trained manpower, and this fact will constrain the rate of growth of this energy-intensive development. Nonetheless, beneficial uses are being created. For instance, about 4 Tcf of gas are being flared in OPEC, and it is reported that Iran and Saudi Arabia, the largest wasters of gas (about 1 quad each), intend to stop by around 1980.[9] Iran has plans to invest $8 billion to harness the country's gas reserves, while Saudi Arabia intends to spend $12-15 billion in gas gathering and treating programs, starting in 1980, to process 1.3 quads of gas.[10]

THE POTENTIAL IMPACT ON WORLD ENERGY MARKETS

Suppose, as the result of these changed economic incentives, that the rest of the world were to find, develop, and use natural gas relative to oil to the same extent as the United States. This would mean, at today's oil usage rate, production and use of an additional 75 quads of natural gas a year, in addition to today's 45 quads level of use. Such an amount would be equivalent to about one-third of total current world energy consumption and about 40 percent of that of the world outside of Communist areas. It would be equivalent to the production of an additional thirty-seven million barrels of oil a day, an amount greater than today's total OPEC output.

Two questions are likely to come quickly to mind: Is the gas really there to be found and developed, and if it is there, how long will it take to develop it to such a level?

The existence of the gas is supported by geological estimates. These estimates have a high variance, but even on pessimistic assumptions, there is a great deal to be found, enough to support a world production rate of 120 quads a year for many years. For instance, current geological estimates center on 10,000 quads of original gas in place (with a spread in estimates from 7,200–16,000 Tcf).[11] Only around 930 quads were produced by 1975.

The possible rate of development of gas presents a difficult problem of forecasting. For the near term we can observe and count a large number of projects and add up the gas utilization associated with them. Even here there are uncertainties, because some of these may be cancelled (as was the recent El Paso gas deal with Algeria as the result of failure to win U.S. government approval) or slipped in time (e.g., the North Slope gas pipeline).

An example of potential slow development is provided by Pakistan, which has substantial proven gas reserves. Current production is low relative to these reserves apparently because the government has set such a low price for gas that the gas utility cannot generate enough capital to expand production at a very high rate.[12]

Estimates of future increases in gas usage vary widely. WAES estimates, which exclude the United States and Communist areas, center on about 32 quads additional usage to 2000.[13] World Energy Conference (WEC) estimates 13 quads more consumed to 2000 (26 quads more including growth in Communist areas). Estimating growth to 1990, the Energy Information Administration comes up with 15 quads, and PIRINC, 22 quads. In a projection made to 1985, OECD estimates 14 additional quads.

There are two features in these projections of particular interest: One is the large variance in projected gas usage—for example, WEC estimated gas use growth by 2000 about equal to that estimated by OECD for 1985. (There was only a two year difference in the base year.) The other is the much smaller absolute increase in gas usage (measured in quads) assumed relative to that of oil in these projections. Of course, the base of natural gas usage is much smaller (around 13 quads in the non-U.S., non-Communist world in the mid-1970s). Even so, a growth rate of 5 percent per year in gas usage could add as much energy output by 1990 (doubling to 26 quads) as Mexico alone might add to oil production over the same period (assuming that it then produces seven million barrels per day—equivalent to 14 quads). The assumption in some studies that increments to oil production might be double or more increments to gas production by 2000 (WAES) or more than triple (WEC) or be more than three times as great by 1990 (EIA) may fail to take adequate account of the changes in relative energy supply costs described above.

Some analyses are more bullish. A report prepared for the World Energy Conference in Istanbul in 1977 estimated global conventional gas availability by 2000 at around 135 quads.[14] The executive vice president of Algeria's Sonatrach has said that OPEC could meet a demand for 21 quads of gas by 2000—if the price were right.[15]

As for the Soviet Union, it is engaged in a large expansion of natural gas production, especially in western Siberia.[16] It is also encouraging the export of large quantities of natural gas, and it seems likely that such exports will replace a substantial part of its oil exports. Despite the large capital requirements for such development, Soviet natural gas production is moving toward equality to its oil production, which is around 20–24 quads.[17] As for China, current evidence supports the assumption that a large expansion in oil and gas production will take place in the next quarter century.

Although there is little reason to doubt the availability of gas resources, there are long lead times in the development of many gas projects. The gas has to be found in the first place. Once found, it may take a long time to discover enough gas to warrant building a pipeline, as the prolonged Canadian exploration in the Arctic islands illustrates. Third, gas transportation projects are highly capital intensive and inflexible. The destination of a pipeline, unlike that of an oil tanker, cannot easily be changed. Not only are these economic risks (How will the project fare if the real price of oil declines?), there are also political risks where national borders are crossed. All of this can cause delays (perhaps of indefinite duration). Some of the countries that can be expected to find natural gas may have poor international credit ratings. Moreover, some of the countries that are abundant sources of natural gas, available at a comparatively low price, may lack other important attributes such as a trained work force or political stability.

These and other uncertanties make the projection of gas usage to the year 2000 and beyond a hazardous business. The wide spread in projection (Table 12-3) illustrates this point.

Table 12-3. Projected Natural Gas Usage, Non-U.S., Non-Communist World.

Year	*High*	*Low*
1975	15 (actual)	
2000	64[a] quads	36[a] quads

[a]The high estimate uses the growth rate of gas usage as in the PIRING report [Palo Alto: 1978]) to extrapolate to year 2000. The low estimate is the same as projected by the WAES study (*Energy: Global Prospects 1985-2000* [New York: McGraw-Hill, 1977]).

On these projections, additions to natural gas usage by 2000 in the non-U.S., non-Communist world could range from 21 to 49 quads. By comparison, additions to nuclear energy capacity in the same regions of the world from now to 2000 might amount to 20 to 30 quads equivalent primary energy. (This assumes the addition of 300-500 GW of nuclear electric power in the non-U.S., non-Communist world.)

NATURAL GAS PROSPECTS IN PERSPECTIVE

These prospects need to be evaluated in relation to projected demand for energy and to the supply of competing fuels. We have not yet made a formal analysis, but some observations may be helpful in thinking about the energy futures.

One is that energy demand worldwide is growing much more slowly than in the several decades before 1973 and more slowly than some of the early projections shortly after the oil price increase after 1973. This lowering of demand projections is the result of assumed consumer adjustment to higher energy prices together with lower forecasts of economic growth.[18] Moreover, there is a good deal of growth expected in certain energy supplies. Even with recent slippage and cancellation of nuclear plants, sizable additions will occur. Using an installed non-Communist world capacity of 500 GW (versus about 100 GW today), added nuclear capacity would contribute around 24 quads (measured in terms of primary energy equivalent). Additions to coal-fired and hydro plants might add 30 quads. Even on a low growth assumption for gas, 21 quads additional would be forthcoming by 2000. The prospect of around 75 quads of additional energy supply, together with slow demand growth, leaves modest growth potential for other fuels, principally oil. On the high growth assumption for natural gas, there might be little additional demand for oil.

This analysis should at least raise questions about the widely held assumption of a rising real price of oil by the year 2000—if not a sharp upward surge sometime between now and then. Because natural gas is an excellent substitute for oil in many applications, increased gas production can be expected to displace oil on virtually a quad-for-quad basis. To the extent that this occurs, it will undermine the ability of OPEC to raise or even to maintain the price of oil. If in addition, new oil sources—for example, Mexico or other developing countries–expand production significantly and old ones more or less sustain production, then the downward pressure on oil prices could become severe.

Of course, if the real price of oil declines, then energy demand would grow more rapidly than is assumed here. This would also slow gas development; LNG developments are particularly sensitive to oil price competition. There would also be an impact on nuclear and coal electric plant growth working through the cross-elasticity of demand on these sectors. These market factors might become evident soon enough in the 1980s, through a declining real price of oil, to substantially slow the rate of natural gas development below the levels suggested here. Such a development should not be surprising.

It might be thought that although there could be a substantial period of energy "surplus" to the year 2000–which is not yet a fashionable view–the world might run into serious energy supply problems early in the twenty-first century. This could happen. However, nuclear capacity can be expected to grow (even without the breeder reactor) and so might natural gas production beyond 2000. But by 2020 or sooner it would be useful to have available a wider array of energy options from which to choose. This analysis suggests that we have a good deal of time to invent and deploy these technologies.

These observations also bear on the appropriate scope of analysis of the world oil market. A good deal has been written about the economics and politics of the OPEC cartel in the past several years, and many of the published analyses have ignored or given little weight to the possible impact of growing gas production on the ability of OPEC to maintain the oil price. The true observation that the OPEC countries also have a large share of the world's gas reserves is not an adequate substitute for an analysis of potential gas usage. For instance, as we have seen, much of the growth in use will take place at home, a phenomenon that may be hard to control in the OPEC bargaining arena.

NOTES

1. Workshop on Alternative Energy Strategies (WAES), *Energy: Global Prospects 1985-2000* New York: McGraw-Hill, 1977); OECD, *World Energy Outlook,* Report by the Secretary General (Paris, 1977); Petroleum Industry Research Foundation, Inc. (PIRINC), *Outlook for World Oil Into the 21st Century,* for Electric Power Research Institute (Palo Alto: 1978); Energy Information Administration, EIA, *Projections of Energy Supply and Demand and their Impacts,* vols. I and II, Annual Report to Congress, DOE/EIA-0036/1,2 (Washington, D.C.: 1977); and CIA, *The International Energy Situation: Outlook to 1985,* CIA Report ER 77-10240 U (April 1977). An incomplete list of recent publications includes those by Shell Oil, Exxon, the Petroleum Industry Research Foundation Inc. (PIRINC), WAES, CIA, and the Energy Information Administration of the U.S. Department of Energy.

2. See PIRINC and EIA reports.

3. Thomas Schelling tells us that several decades ago some of the oil majors in the Middle East calculated that it would pay to pipe natural gas the 2,400

miles to Western Europe but that the political hazards in going through so many countries *en route* made it too risky a project.

4. 1975 production figures are oil: U.S.–3 billion barrels; USSR–3.6 billion barrels; gas: U.S.–21 Tcf; USSR–10.7 Tcf. *Minerals Yearbook* Bureau of Mines, U.S. Dept. of Interior, (1975).

5. The report of large gas reserves in the Deep Basin of northern British Columbia and Alberta raises questions about the likely competitiveness of Arctic gas given its high transport costs.

6. The data in this paragraph are taken from *Petroleum Encyclopedia,* vol. 11 American Petroleum Institute, (1978), pp. 332–34.

7. One trillion cubic feet per year (Tcf) has approximately 10^{15} Btu in heat value or 1 quad of energy.

8. For instance, see *OPEC*, (Vienna-Information Department–*Annual Review and Record, 1976*); and *Hydrocarbon Processing,* June 1978, supplement.

9. See *World Oil,* June 1978, p. 134, for flaring statistics.

10. W.D. Moore, III, "Iran's Mammoth Gas Program to Eliminate Flaring by 1980," *Oil and Gas Journal,* June 26, 1978, pp. 79–83; W.D. Moore, III, "Saudi push exploration, recovery, gas processing," *Oil and Gas Journal,* June 26, 1978, pp. 84–92.

11. Parent and H. Linden, "A Survey of United States and Total World Production, Proved Reserves and Remaining Recoverable Resources of Fossil Fuels and Uranium as of December 31, 1975" (Institute of Gas Technology, Chicago, January 1977). Note that these estimates are not associated with a cost of extraction.

12. Communication from Alan Manne.

13. Base year 1972.

14. W.T. McCormick et al., "The Future for World Natural Gas Supply" (Paper delivered at American Gas Association 10th World Energy Conference, Istanbul, August 1977), p. 2.

15. *Middle East Economic Survey,* 26 June 1978, pp. 6–7.

16. Larry Auldridge, "Soviets pushing projects to boost gas production," *The Oil and Gas Journal,* May 22, 1978.

17. Soviet Union gas production has been increasing at around 1 Tcf per year. The level of 20 quads of gas production might be reached before 1990.

18. A recent analysis of U.S. energy demand based on consumer responses to higher energy prices during the 1970s is William W. Hogan, "Energy Demand Forecasts: Comparisons and Sensitivities" (Prepared for the Non-Proliferation Alternatives Program Summer Study, Aspen, Colorado, August 1978). For a set of price elasticity of demand, economic growth, energy price, and consumer lag variables, this analysis shows a range of year 2000 energy demands for the United States from 62 to 141 quads by comparison with the 1977 demand of 75.9 quads. The central estimate of these parameter values is around 100 quads, a level about one-third greater than that of 1977.

Session C

Identifying Critical Parameters of Resource Availability and Technology

Part I

Fission and Solar Parameters

Chapter 13

High Capital Cost: A Deterrent to Technological Change*

Karl Cohen
Stanford University

In August 1943 construction began at Hanford, Washington, on the water-cooling facilities for the first of three plutonium production reactors. This was eight months after Fermi's demonstration of the first manmade chain reaction and only five months after acquisition of the site. The reactor drawings were released on October 4, 1943. B pile was loaded in September 1944. By the end of January 1945, all three reactors were in full operation, and the first plutonium was produced. In the same period of time nowadays—eighteen months—one might at best hope to have an environmental impact statement completed.

I recall these contrasting examples of the effects of policy on the tempo of technical development not out of nostalgia for the golden age when nuclear energy was actively encouraged by our government, nor yet to imply that the consequences of present energy policies exceed the intent of their framers. The reason is to reassure this audience that my subject, which is the influence of classical market forces on technological change, is treated with full consciousness of the circumstance that we do not live in a pure market economy. Indeed, we are as unlikely to see a pure market economy as we will be to witness the "withering away of the state" in the nonmarket societies. But in market societies, however imperfect, policy and market forces both affect the tempo of technological change.

*The work on which this chapter is based was performed at the International Institute for Applied Systems Analysis in Laxenburg, Austria. It was supported by the General Electric Company. I am deeply indebted as well to Professor W. Haefele, leader of the Energy Project, and to Dr. A.M. Perry, of Oak Ridge National Laboratory, for many ideas and illuminating discussions. None of the above bears responsibility for, nor necessarily concurs with, the opinions and conclusions expressed.

HIGH CAPITAL COST AND TECHNOLOGICAL CHANGE

The technological change feasible in a given period of time depends on the frame of reference. In a wartime mobilization many things are possible that cannot be accomplished by business as usual. Even in peacetime, government development funding, price fixing, regulations, subsidies, loans, taxes, tax incentives, rationing, and import duties can and do affect the direction and tempo of technological change. Nevertheless, the purpose of planning should be to avoid, as much as possible, the necessity for regulations and crash programs to make up for a lack of foresight. We may not be able to attain our goals by business as usual, but it should be our ambition to depart from it as little as possible—to make plans that work with and not against normal economic forces and normal economic rhythms. For such plans will involve the least constraints on the public and the least costs and dislocations to the economy and are most likely to be adopted and to succeed.

With this in mind let us consider the influence of capital cost of energy systems on the rate of technological change. High capital cost contributes to a high cost product. However, if it is accompanied by relatively low operating costs, over a period of severe operating cost increases, either through inflation or resource depletion, it may eventually lead to comparatively lower total costs, even on a present worth basis.

Economic models based on linear programming will typically choose systems that show the lowest present worth total costs over a period of thirty to fifty years, without further inquiry into the capital cost component. In such models the system installed in a given region can flip-flop as soon as some circumstance is calculated to produce a new lowest cost system. This flip-flop, say from fossil to nuclear power, or from light water reactors to breeders, or from fossil to solar power, will occur whether the calculated cost advantage is small or large and is not affected by the capital cost except as it is a component of total cost. The availability of capital for a more capital-intensive system is not seen as a problem. To be sure, constraints can be applied to the growth of new technologies—say a fixed progression of 4, 8, 16, 32 . . . . plants over a period of years—but such constraints are mechanical and do not relate to capital intensity or to the magnitude of the cost advantage.

An illustration of such an analysis is shown in Figure 13-1 which is a reproduction of a page from an AEC report.[1] The report (WASH 1098) was a collaborative effort of five national laboratories and five reactor manufacturers and was published in December 1970. It projects a sudden technological changeover from light water reactors to liquid metal fast breeder reactors. The breeder market share is shown to increase from 6 percent of U.S. power plant installations in 1980-1981 to 50 percent in eight years. I resurrect this bit of early Americana, which contains a number of quaint assumptions about costs and electrical power growth rates, because the pattern of technological change that

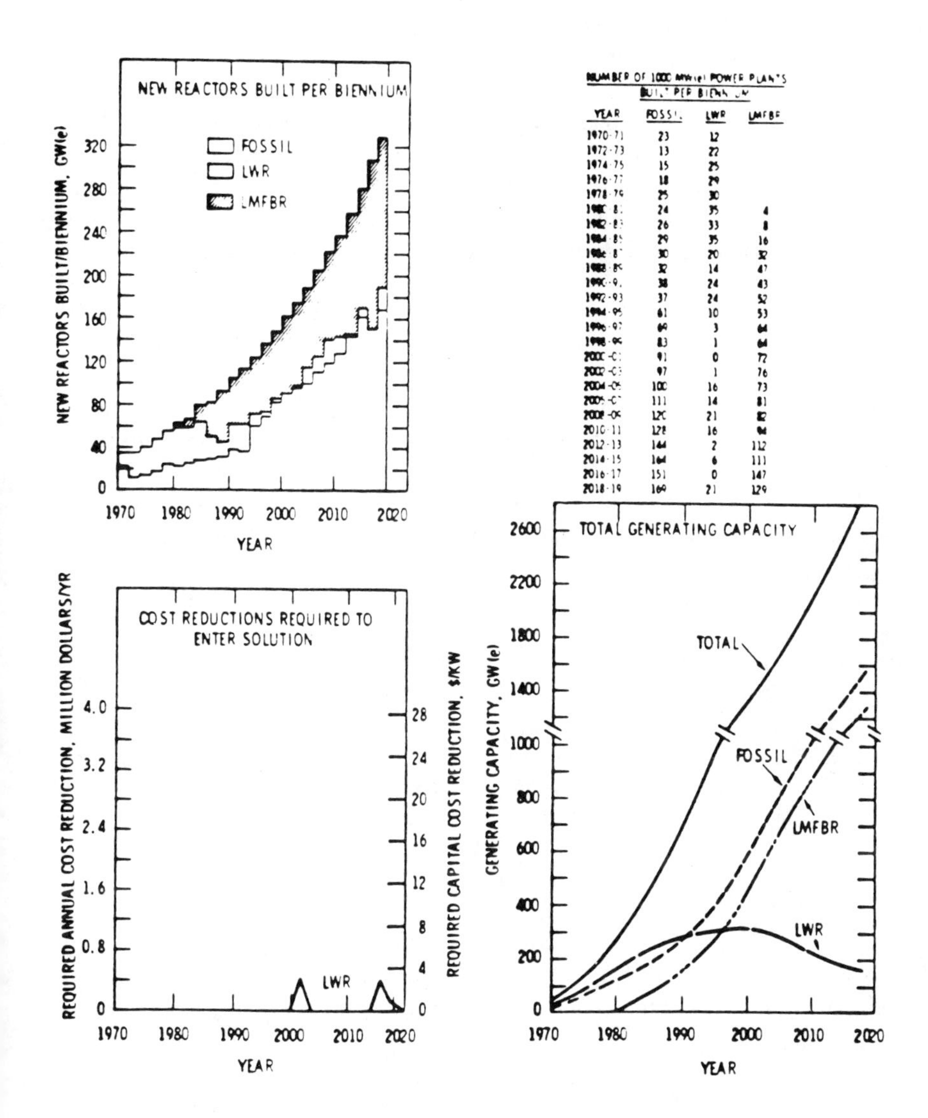

NUMBER OF 1000 MW(e) POWER PLANTS BUILT PER BIENNIUM

YEAR	FOSSIL	LWR	LMFBR
1970-71	23	12	
1972-73	13	22	
1974-75	15	25	
1976-77	18	29	
1978-79	25	30	
1980-81	24	35	4
1982-83	26	33	8
1984-85	29	35	16
1986-87	30	20	32
1988-89	32	14	47
1990-91	38	24	43
1992-93	37	24	52
1994-95	61	10	53
1996-97	69	3	64
1998-99	83	1	64
2000-01	91	0	72
2002-03	97	1	76
2004-05	100	16	73
2006-07	111	14	81
2008-09	120	21	82
2010-11	128	16	94
2012-13	144	2	112
2014-15	164	6	111
2016-17	151	0	147
2018-19	169	21	129

Source: *WASH 1098: Potential Nuclear Power Growth Patterns* (Washington, D.C.: USAEC, Division of Reactor Development and Technology, December 1970).

Figure 13-1. Characteristics of Optimum Generating System with Fossil, LWR, and LMFBR Power Plants.

it exhibits, through many revisions and despite radical changes in assumptions, remains embedded in the thinking of the nuclear industry and of its opponents.

APPLICATIONS OF A MODEL OF TECHNOLOGICAL CHANGEOVER TO A NUCLEAR ECONOMY

A very different model of technological changeover was developed by V. Peterka at IIASA[2] to account for the remarkable regularities in technical substitutions that were described by Dr. Fisher in Chapter 5. He and Dr. Pry found[3] that the substitution of one technology for another follows a particularly simple law. If f_1 is the market fraction of an old technology and f_2 the fraction of a new technology

$$\ln f_2/f_1 = At,$$

where A is a constant, t is time. This generalization was shown to hold for seventeen technical substitutions ranging from synthetic-natural rubber to BOF–open hearth steels.

Marchetti[4] subsequently applied the Fisher-Pry concept to multivariate competition of different primary energy sources (wood, coal, oil, gas) and showed that the same regularities exist (see Figure 4–10).

Peterka, using an observation made in 1961 by Mansfield[5] that the rate of technical change is positively correlated with the profitability of the new technology and negatively influenced by the relative capital investment needed to introduce the new technology, developed a purely economic theory that accounts for the Fisher-Pry observations and relates the substitution constant A to the production costs and capital investments of the competing technologies. The basic equation of the Peterka theory is, if $P_i(t)$ is the rate of production by the ith competing technology (Table 13–1)

$$\alpha_i \dot{P}_i(t) = P_i(t)\{p(t) - c_i\}\ , \quad i = 1, 2 \ldots n\,, \tag{13.1}$$

where α_i is the specific investment for technology i (capital needed to increase the production rate one unit), c_i is the specific production costs (cost, including capital charges, of producing one unit), and $p(t)$ is the market price of a unit of production.

For power plants, P_i is in kilowatts, α_i is in \$/kw, and it is convenient to express p and c_i in \$/yr/kw. This equation asserts that a producer's investment in a technology i on the average will equal his return from production using this technology.

It is plain that in the long run a producer cannot invest in a technology more than his return—he cannot subsidize it indefinitely. Further, if he wishes to retain his market share, he must continue to invest. The equations are intuitively of the right form, the increment of production being proportional to

Table 13-1. Market Penetration Relationships.

Fisher-Pry	$\ln f_2/f_1 = At$
Peterka 13.1	$\alpha_i P_i(t) = P_i(t) \{p(t) - c_i\} \quad i = 1, 2 \ldots n$
13.2	$f_i(t) P(t) = P_i(t)$
13.3	$P_i(t) = P(t) \frac{f_i(t)}{\alpha_i(t)} \{p(t) - c_i(t)\}$
13.4	$\frac{d \ln P(t)}{dt} = p \Sigma \frac{f_i}{a_i} - \Sigma \frac{c_i f_i}{a_i}$
Approximation for $n=2$ 13.5	$\frac{d}{dt} \ln \frac{f_2}{f_1} = \frac{c_1 - c_2}{\alpha_1} = \left(\frac{\alpha_2}{\alpha_1} - 1\right) \frac{d \ln P}{dt}$
Exact solution for $\alpha_2 = \alpha_1 = \alpha$ 13.6	$\frac{d}{dt} \ln \frac{f_2}{f_1} = \frac{c_1 - c_2}{\alpha}$
Better approximation for $n=2, f_2$ small $\alpha_1 \neq \alpha_2$ 13.7	$\frac{d \ln P_2}{dt} \leqslant \frac{a_1 \frac{d \ln P}{dt} + c_1 - c_2}{\alpha_2}$

the present production, and the observations of Mansfield are nicely incorporated.

One could imagine another term on the right-hand side of the equation, representing a subsidy by the producer or the government to get the technology started. When P_i is small, this could be significant, but after a while, unless the subsidy increases continuously as P_i does, it cannot have much effect.

From this set of equations, as is seen from Equation (13.6), the Fisher-Pry relationship drops out immediately and so do the Marchetti relationships for any pair of competitors in multivariate competition. Unfortunately this does not prove that the Peterka theory is correct. All we need to derive the Fisher-Pry formula is an explanation for exponential growth rates of each component individually. The psychology of imitation might furnish such an explanation. The Peterka theory has the advantage over other explanations that the coefficients are determinable without reference to the data that the theory seeks to explain. This enables us to devise objective tests and to use the theory as a predictive tool.

The Peterka formulation is not without its conceptual difficulties. The Fisher-Pry and, particularly, the Marchetti relationships are observed to hold with constant coefficients over extended periods. One does not expect cost relationships as given by Peterka to be so stable. When applied to commodities such as oil and coal, which have multiple end uses and different qualitative properties, it is even hard to define a unique capital cost and price. Similarly, if one compared electricity with oil on a $/Btu basis (as does Amory Lovins), the theory would not predict that electricity would grow at all, much less that it would grow twice as fast as all other forms of energy. This suggests that the theory will be most useful as a predictive tool when comparing technologies with identical end products, such as alternate means of producing electricity.

One way to test the theory is to see what sense it makes out of current situations, such as the ongoing penetration of nuclear against fossil plants in the United States (see Table 13-2). The values given are my estimate of average costs for baseload plants operating on the dates given, mostly completed earlier. They are of course judgmental, but were not chosen to prove a point.

I draw your attention to the nuclear power growth rates predicted for 1977 (10 percent) and 2000 (8.9 percent). There is no fundamental reason that these values should agree: each depends on a momentary competitive situation. Yet the relationships appear stable. Perhaps the reason is that plant construction periods are so long that cost experience is acquired slowly and hence cost relationships persist for extended periods.

Table 13-2. Market Penetration of LWR versus Fossil Base Load Plants in United States.

	Year 1977		*Year 2000*
INPUT			
Capital Cost (plants in being) (1978 $)			
Nuclear	$350/kw		$1000/kw
Fossil	$200/kw		$ 800/kw
Fuel Price Components			
Coal	$1.20/MM btu		$2.50/MM btu
U_3O_8	$20/lb.		$60/lb.
Separative Work	$60/SWU		$100/SWU
Electrical Growth Rate	5.5%/a		5%/a
Nuclear Fraction of Base-Loaded Plants	0.15		0.43 (from 1977 output)
Installed Nuclear Capacity	46 GWe		
OUTPUT			
Nuclear Growth Rate (percent)	10		8.9
Average		9.5	
Installed Nuclear Capacity (2000)	459	404	
Installed Base Load Capacity	1063	1025	
Nuclear Fraction (2000)	0.43	0.40	

The cost assumptions are far too uncertain for us to take the exact values of the growth coefficients literally. Nevertheless, the average of the values predicts LWR capacity in the year 2000 will be 400 GWe. This is unlikely to be far wrong and gives the insight that the present low projected growth of LWRs may reflect economic as well as sociological and political forces.

This exercise shows that the Peterka model is consistent with observed economic trends and suggests that it has some value as a predictive device.

Now let us return to Table 13-1 and examine some of the properties of the theory. Looking at equation (13.4), at any time t all the quantities with subscripts are known. $d \ ln \ P(t)/dt$, the overall market growth rate, may be estimated from long-term trends, and the price p is then determined. We then go back to equation (13.3) and find the growth rates of the individual components. By iteration one can advance through time. Usually our knowledge of the coefficients as a function of time is not precise enough to warrant such a procedure.

Under the Peterka formulation, a rapid growth of a new technology requires a large economic advantage—large compared to the unit capital cost required.

Let us first take the case of equal capital costs for the new and the old system (Equation 13.6). If the product cost difference $c_1 - c_2$ is small compared to the capital cost α, the rate of substitution will be small. To get a feel for this expression, in the year 2000, with yellow cake prices of \$60/lb and separative work at \$100/SWU in 1978\$, LWR fuel burnup costs (including inventory charges) contribute \$35 (5.3 mills/kWh) to the annual operating cost. If we were to introduce a new system, costing in 2000 \$700/kW, which completely eliminates the burnup costs, and if none of the other costs were increased as a result, $\Delta c/\alpha$ would be 5 percent. The ratio f_2/f_1 would increase by a factor e in twenty years. If the initial penetration were 2 percent, it would be 5 percent in twenty years. Only for very large yellow cake prices would the penetration be rapid.

If the new system has a larger capital cost than the LWR, the second term of equation (13.5) is negative, and the market penetration rate is slowed down further. This suggests that advanced converters of a new type are unlikely to become much of a factor in the market or to affect substantial savings in yellow cake.

Now let us look at the technological change in LWR operation from a once through fuel cycle to plutonium recycle. Since there is no experience, we do not really know the economics of Pu recycle. Let us use the official values established for the NASAP (nonproliferation alternative systems assessment program). This shows Pu recycle to be marginally economic for fuel presently being burned and to have a \$5.40/yr/kW advantage for fuel at \$60/lb yellow cake and \$10.20 for fuel at \$100/lb.

If a 1,500 ton/yr reprocessing plant costs \$2 billion, and services 50 GWe, the added capital cost is $40/kWe. Then the relative rates of introduction are, for yellow cake at $60/lb, 13.5%/yr; and for yellow cake at $100/lb, 25.6%/yr. These high rates show that reducing U requirements by reprocessing is economically far more feasible than introducing advanced converters.

The last application of the Peterka theory that we will discuss here is the introduction of the liquid metal fast breeder into the world light water reactor economy. Table 13-3 gives the properties of contemporary fast breeder designs using mixed oxide fuel. The fuel fabrication and reprocessing cost assumptions have been chosen to be favorable to the LMFBR. Further assumptions are that the world nuclear energy generation will increase in the thirty-five-year period beginning in 1995 at an average rate of 4.9 percent/year and that 10 GWe of fast reactors will be operating by 1995.

The highest LMFBR penetration shown on the Table 13-4 is 40 percent. This is the case with the lowest breeder capital costs and the highest yellow cake prices. As capital costs increase—for example the 1000/1000 case—LMFBR penetration is calculated to be only 13 percent. If the LMFBR has a capital cost disadvantage to the LWR, its penetration is further slowed. In the 1000/1200 case, which might be characterized as likely for LWRs and optimistic for LMFBRs, even astronomical uranium prices are not sufficient to produce significant LMFBR penetration in thirty-five years.

This scenario in no way resembles that of WASH 1098, which we saw earlier. WASH 1098 envisioned a rapid substitution of LMFBRs for LWRs. In that scenario, LMFBR growth was ultimately fueled by its own production of plutonium. This is the "plutonium economy" that has so frightened political scientists.

If we follow the Peterka formulation, we would conclude that the LMFBR will not achieve substantial market penetration by the year 2030 under the influence of purely economic forces unless (1) its capital cost is no higher than that of a LWR; and (2) the price of yellow cake, in 1978$, rapidly reaches

Table 13-3. Properties of LMFBR and Economic Ground Rules.

Burnup	75,000 MWD/ton
Thermal efficiency (net)	36 percent
Capacity factor	0.75
Capital charges	12.5 percent
Capital charges (nondepreciating assets)	10 percent
Average fissile inventory	5 g Pu/kWe
Excess fuel produced per year	0.210g U-233/kWe
Value of Pu	0.8 X value of contained U-235 @ 3 percent
Value of U-233	1.1 X value of contained U-235 @ 3 percent

Fabrication and reprocessing of LMFBR fuel set to make LMFBR fuel cycle cost = BWR recycle fuel cost when separation work is $60/SWU and yellow cake = $20/lb.

Table 13-4. Economic Penetration of LMFBRs by 2030 (world).

Initial condition (1995)
900 GWe total nuclear
10 GWe LMFBR

Final condition (2030)
5000 GWe total nuclear

GWe LMFBR installed in 2030

α_1/α_2	*Y = 100*	*200*	*300*
700/700	205	745	2100
700/850	55	180	600
1000/1000	135	350	850
1000/1200	40	100	230
1000/1500	–	–	50

Key: All numbers in 1978$.
α_1 = capital cost of BWRs $/kW.
α_2 = capital cost of LMFBRs $/kW.
Y = price of yellow cake 1978$/lb.
Separative work cost = $100/SWU.

$300/lb. The same conclusion would apply *a fortiori* for advanced converters.

The most likely scenario for the evolution of nuclear fission energy systems under market forces during the first three decades of the next century appears to be:

1. Thermal reactors closely related to present day LWRs, fueled with enriched uranium, will be the predominent reactor type.
2. Plutonium from LWR reactors will be in ample supply to start new LMFBRs. They will not have to breed their own plutonium.
3. Since U-233 is a superior fuel to plutonium in thermal reactors, LMFBRs would produce U-233 to reduce the dependence of LWRs on natural U-235.
4. Chemical reprocessing of both uranium and thorium fuels will serve as the link between these reactor types. It will be practiced on a large scale before the end of this century.
5. Eventually—perhaps in the period 2030 to 2050—enough LMFBRs could be built to effectuate the transition to a renewable resource economy.

CONCLUSIONS

There are two ways government decisionmakers could foster the growth of LMFBRs, to reduce dependence on uranium mining and to hasten the advent of a renewable resource economy.

The first is to subsidize early creation of an LMFBR industry. If 20 GWe of LMFBRs were operating in 1995 instead of 10 GWe, the 1995 industrial base would be twice as large, and successive annual increments would be doubled. Dr. Fisher showed how massive wartime subsidy of synthetic rubber, while not affecting its ultimate market penetration rate against natural rubber, permanently advanced the time scale of its introduction.

Second, a strong effort could be made to reduce the cost of LMFBRs. One way of improving their economics might be to build them larger. This may well be within the range of possibilities, since LMFBRs do not have to contend with size limitations on pressure vessels, and early LMFBR designs are probably not at the specific power limit. If we assume economy of scale to be inversely as the one-third power of the rating, as has been observed for LWRs, a 20 percent cost disadvantage could be overcome by building LMFBRs 1.7 times larger than LWRs.

In this chapter I have tried to apply in a systematic way to nuclear power development relationships derived from historical experience in technological change. I have pointed out some of the conceptual difficulties of the theory, and it is far from my thinking to take the results literally. Nevertheless, the Peterka formulation must be at least qualitatively correct in its identification of capital cost as an inertia that resists technological change. Consideration of this factor drastically influences models of the evolution of nuclear energy systems over the next half century.

NOTES

1. *WASH 1098: Potential Nuclear Power Growth Patterns* (Washington, D.C.: USAEC Division of Reactor Development and Technology, December 1970).

2. V. Peterka, "Macrodynamics of Technological Change," RR 77-22 (Laxenburg, Austria: IIASA, November 1977).

3. J.C. Fisher and R.H. Pry, "A Simple Substitution Model of Technological Change," Report 70-C-215 (General Electric Company R&D Center, June 1970); R.H. Pry, "Forecasting the Diffusion of Technology," Report 73 CRD 20 (General Electric Company, July 1973).

4. C. Marchetti and N. Nakicenovic, "The Dynamics of Energy Systems and the Logistic Substitution Model. Volume I: Phenomenological Part," Administrative Report AR-78-1-B (Laxenburg, Austria: International Institute for Applied Systems Analysis, July 1978); C. Marchetti, "On Strategies and Fate," in W. Haefele et al., Second Status Report on the IIASA Project on Energy Systems 1975, RR-76-1 (Laxenburg, Austria: IIASA, 1976).

5. E. Mansfield, "Technical Change and the Rate of Imitation," *Econometrica* 29, no. 4 (October 1961).

Chapter 14

The Need of the Plutonium-fueled LMFBR

H.H. Hennies, P. Jansen
and
G. Kessler
*Nuclear Research Centre,
Karlsruhe, West Germany*
(Presented by P. Jansen)

INTRODUCTION

The potential for both military and civilian use of nuclear materials (plutonium, U-235, or U-233) was recognized from the beginning of nuclear development. Technical and institutional solutions to prevent the diversion of nuclear material and the proliferation of know-how for military application in other than nuclear weapons states have been proposed from 1944 on. The nonproliferation treaty and the IAEA safeguards and surveillance concepts are important steps toward these aims. Doubts and distrust in these already available institutional concepts have lead again to technical and political proposals during the past two years that culminate in the thesis that any civilian use of plutonium and the deployment of the fast breeder should be abandoned. Alternate, so-called denatured, fuel cycles are being proposed, and the claim is being made that these denatured fuel cycles would have a higher proliferation resistance.[1] We will discuss those arguments, starting especially from the position of West Germany, and will then generalize our conclusions.

THE FUTURE ENERGY DEMAND IN WEST GERMANY

In 1975 the GNP per capita ratio was nearly the same in the United States and in West Germany, but the energy-GNP ratio was two and a half times higher in the United States. This may be a convincing argument that it is more difficult for West Germany than for the United States to improve future energy consumption. Before the oil crisis, the coupling between growth ratios of energy and GNP has been about 1. An attempt to decrease this for the next fifty years to 0.70 calls for a tremendous effort in the case of West Germany.

It is extremely difficult to forecast the GNP growth over a long time period. Based upon present experience we assume that the GNP growth slows down continuously from about 3 percent per year now to about 1 percent per year fifty years from now. Both assumptions lead to an energy demand for West Germany that, compared to 1970, doubles until the year 2000 with 1.3 percent growth per year in that year and triples about seventy-five years from now. We cannot envisage more restrictive scenarios without assuming uncontrolled events for the future (Figure 14–1).

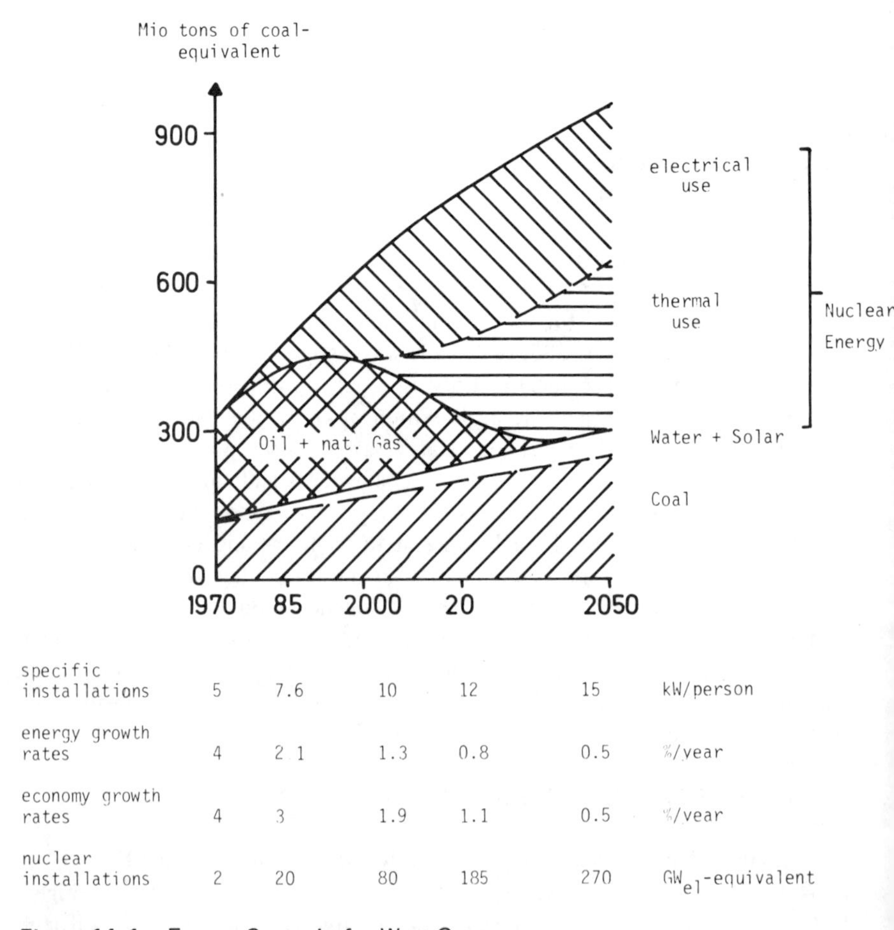

	1970	85	2000	20	2050	
specific installations	5	7.6	10	12	15	kW/person
energy growth rates	4	2.1	1.3	0.8	0.5	%/year
economy growth rates	4	3	1.9	1.1	0.5	%/year
nuclear installations	2	20	80	185	270	GW_{el}-equivalent

Figure 14–1. Energy Scenario for West Germany.

It would be an additional effort for West Germany to hold a coal and solar share in the energy supply of about one-third. We will assume this, despite the fact that West Germany intends to keep hard coal and lignite production on a constant level for the future and that the solar share cannot be more than 6 percent of the total primary energy needs in the year 2050, mainly for home-heating purposes in about 50 percent of the residences. We further assume that in the long run, nuclear energy will be able to supply a second one-third, mainly producing electricity. Under these assumptions our oil and natural gas demand in West Germany—both materials to be imported—will double until the year 2000, and this demand will never slow down. The import of ninety million barrels of oil equivalent per year, however, might be a serious misconception, especially after the year 2020. This leads us to the option to introduce nuclear energy in the nonelectric energy market, beginning in the year 2000. This can in principle be achieved by any reactor type, as mainly low temperature needs must be met. For coal gasification various technical options exist and hydrogen can be produced by electrolysis.[2]

If we define that West Germany will have to decrease the import of oil and natural gas to essentially zero at the latest in seventy-five years, our projections for nuclear energy in West Germany lead to 20 GWe for 1985, 80 GWe for 2000, and 270 GWe-equivalent for 2050. We understand these figures not as forecasts, rather they represent a scenario that is worth considering for a research and development policy of new energy systems. This scenario does not compare to the U.S. situation, as in the United States there is more room for energy-saving measures, coal is available easily and cheaply, and there are domestic oil and natural gas resources.[3]

THE URANIUM DEMAND IN WEST GERMANY

Now, if we try to meet our nuclear energy projection by different reactor strategies, the next problem is the availability of natural uranium, of which we have no domestic resources, either. What can West Germany expect to purchase on the world uranium market? The most recent estimates amount to 5 million metric tons of U_3O_8 within the category assured and estimated up to 50 $/lb U_3O_8.[4] Several million tons additional may be available for higher costs. Including the low grade byproduct uranium in phosphate mining, the total world uranium supply may reach about 18 million tons.[5] As this figure already includes what is called "speculative resources" down to poor grades of 100 ppm, any higher estimates of recoverable uranium resources seem to be unrealistic. In 1976 West Germany shared 6.4 percent of world GNP, 4 percent of world energy consumption, and 1.5 percent of world population. Can we expect 1.5 percent of 5 million tons U_3O_8, or 6.4 percent of 18 million tons?—that is, 75,000 tons or 1.1 million tons?

A responsible policy would not count on the upper limit, but will try to avoid the lower limit. The median may be a good target for West Germany to aim at—that is, about 500,000 tons. What does our scenario now lead to? Table 14-1 gives a summary of our results. In addition Figure 14-2 and Table 14-2 show some details of the basic reactor concepts.

It is easy to realize, that none of the advanced denatured thorium-uranium fuel cycles (2b,c) will meet the conditions for West Germany. Even the introduction of the fast breeders can hardly satisfy our target of 500,000 t of U_{nat}. This is the reason why West Germany is striving to introduce the fast breeder as early as possible. Ten years delay in the commercial introduction will cost about 16 percent in additional uranium needs. Only the use of fast breeders will significantly lower uranium needs down to our supply target. Only with fast breeders can West Germany gain full independence from U supplies.

Table 14-1. Uranium Consumption[a] for West Germany up to the Year 2050 for Different Reactor Strategies.

	Cumulated over the next 75 years in Mio to U_3O_8	*Yearly U_3O_8 consumption at a plateau of 270 GWe in 10^3 to U_3O_8*
1. Open fuel cycles		
a. Light Water Reactors	1.7	40
b. Heavy Water Reactors, slightly enriched, start 1990	1.3	27
c. High Temperature Reactors, mixed Thorium Uranium-fueled (denatured, MEU), start 2000	1.3	27
2. Closed denatured fuel cycles		
a. Light Water Reactors	1.3	30
b. Heavy Water Reactors with U-233 (MEU), start 2005–2010	1.0	11
c. High Temperature Reactors with U-233 (MEU), start 2005–2010	1.1	17
3. Plutonium fuel cycles		
a. Light Water Reactors with Pu recycling, start 1995	1.0	20
b. Fast Breeders, oxide fuel, start 2000	0.7	0
c. Fast Breeders, carbide fuel, start 2005–2010	0.5	0
d. Fast Breeders, oxide fuel, starting with U-235 in 2000	0.5	0

[a]With respect to the scarcity of the uranium resources, 0.1 percent tails assay is assumed.

a) Light Water Reactor - Once Throuah

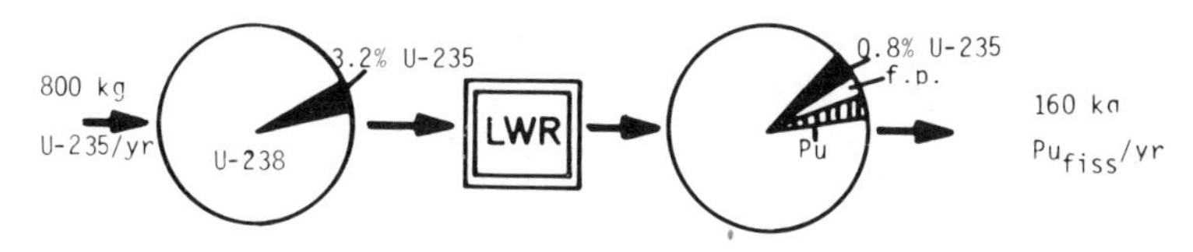

b) Heavy Water Reactor - denatured closed fuel cycle+)

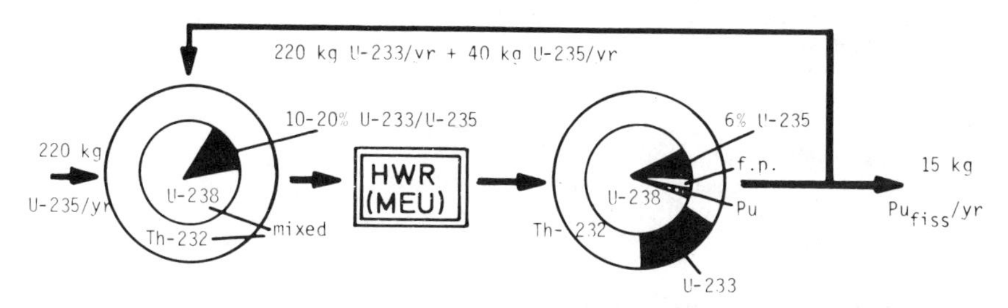

c) High Temperatur Reactor - closed HEU fuel cycle

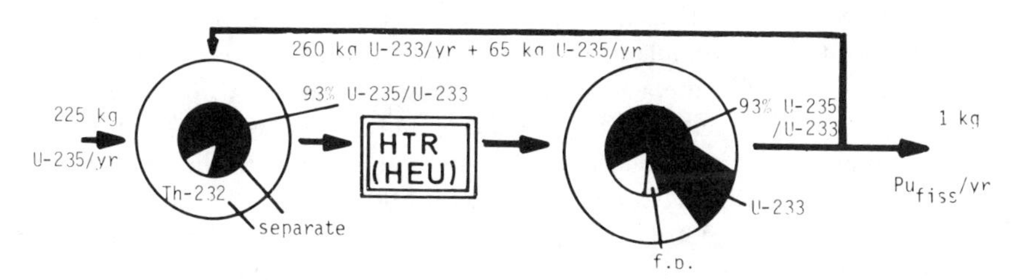

d) Fast Breeder Reactor - closed Pu fuel cycle

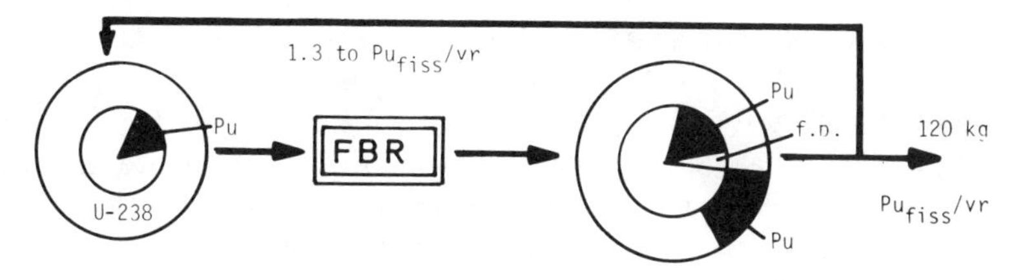

+) The corresponding HTR has similar properties

f.p. = fission products

Figure 14-2. Principal Scheme of Different Reactor Fuel Cycles.

Table 14-2. Sensitive Materials Inventory and Massflow, Uranium Requirement (for a thirty year lifetime of 1 GWe at 75 percent load).

	Uranium Enrichment	*Natural*	*LEU*	*MEU*	*HEU*
1.	Open Fuel Cycles	HWR	LWR/HWR	HTR/HWR	HTR
	Pu_{fiss} in system (kg)	9000	4500	300	30
	MEU/HEU in system (kg)	–	–	500	3500
	MEU/HEU handled (kg)	–	–	15000	15000
	U_3O_8 demand (to)	5000	5000/3000	3000	3000
2a.	Closed Fuel Cycles				
	Pu_{fiss} in system (kg)		1300	300	30
	Pu_{fiss} handled (kg)		7000	300	30
	MEU/HEU in system (kg)		–	1500	2000
	MEU/HEU handled (kg)		–	15000	17000
	U_3O_8 demand (to)		3000	2000	2000
2b.	Fast Breeders				
	Pu_{fiss} in system (kg)	6000			
	Pu_{fiss} handled (kg)	40000			
	Waste uranium consumption (to)	45			
	U_3O_8 when starting with U-235 (to)	1500			

THE WORLD'S URANIUM DEMAND

We will now broaden the analysis to the world situation. As the supply-demand ratio was not very favorable in the case of West Germany, we can assume that it will be similar for the whole world. There are nations with less need for nuclear energy (perhaps the United States?) and nations with even more (like France, Italy, or Japan not having even coal). So we refer to the lower projections made by OECD in February 1978 with 800 GWe for OECD countries in the year 2000 and 1,000 GWe for the world.[6] The world needs are projected for the year 2025 to 2,200 GWe and we extrapolate 3,200 GWe for the time seventy-five years ahead and 4,000 GWe in one hundred years. Here, with fast breeders (Oxide, Pu-fueled, start 2000), we come up with 8 million tons within the next seventy-five years, nothing to be added thereafter. In the best of the alternatives, a heavy water reactor with a closed MEU fuel cycle strategy will need 11 million tons over the next seventy-five years and 15 million tons over the next one hundred years, still growing further. Again we see that in relation to the uncertainties of uranium resources (and having used the lower OECD projection!), the fast breeder is a rational nuclear energy strategy—that is, plutonium is a necessity.

Let us conclude this part of the chapter with a reconsideration of the possible West German share of the world's uranium resources. Roughly speaking the OECD world projection for the use of nuclear energy is twelve times that of West Germany. As long as we follow a strategy based on perpetual uranium needs, it is highly probable that countries without their own resources might not get their equivalent share because nations with uranium resources intend to conserve these for as long as possible. So it would be absolutely unrealistic for West Germany to assume that over 1 million tons would be available—as would be needed if we apply an advanced heavy water strategy as proposed by von Hippel et al.[7] On the other hand, if we follow a strategy worldwide that promises an end of uranium needs, nobody might care about exhausting the resources even if it serves other nations. It is the fast breeder and only the fast breeder that can guarantee this quasi-end of uranium needs. Furthermore, even some additional growth of world's nuclear power supply would be possible with fast breeders. Thus, our analysis leads to a full contradiction to the thesis forwarded in the paper of von Hippel et al.[8] that there would be no need for the breeder.

PROLIFERATION ASPECTS

In the second part of this chapter we analyze the thesis that the use of plutonium in civil nuclear power plants would provide too large a risk for the proliferation of nuclear weapons. In this context it is also claimed that more proliferation resistant reactor systems and fuel cycles do exist and should be applied in the future.[9] The term "latent proliferation" has been defined in addition to vertical and lateral proliferation as a case in which a nonweapon state could divert sufficient amounts of nuclear material (plutonium, U-235, or U-233) from its civilian nuclear power program and apply it in already prefabricated and pretested weapons devices.[10]

In comparing the proposed "more proliferation resistent fuel cycles" with the closed U-Pu fuel cycle of the LWR-FBR strategy, we must discuss three questions:

- What are the differences between the fissile materials (plutonium, U-235, U-233) within the different fuel cycles with regard to their potential use as weapons grade material?
- Which amount of sensitive fissile material is involved within these different fuel cycles?
- What degree of difficulty must be overcome by a country to modify this sensitive fissile material into weapons grade material?

Fissile Materials for Weapons Use

The different reactor systems and fuel cycles discussed in Table 14–1 contain either plutonium in various isotopic mixtures or U-235 and U-233. The so-called sensitive enrichment of both fissile plutonium mixed with other Pu isotopes and U-233 mixed with U-238 is about 12 percent; the one for U-235 is above 20 percent.[11] Above this so-called sensitive enrichment, these fissile materials could be used for crude weapons. For a greater than 93 percent enrichment, the critical mass in spherical geometry with a proper reflector is about 5 kg for plutonium and 10 percent higher for U-233. The critical mass for U-235 is 15 kg.[12] The impurity of several hundred ppm U-232 with its 2.5 MeV γ-radiation in reactor grade U-233 and the radiotoxicity of plutonium accompanied by α-, β-, and γ-radiation as well as the spontaneous neutrons of Pu-238, Pu-240, and Pu-242 are not considered to pose serious technical or handling problems to a country that has decided to use those materials for military application.[13] It remains, however, a question whether such a country would use its civilian nuclear program for atomic weapons material production. Whereas for U-235 and U-233 the gun type and implosion technique can be applied, only the implosion technique is feasible for plutonium weapons. Although spontaneous neutrons of the Pu isotopes Pu-238, Pu-240, and Pu-242 in reactor grade plutonium could cause preignition and therefore decrease considerably the yield and reliability of such weapons, this preignition effect is not considered a large problem. The application of advanced implosion techniques could still result in weapons in the KT range.[14]

In summary, we conclude that despite some differences, all three essential fissile materials—U-233, U-235, and plutonium—are about equally qualified in HEU form as weapons materials. A country that has decided to fabricate a nuclear weapon would use any one of these materials, depending upon the infrastructure of its available technology. This does not necessarily require a civilian nuclear program as a prerequisite. Terrorist or subnational groups, however, would probably have most difficulties with reactor grade plutonium.

Amount of Sensitive Materials in the Fuel Cycle

The fresh fuel of LWRs or HWRs would certainly require the greatest effort to enrich it to weapons grade material. According to our assumption that a country that has decided to produce atomic weapons would use the best and cheapest available technical nuclear infrastructure, we define plutonium containing fuel and 12 percent enriched U-233 and 20 percent enriched U-235 fuels as equally sensitive, because the U-233 or U-235 MEU fuel could be enriched with moderate efforts—that is, by centrifuge techniques or potential future enrichment techniques. We will explain the basis for these definitions in the following section. The amounts of fissile material within the different open or closed fuel cycles and the amounts that have to be reprocessed and refabricated per year are shown in Table 14-2 and Figure 14-2. There is the following clear tendency:

- The U consumption can be decreased only by either higher enrichment (20 percent MEU) or the application of a closed fuel cycle, which means reprocessing and refabrication;
- The amounts of sensitive materials involved within the fuel cycle are increasing with higher fuel utilization—that is, decreasing U consumption;

- The FBRs working in the U/Pu cycle show an amount of sensitive material that is only three times as high as in a LWR cycle but sustain the capability of energy independence for the long range;
- The open fuel cycles would steadily accumulate plutonium and/or MEU fuel. For the proposed strategy of an open MEU fuel cycle, up to 800 tons of U-233 and 100 tons of Pu would be discharged from these denatured reactors up to the year 2050 in West Germany. This is one reason why West Germany follows the concept of reprocessing the irradiated fuel. The second reason is the uranium saving by using the reprocessed plutonium in fast breeders. The Gorleben reprocessing plant is intended to be built in collocation with the Pu refabrication and the waste disposal plant above a salt dome.[15]

Figure 14-3 adds to these results, showing the amount of sensitive material (MEU, Pu) for, denatured heavy water reactors with a closed MEU U-Th fuel cycle added between the years 2000 and 2005 to a LWR stock on a once through fuel-cycle; fast breeders with a closed U-Pu fuel cycle starting at the year 2000; and LWRs only, but with Pu recycling. There is no significant difference in the stock of sensitive material in strategies A and B.

Degree of Difficulty to Modify Reactor Grade Materials

It remains now to classify the degree of technical and handling difficulties to modify the different reactor grade materials into weapons grade materials. The HEU fuel of presently built HTGRs would certainly present the least difficulties for a modification into weapons grade material (Class A). Nonirradiated plutonium MOX fuel for LWR recycle or FBRs together with the nonirradiated denatured MEU fuel would represent the next category (Class B). The promoters of the MEU denatured fuel cycle have claimed a higher proliferation resistance of this fuel, obviously because small-scale reprocessing plants are presently more widespread in the world than enrichment facilities. The technical difficulties to be overcome by a country that has decided to produce weapons grade materials, however, appear to be roughly the same in both cases, as was clearly stated by K. Cohen.[16] The centrifuge or any future enrichment technqiue does not pose any higher technical difficulties than reprocessing techniques.[17] This will especially become true for the time after 1990–2000 when FBRs or other advanced reactor systems are introduced. At least enriched MEU fuel might not pose more additional difficulties as, on the other hand, are posed by reactor plutonium to adapt it for weapons.

The third category (C) would be formed by irradiated MOX U/Pu fuel or irradiated MEU-U/Th:Pu fuel. Again we do not make a distinction between reprocessing and enrichment techniques in this case. For a country that has decided to produce weapons grade material, the difference betwen categories C and B (presence or absence of fission products) will not be relevant.

This leads us to the last point of interest, the question of whether to have a

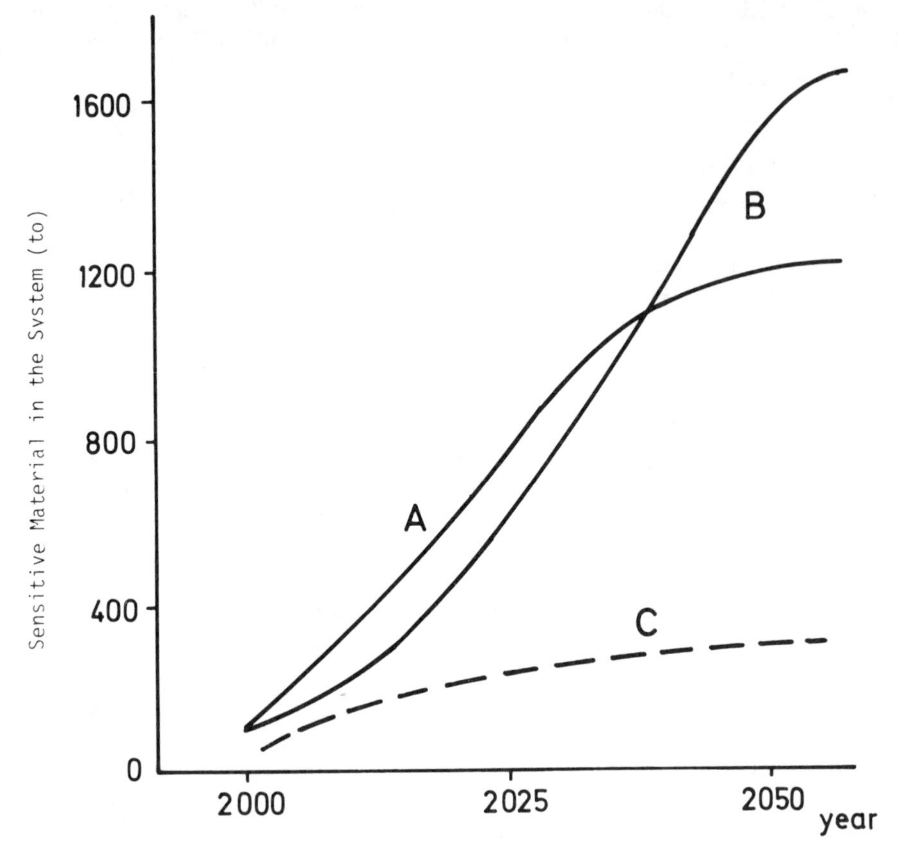

Key:

A. HWRs on a closed denatured MEU U-Th fuel cycle.
B. FBRs on a closed U-Pu fuel cycle.
C. LWRs with Pu recycling.

Figure 14-3. Sensitive Material in Power Reactors and Fuel Cycle for Different Strategies.

national fuel cycle or not to have it, provided that the nation has a nuclear power program, makes any significant difference with respect to proliferation risks. Two facts provide an answer to this question. First it can be stated that the large amount of fissile materials to be handled, as indicated in Table 14-2, does not legitimately distinguish between different kinds of handling categories as far as the possibility of diversion is concerned. We consider the transportation of fissile material as equally proliferation-relevant as chemical handling. Second, we are aware that for a nation that has decided to produce nuclear weapons, the effort to do so would be of such dimensions that there will not be a significant difference whether it already has civilian reprocessing know-how or not. In combining both arguments we do not see any difference in impact on proliferation risks whether a nation only runs nuclear power stations with the fuel cycle internationalized or has a complete fuel cycle of its own.

We conclude from this analysis that we cannot find any essential difference for any of the cases considered in Tables 14–1 and 14–2 and Figure 14–2. Especially, we cannot see a higher proliferation resistance of the MEU open and closed fuel cycles if compared to the LWR-FBR closed fuel cycle. To summarize:

- Pu, U-233, and U-235 are equally weapon-usable;
- The argument that the chemical treatment to get Pu would be easier to be accomplished than to enrich U233/U235 MEU fuel is not true for today's and future technologies;
- The presence of fission products does not make chemical processing on a national basis significantly more difficult;
- As shown in Table 14–2, the amounts of fissile material to be handled (e.g., transported) are large enough to cover any diversion for weapons use, if a nation decides to do so;
- The effort to build nuclear weapons is such that whether a nation has the know-how for chemical reprocessing of nuclear material or not is irrelevant.

This leads to a full contradiction to the second thesis, forwarded in the paper of von Hippel et al.,[18] that the fast breeder with Pu fuel is not acceptable from a proliferation point of view. Rather, looking at this question realistically, we see no difference in proliferation risks, between once through fuel cycles, denatured MEU fuel cycles with internationalized fuel processing, or fast breeders with Pu fuel and the corresponding fuel cycle installed nationally. The hope for a unique technical solution of the proliferation question and the strong connection established between the peaceful civil nuclear fuel cycle and nuclear weapons capability appears to be overestimated. At the same time, the basically political nature of the proliferation problem seems to be underestimated.[19]

ALTERNATIVE FUEL CYCLES AND TECHNICAL REALIZATION

Closing the fuel cycle and developing the breeder option now, without any further delay, appears mandatory for us in view of the limited and very unequally distributed uranium resources and in view of the extremely long time that will appear on the way toward full commercialization of new reactor systems. Figure 14–4 shows an assessment of the technical development status and the technical difficulties still to be overcome for a realization of the alternative

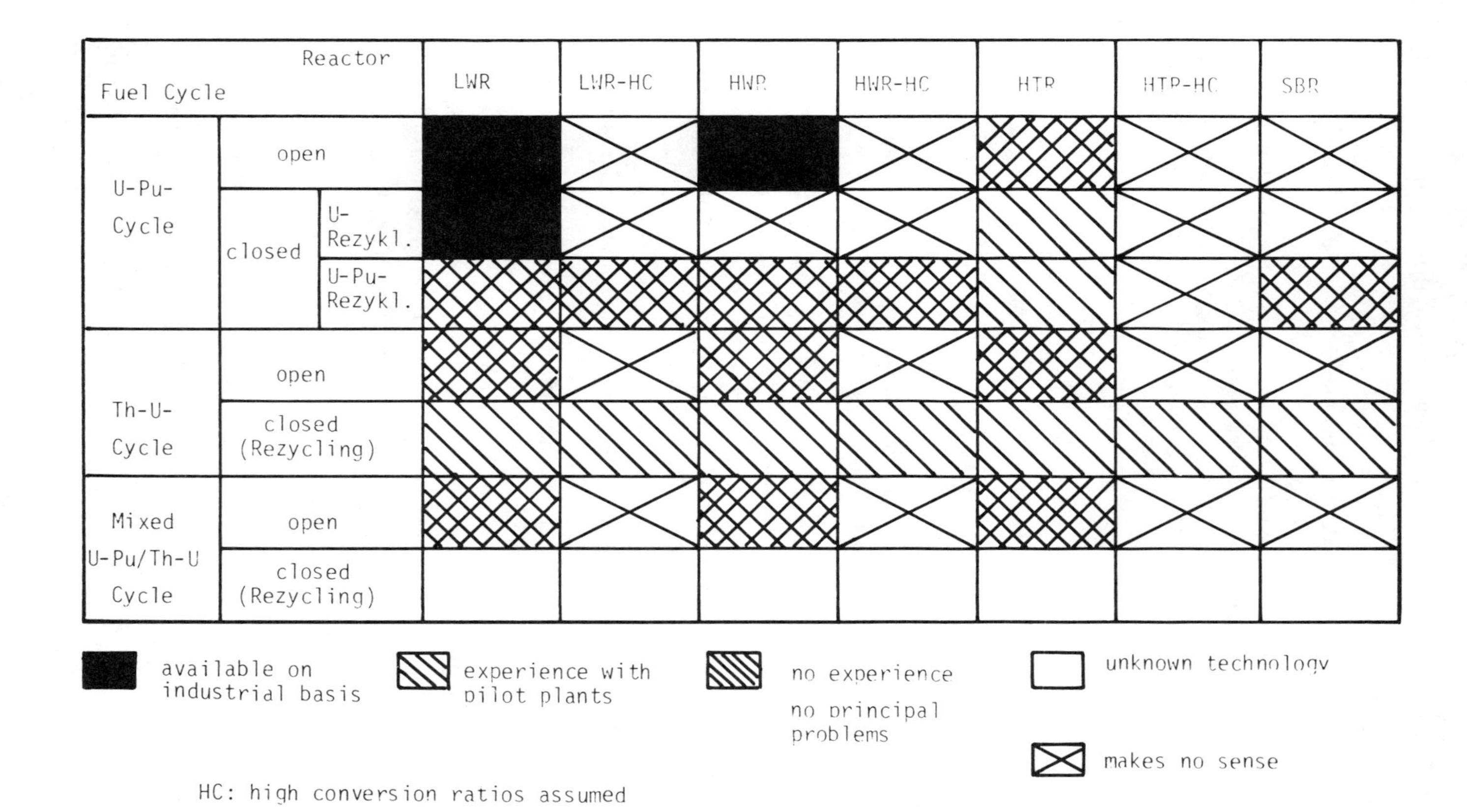

Figure 14-4. Technical Development Status of Various Fuel Cycles.

fuel cycles discussed in Tables 4-1 and 4-2. We have chosen the following classification:

- Technology available on an industrial basis;
- Experience with pilot plants, no essential technical problems expected;
- No experience, but no principal difficulties expected; and
- No experience, broad development programs required since partly unknown technology.

In discussing the closed fuel cycles, one can state that the U/Pu cycle is clearly the most advanced, whereas the Th/U cycle still needs practical experience. The closed mixed fuel cycles for the MEU reactor systems with denatured U-233, U-235, and U-238 fuel are not developed at all.

In view of the requirement that West Germany needs to close the fuel cycle as early as possible in order to gain uranium and energy independence, one would certainly not select the mixed U/Th/Pu cycle of the MEU reactors. Instead we should continue to develop the U/Pu fuel cycle for an LWR-FBR strategy. In addition development work for the Th/U-233 fuel cycle, mainly for special use of nuclear power in nonelectrical, high temperature demand areas, will go on,[20] though for this application the U/Pu fuel cycle could equally be applied.

Finally, to be complete, we should mention the possibility of alternate FBR fuel cycles. Several variants of FBRs with thorium in the radial or in both the radial and axial blankets, even in the core, have been proposed in combination with U-233-fueled LWRs, HWRs, or FBRs.[21] The most interesting of those combinations would be FBRs with thorium in the radial and axial blankets in combination with HWRs on a denatured MEU closed fuel cycle. The FBRs starting at 2000 initially would be supported by Pu-producing LWRs. The FBR then could provide U-233 for the HWRs starting later. In the steady state (around 2050–2075), this FBR-HWR strategy would consist of one-third FBRs and two-thirds HWRs. The uranium consumption of this strategy would be about the same as in the LWR-FBR reference case. However, we do not feel that this strategy would bring any advantages, as:

- The Pu fuel cycle is necessary anyway, and sensitive U-233 would be generated in the FBR blanket (Pu- and U-233 economy);
- There is no advantage in uranium needs;
- The MEU-HWR fuel cycle would not improve the proliferation resistance;
- The institutional arangements for out of nation breeder parks would in our opinion be unfeasible and would not bring advantages with respect to the proliferation risk; and
- The technical and financial measures to provide for this technology and the development and construction of two different reprocessing lines would be immense.

SUMMARY

West Germany does not have its own significant oil, natural gas, or uranium resources. On the long term, we therefore must aim at balancing the depleting oil and gas resources by the application of nuclear energy. Since West Germany must purchase uranium on the free world market, it must aim at the development of reactor systems with low U consumption, which means high converters and the FBR. Both reactor systems require a closed fuel cycle.

Only the introduction of the FBR can guarantee the smallest total uranium requirement of about half a million tons and uranium independence from the year 2050 onwards. The different proposed alternative fuel cycles for reactors with MEU denatured fuel do not have a significantly higher proliferation resistence than the U/Pu fuel cycle, but they would need significantly more uranium than FBRs. Proliferation resistant fuel cycles or reactor systems do not exist. Various smaller improvements on the technical side, being discussed presently in the International Fuel Cycle Evaluation (INFCE) program, can be expected, however. They must be combined with political and institutional solutions on the basis of the NPT and the IAEA safeguards concept.

NOTES

1. H.A. Feiveson; F. von Hippel; and R.H. Williams, *An Evolutionary Strategy for Nuclear Power (Alternatives to the Breeder)* (Princeton, New Jersey: Princeton University, 1978); H.A. Feiveson and T.B. Taylor, "Security Implications of Alternative Fission Futures," *The Bulletin of the Atomic Scientists* 31, no. 10 (December 1976).
2. W. Häfele, "World Energy Resources, Demand and Supply of Energy, and the Prospects for the Fast Breeder," IAEA-SM-225/71 (Report delivered at Symposium Bologna, April 1978).
3. Energy Policy Project of the Ford Foundation, *A Time to Choose—America's Energy Future* (Cambridge, Massachusetts: Ballinger, 1974).
4. Feiveson and Taylor.
5. *Uranium-Resources, Production and Demand* (OECD-NEA and IAEO, December 1977); Die künftige Entwicklung der Energienachfrage und deren Deckung. Bundesanstalt für Geowissenschaften und Rohstoffe, März 1976.
6. *Nuclear Fuel Cycle Requirements and Supply Consideration, Through the Long-Term* (OECD-NEA, February 1978).
7. Feiveson, von Hippel, and Williams.
8. Ibid.
9. Ibid.
10. Feiveson and Taylor.
11. "Primer on the Fuel Cycle," *Review of Modern Physics* 50, no. 1, pt. II (January 1978).
12. Ibid.; M. Willrich, Th. R. Taylor, *Nuclear Theft, Risks and Safeguards* (Cambridge, Massachusetts: Ballinger, 1974); *Nuclear Proliferation and Safe-*

guards (Washington, D.C.: Congress of the US, Office of Technology Assessment, 1977).

13. *Nuclear Proliferation and Safeguards.*

14. Ibid.

15. K. Beckurts, "Nuclear Nonproliferation—A View from Germany" (Report delivered at AIF Conference, New York, October 1978).

16. K. Cohen, "The Science Fiction of Reprocessing and Proliferation" (Report delivered at Fuel Cycle Conference, Kansas City, 1977).

17. C. Starr, "Nuclear Power and Weapons Proliferation—The Thin Link," *Nuclear News,* June 1977, p. 54.

18. Fieveson, von Hippel, and Williams.

19. Beckurts; Starr.

20. Beckurts.

21. M. Schikorr, "Reaktorphysikalische und reaktorstrategische Untersuchungen zur Spaltstoffökonomie des Thorium—und Uranzyklus in Schnellen Brutreaktoren und Hochtemperaturrekatoren" (Dissertation, University Karlsruhe, FRG, 1977).

Chapter 15

On the Thorium Economy

Edward Teller
Stanford University and Lawrence Livermore Laboratory

Briefly, I would like to bring to your attention a few technical changes and a few technical situations that I believe have not received very wide recognition yet. For many reasons (one of them is my incapability of doing so), I will avoid going into details. One of the interesting developing fields is the prospective decrease in the cost of separating isotopes. By means of keeping this subject strictly secret, we managed to hold up progress for several decades. In more recent years this secrecy has been loosened, though unfortunately not abandoned. The figures given for the cost of separative work units are apt to come down by a factor 2 and quite possibly by a factor 10.

This has all kinds of consequences. One is the exploitation of a greater fraction of the U-235 that one gets out of yellow cake. Another consequence is a more complete disproof of the futility of the attempt to forbid reprocessing. Today the South Africans do not need any reprocessing or any reactors. They have reinvented and improved the separation of uranium isotopes, and they can produce atomic bombs if they wish to do so.

Among the various new methods of isotope separation, laser isotope separation is particularly important. This, I think, promises to become very cheap, and there are at least two strong competitive methods that hold promise. Time and secrecy cooperate in forbidding me to tell you details on how this will be done.

Another isotope separation, which is fortunately not secret, is the production of heavy water. The production of heavy water has made practically no progress since Harold Urey. Heavy water has been very successfully used in the CANDU reactors, the Canadian deuterium reactors, which probably is the most economic way to use uranium U-233 derived from thorium as the original fuel. There are competitors: the process of General Atomic and, in Germany, the work in Ulich. There are high temperature, gas-cooled reactors. Furthermore, experience has shown that it is dangerous to eliminate from the race the teacher of the nuclear engineer in the White House, who is working on an epithermal reactor. All of these reactors have good neutron economy. In this manner, U-233 converters could be established reproducing 90 or 95 percent of the fuel in each cycle.

One way or the other, we are going to establish a cycle that will use thorium, which is at least three times as abundant as uranium. We can use all of it and thereby get an available energy source comparable in quantity and in fact bigger than what would be available to the fast breeder.

I want to bring to your attention an approximate comparison of two figures. One is the cost of the separation of uranium (SWU) important in lighter water reactors. The cost of that contributes something like 5 percent of the cost of electricity. On the other hand, the cost of heavy water until recently contributed 20 percent to the cost of the CANDU reactor. I want to say a word (because of the obvious importance of this subject) about a way in which to separate heavy water. We are working on this subject at Livermore, having been supported by the enormous sum of half a million dollars per year. It is on this thread that a new technology is hanging, and recently it was proposed to cut this thread. Actually there are two main proposals.

One is based on laser isotope separation starting with a freonlike substance containing one hydrogen or deuterium, where with very high laser intensities in the infrared, tuned appropriately, you simply tear out the deuterium together with a fluorine. One can reuse the freonlike substance by contact with water so that H and D are exchanged. Therefore, you are not relying on any expensive substance.

The other approach uses an expensive substance, namely hydrogen. Hydrogen, in turn, is being used for all kinds of purposes in industry; when you use it in the Haber process to make fertilizer, you will cheat the poor farmer in that you deprive the fertilizer of its heavy hydrogen content. I do not think he will be badly cheated in this manner. The method of doing so is very pretty. It is based in principle on the quite different solubility of hydrogen in palladium. But because palladium is expensive, we are beginning to experiment with cheaper "isotopes" of palladium. I am using the words with a slight Hungarian distortion. Let me mention one of these isotopes. It is called lanthanum penta cabaltite. Like palladium it has all kinds of incomplete shells, and it has properties not quite as good as palladium, but its cost is quite negligible compared to that of palladium. Then you use essentially chromatagraphic methods in letting the hydrogen go through a bed of particles of this kind; and you admit on one side the mixture of isotopes, and on the other side you get out pure deuterium. I am slightly oversimplifying the process, but the cost of deuterium, heavy water, is apt to come down in a really dramatic fashion if we pursue the subject.

I am an advocate of converters and an opponent of breeders because I claim that the supporters of the breeder exaggerate. In principle, I agree with them; in practice, I don't. Assume, and this is practically present technology, that I can get a CANDU reactor or the like that runs on U-233 and in a cycle it reproduces 90 percent of the U-233. I have to add 10 percent to make up in U-235, and I do not care, because the SWU is inexpensive and the thorium is very cheap. Therefore I can spend for the uranium, if need be, $400 rather than $40, which is the present price of a pound of yellow cake. We may get a practically unlimited amount of yellow cake as long as you can pay $400 without having changed the price structure of electricity. Thus, the supply will suffice.

My final topic is a little uncertain. There is a new competitor in this race that greatly increases my optimism about the thorium cycle. This is the fusion-fission hybrid. I have consistently prophesied, much to the annoyance of an organization called the Fusion Energy Foundation, that fusion will not be economically significant before the year 2000. In spite of great progress that has been recently achieved, in spite of the headlines that will appear in the next few years—"Fusion Energy Accomplished, Energy Problem Vanishes"—I will say that pure fusion will not be economically viable until at least 2000 and probably not until many years later. Technical problems will persist due, for instance, to the bombardment with high energy neutrons of the inner walls of the fusion chamber. These are more efficient than similar problems had been in normal fission reactors. Technical problems will delay the economic realization. There is, however, a way out, and I claim that this way out is excellent and could bring closer the date at which we may have economic pure fusion.

The argument for the fusion-fission hybrid can be simply summarized. Fusion is weak in energy. It produces relatively little energy, but it is strong in neutrons. In fact, it is a beautiful source of fourteen million volt neutrons. Fission is strong in energy, but there are problems (as we all know and have already discussed) concerning the fuel. This drives some people into statements that we must have a fast breeder—and indeed we must unless we get something else. Let me describe my ideal reactor. We have in Livermore a beautiful machine called the mirror machine. We confine the plasma between two magnetic mirrors that are imperfect but good. To extrapolate our result to the point of saying that a pure fusion machine can be made is uncertain. To extrapolate it to the point of a fusion-fission hybrid is I think quite conservative. The best configuration is as follows: You have a pipe approximately thirty feet long. At the end of this pipe we have two regions in which fast ions are stopped. There is a little question for the experts of the transition region between the straight part of the magnetic confinement and the region where reflection of fast ions occurs. We believe that we have enough data to say that this will not be a serious problem. Now into this machine we inject fast deuterium atoms that will be stripped and stopped and that will react with tritons. We run at an energy deficit.

But we are going to surround this arrangement by several layers of other substances. The first layer is uranium. It can be U-238 or natural uranium; uranium is a beautiful neutron multiplier, and from one fourteen million volt neutron you are going to get as many as half a dozen neutrons of an energy that can no longer be used in U-238. I prefer to stay way below the critical mass, and I prefer furthermore to cool the whole thing, like Peter Fortescue would, with helium. This is a good way to convert the energy and the safest way to operate. Once the neutrons are down to somewhere in the neighborhood of an MeV or 1.5 MeV, then you capture them, having first slowed them down. I am not telling you whether to slow them down in heavy water or in graphite. Then you capture lithium to replace the tritium and thorium, giving in the end U-233. The whole apparatus should be self-supporting in energy and might give a little trickle of extra energy to keep a relatively isolated community alive, but should not be a 1,000 megawatt plant. It should instead be a fuel factory. What goes in is some uranium and a greater amount of thorium; what comes out is U-233 that goes into the makeup of reactors like CANDU or similar reactors and enough tritium to replace what was burned up.

I claim that in this way we will have enough energy at least until the next ice age, at which time we will have to fight the battle with the environmentalists whether we will allow the natural process of the ice age to proceed or not.

Before I conclude, let me tell you that for pure fusion as far as I see it, the best apparatus is not what I described but the famous Tokamak. It is possible to make a hybrid with the help of Tokamak, but the geometry is a little more clumsy. But no matter which way we go I do not want to see an "either-or" between pure fusion and the thorium economy. If the hybrid will prevail, we will learn a great amount of the technology on interior walls and similar difficult technological questions, and therefore, the operation of the hybrid will bring us automatically closer to that day when pure fusion will become possible.

Part II

Fusion and Other Parameters

❋ *Chapter 16*

Laser Fusion Hybrids—Technical, Economic, and Proliferation Considerations

G.A Moses, R.W. Conn
and
S.I. Abdel-Khalik
University of Wisconsin
(Presented by G.A. Moses)

ABSTRACT

A possible role for the fusion-fission hybrid in the context of an immediate nuclear future that may not include fuel reprocessing or the LMFBR has been examined. In such a role, the hybrid is used to irradiate fertile fuel assemblies, thereby simultaneously enriching the fuel to the proper fissile concentration and rendering it proliferation resistant by making the fuel highly radioactive. Should reprocessing of spent LWR fuel be allowed, this hybrid concept can be incorporated into an internationally monitored, physically secure fuel production and reprocessing center that meets nonproliferation guidelines.

In the SOLASE-H study, a laser fusion hybrid is conceptually designed to meet the needs of the proliferation resistant fuel cycle. One hybrid operating at a fusion power of 1200 MW can fuel approximately two and a half 1,000 MWe LWRs requiring 4 percent enriched U-233 fuel. (With reprocessing this hybrid can fuel ten LWRs.) The assemblies can be enriched to 4 percent fissile content in 1.9 years in an optimally designed case. The fuel burnup level in the hybrid itself is equivalent to 4,300 MWD/MT. Flat enrichment profiles across

*This work was supported by the Electric Power Research Institute under Contract #RP 237. The results presented herein represent the efforts of the following people: S.I. Abdel-Khalik, R.W. Conn, G.W. Cooper, J. Howard, G.L. Kulcinski, E.M. Larsen, C.W. Maynard, G.A. Moses, M. Ortman, M.M.H. Ragheb, D.L. Smatlak. I.N. Sviatoslavsky, W.F. Vogelsang, R.D. Watson, W.G. Wolfer, and M. Youssef.

the assembly can be achieved at the expense of the U-233 breeding ratio. A figure of merit that maximizes the U-233 breeding ratio subject to minimizing the peak to average enrichment (the hot spot factor) determines the optimum blanket design.

The substantial fusion power required to produce fissile fuel does not allow the laser fusion pellet gain (and hence laser energy) and pulse repetition frequency to be simultaneously relaxed. Laser efficiency can be substantially relaxed due to the blanket energy multiplication. The hydrogen-fluoride laser, studied for the SOLASE-H system, appears to be scalable to an output energy of 2 MJ, and the calculated net efficiency of 2.6 percent indicated HF is an attractive laser candidate for laser fusion hybrid applications.

INTRODUCTION

A fusion-fission hybrid reactor utilizes the 14.1 MeV DT fusion neutrons for breeding fissile material in the hybrid reactor blanket. This bred fuel can be removed periodically from the blanket and burned in conventional fission reactors or it can be burned "in situ" in the hybrid blanket itself. During the last five years there have been many studies of hybrids for a variety of fusion systems (i.e., Tokamak, mirror, laser fusion, electron beam fusion).[1] Such fusion-fission hybrid reactors appear to be attractive because they produce two revenue sources–electric power and fuel for conventional fission reactors–at a fusion performance level that is less than that required for pure fusion reactors. The additional revenue source, fissile fuel, strengthens the economic perspective of the fusion system. The reduced fusion performance is allowable because the 14.1 MeV DT neutron energy is multiplied in the hybrid blanket by the fission process. For the hybrid operating as a fuel factory where fissions are minimized, the blanket energy multiplication is still typically 2–10. In the second option, where the fuel is allowed to burn in the hybrid itself, the multiplication may be as high as 40–50 depending on how close the blanket approaches criticality. It is argued that this relaxation of the fusion energy requirement may allow hybrid reactors to make an impact on the world's energy production problem at an earlier date than pure fusion reactors. However, the hybrid reactor may also appear unattractive if it is considered to have both the disadvantages of complex fusion systems and the radioactive waste, criticality, and proliferation problems of fission reactors.

The SOLASE-H[2] laser fusion hybrid reactor study investigates the possibility of minimizing the perceived disadvantages of the hybrid by operating with a low k_{eff} in the blanket and utilizing a proliferation resistant fuel cycle that allows direct enrichment of PWR fuel assemblies in the hybrid and transfer of the irradiated assemblies to the fission reactor, without intermediate reprocessing. The study established the potential role of the hybrid for a nuclear future that includes no immediate reprocessing or development of the LMFBR. This study is, in fact, a continuation of the SOLASE[3] conceptual laser fusion reactor design, reported by R.W. Conn at this meeting last year.

In the following sections, there first appears a generic discussion of the proliferation resistant fuel cycle. Any fusion system might be used to produce the fuel. This is followed by a description of the SOLASE-H laser fusion hybrid system. The final section summarizes the conclusions derived from the SOLASE-H study.

NONPROLIFERATION POTENTIAL OF FUSION HYBRID REACTORS

Only 0.7 percent of natural uranium is the fissile U-235 isotope. The remaining 99.3 percent is U-238. Other fissile isotopes can be manufactured by the absorption of a neutron in Th-232 and U-238 to produce U-233 and Pu-239, respectively. Once these artificial fissile materials have been produced, they can be mixed with their corresponding fertile material at a 3 to 4 percent concentration and fabricated into fuel assemblies for use in fission reactors. The production of these artificial fissile isotopes is, of course, the purpose of the fusion hybrid reactor. However, the reprocessing of the material produced in the hybrid to remove fission products and the fabrication into cold, clean fuel assemblies exposes the hybrid fuel cycle to the same proliferation considerations as the fast breeder fuel cycle. Because this fuel is easily handled and the fissile material can be removed by chemical rather then physical processes, the fuel is most vulnerable to diversion for the purpose of nuclear weapons development. Feiveson and Taylor,[4] have argued that spent or highly radioactive fuel is self-protecting. Such assemblies weigh nearly half a ton. They argue that stealing such irradiated assemblies would require heavy cranes, tons of shielding containers, and a large vehicle for transporting the stolen, shielded assemblies. Further, the fissile U-233 or Pu-239 must still be separated from the dangerously radioactive fuel.

The fusion-fission hybrid fuel cycle proposed here directly enriches the fertile fuel to 3 to 4 percent fissile concentration in the hybrid blanket. This process also makes the fuel highly radioactive so that it is rendered diversion resistant. The details of this fuel cycle are outlined in Figure 16–1.

The cycle includes four steps:

1. Fertile fuel, ThO_2 or UO_2, is fabricated in a form that is directly usable in a LWR. (Other fission reactors could be included but the LWR is used here because it is the workhorse of the U.S. fission reactor industry.)
2. The cold, clean fuel assemblies, containing only fertile fuel, are placed in the hybrid blanket and carefully enriched to a nearly uniform concentration of 3 to 4 percent fissile fuel as required by the LWR.
3. The enriched, and now highly radioactive, assemblies are transferred as units directly to the LWRs for burning of the fuel.

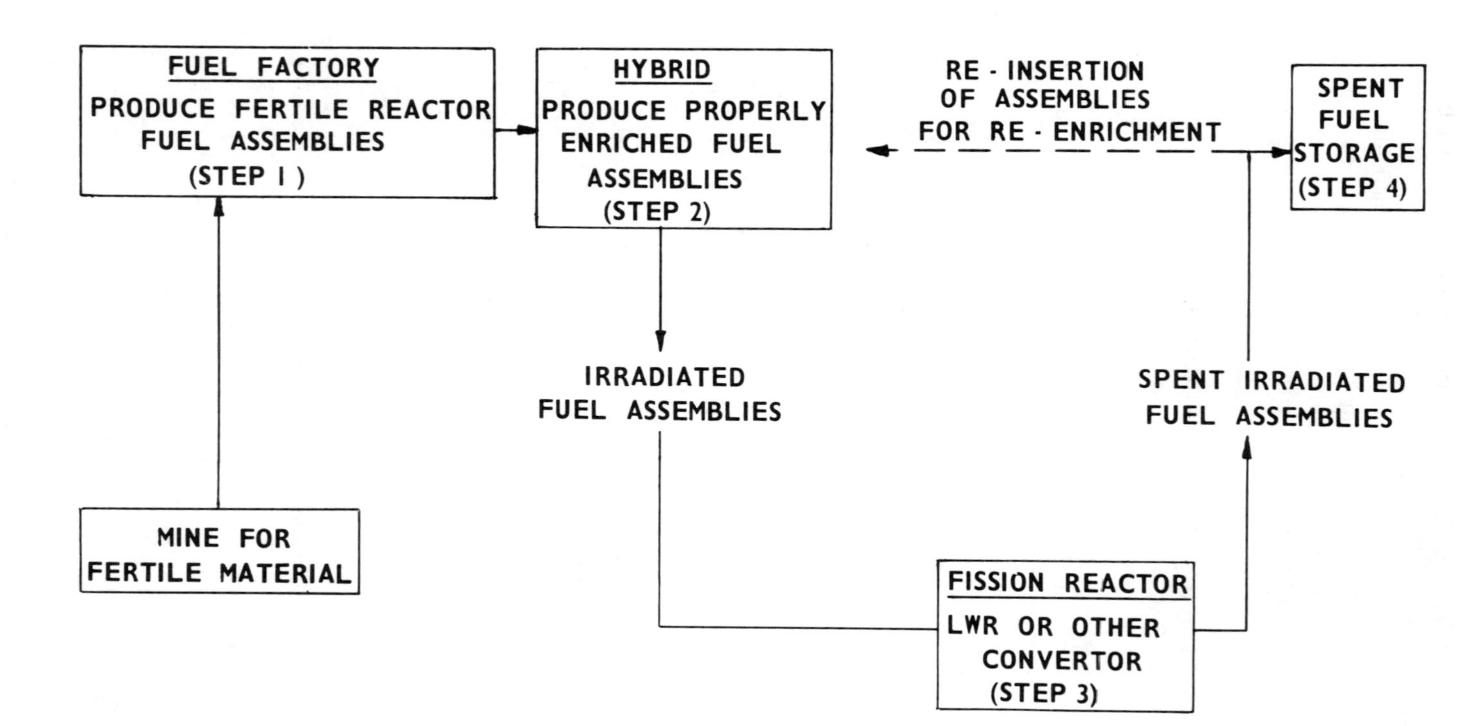

Figure 16–1. Fusion-Fission Hybrid Fuel Cycle Without Reprocessing.

4. The spent fuel from the LWR is stored until a decision is made on reprocessing or storing or both. If feasible, the spent fuel can be reinserted into the hybrid to be reenriched for further burning in the LWR. This possibility depends on both the importance of fission product buildup to LWR performance and the radiation damage to the fuel and cladding.

The attractive features of this cycle are the following:

1. The system is resistant to diversion because fissile material occurs only inside highly radioactive fuel assemblies. Only fresh fertile material is fed to the hybrid, and upon removal, the fuel pellets contain fission products that are highly radioactive, and the pellets themselves are contained in rod assemblies with highly activated cladding. Access to the fissile material is thus very difficult, making the entire cycle proliferation resistant according to the guidelines of Feiveson and Taylor.
2. The fissile fuel reserves are extended substantially. If the average LWR fuel enrichment is assumed to be 3 percent, the fissile fuel reserves are extended by 4.3 × (thorium resources + uranium resources). According to Staatz and Olsen,[5] the occurrence of thorium is widespread but the resources are not well known because present demand is low. The demand in 1968 was for only about 125 tons of ThO_2. Estimates of the thorium content of the earth's crust range from 6 to 13 ppm. Identified world thorium resources recoverable primarily as a byproduct or co-product are about 1.4 million tons, one-third of which occurs in a deposit near Elliot Lake, Canada. The general understanding is that large additional resources would be found with additional exploration. If we assume that the thorium resources are no larger than the uranium resources, the fissile fuel supply is extended by a factor of 4 to 5 without reprocessing.
3. The extension of the fission fuel supply using the hybrid produces additional time that can be used to make deliberate decisions on issues such as internationally controlled, physically secure fuel production and fuel reprocessing centers.[6]
4. The manufacturing of fresh fertile fuel pellets can proceed without the handling problems inherent in the use of a radiation spiking material such as Co-60. This avoids any legal or safety issues associated with the deliberate addition of dangerous materials.

The major disadvantage of this system is that it does not take full advantage of the fertile fuel reserves. To achieve a fuel supply measured in thousands of years, rather than just a few hundred, fuel reprocessing is essential. Without reprocessing, one hybrid reactor is only able to supply fissile fuel to about two and a half LWRs of the same thermal power. This has the economic impact of increasing the effective fuel cost. With reprocessing of the spent LWR fuel, on the order of ten LWRS can be fueled from one hybrid of equivalent power, depending on the conversion ratio of the LWR or other convertor reactor.

The proliferation resistant fuel cycle can be extended to include reprocessing of the spent LWR fuel if one follows the structure outlined by Feiveson and Taylor of internationally controlled, physically secure fuel production and reprocessing sites combined with many national convertor reactors "outside the fence." This process involves the four steps outlined in Figure 16-2.

1. Fresh ThO_2 or UO_2 fuel is fabricated in assemblies that are directly usable in a LWR or other convertor reactor. This step will also involve the fabrication of enriched fuel assemblies at the secure site using fissile fuel from the reprocessing step. We propose that such fuel be only partially enriched (for example, to just 2 percent even though about 3 to 4 percent is required) and that the hybrid be used to produce the required additional enrichment.
2. The fuel assemblies are irradiated in the hybrid blanket to produce the required fissile enrichment.
3. The fuel is transferred directly to the fission reactor and burned.
4. The spent fuel assemblies are shipped back to the physically secure site for reprocessing. The reprocessing plant removes fission products and sends the fissile material to the fuel factory for fabrication into new fuel assemblies.

The advantages of this approach are the following:

1. The fuel supply is measured in terms of the fertile material abundance. All estimates show that such fuel supplies will last for thousands of years.
2. Fuel shipped to and from the convertor reactors is always highly radioactive and would be resistant to diversion and reprocessing for the reasons described earlier.
3. The convertor reactor need not be restricted to a LWR, although using these reactors will minimize the need to develop additional fission reactor technologies.

The potential success of these fuel cycles depends upon two key technical questions: (1) Can the hybrid reactor produce uniformly enriched fuel at an acceptable fusion performance level when the blanket design is constrained to accommodate LWR fuel assemblies? (2) Can a standard LWR burn the irradiated fuel? The first of these questions was the major emphasis of the SOLASE-H laser fusion hybrid study. The second question will be considered in future work.

THE SOLASE-H FUSION-FISSION HYBRID REACTOR STUDY

Introduction

The SOLASE-H study is a coupled set of investigations covering five separate topics: (1) overall proliferation resistant fuel cycles; (2) blanket neutronics

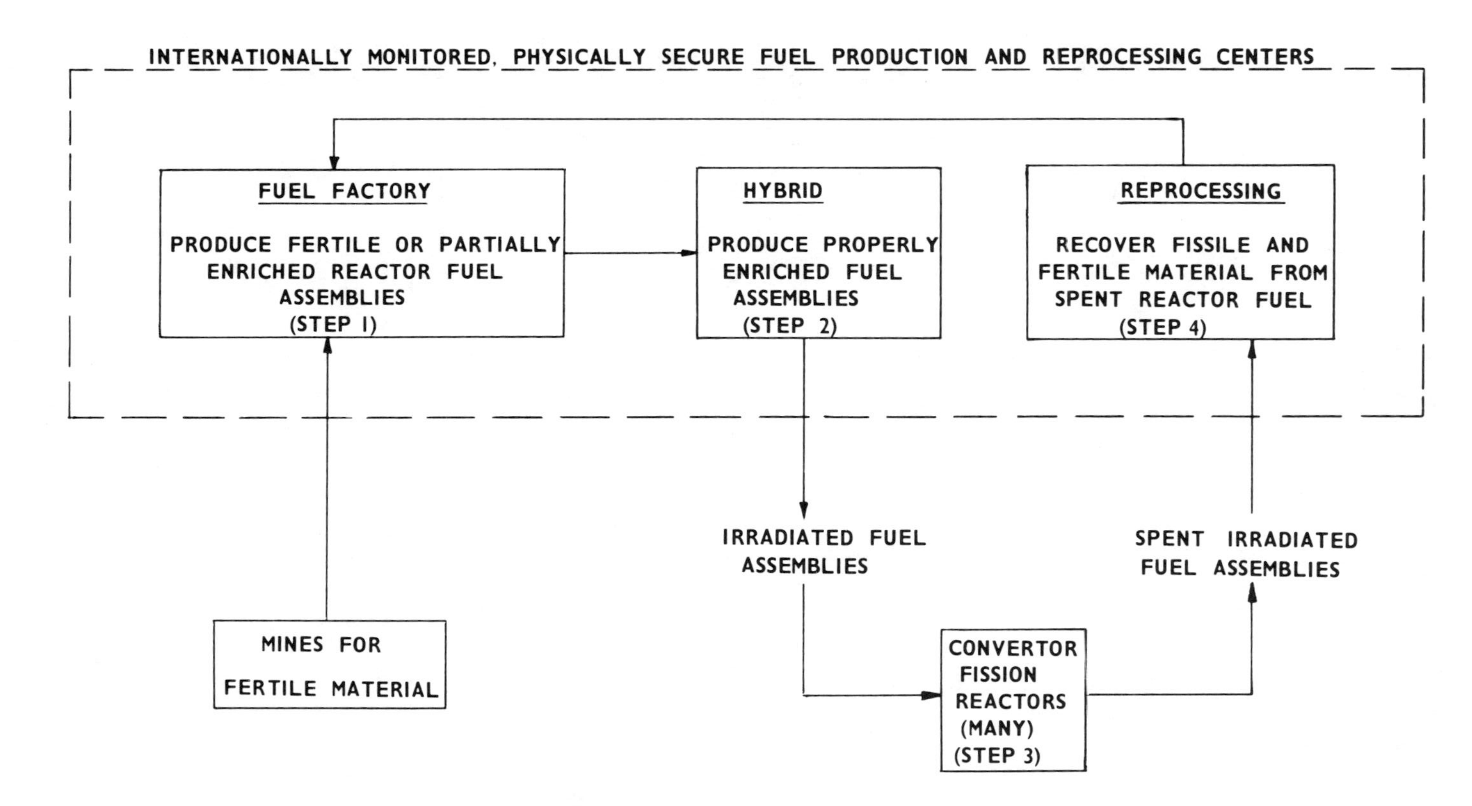

Figure 16–2. Fusion-Fission Fuel Cycle With Reprocessing.

and mechanical design; (3) laser fusion performance requirements for hybrids; (4) first wall protection using xenon cavity gas; and (5) hydrogen fluoride laser design. This study is distinguished from a true conceptual reactor design by the fact that not all systems are treated (e.g., pellet injection, tritium reprocessing), and there is not as much emphasis on a completely self-consistent set of parameters. However, the ranges of parameters studied in each area were chosen to overlap, so that self-consistent sets of parameters can be derived from the study. It is convenient to consider a set of consistent parameters such as those displayed in Table 16-1. Cutaway views of the reactor cavity and blanket are shown in Figure 16-3. The reactor cavity and blanket have a cylindrical geometry to accommodate the fuel assemblies around the circumference. The cavity height allows three assemblies to be stacked on top of one another. The blanket structure is zircaloy to be compatible with the cladding of the fuel assemblies. The zircaloy first wall is protected from the X-ray and ion debris of the pellet microexplosion by 0.5-1 torr of xenon gas that is circulated through the cavity.

The fusion power in this system is 1,200 MW. This is produced by irradiating pellets at the rate of 4 Hz with each explosion yielding 300 MJ of energy. The 14 MeV neutrons are assumed to contain 70 percent of this energy, with the rest partitioned between ions and X-rays. The laser energy on target is 1.5 MJ, thus implying a pellet gain of 200.

The blanket power multiplication varies between two and five during the fuel enrichment process, giving an average thermal power of 2,650 MW. The neutron wall loading at the midplane is 2 MW/m^2. The coolant is sodium. It enters the blanket at 300°C and exits at 350°C. The tritium breeding ratio is 1. The upper and lower blankets, comprising 30 percent of the solid angle subtended at the target, are devoted only to breeding tritium.

The fertile material is ThO_2, clad in 17 × 17 PWR fuel assemblies. The blanket contains 528 assemblies and produces 0.65 Th-232 (n,γ) Th-233 reactions per fusion neutron. This produces 2.5 tons of U-233 per year, enough to fuel about two and a half 1,000 MWe PWRs with no reprocessing. The time to reach 4 percent fertile enrichment is 2.7 years of exposure or 3.8 years of operation at a 70 percent plant factor. The maximum to average fuel enrichment in a fuel assembly is 1.1.

The SOLASE-H study includes a detailed conceptual hydrogen fluoride laser design. The laser energy is 2 MJ and the maximum power is 300 TW. The wavelength of the HF laser is actually a range of wavelengths, 2.7-3.5 μm, because the laser operates on many different lines. The net efficiency of the laser is 2.6 percent. This includes both the electrical efficiency of initiating the chemical reaction and the chemical efficiency of reconstituting the laser gas mixture back into its original constituents. The pulse length is 3 ns and there are twenty final amplifiers, hence there are fifty-six last mirrors. The last mirrors are located at a distance of 22 m from the cavity center, and no attempt has been made to uniformly distribute them around the reactor.

Table 16.1 Solase-H Parameters.

Parameter	Value
Cavity Shape	Cylindrical
Cavity Radius	6 m
Cavity Height	12 m
Structure – Blanket	Zircaloy
– First Wall	2 mm Zircaloy
First Wall Protection	0.5 -1.0 torr Xenon Gas
Fusion Power	1200 MW
Pellet Yield	300 MJ
Neutrons	210 MJ
X-rays and Ions	90 MJ
Pellet Gain	200
Pulse Repetition Frequent	4 s^{-1}
Laser Energy (on target)	1.5 MJ
Average Thermal Power	2650 MWt
Thermal Power Range	2400–2900 MWt
Percent Variation	(19 percent)
Gross Electric Output	925 MWe
Net Electric Output	700 MWe
Recirculated Power Fraction	26 percent
Blanket Power Multiplier	1.5 - 5
Neutron Wall Loading (Max)	2 MW/m^2
Coolant	Na
Coolant Temperatures	300–350°C
Tritium Breeding Ratio	1.0
Fertile Material	ThO_2
U-233 Production Rate	0.65/Fusion Neutron 2.5 Tonnes/yr
Fuel Form	(17 x 17) PWR Assemblies
Number of Assemblies	528
Time to 4 percent Enrichment	2.7 yr
Max/Average Enrichment	1.1
Neutron Multiplier	Pb
Laser Type	Hydrogen-Fluoride
Laser Energy	2 MJ
Net Efficiency	2.6 percent
Electrical Efficiency	24 percent
Wavelength	2.7 – 3.5 μm
Maximum Power	300 TW
Pulse Length (Multiplexed)	3 ns
Number of Final Amplifiers	20
Last Mirror Position	22 m
Number of Laser Mirrors	56
Illumination	Nonuniform

Blanket Neutronics and Mechanical Design

The blanket design for SOLASE-H is shown in Figure 16–3. The reactor cavity is cylindrical, with fissile fuel being bred only in the circumferential blanket. The top and bottom blankets are devoted to breeding tritium. The

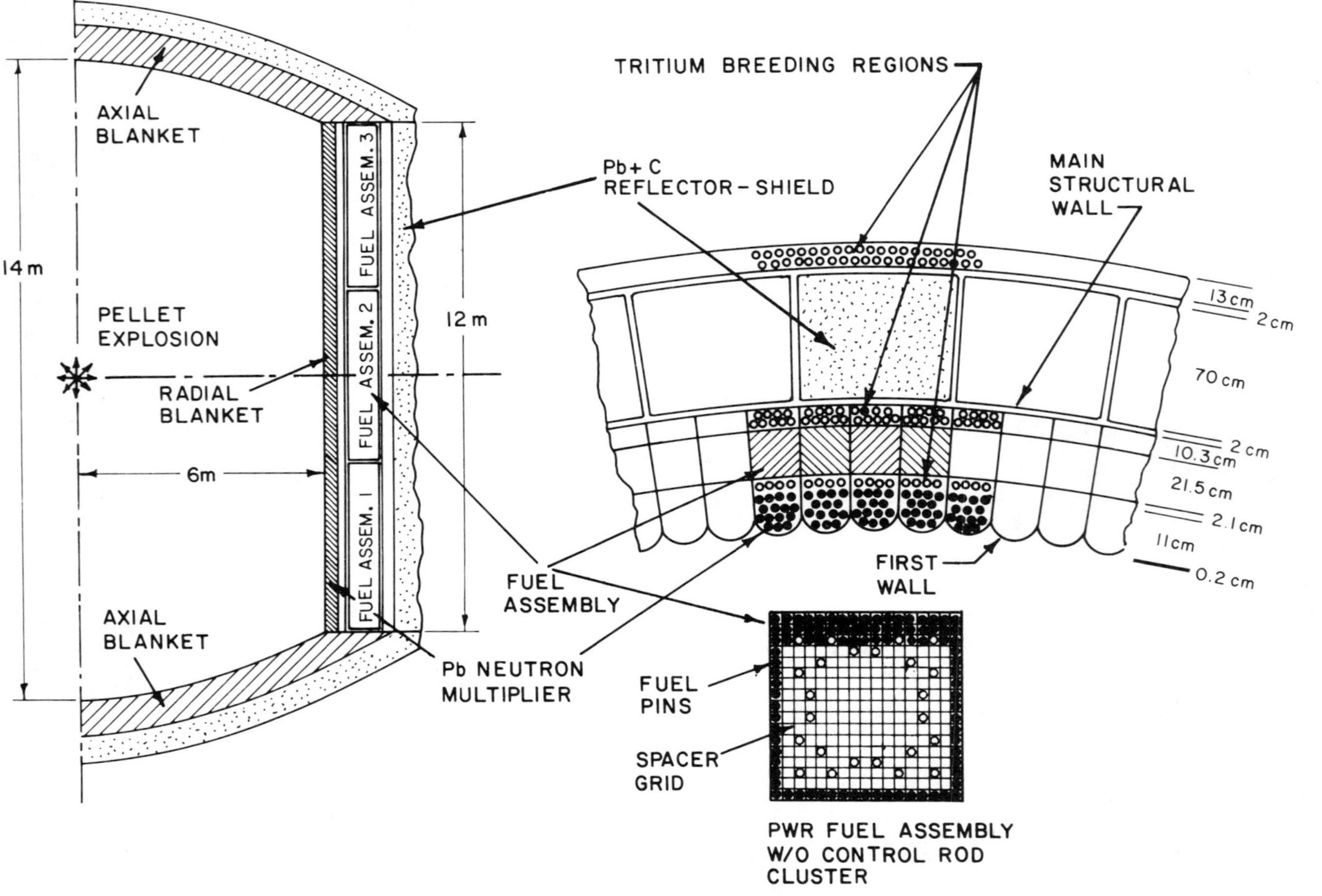

Figure 16-3. Cutaway View of the SOLASE-H Reactor Cavity and Blanket.

radius of the cavity is 6 m and the height is 12 m. This allows three LWR fuel assemblies to be stacked in the blanket. The blanket structure is zircaloy, to be compatible with the cladding of the fuel assemblies. If stainless steel were used as the structure, there is the possibility of carbon transport between it and the zircaloy cladding by the Na coolant. The first wall is 0.2 cm thick and is scalloped as shown in Figure 16-3 to accommodate the Na coolant pressure in the blanket. Directly behind the first wall are pins of Pb, clad in zircaloy. This Pb serves as a neutron multiplier, thus enhancing the fissile production rate. When this zone is removed, the total number of breeding captures per fusion neutron is reduced from 1.63 to 1.46. If the neutron for breeding tritium is subtracted, then the reduction of neutrons available to produce fuel goes from 0.63 to 0.46, a 27 percent effect.

The zone containing LWR assemblies is surrounded in the front and rear with pins containing Li. These Li zones both breed tritium and filter thermal neutrons that might otherwise diffuse into the fuel assemblies and induce fission. By poisoning the thermal flux, they enhance the uniformity of enrichment across the LWR assembly. Behind the LWR fuel zone and its Li filter is a Pb and carbon reflector. The fuel zone is therefore surrounded by fast-neutron-reflecting material and thermal neutron filters. The assemblies behave as a fast neutron flux trap, thus maximizing the fissile fuel breeding rate. The reflector is followed by an outer Li zone to capture any leaking neutrons.

Numerous neutronics calculations using the ANISN neutron transport code were done to optimize this blanket (#13 on the Figure 16-4) such that the uniformity across the fuel assemblies is that shown in Figure 16-4. The maximum to average enrichment is 1.1, with a U-233 enrichment of 4.7 percent at the edge and 3.75 percent in the middle. The time required to reach this enrichment is 2.7 years of exposure. The fuel assembly is rotated 180° at the end of 1.35 years to achieve the symmetric profile. The profile can be made flatter only at the expense of reducing the fissile breeding ratio. The optimized blanket is chosen to be the one having the smallest maximum to average $Th(n,\gamma)$ reaction rate profile, denoted by R, while having a high value of the uranium breeding ratio, UBR. Thus,

$$FM = UBR/R. \tag{16.1}$$

However, since the average $Th(n,\gamma)$ reaction rate is proportional to UBR, we find that

$$FM = (UBR)^2/Th(n,\gamma)_{max}. \tag{16.2}$$

This quantity was chosen to represent the criteria for blanket optimization. A penalty is paid in the LWR for large values of maximum to average enrichment due to hot channel factors. However, a penalty is paid in the hybrid for

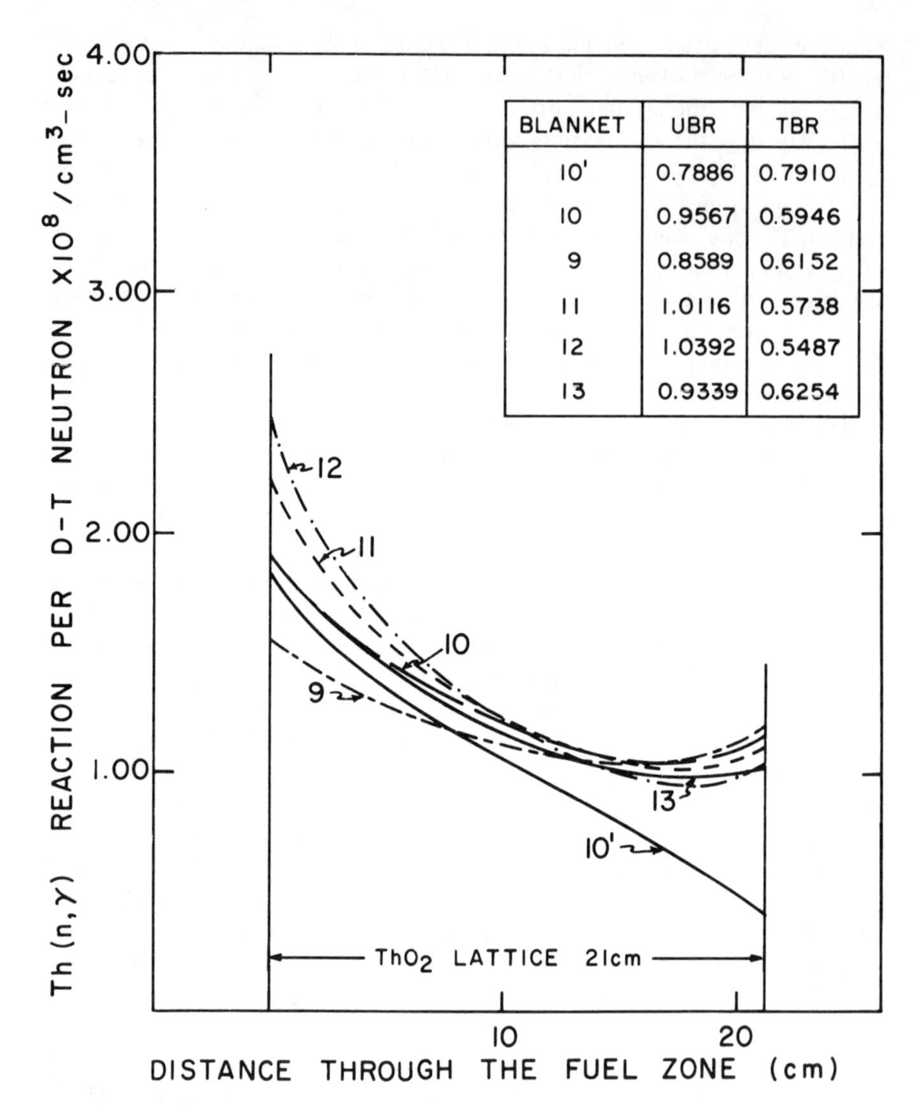

Figure 16-4. Uniformity of Enrichment Across the PWR Fuel Assembly after 2.7 Years of Exposure.

values close to one because these blanket designs have reduced values of *UBR*. Thus our figure of merit tends to minimize each of these penalties. For the base case blanket design, the breeding ratio is 0.65 U-233 per fusion neutron. With this breeding ratio and 1,200 MW of fusion power, the hybrid produces ~2500 kg of U-233 per year, enough to fuel two and a half 1,000 MWe LWRs without reprocessing.

Most of the blanket neutronics analyses were done using ANISN and assuming a one dimensional, spherical blanket. A solid angle weighting of 70 percent is then applied to the results to account for the fact that the fissile fuel is only in the circumferential blanket in the cylindrical reactor. Once a near optimum detailed blanket configuration is determined, three dimensional Monte Carlo calculations are performed on the entire blanket including the upper and lower tritium breeding regions. These calculations are done to determine the enrichment profile in the axial direction and to test the solid angle weighting approximation. This analysis shows that the upper and lower blankets can be strongly neutronically coupled to the circumferential blanket and, hence, that the simple solid angle weighting technique must be cautiously applied. However, the total number of absorptions per fusion neutron is almost constant at 1.65. Therefore, a proper three dimensional design can be established that will give the same results as the one dimensional designs with solid angle weighting. Furthermore, alternate fuel assemblies can be replaced with scattering material and a thermal neutron filter, so that the remaining assemblies are reduced in number and are surrounded by scattering material and thermal neutron filters. The three dimensional analysis shows that this does not seriously reduce the total number of absorptions per fusion neutron but significantly reduces the fuel inventory. The fuel in the blanket is enriched more quickly, reducing the associated carrying charges.

The power generated by the hybrid averages 2,650 MWt. This swings between 2,400 MWt and 2,900 MWt due to the changing blanket multiplication during the fuel enrichment process. These values are for the equilibrium cycle where there are four different batches of fuel in the blanket. Therefore, the blanket contains fuel that is fresh, one-fourth enriched, one-half enriched, and three-fourths enriched. This fuel management scheme limits the power swing in the blanket to 19 percent. A thermal efficiency of 35 percent then gives a gross electrical output of 925 MWe. The laser requires 225 MWe, and thus the net output is 700 MWe.

Laser Fusion Performance

Two economic figures of merit serve as guidelines for the determination of acceptable laser fusion performance. These are (1) the recirculating power fraction and (2) the cost scaling of capital-intensive components, such as power supplies to drive the laser. The first of these considerations is related to the return on investment in the thermomechanical equipment needed to produce the electricity. The recirculating power fraction can be related to the target gain, G, total laser efficiency, η_L, and the blanket energy multiplication, M, by the expression

$$f_R = [\eta_{th}\, \eta_L G(0.3 + 0.7M]^{-1}, \qquad (16.3)$$

where η_{th} is the thermal to electrical conversion efficiency and we assume 70 percent of the fusion energy is in neutrons. If $M = 1$ and $\eta_{th} = 0.4$, then a 25 percent recirculating power fraction implies $\eta_L G = 10$. This is typically taken to be the performance constraint placed on pure laser fusion reactor systems. For instance, the laser efficiency in the SOLASE design was 6.7 percent, and the target gain was 150. In a hybrid, where $M > 1$, the product, $\eta_L G$, can be less than 10, while the system still meets the condition of a 25 percent recirculating power fraction. Furthermore, the blanket multiplication increases the absolute power level; hence, the fusion power necessary to produce a given amount of thermal power is reduced. Both of these effects lead to relaxed laser fusion performance requirements.

The second condition, the cost scaling of capital-intensive equipment, determines the economy of scale associated with such systems. An analysis of power supply costs indicates that high efficiency lasers, which require modest power supply energy coupled with modest target gain, are more likely to operate in an economically satisfactory way than low efficiency lasers and high gain targets. The lower bound on laser efficiency when power supply costs are limited to \$200/kWe of installed capacity is shown in Figure 16–5 for different total electrical power and blanket multiplication factors. The economy of scale is clear because the minimum laser efficiency consistent with this power supply cost is 19 percent for a 100 MWe plant, but only 4.4 percent for a 1,000 MWe plant when $M = 1$. This maximum power supply cost of \$200/kWe is chosen because this would be about 10 percent of the plant cost if the hybrid were to cost \$2,000/kWe. We reason that it is unlikely that the power supplies, only one component of the plant, could be allowed to exceed more than 10 percent of the total plant cost.

The SOLASE-H laser fusion parameters are compared to the SOLASE parameters in Table 16–2. A key relaxation of laser performance is the reduction of

Table 16–2. Comparison of SOLASE and SOLASE-H Laser Fusion Performance Parameters.

Parameter	*Solase*	*Solase-H*
Electric Power (net)	1000 MW	700 MW
Thermal Power	3300 MW	2650 MW
Fusion Power	3000 MW	1200 MW
Blanket Power Multiplication	1.1	1.5–5
Recirculating Power Fraction	25 percent	26 percent
(Target gain) × (Laser efficiency)	10	5
Target Yield	150 MJ	300 MJ
Percent Yield in Neutrons	80 percent	70 percent
Target Gain	150	200
Laser Type	CO_2 Model	HF
Laser Wavelength	–	2.7–3.5 μm
Laser Energy (on target)	1.1 MJ (1.0)	2 MJ (1.5)
Maximum Laser Power	1000 TW	300 TW
Repetition Rate	20 Hz	4 Hz
Net Laser Efficiency	6.7 percent	2.6 percent
Laser Electrical Efficiency	10 percent	24 percent

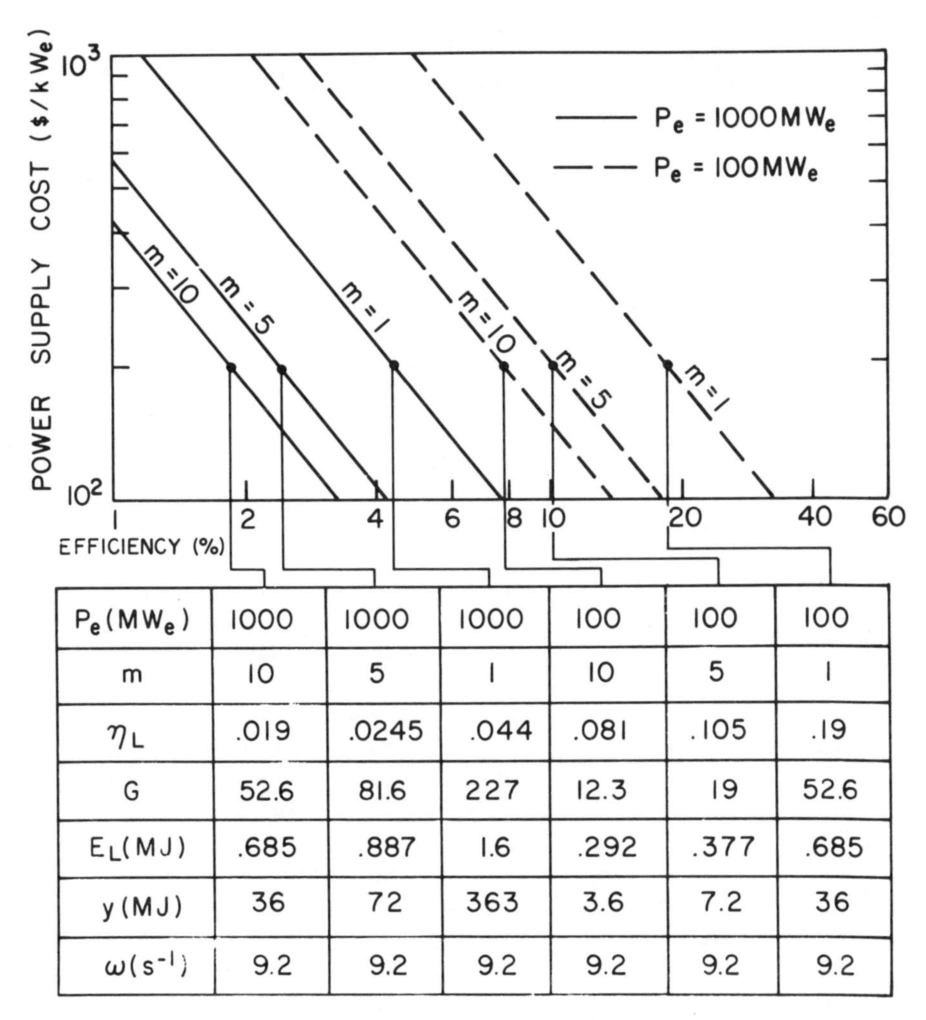

$P_e (MW_e)$	1000	1000	1000	100	100	100
m	10	5	1	10	5	1
η_L	.019	.0245	.044	.081	.105	.19
G	52.6	81.6	227	12.3	19	52.6
$E_L (MJ)$	.685	.887	1.6	.292	.377	.685
y (MJ)	36	72	363	3.6	7.2	36
$\omega (s^{-1})$	9.2	9.2	9.2	9.2	9.2	9.2

Figure 16-5. Laser Parameters for Power Supply Costs that are Limited to $200/kWe Installed Capacity.

repetition rate from 20 to 4 Hz. State of the art power supplies cannot meet the lifetime requirements of 10^8 to 10^9 shots, and although in principle they can be derated in voltage to meet these demands, the prospect that such simple solutions will be successful is quite low. Therefore a relaxation of the repetition rate in the hybrid is important. This is also possible because the blanket multiplication allows a lower fusion power. The electrical efficiency of the HF laser in SOLASE-H is 24 percent, more than a factor of two greater than the electrical efficiency of the SOLASE laser. This allows the laser energy to be increased by a factor of two without increasing the power supply requirements. The larger laser energy is chosen to allow more conservative estimates of the laser beam transport efficiency and the target gain.

For the SOLASE-H study, it was necessary to couple these considerations with the need to produce a copious supply of 14 MeV neutrons. The hybrid blanket is designed to produce a maximum amount of uniformly enriched fuel with a minimum fission rate. This results in a low blanket multiplication. A fusion power of 1,200 MW produces enough fuel for about two and a half LWRs. However, the reactor cavity size is determined by the energy that the first wall can accommodate in a single pellet microexplosion. This favors small explosions at a high repetition rate, yet this is inconsistent with the anticipated scaling of target gain with laser energy. Such considerations lead to the range of parameters chosen for SOLASE-H.

First Wall Protection by Xenon Cavity Gas

As mentioned in the previous section, the cavity volume, and hence the blanket volume, is determined in laser fusion reactors by the size of a single microexplosion rather than by the average fusion power. It is therefore crucial to determine the maximum target yield that can be accommodated on a repetitive basis by the first wall. In most laser fusion reactor concepts thus far, the first wall has been shielded from the pellet blast by some protection scheme. The method of protection has in fact been the fundamental identifying characteristic of these reactor designs.[7] The first wall protection method proposed for the SOLASE-H study involves the introduction of a noble gas, such as xenon, in the cavity. The gas pressure is less than 1 torr to minimize the effects of gas breakdown by the laser beams. This gas absorbs the target ionic debris and X-rays from the pellet explosion and reradiates the energy to the first wall over a time that is long enough to allow the energy to be conducted away. In Figure 16–6 we show the heat flux experienced by the first wall as a function of time for 0.5 torr of xenon. In these calculations it is assumed that 90 MJ or 30 percent of the total 300 MJ yield is deposited in the gas. In Figure 16–7 the transient temperature response of the wall is plotted as a function of time. This analysis indicates that 90 MJ is about the maximum amount of energy that can be withstood by a zircaloy first wall and a 6 m cavity. However, calculations also show that the heat flux at the first wall sensitively depends on the radiative properties of the hot (1-10 eV) xenon gas. These properties have not been accurately computed, so that a final conclusion awaits further analysis.

Hydrogen Fluoride Laser Design

The SOLASE-H study includes the conceptual design of a 2 MJ hydrogen-fluoride laser. This laser is pumped by an electron-beam-initiated chemical reaction.

$$F_2 + H \rightarrow F + HF,$$

$$H_2 + F \rightarrow H + HF. \tag{16.4}$$

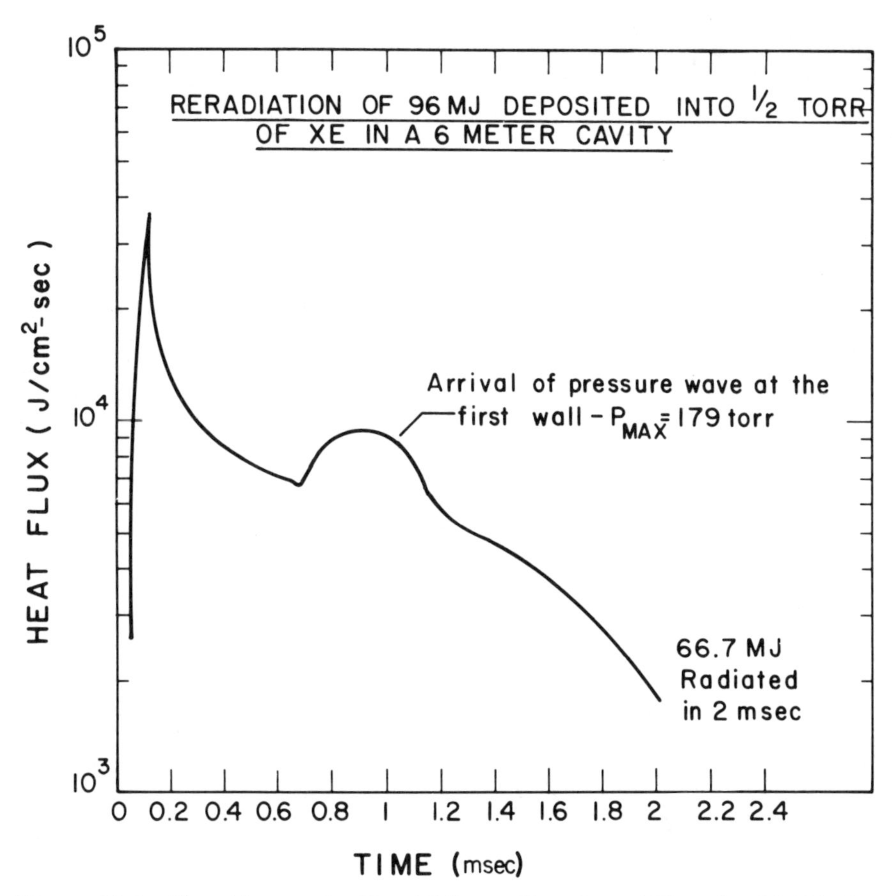

Figure 16-6. Heat Flux at the First Wall as a Function of Time for 96 MJ of Energy Deposited into 0.5 torr of Xenon.

The electron beam pulse is 20 nsec, and the natural pulse width of the laser is 12 nsec. This is likely to be too long for target irradiation, and therefore the twenty final amplifiers are multiplexed. Energy is extracted from eighteen of the amplifiers in a series of three short pulses that are combined on the target in a manner that gives the desired pulse shape. From the remaining two amplifiers, a single long pulse is extracted. The multiplexed pulse from one final amplifier is shown schematically in Figure 16-8. The amplifiers have square optical apertures that are 102 × 102 cm and are 34 cm in length.

The net electrical and chemical efficiencies of the 3,000/900/100 torr mixture of $F_2/O_2/H_2$ laser gas are 24 percent and 4 percent respectively. These are chosen to maximize the overall efficiency, including all recirculating power costs, to 2.6 percent. However, Figure 16-9 shows that the electrical efficiency can be increased, but only at the expense of reducing the chemical efficiency

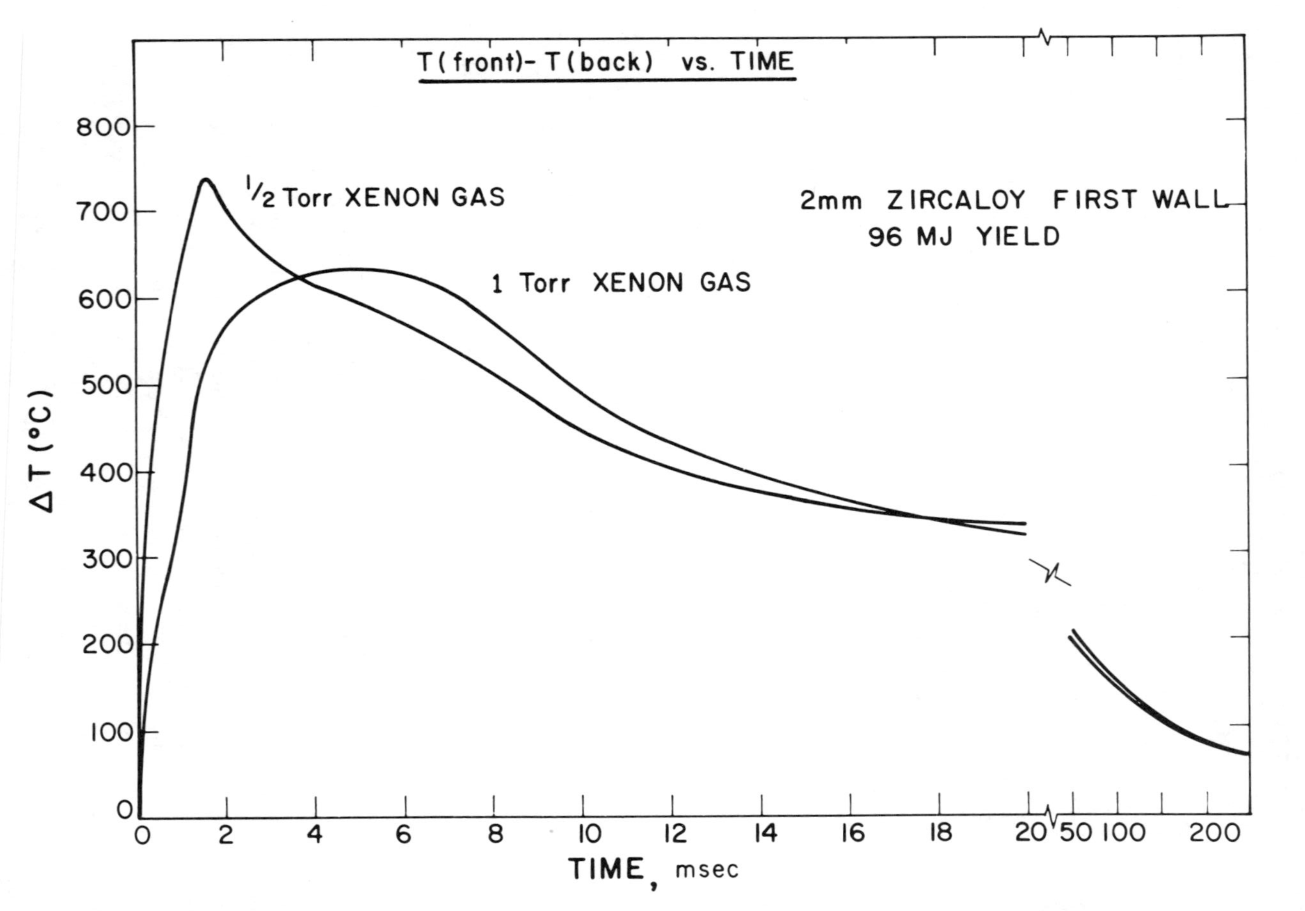

Figure 16-7. Transient Temperature Response of 2 mm Zircaloy First Wall Exposed to the Heat Fluxes in Figure 16-6.

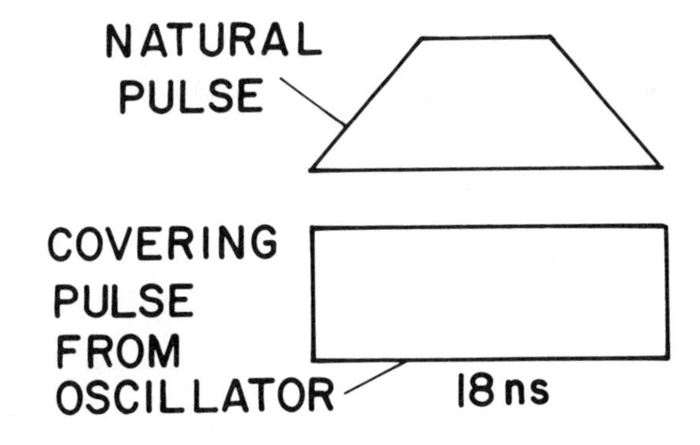

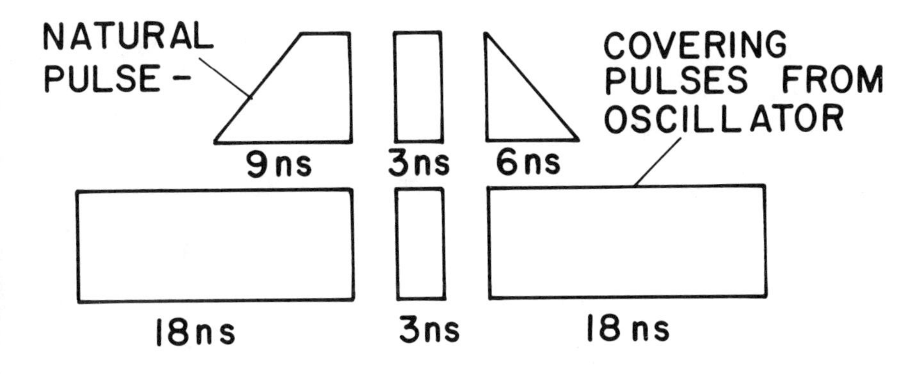

Figure 16-8. Multiplexed Pulse from a Single HF Final Amplifier.

and the overall efficiency. If this could be tolerated, then the power supply requirements could be even further reduced.

There is an increased interest in the HF laser for laser fusion applications because recent experiments have proven that beam quality is good and that amplified spontaneous emission in the very high gain amplifiers can be suppressed. This motivated our study of this 2.7 to 3.5 μm laser. The net efficiency is quite adequate for hybrid reactors and may be used for pure fusion reactors for target gains of 400 to 500 can be achieved with this relatively long wavelength. The amplifiers are compact, making the laser more easily scalable to 2 MJ.

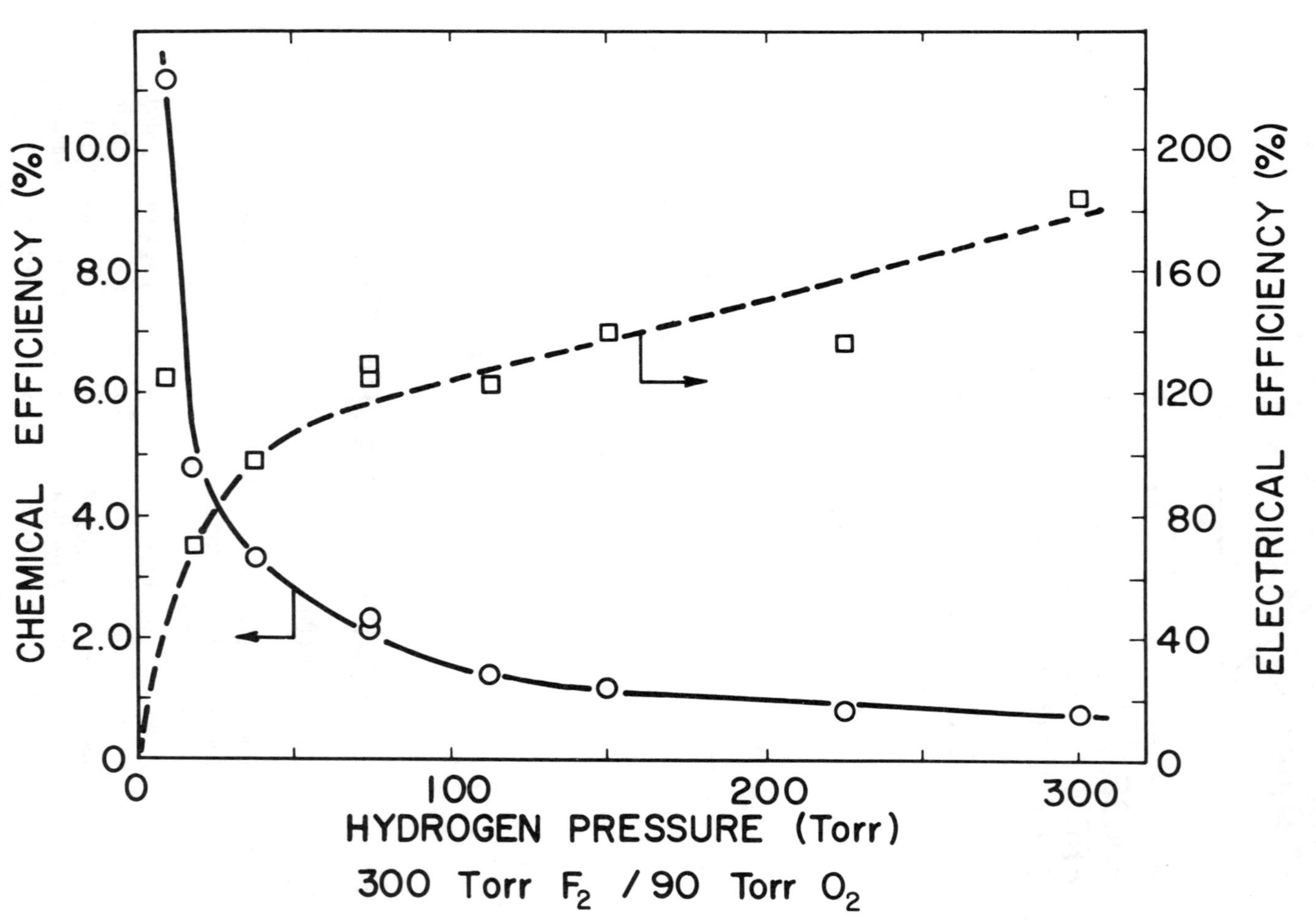

Figure 16-9. Electrical and Chemical Efficiency of HF Laser as a Function of H_2 Partial Pressure.

The HF laser may be repetition rate limited because the laser gas must be reprocessed to convert it from HF into H_2 and F_2. The chemical-reprocessing facility may be the limiting capital cost item in the total laser cost. Just as with the power supplies, there is likely to be an economy of scale associated with the reprocessing plant, but this is not known at this time.

CONCLUSIONS OF THE SOLASE-H STUDY

The SOLASE-H study established the potential feasibility of hybrid reactors for fueling LWR fission reactors in a nuclear future that does not allow reprocessing due to proliferation concerns. This involves direct irradiation of fertile fuel assemblies in the hybrid blanket. This simultaneously enriches the assembly of fuel to the proper fissile enrichment and renders it proliferation resistant by making the fuel highly radioactive. Such a hybrid reactor produces 0.6 to 0.7 U-233 atoms per fusion event while achieving a tritium breeding ratio of one. At this rate, one direct activation hybrid operated at a fusion power of 1,200 MW can fuel two and a half LWRs requiring 4 percent enriched fuel. The hybrid can also be incorporated into a scenario where the spent LWR fuel is sent to an internationally monitored, physically secure fuel production and reprocessing center. This fuel is reprocessed and then inserted back into the hybrid for re-enrichment. This would allow each hybrid to fuel ten LWRs with conversion ratio of 0.75 for U-233 fuel. This scenario also meets the qualifications of the Feiveson and Taylor nonproliferation fuel cycle.

The economic feasibility of the nonreprocessing fuel cycle will be very sensitive to the cost of the hybrid because the support ratio of fission reactors to hybrid reactors is so low. This sensitivity is somewhat reduced for the case of the hybrid and reprocessing center. The first fuel cycle allows time to make deliberate decisions about reprocessing while still maintaining the LWR industry. Without reprocessing, the fissile fuel reserves are extended by about a factor of ten over the U-235 resources. With reprocessing, the fuel resources are measured in terms of the fertile fuel supply, extending the LWR fuel supply to thousands of years.

The potential success of these fuel cycles depends upon two key technical questions: (1) Can the hybrid reactor produce uniformly enriched fuel at an acceptable fusion performance level when the blanket design is constrained to accommodate LWR fuel assemblies? (2) Can a standard LWR burn the irradiated fuel? The first of these questions was studied in detail in the SOLASE-H study. The answer to the second question is currently being pursued.

Using careful blanket design, LWR fuel pins can be nearly uniformly enriched to 4 percent fissile concentration in approximately three years. The spectrum of neutrons incident on the assemblies must be carefully tailored to provide uniform enrichment. A hard spectrum is desired, and this favors Pb, rather than Be, as a nonfissionable neutron multiplying material in the blanket.

The Be has a large (n, $2n$) cross-section, but it also moderates the neutrons, and this is not desired. There is also a serious question of resource availability for Be. The fuel is surrounded by thin zones of Li to filter the thermal neutrons that might otherwise diffuse into the fuel. This serves the dual purpose of breeding tritium and suppressing the fission rate in the fuel assemblies.

Nearly flat enrichment profiles across the assembly can be achieved, but only at the expense of the U-233 breeding ratio. Therefore, a figure of merit is developed that takes account of both minimizing the hot spot factor resulting from nonuniform enrichment and maximizing the U-233 breeding ratio. This shows that the optimum is not necessarily the blanket design that produces the flattest U-233 distribution. The fuel can be rotated at the halfway point in enrichment to provide a symmetric profile. Axial uniformity can be provided by a fuel management scheme in which the fuel spends one-third of its time in each of the three vertical locations relative to the point source. Three dimensional neutronics calculations show that some of the fuel can be replaced by neutron-scattering material and that the remaining fuel still has the same production rate, 0.6 U-233 per fusion neutron. This reduction of fuel inventory shortens the time to 4 percent enrichment from three years to one and a half years of exposure. These three dimensional calculations also show that the circumferential blanket and the upper and lower blankets can be strongly coupled, neutronically. This suggests that blanket design using simple one dimensional calculations with solid angle weighting must be carefully evaluated for validity.

Burnup calculations show that approximately 13 percent of the total fuel generated is consumed before it is removed from the blanket. This burnup is equivalent to 4,300 MWD/MT. The power swing, due to changes in the blanket multiplication during enrichment, is 19 percent. The minimum power is 2,400 MWt and the maximum is 2,900 MWt.

The damage rate to the zircaloy clad during exposure is about 7 dpa over the total three-year period. This low value is the result of the neutron moderation caused by the Pb neutron multiplier zone and the Li filter zone in front of the fuel assemblies.

The direct enrichment fuel factory hybrid requires a substantial fusion power because the rate of fissile fuel production is proportional to the number of 14.1 MeV neutrons that are generated. Fissile or nonfissile neutron multipliers do not affect the rate of net fuel production by more than about ~25 percent. About 2.1 kg of U-233 are produced per megawatt year of fusion energy. Hence, 1,200 MW of fusion power are required to produce the fuel for two and a half 1,000 MWe LWRs. This large fusion power requirement does not allow the repetition rate and pellet gain to be simultaneously relaxed. In SOLASE-H, the pellet gain remains rather high (~200) while the repetition rate is relaxed to 4 Hz. This low repetition rate allows more time to reestablish the cavity's initial conditions before the next target microexplosion. It also relaxes the gas-handling capacity in the HF chemical laser. The laser efficiency can be substantially reduced in the hybrid while still maintaining an acceptable recirculating power fraction. The HF chemical laser has net efficiency of 2.6 percent, and the pellet gain is 200. This leads to a recirculating power fraction of 26 percent. Such a reduction in required laser efficiency will admit more lasers to the "possible laser fusion driver" category than will pure laser fusion requirements.

The cavity and blanket volume are determined by the transient conditions following a single microexplosion rather than the time-integrated fusion power. This leads to the desire for small explosions at a high repetition rate. But this is not likely to be possible because lasers and cavities may be repetition rate limited, and a copious supply of fusion neutrons is needed. However, the repetion rate must be high enough to avoid thermal relaxation in the fuel assemblies between microexplosions. Repeated thermal transients in the fuel lead to thermal ratcheting, which will destroy the fuel integrity. The minimum allowable repetion rate is about 1 Hz.

Gas protection of the reactor first wall from pellet debris and X-rays appears to be applicable to hybrid reactors. The major first wall response to the hot gas at densities of $0.75 - 3 \times 10^{16}$ cm^{-3} comes from a thermal transient due to the gas reradiation to the first wall. The overpressure at the first wall due to blast wave effects is minimal.

The hydrogen-fluoride chemical laser appears to be scalable to at least 2 MJ. Its long pulse nature necessitates multiplexing of beams through the final power amplifiers if pulses shorter than 15 nsec are required. The electrical efficiency of this laser can be as high as 100 percent, but optimum net laser efficiency is associated with an electrical efficiency of 24 percent. This can greatly relax the power supply requirements over those needed for low electrical efficiency lasers. The net efficiency of the HF laser including the chemical efficiency of reconstituting the H_2 and F_2 is 2.6 percent. As mentioned earlier, the HF gas handling and reprocessing limits the laser repetition rate.

NOTES

1. L.M. Lidsky, "Fission-Fusion Systems: Hybrid, Symbiotic and Augean," Chapter 16 (Moses, Conn, Abdel-Khalik) *Nucl. Fusion* 15 (1975): 151; R.P. Rose et al., "Fusion-Driven Breeder Reactor Design Study," Final Report, Westinghouse Electric Corp., WEPS-TME-043, Pittsburgh: Fusion Power Systems, May 1977; R.G. Mills, " System Analyses of Fusion-Driven Fission" Paper delivered at Third ANS Topical Meeting on the Technology of Controlled Nuclear Fusion, May 9–11, 1978. Santa Fe, New Mexico); D.J. Bender (LLL), K.R. Schultz, R.H. Brogli, and G.R. Hopkins (GA), "Performance Parameters for a U-233 Refresh Cycle Hybrid Power System" (Paper delivered at Third ANS Topical Meeting on the Technology of Controlled Nuclear Fusion, May 9-11, 1978, Santa Fe, New Mexico); J.D. Lee, "Nuclear Design of the LLL-GA U_3Si Blanket" (Paper delivered at Third ANS Topical Meeting on the Technology of Controlled Nuclear Fusion, May 9–11, 1978, Santa Fe, New

Mexico); J.A. Maniscalco, "A Conceptual Design Study for a Laser Fusion Hybrid" (Paper delivered at Second ANS Topical Meeting on the Technology of Controlled Nuclear Fusion, September 21–23, 1976, Richland, Washington; and W.O. Allen and S.L. Thomson, "Electron Beam Fusion-Fission Reactor Studies" (Paper delivered at Third ANS Topical Meeting on the Technology of Controlled Nuclear Fusion, May 9–11, 1978, Santa Fe, New Mexico.

2. R.W. Conn et al., "SOLASE-H, A Laser Fusion Hybrid Reactor Study," University of Wisconsin Fusion Design Memo, UWFDM-274, Nuclear Engineering Department, University of Wisconsin, 1978. Also, *Trans. Amer. Nucl. Soc.* 27 (1978): 58.

3. R.W. Conn et al., "SOLASE, A Laser Fusion Reactor Study," University of Wisconsin Fusion Design Memo, UWFDM-220, Nuclear Engineering Department, University of Wisconsin, 1977. Also, R.W. Conn, "Laser Fusion as a Power Reactor Concept," in O.M. Kadiroglu, A. Perlmutter, and L. Scott, eds., *Nuclear Energy and Alternatives* (Cambridge, Massachusetts: Ballinger, 1978).

4. H.A. Feiveson and T.B. Taylor, "Alternative Strategies for International Control of Nuclear Power" (Report Prepared for the 1980s Project of the Council on Foreign Relations, October 1976); and H.A. Feiveson and T.B. Taylor, Security Implementation of Alternative Fission Future," *Bull. of the Atomic Sci.* 32 (1976): 14.

5. M.H. Staatz and J.C. Olsen, "United States Mineral Resources," U.S. Geological Survey Prof. Paper 820 (1973), p. 468.

6. Feiveson and Taylor, "Alternative Strategies . . . "; Feiveson and Taylor, *Bull. of the Atomic Sci.*

7. R.W. Conn et al., "SOLASE, A Laser Fusion Reactor Study"; L.A. Booth, "Central Station Power Generation by Laser Driven Fusion," Los Alamos Scientific Laboratory Report LA-4858-MS (1972); Los Alamos, T. Frank, D. Freiwald, T. Merson, and J. Devaney, "A Laser Fusion Reactor Concept Utilizing Magnetic Fields for Cavity Wall Protection" (Paper delivered at First ANS Topical Meeting on the Technology of Controlled Nuclear Fusion, San Diego, April 1974); J.A. Maniscalco and W.R. Meier, "Liquid-Lithium Waterfall Inertial Confinement Fusion Reactor Concept," *Trans. of the ANS* 62 (1977): 62.

 *Chapter 17*

Progress in the Toroidal Approach to a Fusion Reactor

Harold P. Furth
Princeton University

INTRODUCTION

Fusion reactions, such as D-T reaction of Figure 17-1, give rise to energetic neutrons as well as to energetic charged particles. An "energy multiplication factor" Q can be defined as the ratio of the total fusion power production to the input heating power that is required to maintain the reacting plasma fuel at an appropriate temperature. The range of useful applications for various fusion reactor schemes is defined by their characteristic Q-values.

If a Q-value of ~ 1 can be attained, the corresponding fusion reactor has the potential for efficient conversion of electric input power to high energy neutron output power. Such a facility could serve as part of a "hybrid" power-generating system, where the fusion neutrons breed fuel for a set of fission reactors, which in turn generate net electric power. If the fusion Q-value can be made to exceed five to ten, a stand alone fusion power plant becomes a practical possibility. If in addition the fusion reactor scheme is such as to confine the charged fusion reaction products and allow them to thermalize with the plasma, it is then straightforward to progress from Q = 5–10 to $Q = \infty$—that is, the fraction of fusion power released in charged reaction products (20 percent in the case of Figure 17-1) is sufficient to maintain the plasma temperature without any external heating power input. The plasma is then said to be "ignited." Finally, in still more advanced fusion reactor schemes, it may be possible to utilize nuclear reactions other than D-T, which release their energy mainly, or exclusively, in the form of charged particles rather than neutrons. This operating mode is more difficult to achieve, since higher temperatures and better plasma confinement properties will be required than for the D-T reaction, but it offers an extremely attractive long-range vista for the fusion power approach to the

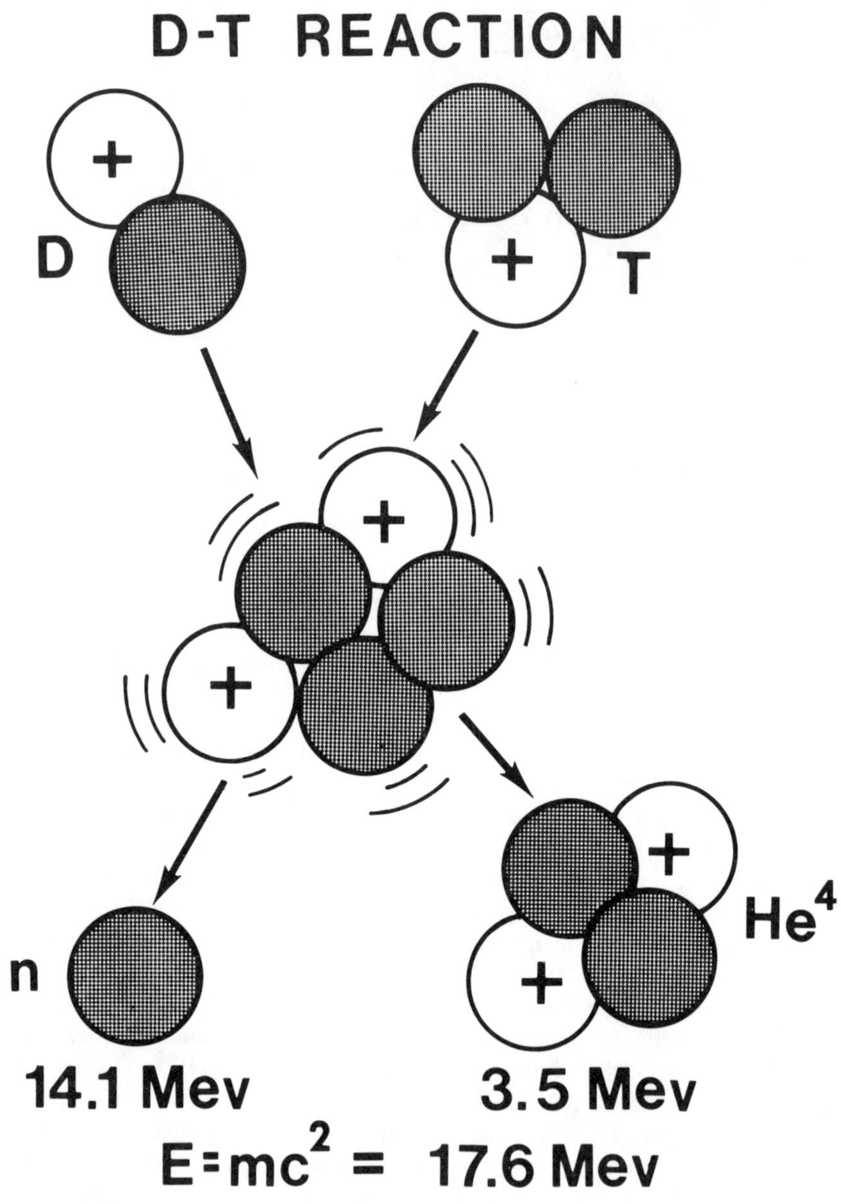

Figure 17–1. D-T Reaction (PPL 761021).

world's energy needs. If charged-particle-producing fusion reactors can be achieved, the residual problem of materials activation will be minimized, and the possibility of exceptionally high thermal efficiency will be introduced.

Historically, the task of achieving even the $Q = 1$ mark in D-T fuel has been difficult and uncertain of success. As shown in Figure 17–2, the required plasma temperatures are very high (1 eV = 12,000 °K), and the measure of energy confinement, $n\tau_E$, calls for confinement times τ_E in the range of seconds if plasma densities of order 10^{14} particles per cm^3 are to be used. During the 1950s and 1960s, it was far from clear that these conditions could be met within the twentieth century.

The exciting development of the past decade has been that the plasma conditions specified by the "Lawson criterion" of Figure 17–2 now seem to be attainable within the next decade or even sooner. Furthermore, new concepts have emerged for reactor operation, which will allow Q-values in the range one to five to be achieved under even less demanding conditions on $n\tau_E$ than those of the Lawson criterion. The net effect has been to shift the focus of fusion energy debates from the traditional issue of whether fusion power was to be taken seriously at all to an entirely new topic: in what form and toward

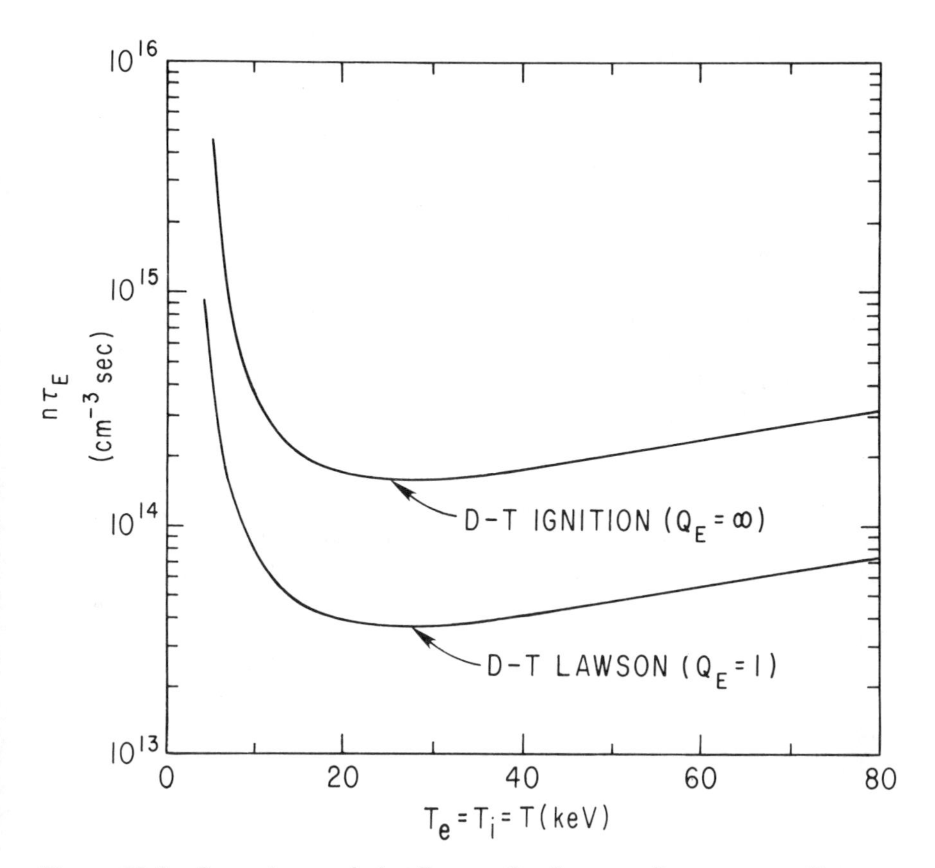

Figure 17–2. Dependence of the Energy Confinement Parameter on Plasma Temperature.

which of the alternative practical goals should our main fusion reactor effort be launched?

As has already been noted, there is a hierarchy of fusion reactor applications, with the most attractive also being the most difficult to achieve. Similarly, there is a hierarchy of alternate fusion reactor schemes, with the potentially most attractive from the point of view of economics being the least developed and the least well understood from the point of view of physics. (These monotonic orderings are, of course, not accidental, since any goal or scheme that appears less attractive as well as less realistic will tend to drop out of competition.) The problem that must be faced by present-day energy planners is thus to select, from a fairly wide range of objectives and means, some concrete strategy that offers the best compromise between what is desirable and what is feasible.

TOROIDAL PLASMA CONFINEMENT

The object of the present chapter is to give a brief account of one particular approach to a reactor, which at the moment is considered to have the best near term prospects of meeting fusion reactor requirements–confinement of a reacting D-T plasma in a toroidal "magnetic bottle." The peculiar advantage of this approach is that the plasma particles as well as charged reaction products can be allowed to undergo many Coulomb collisions without escaping from confinement. As Figure 17-3 makes clear, this feature is important in allowing operation at relatively "low" temperatures of order 10 keV (120 million degrees), where the ratio of the Coulomb cross-section to the fusion cross-section becomes extremely large. In the toroidal geometry, the collision process does not cause direct particle loss, but only gradual diffusion across the magnetic field (Figure 17-4). The confinement time τ_E thus scales up as the square of the plasma dimension. For minor radii a $\gtrsim$ 1 m, which are appropriate for practical fusion reactor designs, there is no difficulty in meeting the $n\tau_E$ requirement for ignition, even taking into account the more complex particle orbits that characterize realistic toroidal confinement geometries.

The historical problem of toroidal confinement experiments has been that the actual rates of plasma transport have greatly exceeded the predicted "classical" rates arising from Coulomb scattering. Until the late 1960s, no experiment was able to exceed the pessimistic "Bohm" prescription for confinement (Table 17-1), and many toroidal experiments seemed to be characterized accurately by this formula. Unlike the classical scaling, the Bohm scaling implies minimum reactor plasma sizes well in excess of near term practical requirements. In the period perior to 1968, the outlook for fusion reactors–as characterized in Table 17-2–was correspondingly bleak.

THE TOKAMAK APPROACH

In 1968, and more definitively in 1969, the I. V. Kurchatov Institute in Moscow reported that the spell had been broken at last and that toroidal confinement

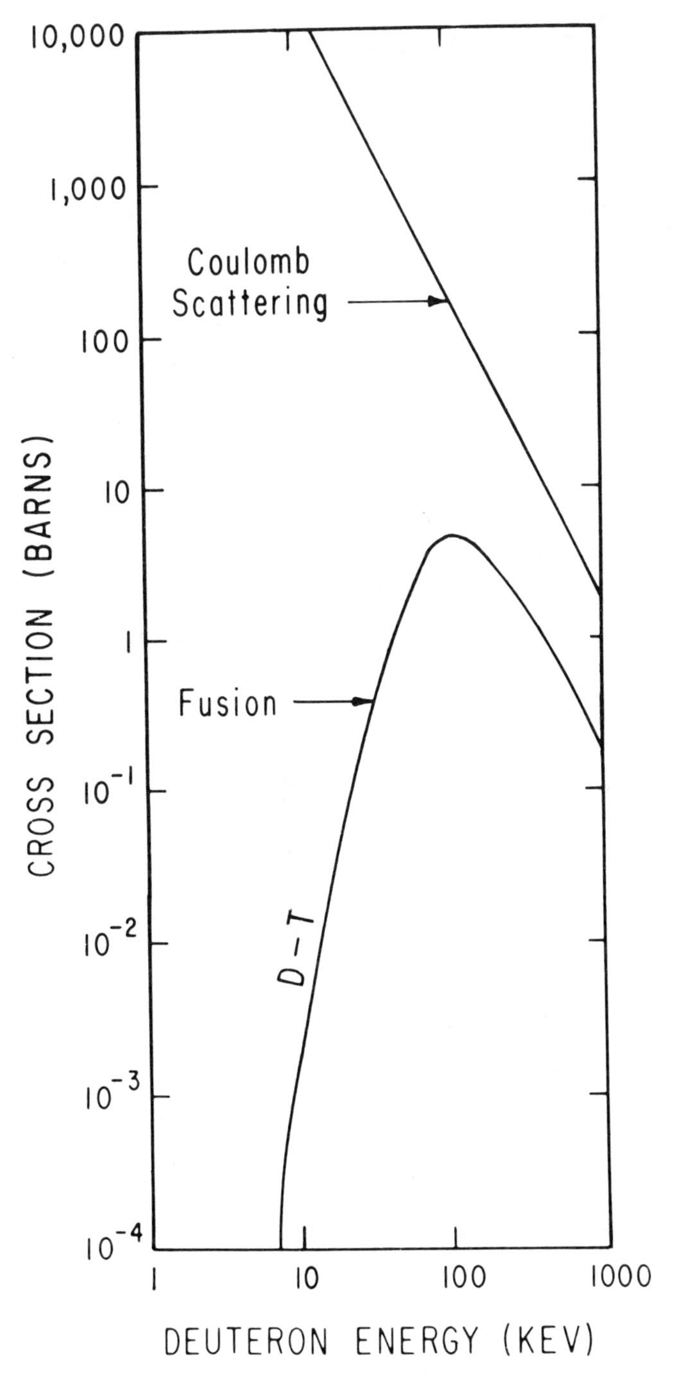

Figure 17–3. Fusion and Scattering Cross-sections. (PPL 753596)

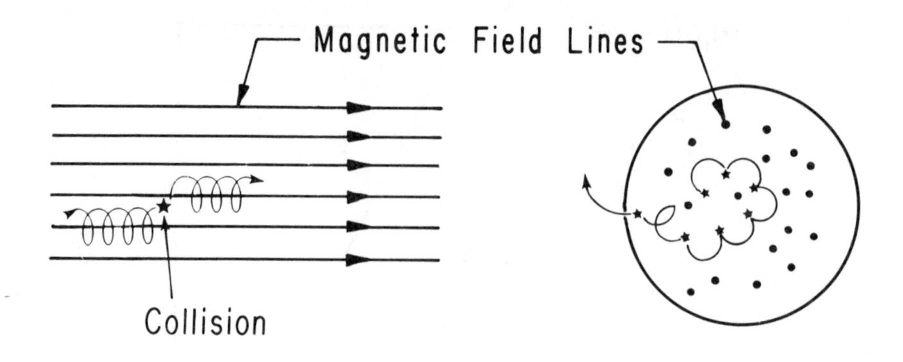

Classical Diffusion: $D = \delta_{\mathrm{orbit}}^{2} / \tau_{\mathrm{collision}}$

Diffusion Time $\sim (\mathrm{Radius})^{2} / D$

$\sim \tau_{\mathrm{collision}} (\mathrm{Radius} / \delta_{\mathrm{orbit}})^{2}$

Figure 17–4. Classical Loss Process of "Straight" Closed System.

Table 17-1. Classical versus Bohm Scaling to a Reactor.

Classical:

$$\tau_E \rightarrow \frac{a^2 B^2 T^{1/2}}{n}$$

A plasma minor radius of a ~ 1m is ample.

Bohm:

$$\tau_E \rightarrow \frac{a^2 B}{T}$$

A plasma minor radius of a > 10 m is needed.

Table 17-2. Toroidal Fusion Reactor Outlook Prior to 1968.

- There are many alternate approaches to a toroidal reactor.
- None of these approaches extrapolates to a practical reactor without supposing large improvements in the physics and innovations in plasma-heating technology.
- The state of nontoroidal approaches is comparable.
- The achievement of a reactor plasma within the twentieth century is questionable.

times approaching the classical prediction were being obtained in their T-3 Tokamak experiment. The tokamak geometry (Figure 17-5) calls for a small poloidal field component to be added to the basic toroidal field. The poloidal field is generated by a toroidally directed induced plasma current, which incidentally serves to provide ohmic heating for the plasma. The 1968 results of the T-3 experiment are shown in Figure 17-6.

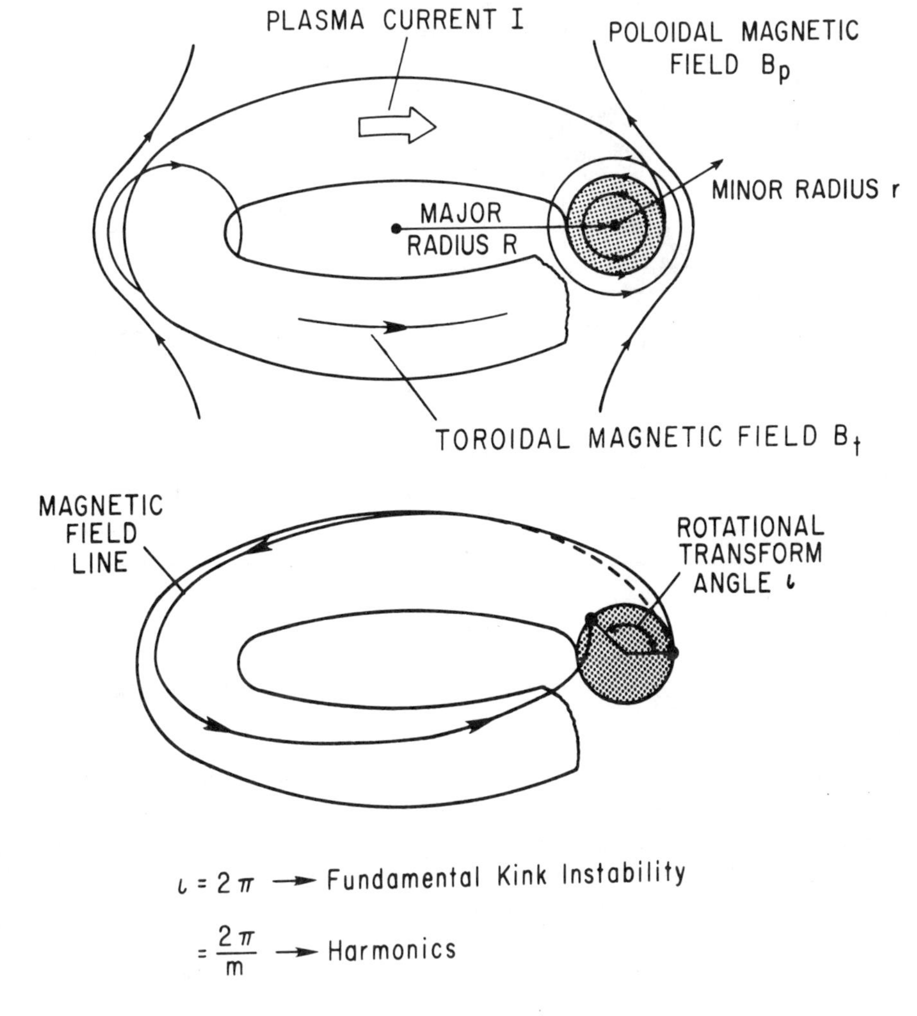

Figure 17-5. The Tokamak Configuration.

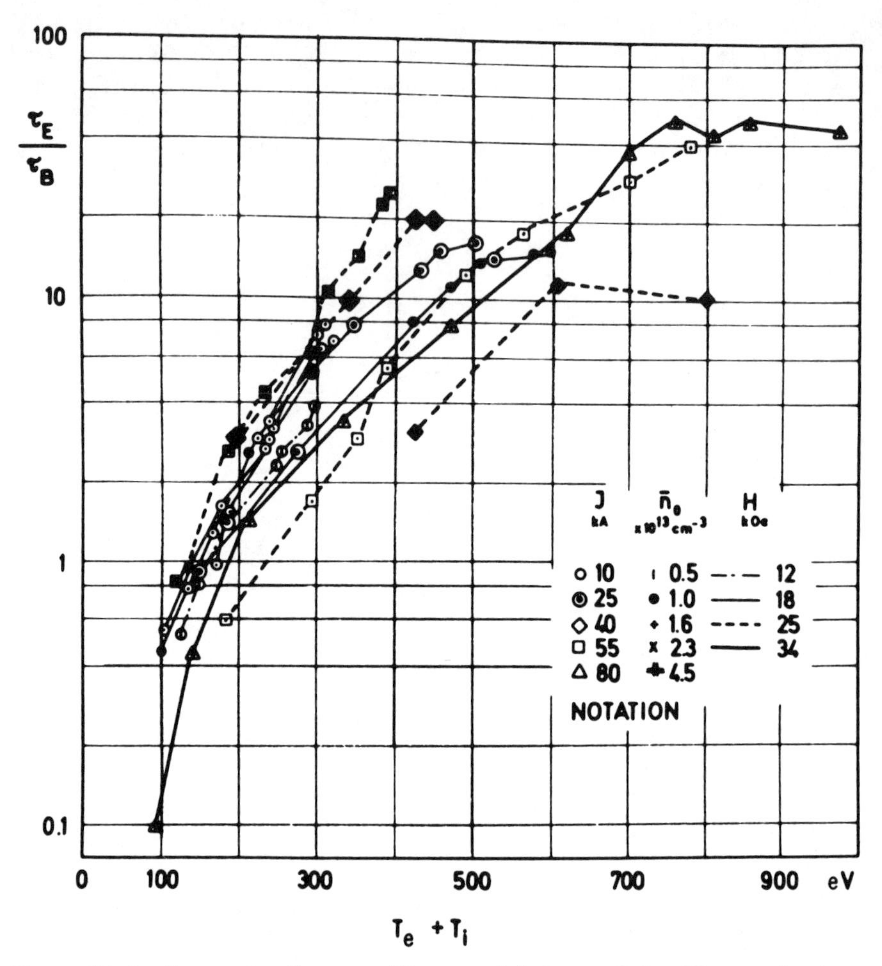

Figure 17-6. Energy Confinement Time τ_E Relative to Bohm Time τ_B in the T-3 Tokamak (1968).

The relatively high temperatures (~1 keV) achieved in the T-3, and the favorable scaling of the plasma confinement, stimulated the initiation of Tokamak research on a worldwide basis. The new Tokamak effort took two forms—attempts to raise the plasma temperature and confinement time still further in ohmically heated plasmas and attempts to develop new auxiliary heating methods that could push the Tokamak plasma parameters into the range of reactor interest.

The most successful experiment in the first category was the Alcator A device at the Massachusetts Institute of Technology. The Alcator concept was to proceed to unusually high magnetic fields (100 kg instead of the 35 kg in T-3) and relatively small plasma size in order to maximize the intensity of the ohmic heating. In this regime, Alcator A was able to produce ten times higher plasma densities (10^{15} particles per cm^3 instead of 10^{14} for T-3) and sixty times higher $n\tau_E$ values ($3 \cdot 10^{13}$ cm^{-3} sec instead of $5 \cdot 10^{11}$ cm^{-3} sec) at approximately the same temperature of 1 keV. A more advanced Alcator device (Alcator C in Figure 17-9) is expected to extend the $n\tau_E$ value to $\sim 10^{14}$ cm^{-3} sec, but the possibilities for raising the plasma temperature by this approach, appear to be limited to about 2-3 keV. The energy confinement scaling that has characterized the Alcator A experiment is shown in Figure 17-8. The inferred reactor size requirement (Table 17-3) is almost as favorable as for the purely classical scaling.

NEUTRAL-BEAM-HEATED TOKAMAKS

The missing ingredient at this point was a heating method more effective than ohmic heating for raising the plasma temperature from a few keV to the desired 10 keV range. During the mid-1970s, initial experiments were carried out at the Princeton Plasma Physics Laboratory (PPPL) and the Oak Ridge National Laboratory (ORNL) on the injection of high-powered neutral atom beams into small Tokamaks (ATC and ORMAK). The basic idea is that a neutral atom can pass freely through the walls of the Tokamak magnetic bottle and is then ionized and trapped in the plasma, where it thermalizes gradually with the other plasma particles (Figure 17-9). The initial experiments were encouraging and led to the recent successful attempt to produce reactorlike Tokamak plasmas in the PLT device at PPPL, using high-powered beam injectors developed at ORNL.

The PLT device is shown in Figures 17-10 and 17-11. It is one of the largest devices currently in operation in the world fusion research effort: the minor radius is $a = 45$ cm, the magnetic field is 35 kg, and the plasma current is $\sim$500 kA. During a typical discharge time of a second, ohmic heating raises the temperature of the electrons to several keV, while the ions lag behind somewhat, at $\sim$1 keV. (This lag is typical for ohmic-heated plasmas, since the energy input is through the electrons.) On injection of 2.5 MW of 40 keV neutral deuterium atoms from the four PLT beam lines, the ion temperature rises to a maximum of 6.5 keV (Figure 17-12), while the electron temperature rises to 3.5-4.0 keV. The central density in this regime is about $5 \cdot 10^{13}$ cm^{-3}.

These PLT results–obtained in July-August 1978–were of particular interest, since they gave the first direct indication of how Tokamak confinement would behave in essentially reactorlike plasmas. Theory had predicted that, at very low Coulomb collision rates, the classical transport would become small, but a new form of anomalous transport by collective plasma fluctuations would make its appearance. Associated theoretical estimates of energy confinement in Tokamak reactors had a highly adverse temperature dependence and pointed to minimum reactor plasma sizes that might be inconveniently large, though not quite so large as the Bohm prediction. Gratifyingly–from the theoretical point

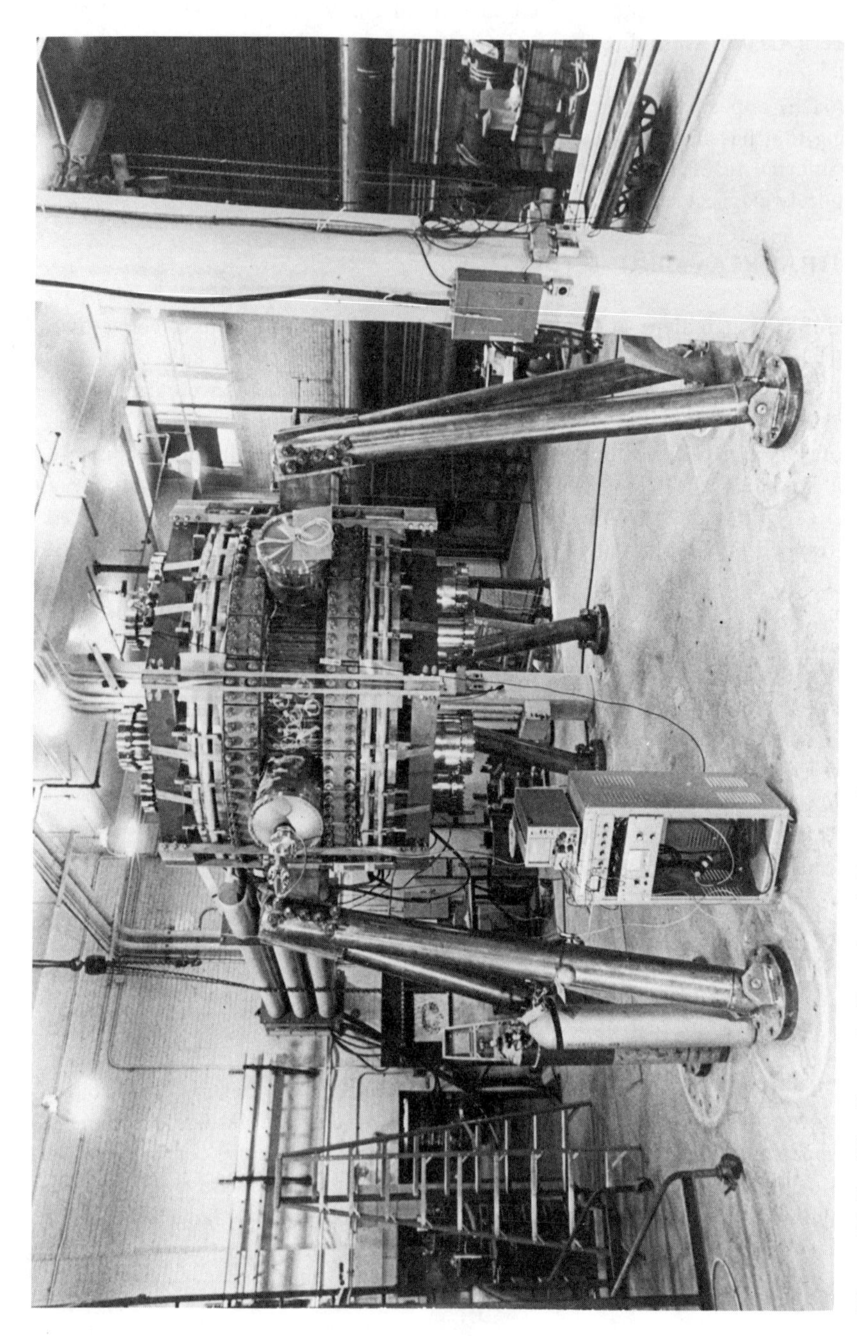

Figure 17-7. The Recently Completed Alcator C Device.

Table 17-3. Alcator Scaling to a Reactor.

$\tau_E \rightarrow na^2$
A plasma minor radius a = 1m is adequate.

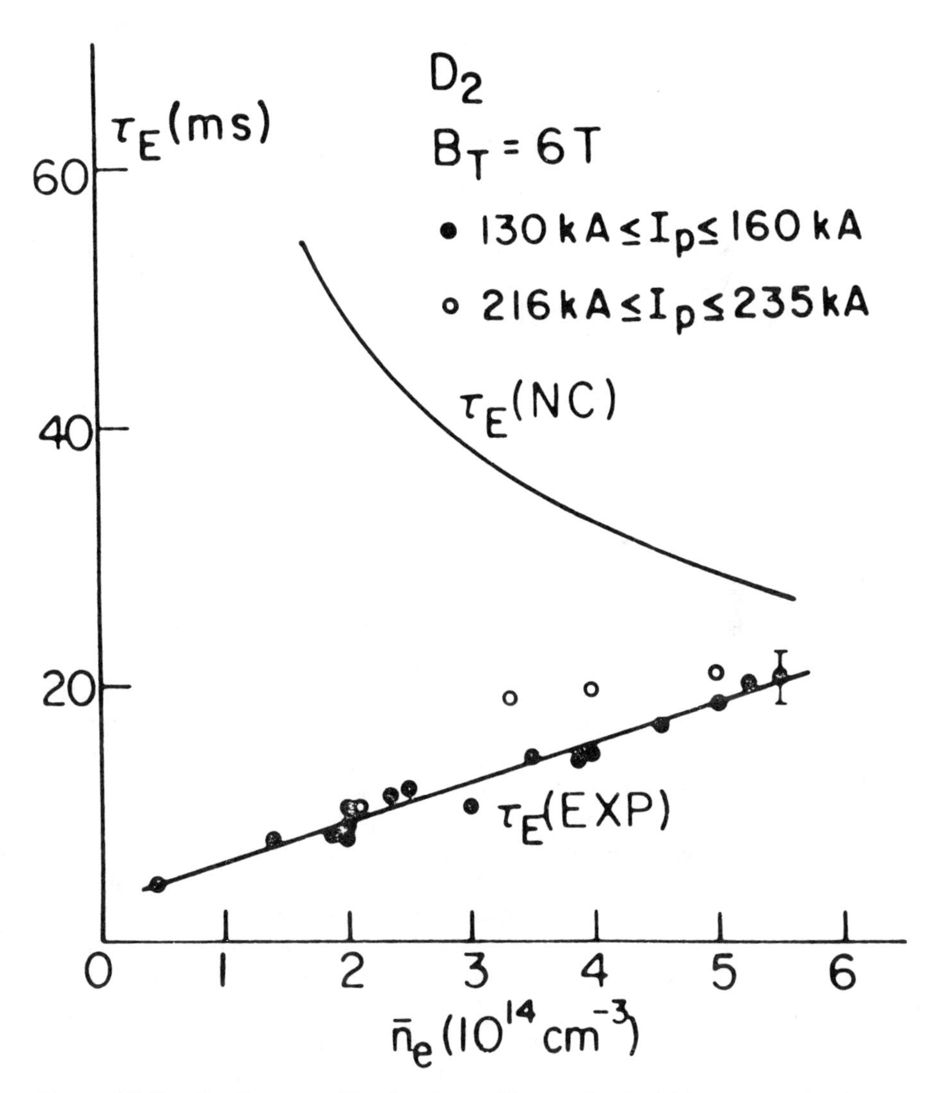

Figure 17-8. Confinement Results from Alcator A. At high average density $n\tau_E$ approaches the classically predicted value.

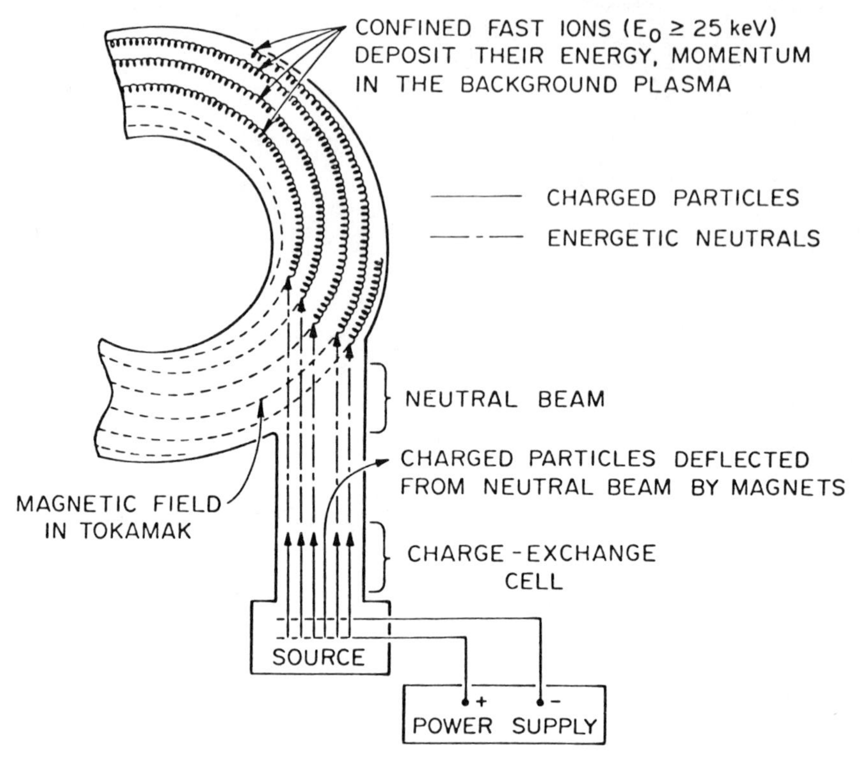

Figure 17-9. How Does a Neutral Beam Heat a Tokamak Plasma?

of view–the PLT experiment did exhibit a new species of fluctuations (Figure 17-13 at the highest plasma temperatures, but also happily–from the practical point of view–the appearance of these fluctuations did not have a measurable effect on the energy confinement, as judged from the scaling of ion temperature with beam power (Figure 17-14). As nearly as could be determined (within a factor of three to five), the ions obeyed precisely the classical prediction for energy transport. Even more surprisingly, the plasma electrons did not exhibit an unfavorable departure from the transport scaling that had been observed in the much more collisional Alcator plasmas, but rather exhibited an apparent improvement in confinement with rising temperature (Figure 17-15).

Thus far, the 6 keV temperature range has been entered in PLT only at moderate $n\tau_E$ values of 10^{12} cm^{-3} sec. At higher densities, $n\tau_E$ values as high as $1.5 \cdot 10^{13}$ cm^{-3} sec have been reached in PLT, but only at 1 keV temperatures. During the next several years, it is hoped to intensify the auxiliary heating capability on PLT (partly by the introduction of new radio frequency wave-heating techniques), in the expectation of realizing higher simultaneous values of $n\tau_E$ and temperature.

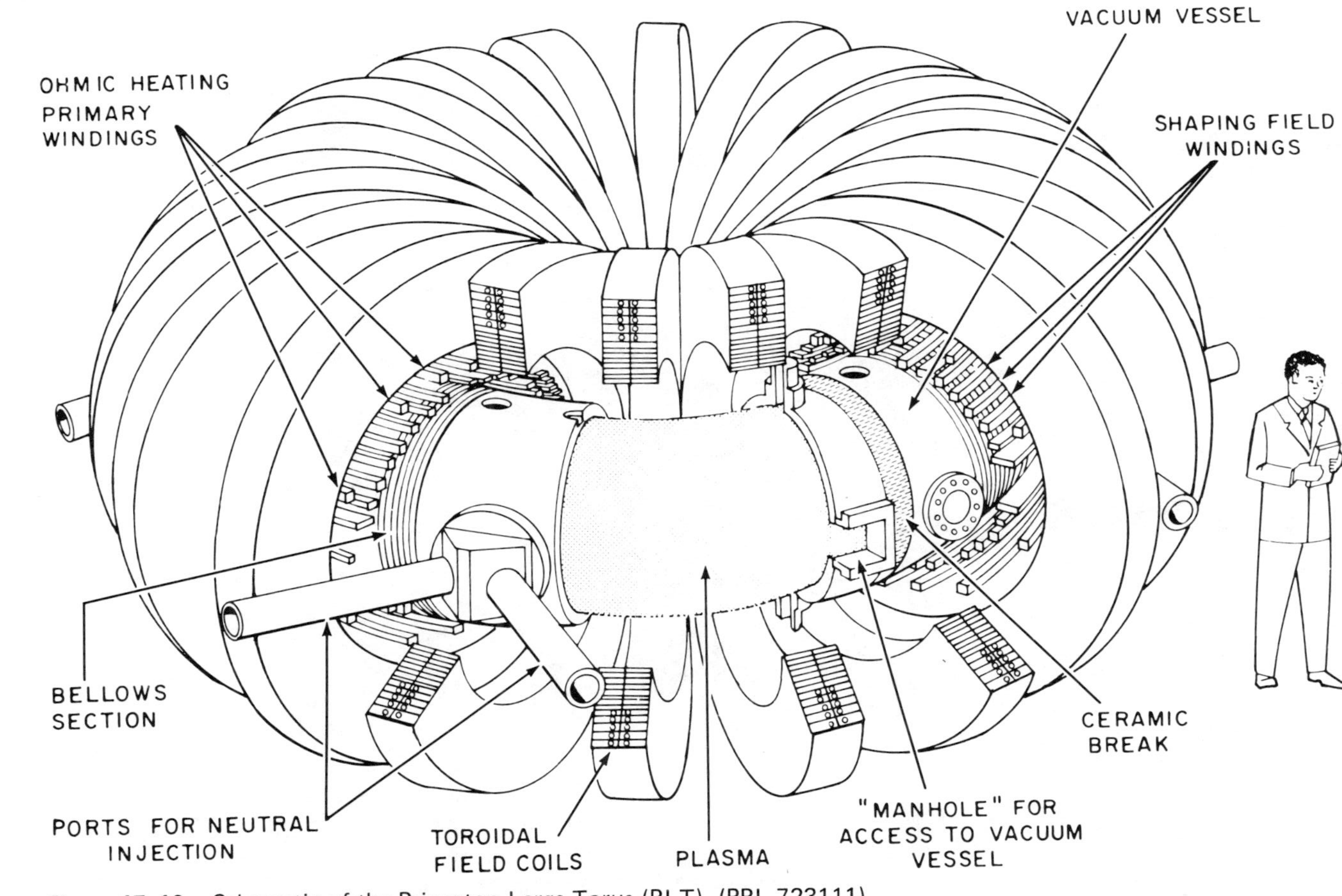

Figure 17-10. Schematic of the Princeton Large Torus (PLT). (PPL 723111)

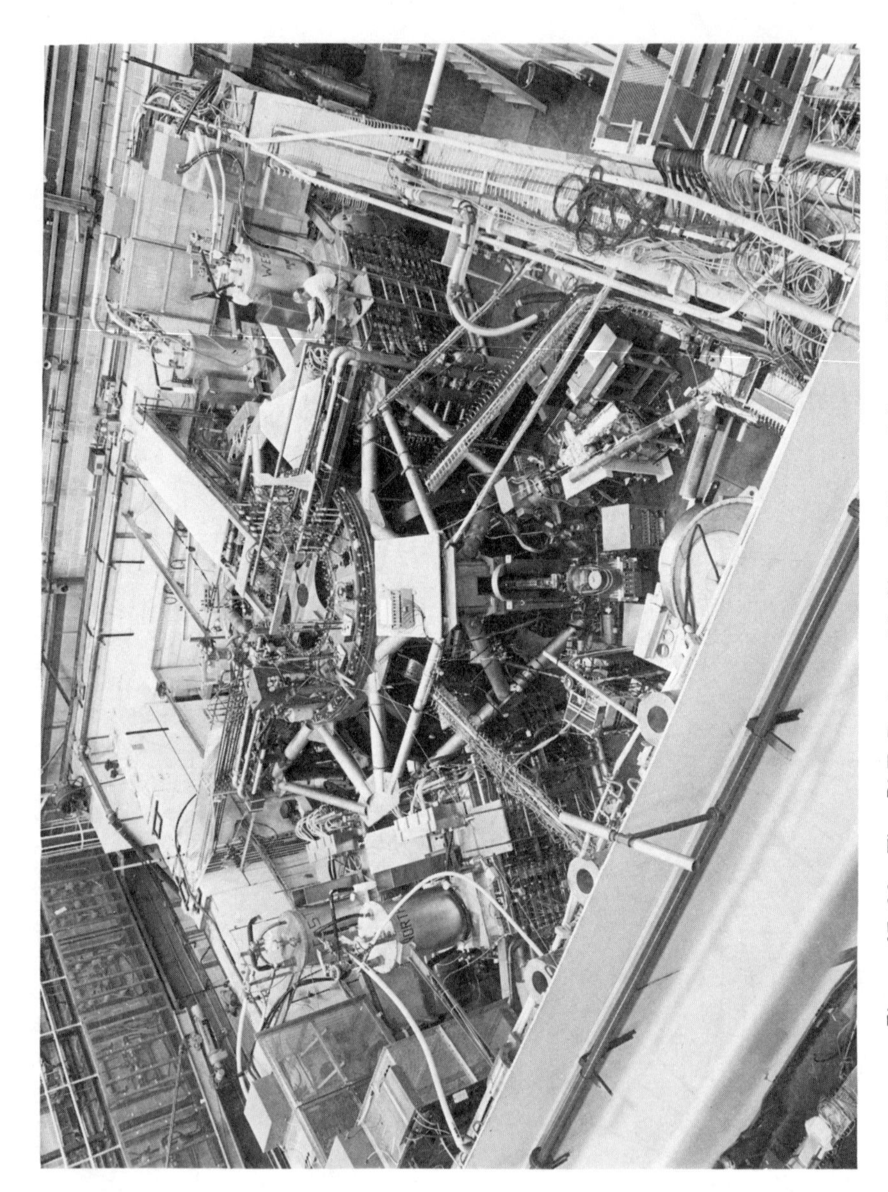

Figure 17-11. The PLT Device with Neutral Beam Injectors. (PPL 783606)

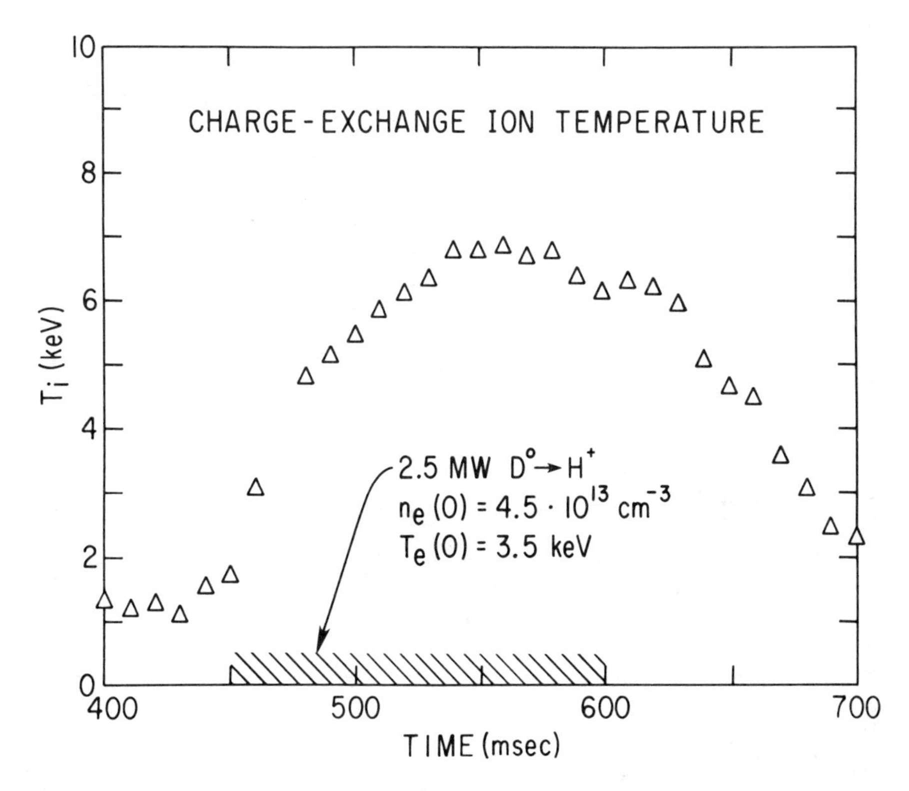

Figure 17-12. Ion Temperature During Beam Heating on PLT (Ohmic heating begins at $t = 0$, beam heating at $t = 450$ msec).

ADVANCED TOKAMAK DESIGNS

Aside from the basic problem of plasma power loss by particle transport, toroidal plasmas have been troubled by line radiation power losses due to nonhydrogenic "impurity" ions–such as oxygen and carbon, from the surface of the vacuum vessel wall, or iron and molybdenum from the wall itself. The fractional population of various impurity ions that can be tolerated in an ignited reactor is shown in Figure 17-16, along with the ranges of impurity levels in present Tokamak experiments. While the best conditions achieved thus far (including the PLT and Alcator A conditions) are compatible with ignition, it seems likely that Tokamak reactors will have to take drastic technological steps to protect themselves from the impurity problem. The most promising technique appears to be the introduction of a magnetic "divertor," which bounds the hot plasma with a magnetic separatrix, rather than with a material limiter of some sort. This approach is about to be studied experimentally in the PDX device at PPPL (Figure 17-17).

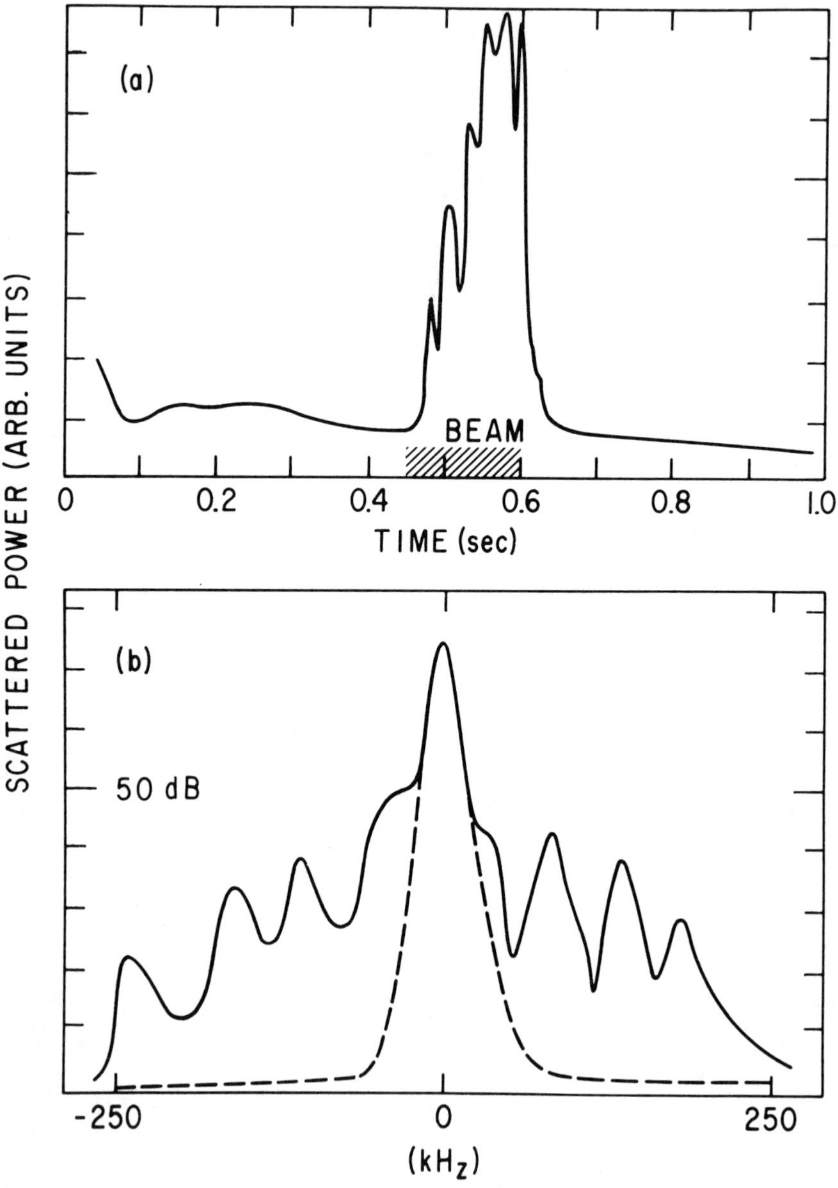

Figure 17-13. Plasma Density Fluctuation Level and Spectrum, as Inferred from Microwave Scattering During Beam Heating on PLT.

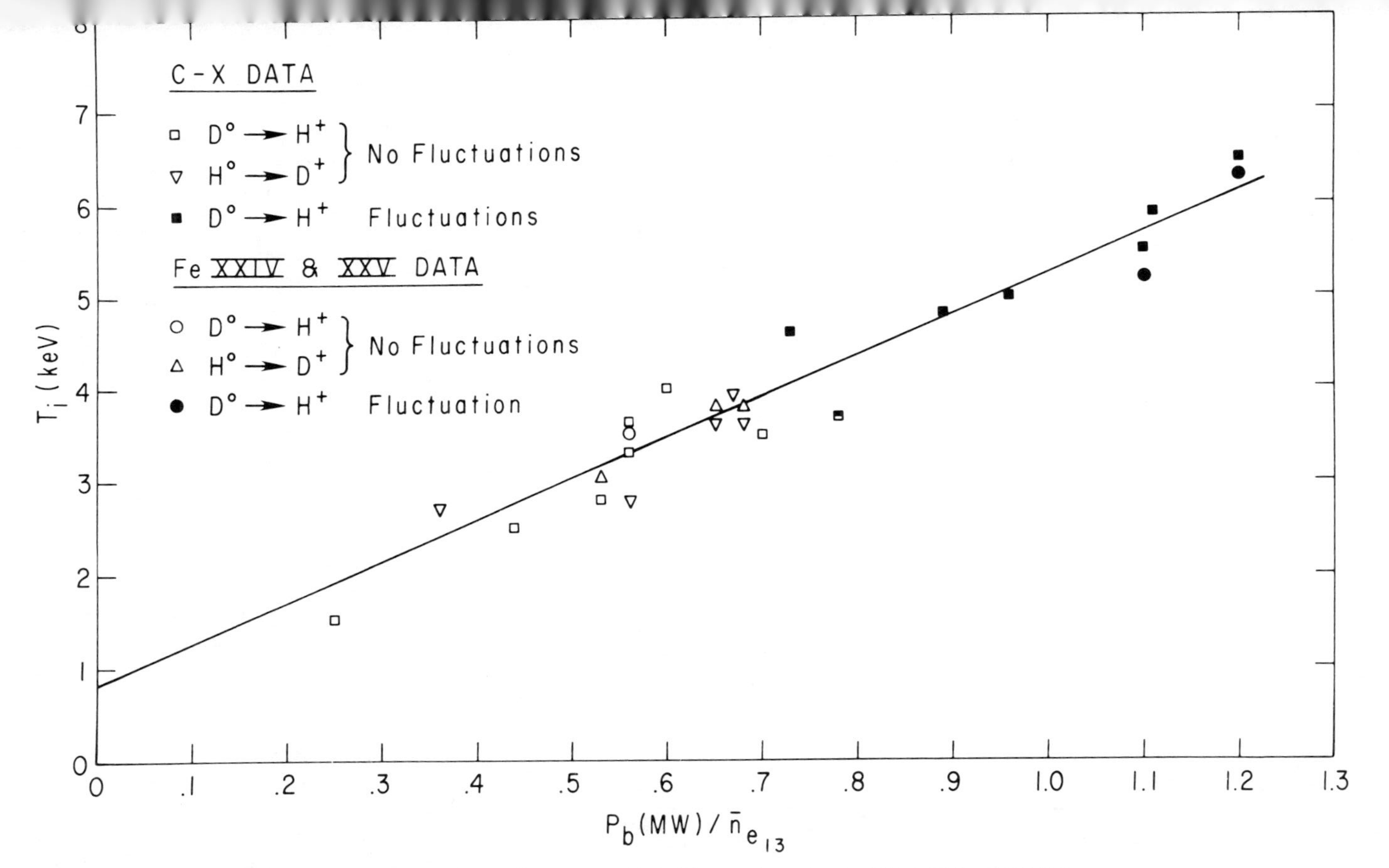

Figure 17-14. Ion Temperature versus Beam Power per Unit Plasma Density in PLT. (PPL 786285)

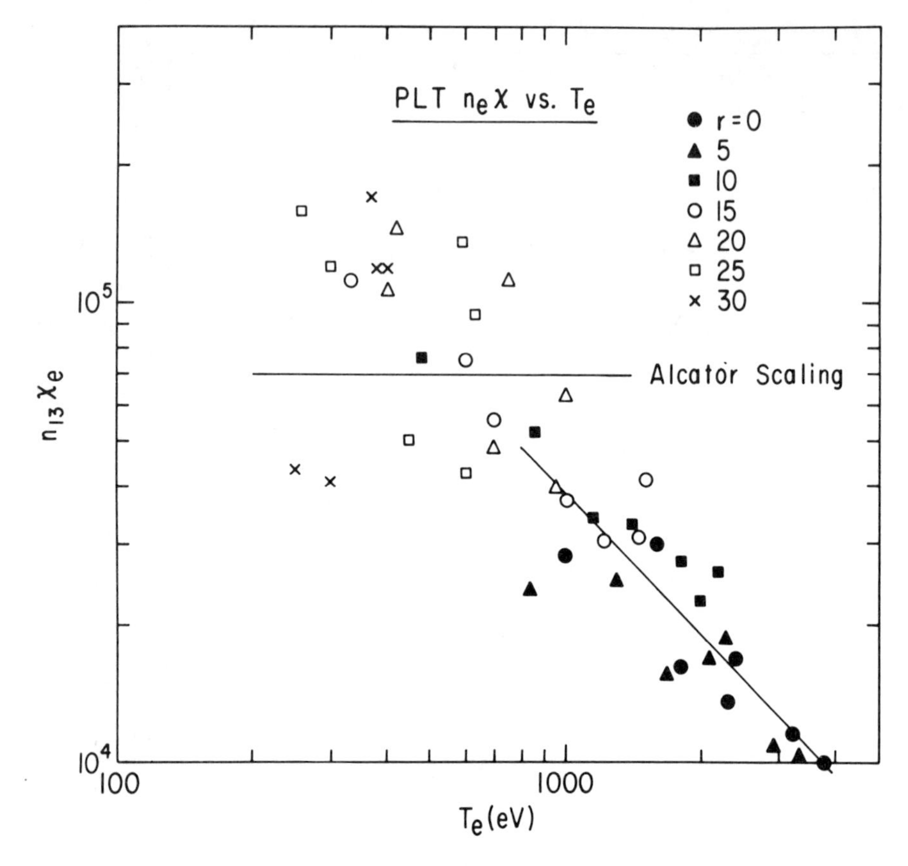

Figure 17-15. Electron Heat Conductivity χ_e times Plasma Density (in units of 10^{13} cm^{-3}) versus Electron Temperature During Beam Heating on PLT.

Departure from the conventional circular minor cross-section of the Tokamak has other benefits. The maximum tolerable ratio of plasma pressure to magnetic field pressure—the so-called beta value—depends on the plasma shape and is predicted to be particularly favorable for D-shaped cross-sections, such as that in Figure 17-17, or peanut-shaped cross-sections, such as that in Figure 17-18), which has recently entered experimental operation at the General Atomic Company. If the beta value can be raised from the level of ~3 percent, which is theoretically stable in the conventional Tokamak, to levels of 5 to 10 percent, the fusion power output rises roughly as β^2, and the impact on Tokamak reactor economics is correspondingly favorable. In previous Tokamak experiments, plasma heating has never been sufficiently intense to push the plasma pressure into the unstable range; accordingly, this important question remains to be documented experimentally.

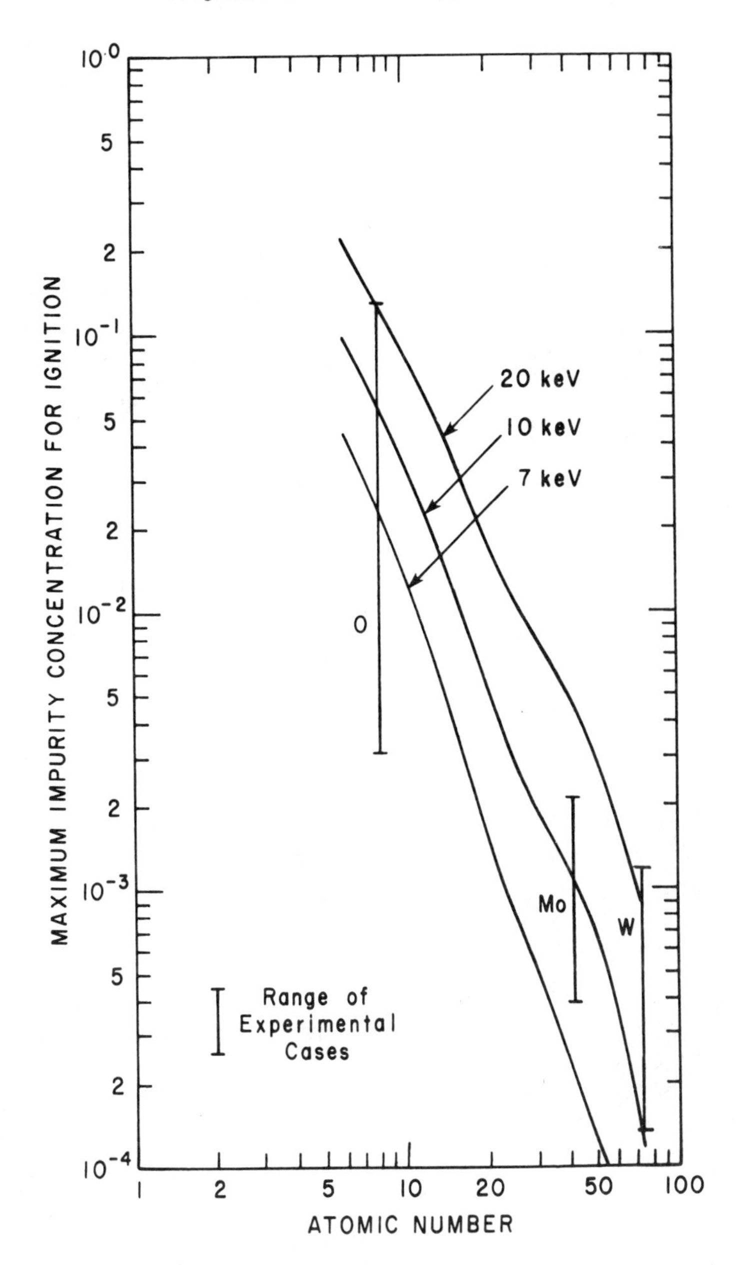

Figure 17-16. Calculated Maximum Tolerable Impurity Ion Fraction for Ignited Plasmas of Various Temperatures. The ranges of present-day impurity content are indicated; the most advanced Tokamaks operate at the bottom ends of these ranges. (PPL 783250)

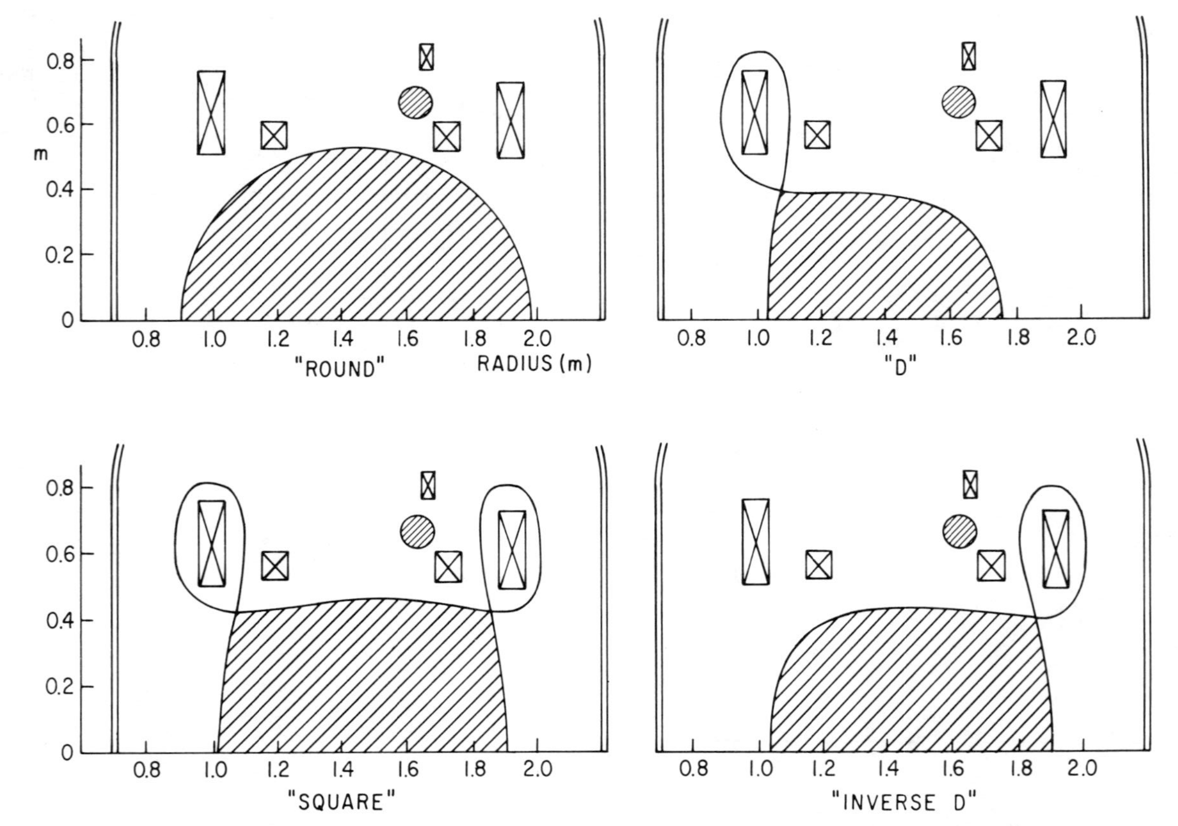

Figure 17-17. Various Divertor Configurations can be Realized on the PDX Device. The edge plasma flows along the separatrix to special collectors on the backsides of the divertor coils. (PPL 753910)

Figure 17-18. The Recently Completed Doublet III Device which Will Study the Maximization of Plasma Pressure (Beta Value) in a Tokamak Plasma of Specially Shaped Cross-Section. (PPL 783242)

FUSION POWER PRODUCTION

The typical neutron yield from the PLT experiment is shown in Figure 17-19. This yield, of about 10^{13} per pulse, is particularly large because not only the heated plasma ions but also the injected beam ions contribute to the reaction rate. The importance of an energetic ion "tail" is illustrated in Figure 17-20. For the 5 keV Maxwellian plasma in the illustration, the greater part of the reactions comes from a small percentage of the particles having energies of several times kT. (The presence of the bulk plasma may, indeed, be considered detrimental to a fusion reactor, since it contributes predominantly to the plasma pressure, without contributing comparably to the output power.) Under these circumstances, it is obviously most effective to seek to distort the plasma energy distribution function as much as possible in the direction of an enhanced ion tail. The maximum that can be accomplished along this line is to heat the bulk plasma through injection of energetic ions—exactly as in the neutral beam-heating approach. The optimum injection energy from the point of view of maximizing the Q-value lies in the range 120–240 keV.

The beneficial effect of heating the reactor plasma by means of an energetic ion beam is illustrated in Figure 17-21. The benefit is most spectacular for very low $n\tau_E$ values. As ignition conditions are approached, of course, the auxiliary heating power requirement dwindles, and the nature of the heating method becomes unimportant.

REALIZATION OF REACTOR CONDITIONS

The gradual improvement in toroidal plasma parameters is illustrated in Figure 17-22. At the present time, only a modest further step is required to reach the Lawson condition, and a much smaller step (i.e., $T \gtrsim 5$ keV at $n\tau_E > 10^{13}$ cm^{-3}sec) will be sufficient to achieve $Q > 1$ in the beam-heated reactor regime.

In 1974, when the near term Tokamak reactor outlook began to appear seriously promising, construction of the TFTR device (Figure 17-23) was undertaken at PPPL. The initial objectives of the TFTR are listed in the first column of Table 17-4. The selection of the TFTR parameters and objectives represents a previous instance where a compromise had to be struck between desirable and feasible strategies; the outcome has proved to be gratifying thus far. As Tokamak experimental results have accumulated, the TFTR base plan has advanced from an initial phase, where it was viewed as audacious and a bit fantastic, to its present status of being generally regarded as very sound and a bit modest. This possible development was foreseen in the original TFTR design, and a TFTR Improvements Project (TIP) is now under consideration, which is aimed at the second column of objectives in Table 17-4. If present favorable trends continue, it appears possible that the $Q = 1$ point may be exceeded substantially on TFTR before the mid-1980s and the phenomenon of

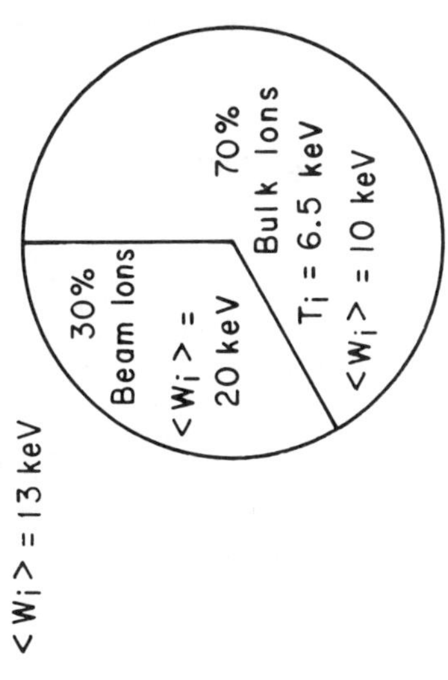

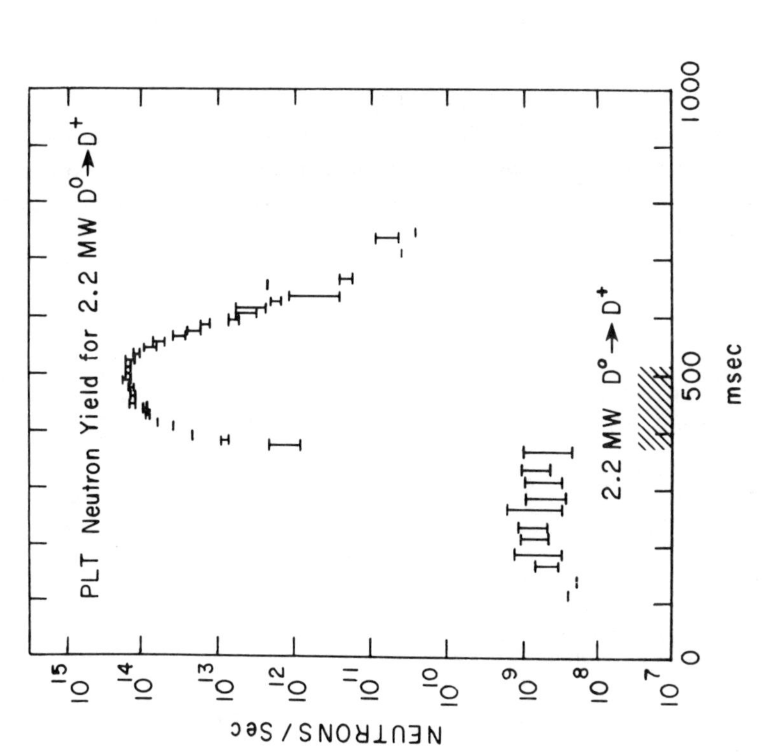

Figure 17–19. Fusion Yield from D-D Reactions in PLT. If D-T were used, the ideal *Q*-value (not counting injector inefficiencies and magnet coil power requirements) would be 2 percent.

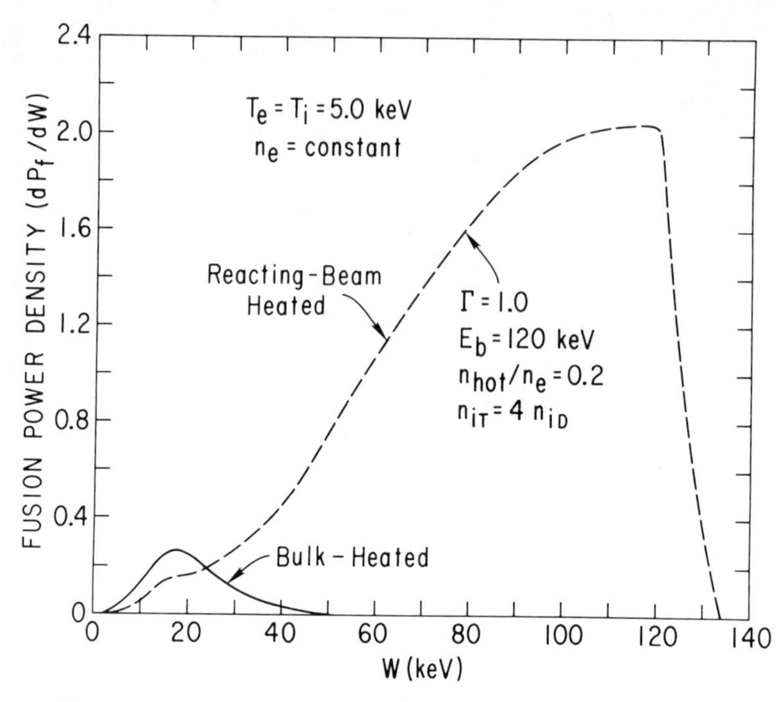

Figure 17-20. Fusion Power Density Distribution Relative to Plasma Particle Energy for a Maxwellian 5 keV Plasma Compared with a 5 keV Plasma Heated by a 120 keV Beam. The presence of the injected ions serves to double the total plasma pressure ($\Gamma = 1.0$), while greatly increasing the total fusion power. (PPL 783232)

Table 17-4. Objectives of the TFTR Project.

	Basic Operation	*Extended Operation*
1. Test tokamak performance in reactorlike plasma	$T_i \sim T_e = 5 - 10$ keV	$T_e = 10$ keV $T_i = 20$ keV
2. Control high current equilibria	$I = 1.0 - 2.5$ MA	$I = 3.0 - 3.5$ MA
3. Study confinement at high $\eta\tau_E$ levels	$n\tau_{E\epsilon} \sim n\tau_{Ei} \gtrsim 10^{13} \mathrm{cm}^{-3}\mathrm{sec}$	$n\tau_{E\epsilon} \sim 3 \cdot 10^{13} \mathrm{cm}^{-3}\mathrm{sec}$ $n\tau_{Ei} \gtrsim 10^{14} \mathrm{cm}^{-3}\mathrm{sec}$
4. Study α effects at reactor-like fusion power density	$P_F \sim 1\ \mathrm{Wcm}^{-3}$	$P_F \sim 3\ \mathrm{Wcm}^{-3}$
5. Demonstrate Q-values of practical significance	$Q \sim 1$	$Q \gtrsim 2$

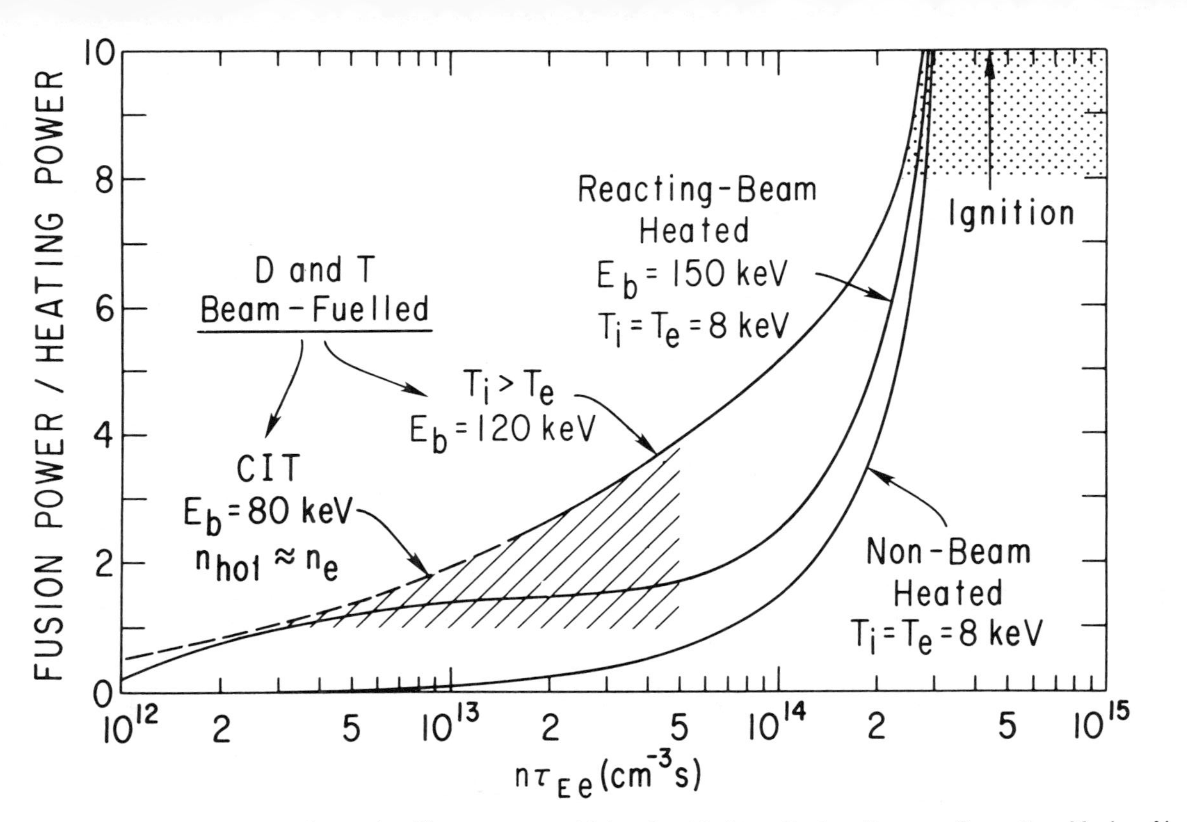

Figure 17–21. The Dependence of Q on the Electron $n\tau_{Ee}$ Value for Various Fusion Reactor Operating Modes. Non-beam-heated plasmas have Q-values below unity until ignition is approached. Beam reactions can extend the $Q > 1$ regime down to low values of $n\tau_{Ee}$. If $n\tau_{Ei} >> n\tau_{Ee}$, and the plasma ions are heated preferentially, as by an energetic beam, appreciable Q-values can be reached for $n\tau_{Ee} < 10^{14}$ cm^{-3} sec. (PPL 783440)

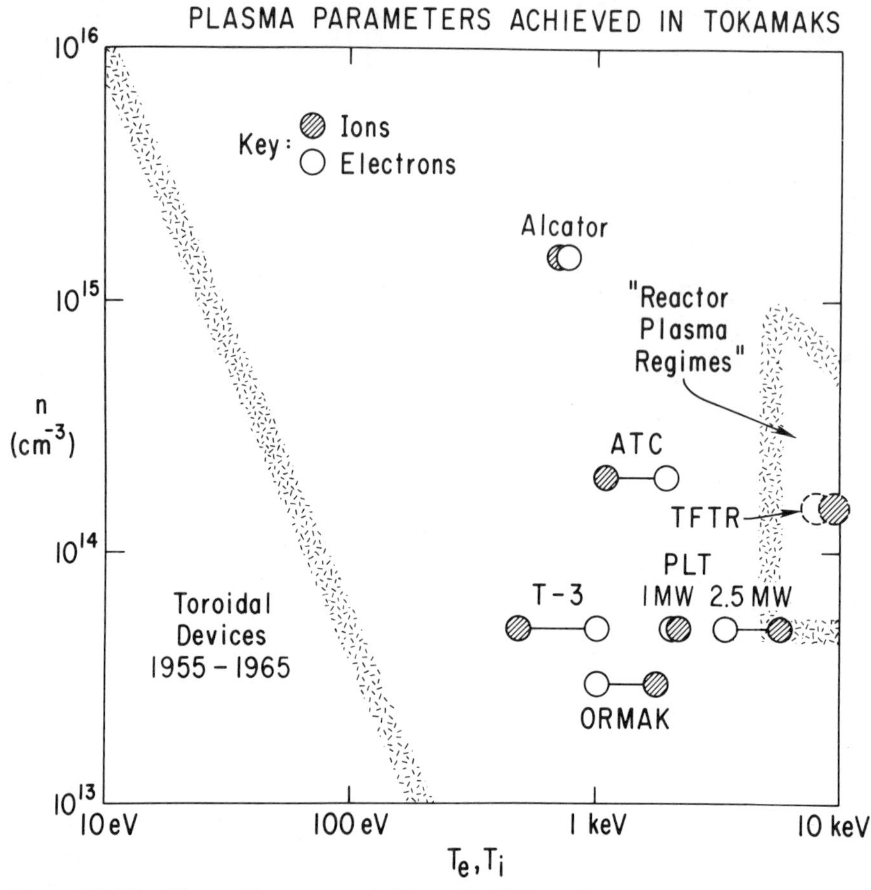

Figure 17-22. Plasma Parameters Achieved in Tokamaks.

self-heating by the fusion reaction products may become a significant factor in the plasma power balance.

THE TOROIDAL REACTOR OUTLOOK

The multiplicity of possible technological approaches has always been a distinctive feature of fusion research. During the period before 1968, this multiplicity was generally taken as an unfavorable sign. Since every one of the diverse fusion reactor approaches seemed to be encountering experimental limitations at a low level of performance, the situation was a little reminiscent of other "impossible" historical research undertakings, such as the transmutation of the elements or the achievement of perpetual motion—which have characteristically stimulated a great diversity of approach.

In the pre-1968 era of fusion research, the cumulative weight of negative

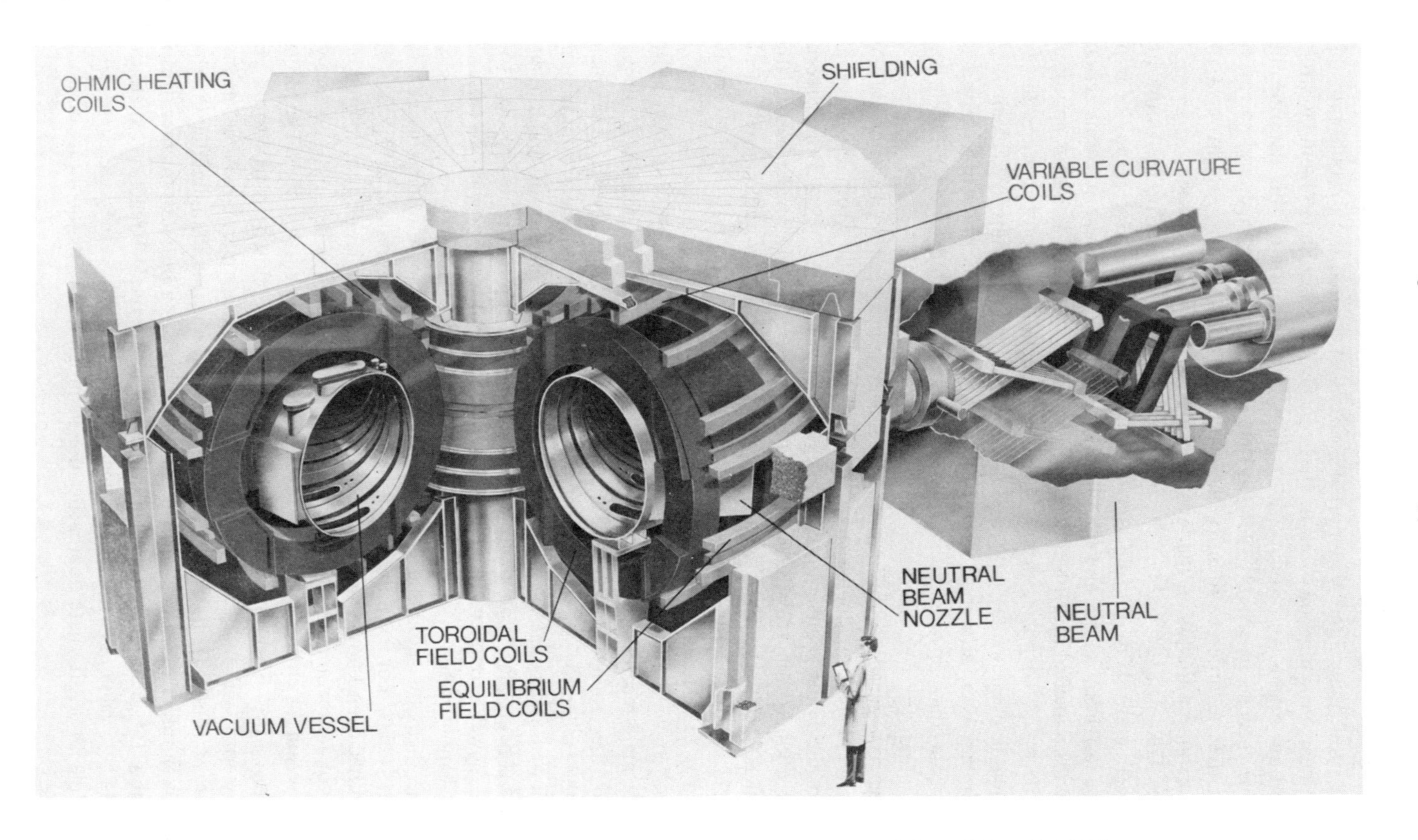

Figure 17–23. The Tokamak Fusion Test Reactor (TFTR) Under Construction at PPPL. (PPL 764831)

experimental results led to a kind of "conventional doctrine"—not based on logical proof, but broadly supported by experience—that there would probably be no near term solution at all or that conceivably some single fabulous stroke of fortune might yet produce a solution. In the spirit of this doctrine, the funding agencies persistently advised that the fusion program should narrow its alternatives in order to identify the successful approach—or else arrive at the conclusion that the fusion goal was beyond present-day capabilities.

The Tokamak successes of the late 1960s were at first interpreted from within the framework of the old conventional doctrine: a single clear ray of hope had at last been found, and the obvious course was to pursue it. Some serious questions remained in 1968 concerning the feasibility of the Tokamak approach, but the predicted obstacles have faded with remarkable consistency as the experimental program has pushed forward. If the present state of Tokamak research had been anticipated ten years ago, it would certainly have served to strengthen the conviction that the right solution to the fusion problem had been found.

The past decade of fusion research, however, has been marked not only by an increasingly firm documentation of Tokamak feasibility, but by an evolution of the whole "conventional doctrine" about fusion prospects. During recent years, a kind of doctrinal inversion has taken place—largely on the basis of the Tokamak results themselves, but partly on the basis of favorable developments in other approaches: the multiplicity of fusion alternatives now tends to be regarded as a resource rather than as a liability.

The conventional doctrine of the late 1970s is that many (or even most) of the main reactor alternatives have prospects of near term feasibility. In particular, the tandem mirror machine, the bumpy torus, the reverse field pinch, and the stellarator—as well as a number of inertial confinement approaches—are viewed as having favorable prospects. In the new climate of opinion, emphasis is being placed on identifying the ultimately most desirable approach to a fusion power economy, rather than on proceeding directly toward an early reactor demonstration.

The focus on economic desirability has stimulated some interesting new research trends. For example, the spherical plasma configuration of Fig. 17-24, which had been known since the 1950s but had been left largely unexplored because of its theoretical and experimental inaccessibility, is beginning to receive more serious attention from both toroidal and mirror research groups. The special feature of the "spheromak" is that the field lines are closed, as in a torus, yet the external coil and blanket system avoids linking the plasma volume, as in a mirror machine. If the problem of generating and maintaining the internal plasma currents of the spheromak can be solved in a practical manner, this is probably the ideal reactor configuration, at least for the D-T reaction, where the plasma beta value need not be extremely high.

While these alternate approaches are in progress, the major thrust of world

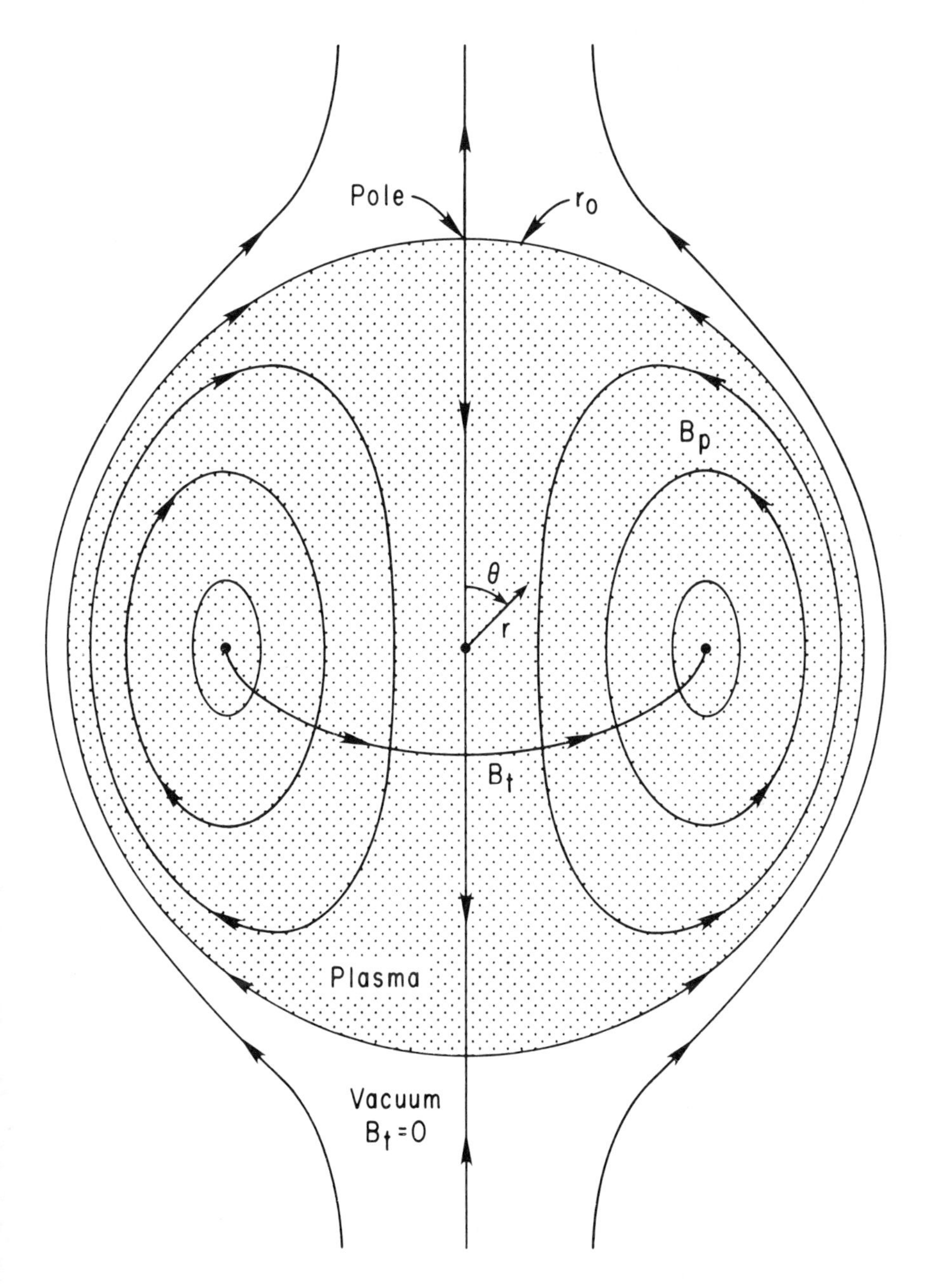

Figure 17-24. The "Spheromak" has an Internal Toroidal Field, Maintained by Poloidal Plasma Currents, and also a Poloidal Field Maintained by Toroidal Plasma Currents and External Axisymmetric Coils. (PPL 783065)

Table 17-5. Toroidal Fusion Reactor Outlook in 1978.

- The Tokamak approach extrapolates to a practical reactor without supposing any improvements in the physics or innovations in plasma-heating technology.
- The achievement of a Tokamak reactor plasma within the next decade is assured.
- Nontoroidal alternates and alternative toroidal approaches lag behind, but appear promising.
- The timing of fusion power development has now become a question of strategy.

fusion research continues to follow up the Tokamak initiative. Increasingly detailed reactor technology studies, based on increasingly credible plasma physics assumptions, are being carried out, and the practical outlook appears favorable. The central problem at present is to select the wisest strategy for launching fusion power on a prosperous initial course (Table 17-5).

Chapter 18

Terrestrial Sources of Carbon and Earthquake Outgassing*

Thomas Gold
Cornell University

ABSTRACT

The earth has replenished its surface with carbon throughout geological time. The supply may have come from hydrocarbons originally included in the body and outgassing largely through faults in the crust. Combustible gases are frequently released in earthquakes and seem to be an essential part of that phenomenon. New sources of fuel and improved earthquake prediction may come from a better understanding of the processes.

INTRODUCTION

Volcanic processes are known to facilitate the outgassing of some regions of the lithosphere or mantle of the earth. It is not known, however, how much outgassing occurs in nonvolcanic regions, either as a general diffusion through the ground or associated with major fault lines and with earthquakes. Limits can of course be placed on the rates at which certain gases could be added to the atmosphere at the present time or in geologic history, but such limits would admit the possibility of a range of interesting effects. In particular, a set of

* The author wishes to acknowledge many helpful discussions with colleagues, including Sir Fred Hoyle and Dr. G. F. MacDonald, that have contributed to the viewpoint presented. The extensive help and keen interest of Dr. S. Soter is gratefully acknowledged; he is responsible for finding much of the supporting material cited and for many improvements of the text. A fuller joint account with him concerning the relation of gas emission to earthquakes is under preparation. The work was supported by a grant from the National Science Foundation (AST-17838 A02).

phenomena that are now well recognized to be associated with earthquakes have no adequate explanations without the assumption that large quantities of high pressure gas escape from the ground into the atmosphere at such times. These phenomena include sudden and surprisingly large changes in ground conductivity preceding and during earthquakes, even hundreds of kilometers from the epicenter; changes in ground water level; changes in the ratio of compressional-to-shear wave velocities; sharp increases in the radon concentration in ground water and in the atmosphere; and in the case of large earthquakes, frequently the appearance of startling luminous effects in the sky fanning out from the surface.

The radon phenomenon clearly implies the presence of a carrier gas, and in adequate amounts, such a gas can account for all the other phenomena. Gases escaping from deep below and at very high pressures can pervade the porosity of the shallower ground, rapidly displace ground water, change the pressure in microcracks, and of course transport radon to the surface in a time short compared with its 3.8 day radioactive half life.

No general studies of the possibility of earthquake-related gas emission seem to have been undertaken, and there is little direct evidence concerning the chemical nature of such gases. Gases that are known to be associated with deeper layers of the earth's crust are CO_2, CO, CH_4, H_2, H_2O, and H_2S, as major components, as well as small quantities of noble gases. It is the phenomenon of luminosity in the air that gives the only direct clue—namely, that the escaping gases generally are combustible.

Although attempts have been made to discuss the luminosity effects in terms of electrical atmospheric phenomena, the high conductivity of the ground really rules out any such explanation. We have pursued many detailed eyewitness accounts of the luminosity phenomena, and these leave little doubt, as we shall see, that one is concerned with combustion of gases coming out of the ground. Methane, carbon monoxide, hydrogen, and hydrogen sulfide are candidates, and some of those must be involved in combustible concentrations. We believe that ignition of any combustible gases coming out of the ground from a high pressure source will always tend to occur through the frictional electricity of dust particles carried in the stream. Of the combustible gases, we consider methane the one that is likely to be the most abundant, and in the following discussion, we concentrate attention on this; however we recognize that any of the other combustible gases may also be involved and that noncombustible gases may dilute the medium, but generally not to the extent of making combustion impossible.

Continuous outgassing may have important consequences for tectonic processes. Thus it has been stressed by several investigators of seismic phenomena that the deeper earthquakes cannot be understood except with the hypothesis that a pore fluid is present and at a pressure of at least the lithostatic value.[1] Only then can a shear stress result in a sudden slippage along a surface;

without such a pore fluid, the high rock pressure would cause shear stresses to be discharged by a gradual and distributed deformation of the rock only. What the rate of loss of such a pressurized fluid would be, and how it is replenished, does not appear to have been discussed.

Since any such fluid—H_2O, CO_2, CO, CH_4, N_2—would always be lighter than rock, one has to suppose that in fractured rock it will generally move upwards. Even without the fractures caused by an earthquake, the tendency for upward migration must always be there, as soon as the pore spaces interconnect over some distances, for then the pressure head in the fluid cannot be balanced everywhere by the rock pressures. In particular, the upper zone of an interconnected domain will generally have an excessive, and the lower part an insufficient, pressure to balance that of the rock; the lower pores will thus tend to close and the upper ones open, making for a general upward migration of the fluid. A downward transport of fluids, to make up any losses, can only occur with a subduction of a mass of rock. Thus, subducted sedimentary rocks may supply water, or CO_2 from the dissociation of limestone, or other contained volatiles. But deep earthquakes occur in areas where there has been no large-scale subduction and where such a source of fluids cannot be expected. If fluids are present there, a supply of volatiles in the primeval material of the earth must be suspected.

A continuous (though by no means steady) outgassing process may account for the ability of rocks to fracture suddenly and discharge shear strain by slipping along a surface, even at great depth (earthquake foci occur down to 700 km). But several other features of earthquakes are now known that may well depend upon the supply of gases. Thus, the usual occurence of aftershocks following major earthquakes requires a time constant for a redistribution of stresses; but this cannot be found in plastic flow, since this would generally allow the gradual discharge of the strain in the first place. If pore fluid pressure is involved, the possibility suggests itself that after a major earthquake and the subsequent escape of fluids through the cracked rocks, the remaining fluids will redistribute themselves with speeds dependent on the porosity. Subsequent shocks may represent the consequences of this gradual redistribution.

The range of other earthquake-related phenomena that have been identified (changes in the water table, escape of radon from the ground, changes in the seismic velocities, flaming and other pseudovolcanic phenomena, and also the well-documented strange behavior of some animals preceding an earthquake) may well all be caused by the gases from the deep ground. First, there may be a slight leakage at the top as a gas mass makes its way upward at some depth; and later, if an earthquake has indeed been facilitated by this gas, there will be the more violent effects of gas escaping through the new cracks generated. The precursory effects will involve frequently no more than the slow displacement of the gases normally present in the porosity of the shallow ground, but this may be a process that is readily perceived by the sense organs of some animals (i.e., sense of smell of dogs and pigs, low frequency acoustical effects, asphyxiation for ground-dwelling animals, etc.).

Much of the hydrocarbon supply of the earth that is at present being exploited shows strong evidence of biogenic origin. Petroleum deposits and gases associated with such deposits must have derived, at least in part, from biological materials. The evidence for this is strongest for the youngest petroleum, but gets weaker for the oldest.[2] Hydrocarbon gases unassociated with petroleum and coming from depth carry no definitive information to identify their origin. Such nonassociated natural gas is known in many regions, and in exploitable quantity, where no ready explanation for a biogenic source is at hand. Even in cases where hydrocarbon gases are associated with petroleum, it is not certain that they derived from it or from a common biogenic source. It is also possible that petroleum reservoirs become augmented through absorbing and polymerizing hydrocarbon gases that get into the reservoir from below. The disposition of oil fields along major fault lines can perhaps be understood in that way and also the fact that the older oils are generally the more hydrogen-rich. If, indeed, some outgassing of primeval materials is taking place, one may suspect hydrocarbon gases to be a component of these, since carbon in the early solar system was probably in hydrocarbon compounds to a major extent and is still found in that form in the most primitive meteorite material. Thus, terrestrial hydrocarbons may indeed have two different types of origin, as Robinson suggests.[3]

Information that combustible gases are a widespread constituent in the crust of the earth cannot fail to be of interest from the point of view of the availability of natural fuels. The quantities that may have escaped in this way and contributed to the carbon on the surface of the earth in all of geologic time may be very large when compared with fuel requirements. Thus, if primeval methane had been the chief source of carbon, the amount necessary to produce all carbon in the sediments would be the equivalent of twenty million years of present-day fuel consumption. There is no clear indication that the outgassing of carbon has come to an end, and the amounts remaining in the earth may still be very large compared with any foreseeable human requirements. No doubt most of it is too deep and probably too diffusely distributed to be accessible to exploitation; but even a very small fraction that may have become concentrated in the vicinity of faults and temporarily stored at a shallow depth may still be a major item when compared with the quantities of known fossil fuels. The detailed investigations of earthquakes and of many fault lines will therefore be of interest.

THE SOURCES OF TERRESTRIAL CARBON

The derivation of all the carbon on the surface of the earth and in the biosphere is not known with any certainty. The bulk of it is in the form of carbonates, chiefly $CaCO_3$, whose derivation as an ocean precipitate is clear, with atmospheric CO_2 being the source material for the carbon. The two basic schemes that can be discussed are (1) that all the biospheric carbon came into the atmosphere early in the history of the earth, that it was then quickly precipitated chiefly as carbonates, and that these carbonates subsequently were reworked by erosion and by volcanic heating to produce a continuous (but by no means constant) supply of CO_2 to the atmosphere ever since; or (2) that the source material for the atmospheric CO_2 has remained locked up in the body of the earth, being released gradually at a slight but significant rate. The evidence of the deposits seems to favor the second alternative. However, the method by which carbon from the interior of the earth slowly became concentrated on the surface is still quite uncertain.

If the meteorites are in any way representative of the materials that contributed to the construction of the earth, then there are two major possibilities: either the carbon was supplied in comparatively high concentration as hydrocarbon compounds, as in the carbonaceous chondrites, and then outgassing from a comparatively small amount of such material and from shallow depths would suffice to produce the observed amount; or the carbon was supplied in the form of carbonates, carbides, and elemental carbon, as present in many meteorites but in much smaller concentration, and then outgassing from a volume approximating the entire mantle of the earth would have been required.

Such extensive outgassing would have required an epoch during which the mantle of the earth was largely molten. One would have expected that most outgassing would have occurred then and very little after it solidified. Yet the evidence of the deposits of carbon seems to favor a much more continuous supply.[4]

The case for an early complete outgassing and a subsequent reworking of the sediments resulting from an early massive CO_2 atmosphere has been argued.[5] But there is no evidence for such large quantities of very early carbonate deposits, and they would need to have been rather completely subducted to escape detection. Also, a massive CO_2 atmosphere would need to be coupled with a substantially weaker sun in order to avoid a high temperature on the earth that would preclude the deposition of carbonates. Still, these are possibilities that cannot be ruled out—but there is no case here that is so strong that the alternative does not need to be considered.

The atmospheric content of the noble gases has been used to estimate what fraction of the earth has been outgassed. This is dependent upon the critical assumption that the building material of the earth was similar in its rare gas content to that of the small sample of apparently primitive material that present-day meteorites provide. If the earth was constructed from material that had a more complex history, and accreted inhomogeneously, these considerations would not be applicable. Even using such assumptions, the range of meteorite rare gas concentrations[6] would allow estimates ranging from 4 to 100 percent of the earth having been outgassed.

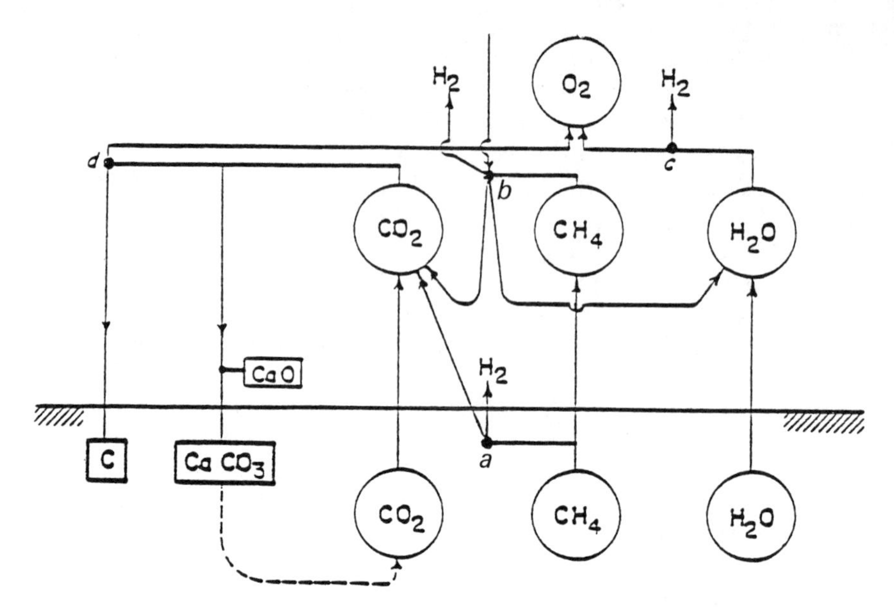

Figure 18-1. Schematic Flow Diagram for Carbon Supply. CH_4, CO_2, and H_2O are shown as major volatiles below the surface. Chemical processes are indicated by small letters. Biological cycles are omitted. The diagram indicates that CH_4 may reach the surface without modification or, if the pathway favors oxidation (indicated as process a), that the carbon may emerge as CO_2 (or as CO, not indicated). The necessary oxygen may be supplied either by water or by metal oxides in the rocks. (The hydrogen may emerge as molecular hydrogen into the atmosphere or oxidize to water in the ground or form hydrides in the rocks.)

Atmospheric CH_4 will be oxidized and the carbon will finally end up as CO_2 (process b) using atmospheric oxygen. The hydrogen may escape from the atmosphere or make more H_2O. Atmospheric H_2O suffers photodissociation to produce a fresh supply of O_2 (process c). The hydrogen must escape for this process to yield a net oxygen supply. Atmospheric CO_2 is finally lost from the atmosphere by forming carbonate deposits, with calcium oxide eroded from rocks, or by deposition as unoxidized carbon, having been reduced by photosynthesis (process d) in plants.

Urey[7] gave reasons for considering that the original condition of carbon in the accretion of the earth was in the form of hydrocarbons, and many authors have since followed this line of reasoning. Whether this implies that the supply to the surface over geologic time was largely methane or that earlier processes had oxidized this is not clear. In each case CO_2 would be made available in the atmosphere very quickly and provide the source material for laying down the carbonate deposits that form more than two-thirds of all surface carbon now. Methane is estimated to be destroyed in the atmosphere in a period of between four and seven years,[8] with essentially all the carbon ending up as CO_2. If methane from the interior formed the main source of a continuous carbon supply to the surface, because of its short atmospheric residence time, it would still imply only a small methane concentration in the atmosphere, and as we shall see, the present atmospheric composition does not rule this out.

The view that carbon had emerged as CO_2 and not as hydrocarbons appears to have been based chiefly on the observation that CO_2 is abundant and hydrocarbons rare in volcanic gases. This observation has nothing to say as to the original form in which the carbon was held in the earth before emerging through a volcanic zone; one understands that the equilibrium between hydrocarbons and CO_2 in the presence of potential oxygen donors, such as water and metal oxides in the magmas, would greatly favor almost all carbon to emerge as CO_2 irrespective of the original form. Outgassing by other pathways, where the temperature is low when the pressure is low, would retain hydrocarbons if they were present in the original source, and among those, methane would be favored as being the most stable against dissociation.

Discussions of the cosmochemistry of the early solar system make it seem probable that carbon was initially much more in the form of hydrocarbons than in oxidized form. If the meteorites are in any way representative of materials in the early solar system, then it is relevant that the ones that are rich in carbon—namely, the carbonaceous chondrites—have it mostly in hydrocarbon form. At temperatures and pressures occuring down to several hundred kilometers, some hydrocarbons, including methane, would still be stable, and one has every reason therefore to inquire whether hydrocarbon outgassing is taking place.

The total quantity of carbon that has been supplied to the surface has been estimated by various authors. We shall consider the range given by Rubey's estimate[9] of 2.5×10^{16} tons and that of Galimov, Migdisov, and Ronov[10] of 7.4×10^{16} tons of total carbon in sediments (the atmospheric, oceanic, and biomass carbon is a small quantity by comparison). For this to be supplied steadily over four and a half billion years, between 1 and 3×10^{10} cubic meters of CO_2 or CH_4 would need to be provided on an average per year.

One method of estimating at least limits on the amount of primeval methane that may be entering the atmosphere is to observe the proportion of ^{14}C in atmospheric methane and to compare that with the proportion in atmospheric CO_2. Since ^{14}C has a half life of 5,730 years, this will define what contribution is derived from sources that have not interchanged carbon with the atmosphere for many thousands of years and therefore lack ^{14}C. Methane from petroleum operations and industrial processes is in that category, as is also the product of decay of some buried biogenic deposits, in addition to any primeval methane.

The estimates given by Ehhalt[11] are that 20 percent of the atmospheric methane lacks sufficient ^{14}C to be of recent biological origin, but that only between 7.4 and 38 percent of that can be accounted for by industrial and known natural sources. His estimates would entitle one to consider that unknown sources of ^{14}C-free CH_4 amount to between 2.7 and 1.1×10^{11} cubic meters per year at the present time. This may be compared with the estimates of the time average of the total carbon supply. According to those, if methane were the chief source, an average of between 1 and 3×10^{10} cubic meters per year would need to be supplied. This is still only a small fraction of the "unaccounted for" atmospheric ^{14}C-free methane.

We cannot judge at the moment whether there indeed is such a primeval methane supply that is at present a little above its long-term mean, or whether the discrepancy is due to uncertainties in the estimates. But the figures serve to show that contributions of primeval methane, very significant for the total surface carbon content, would still not have produced any very prominant effect in the atmospheric methane budget.

The other carbon isotope determination, the ratio of ^{12}C to ^{13}C, is also not conclusive in deciding whether any methane that is found constitutes a primeval supply or is the result of a biogenic deposit. Carbon that is biogenically deposited is generally isotopically light—that is, contains less ^{13}C—by between ten and thirty parts per thousand compared with marine limestone (this is usually expressed on the PDB carbon isotope scale[12] as $\delta^{13}C$, which is the deviation, expressed in parts per thousand, from the adopted standard $^{13}C/^{12}C$ ratio of 1123.72×10^{-5}). In almost all circumstances, however, the oxidized carbon is found to be heavy and the reduced carbon light, and this can be understood, at least in part, as an equilibration process, dependent on the temperature at which reactions take place.[13] The fact that most methane coming from depth is isotopically light has been interpreted as implying a biogenic origin, since biological processes are known to select in favor of the light isotope for reduced carbon. However, a range of light carbon exists in the hydrocarbons of carbonaceous chondrites also; moreover, the process of cracking that would be responsible in the crust for generating methane from complex hydrocarbons is known to favor the light isotope, and the process of slow diffusion of methane through micropores in the rocks may further contribute to such selection. (For the heavy CO_2 molecule this effect would be small.) It is not at all clear that biology is unique in its ability to shift the $\delta^{13}C$ value by ten or twenty parts per mil towards the light side,[14] which is all that is involved in the case of sources of methane that may be candidates for a primeval supply. (Fuex[14] gives a range of -25 to $-30°/00$ for the $\delta^{13}C$ value of "geothermal methane," while the mean of all terrestrial carbon appears to have a value around $-10°/00$.)

The meteorites demonstrate a large amount of isotopic fractionation, with the carbonates heavy and the reduced carbon light, but with a range between them still larger than generally occurs in terrestrial carbon. Carbonaceous chondrites[15] have a range of $\delta^{13}C$ values from +60°/00 for the carbonates to −30°/00 for the reduced carbon, with the mean for most samples being around −10°/00, similar to the terrestrial average. In this case equilibration and diffusion are generally discussed as responsible, and not biological, processes.

Terrestrial igneous rock is reported to have a rather evenly distributed concentration of 200 ppm of reduced carbon with a $\delta^{13}C$ value around −25 to −28°/00. Hoefs[16] considers that this uniform distribution would be difficult to understand on a basis of biogenic sources and thinks it more likely that it is of primary origin. He concludes that such isotopic selection may arise abiogenically. This range of $\delta^{13}C$ values is very small compared with most other materials, and it is just the same as the small range quoted by Fuex[17] for geothermal methane. One may wonder whether a generic connection exists between these two forms of reduced carbon.

There are many other problems in the interpretation of the terrestrial carbon isotope variations that we have not touched upon here; they leave an uncertainty whether the supply isotope ratio has been constant over geologic time,[18] and they make a suggestion that the present supply is lighter than recycled archaen carbonates would have made it.

It has sometimes been stated that hydrocarbons cannot be present at great depth in the crust, since the temperatures there would be high and cause decomposition. However, the stabilizing effect of the pressure is very important. We may assume that a temperature of 1300°K is reached generally at a depth of approximately 100 km and therefore at a pressure of 33 kilobars.[19] It appears that some hydrocarbons are quite stable under these conditions. Detailed thermodynamic calculations have been carried out by French[20] and by Karzhavin and Vendillo[21] of the equilibrium between CO_2, CO, CH_4, H_2, and H_2O for temperatures up to 1500°K and pressures up to 5 kilobars, and these show the important stabilizing effect of pressure, in particular on methane, which would be the major component in the 1500°K and 5 kilobar condition and probably even at much higher temperatures still. If methane were the supply at depth, a wide range of temperature-pressure conditions on the way to the surface would maintain the gas in that form.

Oxidation of methane to CO and CO_2 would take place enroute in circumstances where, first, an oxygen donor was available and, second, the temperature was high when the pressure was low. These conditions are most likely to be met along a volcanic pathway. In addition to providing the high temperature near the top, the process of bubbling gases through liquid rock will allow oxygen to be extracted from dissolved water or from oxides. In pathways through pores in solid rock, the available oxygen from surfaces is much more limited. On this basis one could understand that the gas emission from volcanic vents contains most of the carbon in oxidized form, even if it originated from hydrocarbons. (Of course there is always the possibility that some volcanic regions provide CO_2 and CO that was derived from the decomposition of subducted limestone, rather than from a primeval source of carbon.)

Karzhavin and Vendillo[22] also provide calculations for the comparative stability of methane, ethane, propane, and butane. These show that methane would be greatly favored as the gas to escape on top, at low temperature and pressure, even if the other hydrocarbons were present at high temperatures and pressures. This can account for the finding that the deepest sources of hydrocarbon gases appear to supply a very high proportion of methane.[23]

Out of all this, no compelling reason emerges for identifying the gas that has been responsible for a continuous supply of carbon to the surface. Many other considerations can be introduced, the most obvious being the question of the oxygen supply.

The two sources of supply of oxygen to the atmosphere (other than the mere recycling of oxygen in the biosphere) are the photosynthetic reduction of the carbon that emerged as CO_2 but was laid down as organic carbon and the photodissociation of water in the upper atmosphere, coupled with the escape of hydrogen. Brinkmann[24] discusses the earth's oxygen balance, and in contrast to the earlier work of Berkner and Marshall,[25] he concludes that a very substantial sink for oxygen over geologic time has to be found for the atmospheric oxygen to be restricted to the low values in precambrian times that most investigations consider to be indicated by the biological evidence. He notes that the oxidation of methane might constitute such a sink, but considers that the quantities that would have been required are excessive.

We can draw up an oxygen balance sheet for the atmosphere and the sediments, for the period before the widespread release of oxygen by the photosynthesis, and for various assumptions concerning the quantities of the sediments and the nature of the supply of carbon. Brinkmann's calculations then give for each case the approximate value of the oxygen pressure that would have had to have existed over most of geologic time (4.6 billion years) in order for photodissociation of water to have provided the required oxygen. (The photodissociation rate of water in the atmosphere is high when the oxygen pressure is low.)

Case I: Adopting Rubey's estimate[26] of the quantities of the carbon deposits and assuming, arbitrarily, that half the carbon emerged as CH_4 and half as CO_2.

Case II: Like Case I, but adopting estimates for the carbon deposits three times as high as those of Rubey, approximately in accord with some more recent estimates.[27]

Case III: Adopting again the higher estimate for the deposits, but including the possible oxygen demand of the lithosphere, estimated by Brinkmann[28] as due to the oxydation of all lithospheric FeO to Fe_2O_3.

Case IV: Rubey's values for the deposits, no lithospheric oxygen demand, but all carbon supplied unoxidized as CH_4.

Case V: Like Case IV, but with the high value for the deposits.
Case VI: The high values for the deposits and the lithospheric oxygen demand, again with CH_4 as the supply. This represents the highest oxygen requirement to be supplied by photodissociation of water within any of the possibilities discussed.

From this tabulation it is evident that methane may well have been a major source of all surface carbon and that the oxidized state of precambrian deposits can be accounted for by the photodissociation of water. Indeed, if the precambrian atmosphere had little oxygen, as has frequently been suggested based on the biological evidence, then a very substantial methane supply is indicated.

In the course of geologic time there may well have been large changes in both the rate of carbon outgassing and in the proportions supplied as CH_4 and CO_2.

The composition of the present atmosphere is difficult to account for if all carbon had come up as CO_2. The oxygen resulting from the photosynthetic reduction of CO_2 (in the period since widespread photosynthesis came into existence) and from the permanent deposition of reduced carbon in sediments amounts to much more than appears to be present in the atmosphere or in additional oxidation of sediments. While the quantitative estimate of the various deposits is perhaps not very certain, it is at least clear that the data favor rather than deny the possibility that some of the carbon has emerged in unoxidized form.

The next question we shall turn to concerns the processes of outgassing through faults in the crust and upper mantle. It has sometimes been stated that at a depth of more than 10 km or so all porosity in the rocks must be crushed out, suggesting an absense of any pathways for gases. This is not so if any gases or liquids are present at a pressure equaling or exceeding that of the local rock. One presumes that shear faults would greatly facilitate the movement of such high pressure gases and that particularly favorable circumstances for their escape would occur at the instant of a major earthquake. We shall now discuss the evidence that large quantities of gas, and usually of a combustible gas, do indeed escape during major earthquakes.

THE EARTHQUAKE EVIDENCE FOR THE ESCAPE OF GASES

Many features of major earthquakes are known that seem puzzling and for which no completely satisfactory explanations have yet been offered. A list of these phenomena might be tabulated as follows:

1. Lights in the air associated with earthquakes;
2. Large changes in the electrical conductivity of the ground preceding and following earthquakes;

Table 18-1. Oxygen Pressure in Period Before the Photosynthetic Production of Oxygen, as deduced by Brinkmann, for a Range of Assumptions Concerning the Quantities of the Sediments.

	Case I	*Case II*	*Case III*	*Case IV*	*Case V*	*Case VI*
Total carbon in carbonate deposits	1.8×10^{16}	5.4×10^{16}	5.4×10^{16}	1.8×10^{16}	5.4×10^{16}	5.4×10^{16}
Total carbon in reduced carbon deposits	6.8×10^{15}	2.0×10^{16}	2.0×10^{16}	6.8×10^{15}	2.0×10^{16}	2.0×10^{16}
Total oxygen demanded for carbonates	4.8×10^{16}	14.4×10^{16}	14.4×10^{16}	4.8×10^{16}	14.4×10^{16}	14.4×10^{16}
Total oxygen demanded by lithosphere	0	0	5.6×10^{16}	0	0	5.6×10^{16}
Total oxygen supplied by primitive CO_2	3.3×10^{16}	9.9×10^{16}	9.9×10^{16}	0	0	0
Extra oxygen required from photodissociation	1.5×10^{16}	4.5×10^{16}	10.1×10^{16}	4.8×10^{16}	14.4×10^{16}	20×10^{16}
Average atmospheric oxygen pressure	p.a.l.	0.3 p.a.l.	0.1 p.a.l.	0.3 p.a.l.	0.1 p.a.l.	0.07 p.a.l.
Mean annual supply of CO_2	9.9×10^{6}	3.0×10^{7}	3.0×10^{7}	0	0	0
Mean annual supply of CH_4	3.6×10^{6}	1.1×10^{7}	1.1×10^{7}	7.1×10^{6}	21.4×10^{6}	21.4×10^{6}

All quantities are in metric tons; p.a.l. means present atmospheric level. The averages over geologic time of the oxygen pressure are shown that would follow from the range of assumptions quoted, Brinkmann's calculations, and the long-term mean rates of supply to the surface of CO_2 and CH_4 that would be implied.

3. Changes in ground water levels;
4. Radon222 excesses in the atmosphere preceding or during earthquakes;
5. Changes in the ratio of the seismic velocities;
6. Aftershocks;
7. The "visible waves" phenomenon; and
8. Large volumetric changes associated with earthquakes.

All these phenomena are very well documented, and we shall discuss their interpretations in terms of the escape of gases.

Earthquake Lights

Most major earthquakes that occur at night appear to have been accompanied by a display of luminescence of the air near the epicenter (and sometimes even as much as a hundred kilometers away) described by most observers as light fanning out from the ground and getting weaker with height. The patterns of light are sometimes described as a set of beams radiating from a point on the horizon, like searchlights turned to the sky, and the color is usually described as blue or white. The phenomenon is also seen at sea and has there been described to look like a burning ship on the horizon. Such displays in general seem to have lasted some minutes, usually following but sometimes even preceding the major earthquake. Richter[29] states that "rarely have they [earthquake lights] been missing from reports of any large earthquake in a populated area." Derr[30] gives a review of such observations that leaves little doubt that the phenomenon is real and, indeed, frequent. Photographs of such events are available, chiefly from Japan in the mid-1960s, showing a bright hemispherical luminescence based at ground level.

The number and the consistency of such reports from most major earthquakes is quite remarkable. There is no question that the explanation of earthquake phenomena must encompass this effect. Some attempts at an explanation are in the recent literature, mostly considering the phenomenon as a manifestation of atmospheric electricity. These have been criticized, and it is indeed clear that no electrical phenomena causing a breakdown in the atmosphere can arise from the ground in any circumstances in view of the values of ground conductivity. Before any atmospheric breakdown could be generated, earth currents would reach values at which very large magnetic effects would be a major consequence: yet no significant magnetic disturbances are reported in association with earthquakes, even from nearby magnetic observatories. The well-documented occurrence of similar phenomena at sea makes the same point even more clearly. Electrical breakdown in the atmosphere has also been attributed to some unspecified consequence of atmospheric pressure waves generated by the earthquake, but no such mechanism is known; furthermore, the purely mechanically generated disturbances in the atmosphere are quite small compared with the normal meteorological disturbance level.

Combustion of gases is the obvious alternative, and there is a large amount of evidence, in many cases quite decisive, that this was involved. Very detailed descriptions exist of flames seen issuing from the ground during great earthquakes in different parts of the world. For example, during the Owens Valley (California) earthquake of 1872, the *San Francisco Chronicle* reported that "people living near Independence, at points where they could see plainly the sides of the mountains on either hand, [said] that at every succeeding shock they could plainly see in a hundred places at once, bursting from the rifted rocks great sheets of flame apparently thirty or fifty feet in length, and which would coil and lap about a moment and then disappear."[31] Or, from the *Inyo Independent* "Immediately following the great shock, men, whose judgment and veracity is beyond question, while sitting on the ground near the Eclipse mine, saw sheets of flame on the rocky sides of the Inyo mountains but a half a mile distant. These flames, observed in several places, waved to and fro apparently clear of the ground, like vast torches; they continued for only a few minutes."[32] Similar observations were reported during the great earthquakes at Lisbon in 1755, at Calabria in 1783, and at Cumana in 1797.[33]

Not only are there many eyewitness reports describing the phenomena in terms of flames of a burning gas, but evidence of burning could subsequently be found along fault lines. For example, Goodfellow describes the great Sonora earthquake of 1887:

> The Sierra Madre fires were, beyond question, synchronous. . . . The evidence of gaseous irruption were few but striking. Primarily were the statements of many who claim to have seen streaks of flame at different points, in the course of the first night in particular, and several times thereafter during succeeding days and nights while the heavy shocks continued. . . the evidence [for ignited gas] was found in several places, both in the river beds and in the hills along the line of faulting. This consisted of cinders about the margins and on the walls of the river fissures, and the discovery of burnt branches overhanding the edges of such places, as well as the same testimony on some of the hills and mountains near the main fault.[34]

In the Owens Valley earthquake there are also reports of fires having been started in the mountains by such sources.

Since unstable burning of gases frequently results in explosions, one might expect loud airborne explosive noises to have been associated with the phenomenon. Indeed there are many reports of this kind also. For example, Kingdon-Ward reported of the great 1950 Assam earthquake: "From high up in the sky to the northwest (as it seemed) came a quick succession of short, sharp explosions—five or six—clear and loud, each quite distinct, like 'ack-ack' shells bursting."[35] Kingdon-Ward was a few miles from the epicenter but the sounds were heard as far as 750 miles away.[36] Other accounts of earthquakes have referred to airborne noises "like the whistling and rush of wind[37] or "like the escape of steam from a boiler."[38]

If the cause of these phenomena is indeed the escape of combustible gases at high pressure out of fissures in the ground, one can give a description of the events that would be expected. The very high velocity gas in the cracks would pick up particulate matter, and as in practically every industrial dust-pumping process, electric sparks would be generated throught the frictional charging of such particles and the transport of this charge against the electric field set up (Van de Graaf action). These will serve to set fire if a combustible mixture exists. If the velocity of flow is high, the flames would usually be disconnected from the orifice and continue burning only at the level above the ground where mixing with atmospheric oxygen is sufficient to produce a flame that can burn back at the local stream speed of the gases, so as to remain stationary (just as in the case of a bunsen burner, whose air intake is closed). It is to this mode of burning that we attribute the evidence that although flames are seen coming from the ground, in many cases there is an absence of evidence of scorching of material on the ground, although in other cases, as we have said, there clearly is such evidence.

In the descriptions of great earthquakes in cities, there have been many reports of fires shooting out of fissures in the streets. Usually this has been interpreted as due to gas mains bursting, and this interpretation may indeed be correct. It is also possible, however, that some of these fires have to be attributed to the same causes as we have just discussed.

Many reports of flames or light make a point of stressing the preponderance of the phenomenon on the hillsides and in regions of bare rock. One can understand that in terms of the outflow of gases, since fissures in the brittle rock would provide much better escape routes where they are not overlaid by a thick alluvial deposit that does not crack nearly so readily.

Fires of combustible gases associated with earthquakes were regarded as well-documented occurrences in much of the earlier literature on the subject; von Humboldt, for example, cited several examples from different areas.[39] It is strange that this interpretation of the "earthquake lights" has been forgotten to the extent that it is not even mentioned among the possibilities in modern articles on the subject. In fact, the general association of earthquakes with gas emission seems to have been generally recognized as far back as classical times.[40] Since the beginning of the present century, however, little attention seems to have been given to this type of evidence and to the earlier discussions.

Earthquake Precursory Effects

A number of different physical measurements of the ground have been identified in recent times as showing precursory effects for earthquakes and usually large changes at the time of the quake.[41] These measurements concern the electrical conductivity of the ground, ground water levels, the transport of radon gas through the ground, and variations in the ratio of the compressional to the shear wave seismic velocity. All these effects are in a sense remarkably large. For example, precursory changes in electrical conductivity exceeding 20 percent have been measured.[42] And in a region in which changes in strain of the ground amount to no more than 10^{-9}, changes in electrical conductivity of 10^{-5} are observed;[43] such effects are seen as clear precursors to earthquakes that may be as much as several hundred kilometers away, occurring usually a few hours after the onset of the conductivity change. Similarly, the precursory effects of a decrease of the ratio of compressional to shear velocity occupies a very large region and not just the one that will subsequently be involved in the fracturing process of the earthquake. Equally, ground water level changes are seen over a large region, most of which will not take any active part in the generation of the earthquake.

If, as has been proposed,[44] all these effects were due to a dilation of the rock due to the widening of microcracks when the shear strain approaches the critical value for fracture, then the phenomena should be confined to the region that initiates the eventual fracture. It would seem most improbable that all rock in a very large region would reach the critical stress for fracture at so nearly the same time in the case of a very slow buildup of stress and in view of the scale of unevenness of the material, the temperature, the topography of the overburden, and many other factors. In any fracture phenomenon it is more usual for a small region to reach the critical stress, but for the resulting crack to propagate into regions in which the stress was originally much below critical, but which became locally stressed through the propagation of that crack. In any such picture of fracture, we would not understand how regions of hundreds of kilometers would be simultaneously affected just shortly before the earthquake.

The rock dilatency model clearly gives a good explanation for the relationship between all the observable quantities mentioned in terms of a change of the amount of porosity of the ground or of the content of the pores. Thus, sudden changes in ground water level can readily be correlated with sudden changes in ground conductivity or with a change in the ratio of the sonic velocities. It is only the origin of these porosity changes over wide areas and the sudden onset that are not satisfactorily accounted for.

The large changes of the radon content of ground water, or of the air above,[45] cannot be understood without another gas serving as the transport agent. Such changes have been observed a few days before an earthquake and at distances of the order of one hundred kilometers from the epicenter. Radon^{222} has a halflife of only 3.8 days, and the amount of diffusion that it could suffer in periods of that order in the ground would be restricted to a few meters. This is true for any realistic value of the porosity of the ground, and therefore a mere change in the dimensions of the pores is not the explanation. Even the mere transport of ground water, which may change its level in the same time by distances usually not exceeding one meter, cannot be held responsible for the enrichment of radon in the water, let alone for its escape into the air above. Mogro-Comparo and Fleischer[46] discuss this problem, but confine themselves to the gaseous transport provided by a thermally driven convective motion. The distances over which such convection could drive radon in a few days are very small (the authors quote a hundred meters in twenty days in loose sand). In compacted soil at some depth this can certainly not account for such large changes as are observed in a few days as precursors to earthquakes. It is clear that a gas driven by a large pressure gradient has to be the vehicle.

Faults in the crust will assist in the outgassing process from great depths by letting high pressure gases migrate upwards along them. Any crack held apart by gas and spanning some interval of height then possesses the hydrostatic pressure gradient in the rock over this height, but a very much smaller pressure gradient in the gas, corresponding to its lower density. If the rock can deform, the tendency would therefore be to widen and extend the crack at the top and to close it at the bottom. In the course of this upward migration, in a deformable rock, the gas will be maintained at approximately the ambient rock pressure until it reaches the level at which there is naturally a porosity in the ground. At that level, generally between two and ten kilometers below the surface, the gas can spread out and move rapidly through the preexisting porosity over large regions. If, for example, this general porosity becomes momentarily connected to a fissure that extends to a depth a kilometer lower, the gas in it may supply a pressure of 300 atmospheres above that of the ambient rock. On this basis, one could understand that it would move over large distances rapidly and that it would dilate the pores. Where there is water, it would tend to displace it and to temporarily raise the water table until the hydrostatic instability causes the lighter gas to get above the heavier water. In the fine pores, surface tension and viscosity will delay this instability from acting, and we estimate that it could easily be much slower than the period of days that is involved in some of the precursor effects. Only such a gas can be held responsible for transporting the radon over large distances in a short time, and since it would effectively collect the radon from a large surface area of porosity below, it must in general cause an increase in the surface radon concentration. The radon thus merely acts as a convenient tracer, showing that an event of high speed gas flow through the rock pores has taken place.

The mechanical consequences of a gas invading a body of rock under high pressure are likely to be very important. At a certain depth, in the absence of gas, a rock will have insufficient strength to maintain any vacant volumes. While such a rock would acquire a porosity and dilate if subjected to a certain shear stress under low hydrostatic pressure, at a sufficiently high hydrostatic pressure, no value of shear stress can create porosity. Under those circumstances, the phenomenon of the propagation of cracks must be greatly impeded; rocks are known to become ductile.[47] Evison[48] discusses the serious problems this raises for the widely held view that brittle fracture and elastic rebound provide the mechanism for earthquakes. He concludes that another mechanism is required. The occurrence of earthquakes down to a depth of 700 km is cited, where the frictional forces in a crack would be too large by a factor of 1000 to allow crack propagation. He also considers that the striking similarity of seismic signals of earthquakes with those of subterranean atomic bomb blasts provides a further argument against that theory.

In the presence of a gas at the hydrostatic pressure of the rock, the situation is completely changed. Any pores that the gas can reach can open up, with the displacement energy being provided almost entirely by the gas, because of its "soft" equation of state (i.e., large volume change corresponds to a small pressure change). If the rock is under a large shear stress that would have caused the development of porosity under zero hydrostatic rock pressure, the gas will effectively bear all the pressure in any pores it can invade and again allow the growth of porosity. One would assume that a connected pattern of pores would then grow, pervading such a body of rock, and that the effects will be a general volumetric expansion accompanied by a rapid change in the mechanical properties from a ductile to a brittle material. The invasion by a gas is then a precursor of an earthquake, because it is this that suddenly converts the rock into a material capable of shear fracture and elastic rebound. If, then, the shocks so created are strong enough to fracture the ground up to the surface, this same gas will find a rapid escape route and produce the large surface effects of gas emission that are reported for almost all major earthquakes. (In somewhat different discussions of the problem Orowan,[49] Mogi,[50], Griggs and Handin,[51] and Evison[52] have also stressed the importance of pore fluid pressure to facilitate sudden slip.)

Aftershocks

Large earthquakes generally have a series of aftershocks decaying over a period of months or years. This is usually attributed to a gradual "settling down" of the stress distribution in the ground that takes new areas to breaking point in the general vicinity of the original epicenter. The nature of the time constants, of the order of months, for this settling effect is not clear. If these are time constants derived from plastic deformation, they are hard to reconcile, at least in a sensibly homogeneous rock, with the occurrence of brittle fracture. The amount of redistribution of the stress has to be very large so that rock will later initiate a fracture, when earlier at the time of the main quake it did not break and relieve its stress, despite the severity of the shock. If the phenomenon is to be interpreted in terms of some combination of rheologic properties of rock, this explanation must suffice for many different types of rock under different pressure and temperature conditions. We think it unlikely that sufficiently universal properties of rock leading to aftershocks can be found.

On the basis that the escape of gases is a major component of the earthquake phenomenon, one can invoke time constants of the gas flow through pores. The main earthquake presumably corresponds to the escape of gas from underground chambers, released by the sudden opening of fracturing rock. After this release of a gas mass, the internal gas pressure distribution in the remaining porosity is changed, and probably a fairly good exit path continues to exist for some time. One may imagine that this will lead to the draining of gas from pockets over a gradually widening zone and in turn that the withdrawal of pressure in such pockets will cause local fractures and collapse. We would attribute the aftershock phenomenon to this type of gas migration and consider, therefore, that it should be possible to detect a continued outflow of gases in the general region for the entire duration of the aftershock sequence.

The "Visible Waves" Phenomenon

In detailed reports of many major earthquakes, a very remarkable and unexpected phenomenon is described whereby large waves are seen to progress on the surface of the ground with a height sometimes declared to be as much as one or two feet, a wavelength ill-defined but of the general order of ten to a hundred meters, at a speed of propagation also perhaps ill-defined but at any rate low enough for the progress of individual wave crests to be clearly seen. Eyewitness accounts generally describe this phenomenon as quite distinct from the "hammer blows" of the sharp shocks, and some descriptions compare the land surface with that of waves in the sea.[53] Fuller quotes one eyewitness to the New Madrid earthquake who says that "the Earth rolled in waves several feet high with a visible depression between the swells, finally bursting and leaving parallel fissures extending in a north-south direction for distances as great as five miles in some cases."[54] Richter writes that "visible waves are most commonly reported from the meizoseismal areas of great earthquakes, particularly on soft or alluvial ground. Consequently many of the clearer and more accessible accounts come from India."[55] He quotes one:

> The ground rocked violently and we were both thrown down. . . We saw a series of earthwaves approaching over the surface of the ground, exactly like rollers on the sea. As these passed us we had some difficulty in standing, but none of the waves reached the intensity of the first which had overthrown us. . . . As the waves above subsided the ground began to crack at our feet. . . This was immediately followed by the emergence on the spot of earthquake fountains. . . This occurred in an alluviated region. . . .[56]

Another observer of the same earthquake "saw the ground in every direction shaking like soft jelly."[57]

Similar observations were made during the earthquakes of 1886 at Charleston, South Carolina[58] and of 1906 in San Francisco.[59]

It is not the magnitude of the displacements that are so remarkable, but rather the apparently slow wave speed. If all phenomena occurred at the speeds of the appropriate seismic waves, these speeds would be far too great for an observer to be able to see the approach of individual wave crests. Richter considers the phenomenon real, although frequently somewhat exaggerated in the reports. Just as in the case of the earthquake lights, it is the consistency of the descriptions from different parts of the earth that are most persuasive of the reality.

In terms of the gas release interpretation of earthquakes, this phenomenon can be accounted for in the following manner. As a result of the major fissure in the underlying rock, a large amount of high pressure gas comes up with pressures that may well be of the order of hundreds or thousands of atmospheres. The alluvial fill covering the fractured bedrock is generally less brittle and will not so readily open large fissures. It therefore acts as an extra impediment to the outflowing gases whose pressure is easily sufficient to lift it entirely. When so lifted from the bedrock, the rigidity of this material is low, and it is of course quite unstable. A phenomenon of the ground "shaking like a jelly" or "rolling like waves at sea" then seems entirely possible. (We find that Michell[60] already offered a similar explanation.) One may well presume that the concomitant fissures were the exit paths that the high pressure gas had made. In the earthquakes of 1783 in Calabria[61] and of 1887 in Sonora,[62] both the "visible waves" and the flaming phenomenon were reported, showing that a combustible gas was involved.

We do not know at present how much of the devastation of earthquakes is generally caused by this phenomenon and how much by the sharper shocks transmitted at the seismic speeds. Perhaps cities that are built on alluvial fill of no very great depth could take precautions against the phenomenon by digging a pattern of trenches down to the bedrock, thus allowing a rapid escape of gas from there.

A quantitative estimate can be made for the mass of gas necessary to cause the phenomenon. Thus, for example, an area of 10×10 km and an alluvial fill thirty meters deep to the bedrock would require a minimum of 50,000 tons of gas to raise, so waves 30 cm high would form. Probably several times more is actually required, allowing for some escape. In any case, these amounts, in evidence in large earthquakes every few years, would amount to only a small fraction of the mean terrestrial outgassing rate, which we estimated as between seven and twenty-one million tons per year (ten to thirty billion cubic meters), if in the form of CH_4. We have no estimate of the amounts of gas that may be escaping at the same time without causing this type of phenomenon.

Volumetric Changes

Large volumetric changes of the ground are related to earthquakes. Very long period precursory phenomena have been observed, involving raises of hundreds of square kilometers of ground by several centimeters, taking place over periods of many years,[63] sometimes followed by a sudden collapse at the time of the earthquake. In other cases, sudden raises or sudden falls of the land occurred at the time of the earthquake,[64] in some cases by as much as several meters.

These volumetric changes have been attributed in recent times mainly to dilation phenomena due to opening of microcracks under shear stress or to the rapid closing of them after the annihilation of such a stress. With the evidence that high pressure gas is frequently involved, one may consider that the porosity increases and the consequent surface lifts are due to the migration of a gas through layers of rock, expanding as it ascends, and that a surface fall then corresponds to the escape of that gas. Such an interpretation has the advantage that it can account for volumetric changes even in cases where the phenomena are taking place at a depth at which the maximum shear strain energy would be insufficient to create porosity in the presence of the pressure of the overburden; also, a wider range of temperature and material properties can be accomodated by such an explanation.

The large uplift in Southern California, referred to as the "Palmdale bulge,"[65] appears to have taken place mostly over the last fifteen years and may be a similar phenomenon to other precursor uplifts. If it is, it would be a substantially larger effect than has been reported in the other cases (25 cm maximum rise, and lateral dimensions of 200×100 km). If it is interpreted as an expansion of rock due to the invasion by gas, then quantities of the order of 10^9 tons would be implied. If CO_2 and CH_4 are a major component, then their escape would correspond to several hundred years of the mean supply rate. While the nature of the Palmdale bulge is not clear, an interpretation along these lines would make it an unusually large phenomenon.

The production of tsunamis ("tidal waves") in the ocean by large earthquakes is common and is indicative of large and rapid volumetric changes. A vertical displacement of the sea floor is usually considered as the cause.[66] Good estimates of tsunami energy can be made, and these have been compared with estimated earthquake energy. For large earthquakes, the tsunamis seemed to contain approximately one-tenth of the total earthquake energy and a somewhat smaller fraction for smaller earthquakes. However, the calculation of the volumetric displacement necessary to produce the wave leads to values that are quite outside the range of known land-based displacements.[67] In turn, such sea floor displacements as would be required would imply very much larger values for the earthquake energy. This can be seen by the following simple consideration.

The energy of the tsunami is given by the volume of water in the wave, multiplied by its height, or $E_w = Ah^2 \rho_w g$, where A is the area, and h the height of the water displaced, ρ_w the density of water, and g the gravitational acceleration. For sea floor displacement as the cause, the displaced volume Ah has to be the same. But the energy in the ground implied by this is then $E_r = Adh \rho_r g$, where d is the depth of the plug of rock that has been displaced, and ρ_r the rock density. d is presumably similar to the depth of the epicenter, a dimension usually of some tens of kilometers, while h, the wave height in the deep ocean, is of the order of one meter. Thus the ratio of the energies is

$$\frac{E_r}{E_w} = \frac{d\rho_r}{h\rho_w},$$

a quantity of the order of 30,000 or more and not 10 as otherwise estimated. Not only is the required sea floor displacement much too large to compare with other earthquake displacements, but the energy implied by such displacements would be immense compared with other estimates of earthquake energy.

The conclusion is that a sea floor displacement is unlikely as the cause of tsunamis; instead, a large volume displacement has to be responsible that is connected with much less energy, and a release of gas will satisfy this condition. The wave volume can then be equal to the gas volume when the gas has expanded almost to atmospheric pressure. In this interpretation, the tsunami at sea is just the equivalent of the "visible wave" phenomenon on land.

An estimate of the wave volume, and hence of the gas masses involved, can be made for tsunamis. Taking the observed period, the wave propagation velocity and the total wave energy quoted by Iida for the largest event recorded in seventy years (Sanriku, March 3, 1933), we obtain a volume of the order of $10^{12} m^3$ and hence a gas mass of the order of 10^9 tons. (For all other events the figures would be much smaller.) If it is thought that this is a large figure, it is still the method of displacing the required amount of water with the least amount of energy possible; the energy supplied by the gas may be only marginally larger than the wave energy produced. An earthquake energy only ten times larger than the tsunami energy is then a possibility.

Detailed studies of this phenomenon will allow much better estimates to be made of the gas masses involved in any one case; also, the nature of the gases could be determined from a subsequent analysis of the dissolved components in the ocean water, where dispersal would be much slower than in the air.

The other gas-related phenomenae we have discussed are not easily used for a quantitative estimate. Sampling of the upper atmosphere over the region and soon after an earthquake may yield some such data. At the present time, all that can be said is that the quantities of gas escaping at earthquakes may be a significant item for the mean carbon supply to the surface. Of course, the rate of this supply at the present time may deviate substantially in either direction from the mean rate over geologic time that we have quoted; as happens often in the geologic record, the short-term mean may be quite misleading.

The fact that earthquake-related gas seems to be frequently or usually combustible, even where the site is unrelated to any known deposits of natural gas, suggests that hydrocarbons are widespread in lower levels of the crust. Methane, because of its pressure-induced stability, is then likely to be the major component to emerge. Noncombustible gases like nitrogen and CO_2 must generally be present in insufficient concentration to prevent burning. Water vapor also cannot generally be so abundant that it would form clouds above exit points due to decompression and cooling, since this is not a part of the usually observed set of phenomena (though this may fit some of the accounts). Hydrogen sulfide smells seem to have been reported often from earthquake sites. The frequent reports that animals, especially dogs, seem to have been disturbed for some period preceding a major earthquake can be attributed to the escape of gases from the ground, detected by the superior sense of smell of these animals. Global surveys of atmospheric methane[68] have been carried out and show a significantly higher concentration in northern high latitudes than in southern. There is a suggestion of an unexplained source in the northern Pacific Ocean. One may wish to relate this to the generally high tectonic activity of the northern high latitudes compared with the stability of the southern; and the survey data could be substantially influenced by individual events of a release of gas.

Systematic and long duration observations of lithospheric outgassing in relation to faults and earthquakes have been carried out in southern Dagestan, in the fault system flanking the northeastern part of the Caucasus, by Kravtsov and Voitov.[69] The reports of the observation at many sites there indicate that deep faults are the escape routes of gases; that the hydrocarbon component is sometimes remarkably pure methane (98.5 percent); and that the effect of earthquakes is to change the composition of escaping gases over wide regions. Of particular interest is the observation that the carbon component of gas may change from CO_2 to CH_4, with the flow of other gases remaining nearly constant. This would be the situation expected if the supply started as CH_4 and if the oxygen supply of the pathway had become depleted.

The authors of that study give reasons why they consider the gases to come principally from high temperature zones, particularly those gases that appear in fault zones and are characterized by a predominantly methane composition. Detailed studies of the isotopic composition and of the rare gas content all confirm a high temperature, deep source. The depths of the Dagestan earthquakes are mostly 30–35 km.

Kravtsov and Voitov also noted that at the epicenter of the earthquake of May 14, 1970, an intensive release into the atmosphere of H_2, CO_2, and He was observed for a period of forty days. Thus, the observations in southern Dagestan give strong support to the view that there are deep and presumably abiogenic sources of hydrocarbons in the crust; that faults and earthquakes play a role in the escape of gases from great depth; and that gas emission over regions of hundreds of kilometers can all be affected by a single event. Indeed, the authors suggest that lithospheric outgassing of carbon through faults may be a major factor in the supply of surface carbon.

Prior to the February 4, 1975, earthquake in Hai-cheng, China, elevated temperatures were observed in the entire fault region for several weeks, and for a few hours before the quake a low-lying, foul-smelling fog was observed.[70]

CONCLUSION

The importance of understanding the sources of the carbon on the surface of the earth is great, both for a general understanding of the geologic and biologic past and also for many purposes related to mineral prospecting. If some or much of this carbon originates from unoxidized sources at some depth in the crust, as many of the present data suggest, then one will be interested in the possibility of exploiting this reservoir for fuel supplies. The largest mean rates of escape of hydrocarbons that are permitted by the considerations of the present atmospheric composition and processes are not large and are certainly smaller than 1 percent of the industrial fuel consumption.[71] On the other hand, the size of the original reservoir, if it supplied the entire surface carbon, is of the order of twenty million years of present-day usage of natural fuels. One may wonder, therefore, whether some small fraction of this large reservoir is still available and is at such depth, or communicating by natural channels to such depths, that recovery is possible. In any case, the information that deep and presumably abiogenic hydrocarbons exist would affect many aspects of the search for natural fuels. Entirely new techniques for both search and recovery would need to be invented. Igneous rocks and fault systems would come under investigation, not just sedimentary regions. A new significance would be seen for the cases where a reducing process has appeared to be associated with faults or with igneous processes, without the recognizable intervention of any organic reducing agents. An example of this is the general distribution of reduced carbon in igneous rocks that we have mentioned;[72] others are reports of reduced minerals without any apparent reducing agent, such as the native iron on Disco Island, Western Greenland.[73] There is also the interesting relation between methane and a rift provided by the anomalous large source of methane of Lake Kivu,[74] a lake of the great African rift valley.

The supply of carbon by tectonic releases of gases may have been a very unsteady process in different geologic eras. Both the primeval supply and the recycled one may have undergone substantial changes, and consequently, the composition of the ancient atmosphere and the climate may have greatly depended upon these.

Many of the earthquake phenomena can be accounted for by large quantities of gases in the crust and upper mantle. Their presence must greatly modify the rheological properties of the rocks. The escape routes of such gases would be favorable locations for earthquakes, and earthquakes in turn will facilitate the escape. Although large amounts of gas may have escaped undetected in the past or been noticed only in a qualitative manner, there can be no serious difficulty in subjecting this entire process to good quantitative analysis. While a range of gas detectors may not be available at the site of the next major earthquake, they can certainly be deployed in the period of the aftershocks, and the quantities and chemical nature of the gases can be established. Permanent gas detection equipment can be emplaced at known active faults. Seawater gas analyses and upper atmosphere gas analyses can be carried out after major earthquakes. The indications of the masses involved that are obtained from volumetric changes on land or at sea tend to give figures somewhat larger than the estimates based on the carbon budget; either the present epoch is particularly active; or the estimates are inexact; or a large proportion of the gases appearing at the present time are not primeval, but recycled deposits such as CO_2 derived from limestone that has been subducted and heated.[75] Water vapor, the fluid whose supply can most easily go unnoticed, may also play a larger role than we have discussed.

The process of earthquake prediction can be sharpened up greatly if indeed the escape of gas from deep levels is a precursory effect. Much more direct methods can than be used to detect gas flow in the ground, and perhaps the entire chain of events can be understood better. "Defusing" an impending earthquake is not out of the question.

NOTES

1. D. L. Anderson, and J. H. Whitcomb, "Time-dependent seismology," *J. Geophys. Res.* 80 (1975): 1497–1503; D. Griggs and J. Handin, "Observations on fracture and a hypothesis of earthquakes," *Geol. Soc. Amer. Memoir* 79 (1960): 347–64; F. F. Evison, "Earthquakes and faults," *Bull. Seismol. Soc. Amer.* 53 (1963): 873–91.
2. R. Robinson, "The origins of petroleum," *Nature* 212 (1966): 1291–95.
3. Ibid.

4. W. M. Rubey, "Geologic history of seawater," *Bull. Geol. Soc. Amer.* 62 (1951): 1111–48.

5. F. P. Fanale, "A case for catastrophic early degassing of the Earth," *Chem. Geol.* 8 (1971): 79–105.

6. J. T. Wasson, "Primordial rare gases in the atmosphere of the Earth," *Nature* 223 (1969): 163–165.

7. H. C. Urey, "On the early chemical history of the Earth and the origin of life," *Proc. Nat. Acad. Sci.* 38 (1952): 351–63.

8. D. H. Ehhalt, "The atmospheric cycle of methane," *Tellus* 26 (1974): 58–70.

9. Rubey.

10. E. M. Galimov, A. A. Migdisov, and A. B. Ronov, "Variation in the isotopic composition of carbonate and organic carbon in sedimentary rocks during Earth's history," *Geochem. Internat'l.* 12 no. 2 (1975): 1–19.

11. Ehhalt.

12. H. Craig, "The geochemistry of the stable carbon isotopes," *Geochim. Cosmochim. Acta* 3 (1953): 53–92.

13. Y. Bottinga, "Calculated fractionation factors for carbon and hydrogen isotope exchange in system calcite-carbon dioxide-graphite-methane-hydrogen-water vapor," *Geochim. Cosmochim. Acta* 33 (1969): 49–64.

14. A. N. Fuex, "The use of stable carbon isotopes in hydrocarbon exploration," *J. Geochem. Explor.* 7 (1977): 155–88.

15. H. R. Krouse and V. E. Modzeleski, "C^{13}/C^{12} abundances in components of carbonaceous chondrites and terrestrial samples," *Geochim. Cosmochim. Acta* 34 (1970): 459–74.

16. J. Hoefs, "Is biogenic carbon always isotopically "light", is isotopically "light" carbon always of biogenic origin?" *Adv. Org. Geochem.* 1971 (1972): 657–63.

17. Fuex.

18. Galimov, Migdisov, and Ronov; R. Eichmann and M. Schidlowski, "Isotopic fractionation between coexisting organic carbon-carbonate pairs in precambrian sediments," *Geochim. Cosmochim. Acta* 39 (1975): 585–95; J. Veizer and J. Hoefs, "The nature of O^{18}/O^{16} and C^{13}/C^{12} secular trends in sedimentary carbonate rocks," *Geochim. Cosmochim. Acta* 40 (1976): 1387–95.

19. H. Jeffreys, *The Earth* 4th ed. (Cambridge University Press, 1959), p. 305.

20. B. M. French, "Some geological implications of equilibrium between graphite and a C-H-O gas at high temperatures and pressures," *Rev. Geophys.* 4 (1966): 223–53.

21. V. K. Karzhavin and V. P. Vendillo, "Thermodynamic equilibrium and conditions for existence of hydrocarbon gases in a magmatic process," *Geochem. Internat'l* 7 (1970): 797–803.

22. Ibid.

23. E. N. Tiratsoo, *Oilfields of the World* (Beaconsfield, England: Scientific Press Ltd., 1976).

24. R. T. Brinkmann, "Dissociation of water vapor and evolution of oxygen in the terrestrial atmosphere," *J. Geophys. Res.* 74 (1969): 5355–66.

25. L. V. Berkner and L. C. Marshall, "Limitation on oxygen concentration in a primitive planetary atmosphere," *J. Atmos. Sci.* 23 (1966): 133–43.

26. Rubey.

27. Galimov, Migdisov, and Ronov.

28. Brinkmann.

29. C. F. Richter, *Elementary Seismology* (San Francisco: W. H. Freeman & Co., 1958).

30. J. S. Derr, "Earthquake lights: A review of observations and present theories," Bull. Seismol. Soc. Amer. 63 (1973): 2177–87.

31. *San Francisco Chronicle.* April 2, 1872.

32. *Inyo Independent,* April 20, 1872.

33. Mr. Stoqueler, "Observations, made at Colares, on the earthquake at Lisbon, of the 1st of November 1755, by Mr. Stoqueler, Consul of Hamburg," *Phil. Trans. Roy. Soc.* 49 (1756): 413–18; W. Hamilton, "An account of the earthquake which happened in Italy, from February to May 1783," *Phil. Trans. Roy. Soc.* 73 (1783): 169–208; A. von Humboldt, *Personal Narrative of Travels to the Equinoctial Regions of America, during the Years 1799 to 1804,* vol. 1 (London: George Bell & Sons, 1881), pp. 163–64.

34. G. E. Goodfellow, "The Sonora earthquake," *Science* 11 (1888): 162–66.

35. F. Kingdon-Ward, "Notes on the Assam earthquake," *Nature* 167 (1951): 130–31.

36. S. M. Mukherjee, "Landslides and sounds due to earthquakes in relation to the upper atmosphere," *Indian J. Meteorol. Geophys.* 3 (1952): 240–57.

37. A. Thomson, "Earthquake sounds heard at great distances," *Nature* 124 (1929): 687–88.

38. M. L. Fuller, "The New Madrid Earthquake," *U.S. Geol. Survey Bull.* 494 (1912).

39. A. von Humboldt, *Cosmos,* vol. 1 (London: Bohn Edn., 1849), p. 209.

40. F. D. Adams, 1938. *The Birth and Development of the Geological Sciences* (Dover, 1938), p. 399.

41. C. H. Scholz, L. R. Sykes, and Y. P. Aggarwal, "Earthquake prediction, a physical basis," *Science* 181 (1973): 803–10.

42. A. Mazella and H. F. Morrison, "Electrical resistivity variations associated with earthquakes on the San Andreas fault," *Science* 185 (1974): 855–57.

43. Y. Yamazaki, "Precursory and coseismic resistivity changes," *Pure & Appl. Geophys.* 113 (1975): 218–27.

44. Scholz, Sykes, and Aggarwal.

45. P. J. Smith, "Radon to predict earthquakes," *Nature* 261 (1976): 97–98; A. Mogro-Campero and R. L. Fleischer, "Subterrestrial fluid convection: a hypothesis for long-distance migration of radon within the Earth," *Earth Planet. Sci. Lett.* 34 (1977): 321–25.

46. Mogro-Campero and Fleischer.

47. Griggs and Handin.

48. Evison.

49. E. Orowan, "Mechanism of seismic faulting," *Geol. Soc. Amer. Memoir* 79 (1960): 323–46.

50. K. Mogi, "Earthquakes and fractures," *Techonophys.* 5 (1967): 35–55.

51. Griggs and Handin

52. Evison.

53. H. Sloane, *Phil. Trans. Roy. Soc.* 18 (1964): 78–100; B. MacDonald, "Remarks on the Sonora earthquake–its behavior at Tepic, Sonora, etc.," *Bull. Seismol. Soc. Amer.* 8 (1918): 74–78; G. Thomas and M. M. Witts, *The San Francisco Earthquake* (New York: Stein & Day, (1971), p. 69.

54. Fuller.

55. Richter.

56. Ibid. See also R. D. Oldham, "Report on the great earthquake of 12th June 1897," *Mem. Geol. Survey India* 29 (1899): 1–379.

57. Ibid.

58. C. E. Dutton, 1889. "The Charleston earthquake of August 31, 1886," *Ninth Annual Rept. U.S. Geol. Survey* (1899).

59. A. C. Lawson et al., *The California Earthquake of April 18, 1906.* vol. 1, pt. 2 (Washington, D.C.: Carnegie Institute 1908), pp. 380–81.

60. J. Mitchell, "Conjectures concerning the cause and observations upon the phenomena of earthquakes," *Phil. Trans. Roy. Soc.* 51 (1761): 566–634.

61. Hamilton; C. Lyell, *Principles of Geology,* vol. 2, 11th ed. (New York: Appleton & Co., 1892).

62. Goodfellow; MacDonald.

63. M. Wyss, "Mean sea level before and after some great strike-slip earthquakes," *Pure & Appl. Geophys.* 113 (1975): 107–18.

64. Lyell; T. Rikitaki, *Earthquake Prediction* (Amsterdam: Elsevier, 1976).

65. R. O. Castle, J. P. Church, and M. R. Elliott, "Aseismic uplift in Southern California," *Science* 192 (1976): 251–53; Smith.

66. F. Press, "The seismic source," *Int'l Union Geodesy Geophys., Monogr.* 24 (1963): 7–18; K. Iida, "On the estimation of tsunami energy," *Int'l Union Geodesy Geophys., Monogr.* 24 (1963): 167–73.

67. A. I. Kravtsov, and G. I. Voitov, "Evaluation of the role of faults in the gas exchange between the lithosphere and the atmosphere (illustrated by southern Dagestan)," *Izv. Vyssh. Uchebn. Zaved., Geol. Razved.* 19 no. 4 (1976): 18–26 (in Russian).

68. Ehhalt.

69. Kravtsov and Voitov.

70. Liao-ling Province Meteorological Station, "The extraordinary phenomena in weather observed before the February 1975 Hai-cheng earthquake," *Acta Geophysica Sinica* 20 (1977): 270–76 (in Chinese, but see *New York Times,* May 29, 1978).

71. H. Perry and H. H. Landsberg, "Projected world energy consumption," in *Energy and Climate* (Washington, D.C.: Nat'l Acad. Sci., 1977), pp. 35–50.

72. Hoefs.

73. J. M. Bird and M. S. Weathers, "Native iron occurrences of Disko Island, Greenland," *J. Geol.* 85 (1977): 359–71.

74. W. G. Deuser, E. T. Degens, G. R. Harvey, and M. Rubin, "Methane in Lake Kivu: New data bearing on its origin," *Science* 181 (1973): 51–54.
75. A. T. Anderson, "Some basaltic and andesitic gases," *Rev. Geophys. Space Phys.* 13 (1975): 37–55.

Session D

Progress in Resolving Critical Environmental Problems, Issues of Risk-Benefit, Perceptions of Public Acceptance, and Institutional Constraints

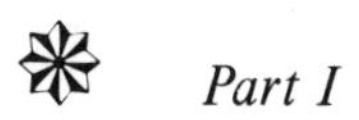

Part I

Risk-Benefit Ethics and Public Perception

Chapter 19

Risk-Benefit Analysis and Its Relation to the Energy-Environment Debate

Chauncey Starr
Electric Power Research Institute,
Palo Alto

INTRODUCTION

Ten years ago one could speak of the energy-environment problem; today it is a political and ideologic conflict. Whether it is a problem or a conflict, the commonly accepted social goal remains unchanged: optimize the public welfare by balancing the benefits of energy availability, the impacts of energy cost, and the risks and environmental insults of energy production and use. The criteria are also the same; they are the social values that define public welfare. The pragmatic question is whether a balance that optimizes the public welfare can be achieved by our present processes of decisionmaking.

RISK-BENEFIT ANALYSIS

The task of balancing social costs and benefits would be straightforward were there a common unit of comparison. Well-established methodologies such as cost-benefit analysis (with dollars as the common currency) or its well-marketed descendant, decision analysis (which frequently converts all factors into non-dimensional units called utils, based upon measurement of utility), are capable of defining this balance even under conditions of uncertainty. But the difficult part of the problem remains—that of converting multidimensional social impacts to comparable units. If these impacts are to be expressed as common currency, then the rates of exchange must be derived from social values. The counterargument, that certain social costs are noncomparable, stems from the distasteful and politically unpopular task of assigning monetary values to life, health, or clean air. The fact is that decisions that require tradeoffs of risk, cost, and benefit are being made constantly by key groups, and the public deserves a full disclosure of the values upon which these decisions are based. Although it is a popular notion that we cannot analytically compare apples and oranges, the fact is that we do choose between them in terms of personal satisfaction.

In addition to the difficulties posed by this lack of a single social cost scale, the values with which social costs are evaluated vary with both the evaluator and the specific circumstances. At an individual level, everyone makes intuitive tradeoffs between cost, safety, convenience, and a number of subjective factors, but not everyone arrives at the same result. The degree to which risk aversion varies is obvious–some individuals are willing to ride motorcycles; some are not willing to ride in airplanes (undoubtedly including some motorcyclists).

The reasons for this diversity are complex. Much of the risk-benefit literature focuses upon specific characteristics of risk and upon the distribution of perceptions of benefits and risks. Among these various factors are:

Discount rates: Individual perceptions of the present value of future benefits and risks vary. This explains why some individuals save and others consume, and it may also partially explain the contrasting individual attitudes toward the risks of smoking (future completely discounted) or nuclear waste (future fully present–valued).

Intangibles: Individual perceptions of the value of benefits and risks vary widely. For example, small percentages of the population will engage in mountain climbing or hand gliding, while the majority judge these risks excessive.

Perception versus analysis: Individual decisions are governed by perceptions of risk and benefit. Perceptions in general depend upon individual experiences and specific risk characteristics to a greater degree than upon actuarial data. Most individuals "feel" about issues rather than "think" analytically. It is this difference that defines the "professional" in any field.

Direct control: Among the risk characteristics that influence the perception of risk magnitude is the degree to which the individual believes he or she has direct control over the system exposing him or her to risk.

Voluntary and involuntary risks: Risk acceptance from activities undertaken voluntarily characteristically are much higher than socially accepted levels of imposed risks.

Catastrophies and societal resilience: This concept deals primarily with the survivors' view of their ability to recover from the effects of events. Two types of societal resilience can be examined, relative to two types of damage. First is the large accident, such as an earthquake or flood, that causes a great deal of immediate and concentrated damage. The threat associated with this type of event is that the loss of substantial portions of a population or resources will require substantial effort and time for recovery. The second type of risk is that which threatens the societal structure and culture, rather than large losses of life. This second type of risk is apparently of great importance, for nations have gone to war to avoid such challenges to their social stability.

SOCIAL VERSUS INDIVIDUAL DECISIONS

For the individual, these complexities in valuing costs, risks, and benefits are usually handled intuitively, and decisions require little external justification. When the issues are public and require arriving at consensus values, the interpretation is significantly more difficult. The decisionmaker is constrained to the values of an organization (a corporation or a government agency, for example), and frequently the criteria must be made explicit. As an example of how the value system changes between individuals and large organizations, a simple case can be considered—that of the cost of successive days of disability. As illustrated in Figure 19-1, the societal value system usually assumes a constant daily disability cost. This approach is taken by the U.S. Public Health Service in decisions regarding allocation of their budget. The goal for this and other similar organizations is to minimize disability days, because the cost to society is assumed proportional to the total number of disability days of the population.

The second curve on Figure 19-1 represents the usual individual's value system, in which an illness or injury lasting only a few days is considered as merely a minor inconvenience. However, for the individual, the disbenefit associated with the long-term disability increases at a rate faster than linear—that

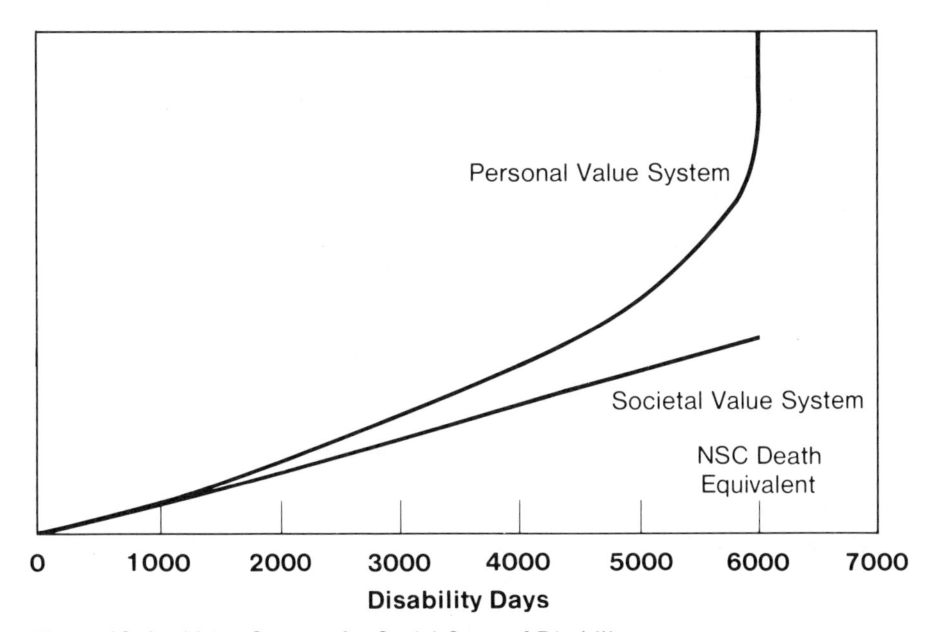

Figure 19-1. Value Systems for Social Costs of Disability.

is, a year of disability is more than fifty times as bad as a week. And, of course, death is considered an extreme disbenefit.

This is one explanation of the popular support for the charitable organizations that raise money for research to develop cures for multiple sclerosis or muscular dystrophy. In terms of disability days to society, these diseases are far less significant than the common cold. The catastrophic nature of the effects upon the individual results in an emphathetic personal perception of a very high cost.

Autocratic countries and military organizations are more likely to use a societal scale, because for these agencies the general welfare of society is placed above that of an individual. The decision to have a conscript army is representative of this viewpoint, because the values of the individuals are totally subordinated to the larger organization. Democratic governments are likely to use a mix of both scales. The congressional decision to spend extensively for cancer research is probably a result of the individual values of older voters, while the fifty-five mile per hour speed limit is designed to satisfy societal rather than individual objectives.

The incorporation of risk estimates into national decisionmaking is in its early stages of methodology development. The creation of a societal value system for public risk acceptance must rank as the major unresolved issue. One approach is to infer these public values by the method known as revealed preferences. This assumes that society has arrived at an acceptable balance between risks and benefits through an iterative process of trial and error and that by analysis of the risks and benefits of existing activities, the acceptable consensus level is revealed. The application of these results, displayed in Figure 19-2, should be considered very tentative and primarily useful for illustrative purposes. Due to the distinction between public and individual evaluations described above and to the distinction between voluntary and involuntary risks, Figure 19-2 is intended to illustrate only the acceptability boundary for involuntary public risks. Further caveats are in order, for the underlying data from which this was drawn were the actuarial or scientific estimates of risk and benefit, when in fact, the public decisions were based upon public perception of risk and benefit. This figure is nonetheless intuitively satisfying in that it indicates that the range of risk is bounded; that an upper limit exists above which public risks are unacceptable regardless of benefit; and that a lower limit, or noise level, can be defined. This lower level, corresponding to unavoidable natural risks such as lightning or insect bites, is representative of a regime in which residual risks are likely to be unremovable.

ENERGY-ENVIRONMENT BALANCES UNDER CONFLICT

Under conflict conditions, cost-benefit analysis or decision analysis are merely sideshows; the primary mechanism for balancing costs and benefits is political

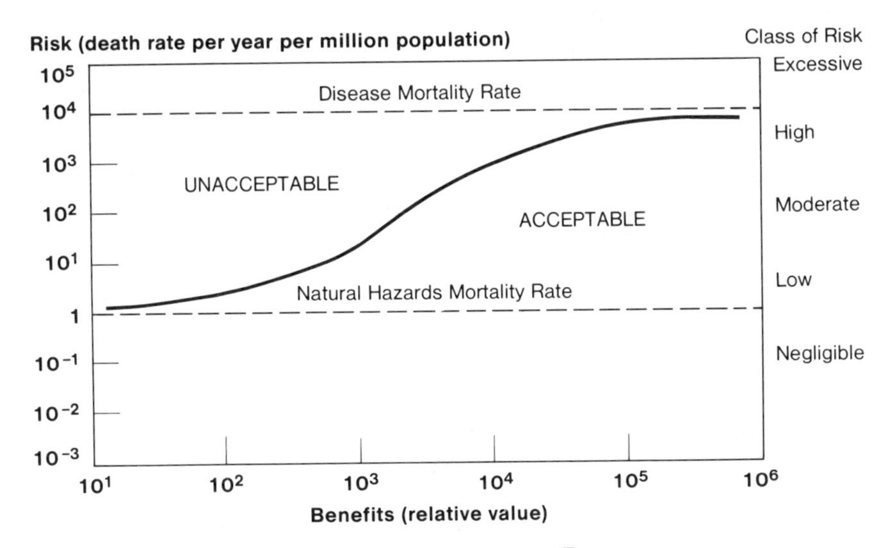

Figure 19-2. Benefit-Risk Pattern of Involuntary Exposure.

pragmatism, a method that generally results in decisions incurring a high social cost. The administrative agencies that are traditionally responsible for defining the balance of social cost and benefit have generally become political battlegrounds for social policy. With increasing frequency, the actions of these agencies are subject to intense group pressures. Interest groups see in these agencies an opportunity to influence policy by challenging decisions counter to their philosophies. The courts are asked to review agency decisions for detailed consistency with legislation, and Congress, as a final arbiter of political debate, often intervenes in regulating agency actions. Unfortunately, legislation, regulation, and the legal process do not ensure arriving at the optimum benefit for society.

Decisionmaking under these conflict conditions is a mixed blessing. On the positive side are the obvious advantages of having critical reviews of the assumptions and analysis upon which public decisions are based. Equally important is the openness that comes for public scrutiny of regulatory or policymaking processes. The disadvantages of increasing conflict are primarily operational: decisions are delayed, often for years.

From the viewpoint of the unregulated, probably the worst feature of this mode of decisionmaking is the problem of moving targets—that is, decisions are revised at a rate faster than industry can adapt. The viewpoint taken by industry is that it can adjust to a wide range of policies, even those seen as costly, provided the rules are clear and specific and that policies are predictable for periods roughly comparable to the planning and investment horizons of the industry. In the energy supply sector, these horizons run thirty to fifty years or longer, yet predictions of national energy policy for even the next several years are speculative.

There seems to be a parallel between the current growing awareness of regulatory costs and the similar awareness that came about a decade ago that environmental impacts were social costs paid by the public. In dealing with these environmental effects, an attempt has been made to internalize the costs. For example, Figure 19-3 represents conceptually a typical cost versus impact relationship, with the horizontal axis representing the level of an externality such as risk or the quantity of a pollutant. The upper curve is simply a sum of the lower curves and represents the sum of control costs and internal costs. Through such internalization, an operating point may be determined that minimizes total social cost. By including the cost of conflict as shown in Figure 19-4, in principle they may also be internalized.

Figure 19-4 represents the social cost of conflict for a technology in which it is assumed that the level of conflict is roughly proportional to the other external costs. That is, as the level of risk or pollution increases, so does the conflict. This case, which roughly describes the regulations governing the use of coal, illustrates the effects of the cost of conflict. The result is, first, to raise the total cost and, second, to permanently add the burden of excessive precaution.

Knowledge of this cost relationship by producers and regulators has traditionally resulted in a kind of large-scale poker game. For example, it is far cheaper for the producer if a preemptive decision to overcontrol externalities (that is, to move to excessive precautions) could prevent the onset of conflict. The regulator, who generally wants technical systems cleaner or safer than does the

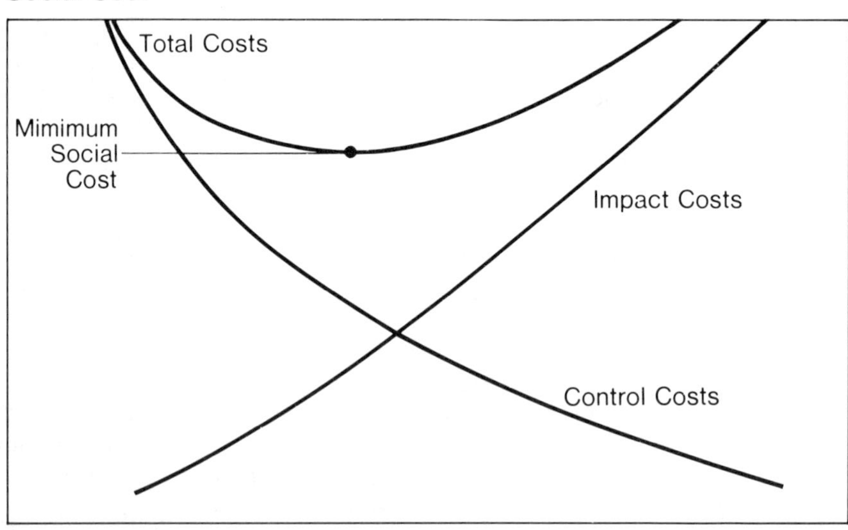

Figure 19-3. Cost Cuves for a Typical Technology.

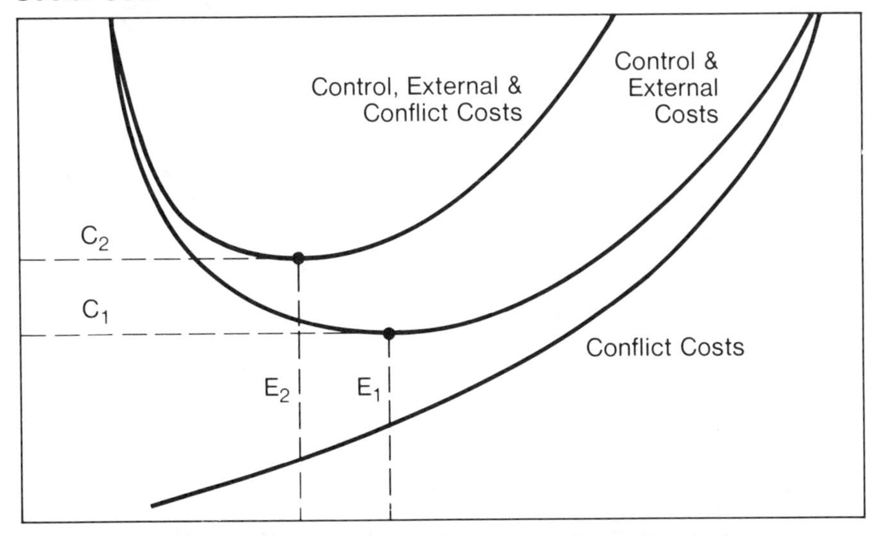

Figure 19-4. Control, Conflict, and Externality Components of Social Cost—Hypothetical Coal Example.

regulated, makes it clear that the time for approval of a project will depend strongly upon the degree to which these externalities are reduced. The result is a higher cost to society than would result solely from a benefit-risk optimum. Industry contends that this distortion is the result of overregulation of drugs, foods, coal, and so forth.

Figure 19-5 illustrates, again in a hypothetical sense, the cost of conflict for a different type of technology. In this case it is assumed that the conflict arises independently of any specific safety or pollution control decision, a description applicable to nuclear power. In both cases, the effect of conflict is to increase the total social cost.

Even with growing public awareness of the cost of delay and litigation, there remain a number of groups and individuals with a vested interest in the continuity of conflict. The "conflict industry" includes many in the legal profession, environmentalists who find conflict an effective means of livelihood or of achieving their social objectives, one issue politicians (as opposed to statesmen), and the media. As a consequence, proposed revisions in decisionmaking systems will not be evaluated solely by their impact upon achieving a public welfare balance. A more important consideration will be the effect of system changes upon the ability of the "conflict industry" to counter initiatives by the energy supply industry.

The identification of vested interests opposing many proposed energy

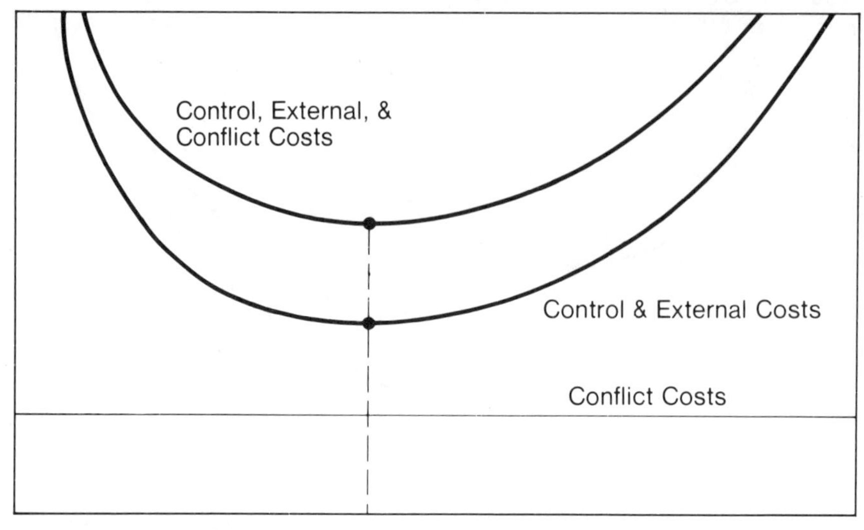

Figure 19-5. Control, Conflict, and Externality Components of Social Cost—Hypothetical Nuclear Example.

projects is not intended to dispute the legitimacy of their role in the conflict. Rather, it is intended to point out a popular misconception–that these groups are somehow purer than the advocates of supply expansion. In this vein, the use of the term "public interest group" is particularly misleading, because the value systems represented by these groups are minority values, and they do not represent a public consensus achieved by a democratic process.

RISK MANAGEMENT CRITERIA

The core technical and economic criteria for assessing a specific system include estimates of aggregate risk and benefit to within a plausible range of uncertainty and the distributions of these risk and benefits to various public sectors. A second general category of criteria is the public input. This includes both social values and perceptions. These values and perceptions are not fixed points that can be measured, but vary from person to person and group to group. The interest groups at the extreme ends of the distribution of values are generally the most visible and vocal, and understandably so, because the members of these groups face stronger challenges to their values than do individuals with "average" values. It is the distributions of values and the balance achieved between these values that establish the political nature of public risk management.

The third category of decision factors is based upon the requirements that the institutions responsible for risk management meet certain basic criteria, characteristic of their defined roles. In general, such institutions tend to optimize their own welfare and survival, and this may not necessarily be the same as the public welfare. The balancing of multi-institutional interests is the key to approaching sound public decisionmaking.

Finally, for a risk decision process to be practical, the requirements it places upon the regulators must be considered. If the structure is such that the regulator inevitably is exposed to criticism for being arbitrary or biased, a natural response of the regulator is to avoid making decisions (as through delay) or to slant decisions in favor of the most vocal or powerful of the interest groups. In actual practice, isolation from this effect is unlikely to be totally achieveable, but a clearer understanding of the decision criteria and process by all interested groups would minimize its influence.

RISK MANAGEMENT SYSTEMS

The decision systems in governmental use do, however, tend to follow a limited number of simplistic philosophies in order to escape from the difficult task of comprehensive assessments. Examples of these are outlined in Table 19-1 and are discussed below:

Best Available Control Technology

This approach, usually shortened to BACT or BAT, satisfies the institutional criteria. Regulation under this approach is achieved through a clearly stated set of rules that are easily understood. Perhaps the greatest attraction for this approach is due to the degree to which it minimizes regulator regret and guilt: if a serious accident occurs—an airplane crash, for example—the regulator is comforted by the knowledge that "we made them use the best system available." Additionally, it offers the benefit of providing incentives for technological

Table 19-1. Risk Management Systems.

Value Systems	*Approaches*
Technical and Economic	Minimum Cost Balanced Risk Economic Incentives
Public	Acceptable Levels Minimize Regret Judgmental
Institutional	BACT Zero Risk Zero Uncertainty

improvement, although often in industries secondary to that engaged in the regulated activity, because it is preordained that a safer technology will be bought to replace current technologies.

But it often achieves these admirable goals at a high price. Under this approach, the ultimate goal, that of balancing social costs with control costs, is forgotten. At times this approach leads to a preoccupation with only one of many social goals and as a result produces undesired side effects. This occurred when the impact of automobile emission standards on fuel economy was neglected. The BACT approach also produces a distorted view of alternatives. As an example, the use of coal to replace oil for electric power generation is a widely accepted national goal, but the BACT regulations applied to coal use are substantially slowing down the rate of substitution. The social value implied in BACT is that risk and environmental degradation have infinite costs, and the resultant resource allocations to reduce risk reflect this fact.

Zero Risk Approaches

For those who view risk as discrete (either present or absent), rather than continuous, the zero risk philosophy appears sensible. This approach is analytically unsupportable, because risk is continuous, and unless the measurement techniques applied to risk are extremely crude, no activities would be permitted. The most visible example of this philosophy is the Delaney Amendment, which prohibits food additives that are carcinogenic at any concentration. As was the case for BACT, the regulations are clearly understood; the regulator is free from guilt, regret, and punishment, if not from embarrassment, due to the occasional extreme results.

As a corollary, there is the "zero uncertainty" goal exemplified by the call for "one-armed scientists" in reference to scientists who, when asked for exact answers where none exist, resort to saying "on the other hand." This notion creates an unclimbable barrier when the two approaches, zero risk and zero uncertainty, are combined. The extreme statement that "the project can go ahead only if you prove beforehand that it will be absolutely safe" summarizes this philosophy. Such glib doctrines are dangerously simplistic, but unfortunately are politically attractive by providing a nobility of purpose to an escape from reality and responsibility.

Minimum Cost Criteria

The minimum cost approach to risk management is the theoretically exact method for satisfying the technical-economic criteria. The minimum cost as used here is the same as the expected value of probability theory. Under this method, resources are allocated to minimize the social cost of risk to society as a whole, usually by converting all kinds to some common unit, such as disability days or dollars. Unlike the approaches based upon the institutional criteria, the use of the minimum cost based on analytic estimates does balance social costs and benefits.

The drawbacks to this method are due to its failure to satisfy public and institutional criteria. By converting all risks and other social costs to a single scale, the social values relating to the multidimensional character of perceived risks are lost. On a basis of disability days alone, it is likely that efforts and resources expended to reduce social cost would be directed in a significantly different way than is current practice. The common cold, undoubtedly a major cause of disability days, would assume an extremely high priority. Rare, but extremely disabling, diseases would be low priority risks because their aggregate impact on public disability is small. But it seems quite clear from examples given previously that a simple measure of the total days of disability is not the criterion by which risks from disease are valued publicly.

Institutionally, minimum cost criteria place greater emphasis upon risk calculations than do the other approaches mentioned above. While it intuitively seems that a system that utilizes the best risk information available would be preferable to other systems that do not, under minimum cost approach, the debate between the proponents and opponents is intense, and each generally produces his own calculation of expected costs and benefits. As a result, the opportunities for regulator criticism are great if a balanced decision does not satisfy any of the adversaries.

Balanced Risk Approach

An extension of minimum cost decision criteria is a system that, as its basic measuring tool, uses the comparative cost effectiveness of alternative substitutional and control strategies across a wide range of activities. The appeal of this approach is that comparisons (and equalizations) of safety expenditures can be made rather easily by using a cost per "life saved" unit. Not surprisingly, this approach is favored by industries that are required to meet higher than average safety standards.

While this approach is a useful indicator of the balance in current regulation, it does not take under consideration the absolute magnitude of risk and benefit of a specific activity, nor is it useful in comparing the aggregate impacts of alternatives. Nonlinearities in social values are not generally included, although it seems likely that they could be.

Minimize Public Conern

While no explicit method is in use that bears this title, the common practice by which regulatory budgets are increased to reduce specific risks following newsworthy accidents (the actions of the California legislature following a major school bus accident or earthquake serve as an intuitive model) leads to a situation in which the politically desirable goal of minimizing the press coverage of risk can be realized. Under unusual circumstances, an accident need not occur. The movie *The Towering Inferno* triggered a review of high rise fire standards.

Acceptable Levels Approach

The concept that the level of acceptable risk varies with the degree of benefit (see Figure 19-2) is both intuitive and in agreement with the economic cost-benefit analysis that underlies minimum cost criteria processes. Unlike the minimum cost criteria, the acceptable levels approach does not presume a linear relationship between risk and benefit; in fact, most efforts in developing this method have been dedicated toward defining the relation between risk acceptability, benefit, and a number of risk attributes, such as the chronic or catastrophic nature of a risk or whether exposure is voluntary or involuntary.

An approach based upon acceptable levels, modified for the public response to specific risk characteristics, comes closest to simultaneously satisfying the technical-economic and public criteria, but thus far the institutional problems associated with doing this in a workable way have not been overcome.

Reasonable Judgment

Subjective regulation has a strong historical basis. The concept that a group of people certified honest, competent, and reasonable by congressional approval can form a useful regulatory commission has historically found acceptance. This approach cannot be defined as having a basis in meeting the technical-economic, the public, or the institutional criteria, because it has historically represented the democratic compromise of all three. The drawback is that public trust is required, a trust that now seems lacking, perhaps due to the fact that most regulators are political appointees and politically approved for office. Further, to have the continuity and predictability of regulatory actions that regulated industries require, the joint management relationship that inevitably develops gives the appearance of collusion between the regulators and the regulated. The alternative (of which examples can be found) is a commission openly hostile to the regulated parties. In either case the exposure of regulators to second guessing and charges of bias and incompetence is intense.

Economic Incentive Approaches

One alternative to the adversary process of risk-benefit decisionmaking that has not been applied on a significant scale is the substitution of economic incentives and penalties for mandated technical performance standards. Under an economic approach, specific social goals, such as the reduction of a certain pollutant, would be required through taxes on emissions. In this way, industries that failed to reduce emissions would be penalized relative to those that took control actions. As a hypothetical example of a risk application of this approach, the list of required coal mine safety regulations and expensive inspection programs could be replaced by a single requirement—that mining companies buy insurance to pay a specified amount, perhaps $500,000 to $1,000,000, for each mining fatality and a lesser amount for injuries to coalminers. In this way the insurance industry would expand their role in the regulatory process, for their rates would undoubtedly vary with the safety features and performance of each mining company. While there is much room for disagreement in the implementation of an economic approach (for example, upon the rates at which pollutants would be taxed), this approach would focus many of the issues. A similar approach has been used in the regulation of low level radioactive emissions from nuclear power plants. At one time NRC regulations allowed the use of $1,000 per man rem cost-benefit analysis of devices to control low level radiation emissions. A secondary benefit of this approach is that it encourages technical innovation in control systems, because any system that satisfies the economic criteria has a guaranteed market. The limitation of this approach is that it is incompatible with a zero risk philosophy. As long as environmental and safety costs are viewed as simplistic, ideologic, and political issues, the conflict is likely to continue.

CONCLUSIONS

Over the past decade, increasing public awareness of the externality costs of energy system environmental impacts and risks has led to careful scrutiny of these effects through environmental impact assessments and regulatory overview. Despite, or in part due to, this increased concern, the present mix of energy use costs, benefits, and regulatory costs is far from optimal. This is largely because of the use of risk management systems that are based upon a goal of eliminating risk, rather than controlling risk cost. Such a decisionmaking framework leads to insufficient consideration of both the technical and economic factors determining the costs of risk and risk control, as well as the public values that underlie risk taking.

A primary objective of changes in these decision systems should be to bring the costs of the process under control. The external costs of delay, litigation, and rapidly changing performance standards should be considered along with the tradeoff between environmental or risk costs and the costs of control. With a goal of balancing these regulatory costs with regulatory benefits, methods have been described that can do a better job at lower cost. Foremost among these is a regulatory philosophy that recognizes that economic incentives are more efficient than statutory performance standards in the application of risk management goals to energy systems. This approach, coupled with the ongoing improvements in risk estimation and the interpretation of public values, can meet the objectives of minimizing all public costs associated with the supply, conversion, and delivery of energy.

REFERENCES

Fischoff, B.; P. Slovic; S. Lichtenstein; and S. Read. "How Safe is Safe Enough? A Psychometric Study of Attitudes Towards Technological Risks and Benefits." Draft report, Decision Research, Eugene, Oregon, October 1976.

Morgan, M. Granger. Editorial. *Wall Street Journal*, August 26, 1978.

Schwing, R. "Expenditures to Reduce Mortality Risk and Increase Longevity." General Motors Research Laboratories Report GMR-2353, February 1977.

Slovic, P.; H. Kunreuther; and G. White. "Decision Processes, Rationality, and Adjustment to Natural Hazards." In *Natural Hazards: Local, National, and Global,* G.F. White, ed. New York: Oxford Press, 1974.

Starr, C. "Risk/Benefit Analysis and Full Disclosure," Presented at AAAS meeting, Washington, D.C., February 13, 1978.

Starr, C. "Social Benefit Versus Technological Risk." *Science* 165 (September 1969).

Starr, C.; R. Rudman; and C. Whipple. "Philosophical Basis for Risk Analysis." *Annual Review of Energy* 1 (1976).

Starr, C., and C. Whipple. "Application of the Risk/Benefit Approach to the Energy/Environment Conflict." Presented at the Conference on Environmental Risk/Energy Benefits and the Implications for Education, Lake George, New York, October 18, 1978.

Tversky, A., and D. Kahneman. "Judgment under Uncertainty: Heuristics and Biases." *Science* 185 (September 17, 1974).

Whipple, C. "Methodologies for Judging Risk Acceptability." Presented at the Joint National Meeting of the Institute of Management Sciences and Operations Research Society of America, New York City, May 2, 1978.

Wilson, R., and W. Jones. *Energy, Ecology, and the Environment.* New York: Academic Press, 1974.

10CFR50 (Code of Federal Regulations; (issued by NRC), Appendix I. Numerical guides for design objectives and limiting conditions for operation to meet the criterion "as low as practicable" for radioactive material in light-water-cooled nuclear power reactor effluents, Sec. II, Para. D.

❋ *Chapter 20*

One Approach to the Study of Public Acceptance

Maurice Tubiana
Institut Gustave-Roussy
Villejuif, France

During these past few years, it has become evident in Western democratic countries, that public acceptance is as important as, and probably more important than, any technical consideration for the future of the world energy crisis. However, up until now, relatively little research has been devoted to this subject, probably because most administrators, physicists, and engineers have not yet realized that the understanding of psychosociological phenomena requires studies as complex and as elaborate as any of those needed to solve a physical or engineering problem. One of the difficulties of psychological reactions is that almost everybody feels competent to discuss them. Furthermore, many still believe that the human mind works in a logical way. These simplistic views of mental processes explain why so few efforts have been made to understand the real nature of opposition to nuclear energy and why the reaction has often been that the problem could be solved by good public relations. It is only recently that the obvious failure of all information campaigns has led us to reconsider the psychosociological reaction provoked by the development of nuclear industry.

THE IRRATIONAL PERCEPTION OF RISK

Studies since the beginning of the century have shown that behavior, value judgment, and decisionmaking are the results of complex processes that involve the entire mental history of the individual as well as genetic factors. In fact, if risk versus benefit analysis is useful for decisionmaking by technologists, it is only of limited use in understanding an individual's decisionmaking process. Moreover, reasoning based on risk versus benefit is not only of little use in influencing public acceptance but may even be detrimental in that it arouses fears.

Most of us know from daily experience that logic is more often used to justify a decision after it has been made than to actually make the decision. I would like to illustrate this statement with examples from my personal experience as a physician. The first example has nothing at all to do with the energy problem.

Let us consider a lady with a breast cancer for which there are two possible treatments: (1) radical mastectomy, or (2) tumorectomy plus breast irradiation. Let us assume that in the latter treatment there is a greater probability of local recurrence (for example 35 percent versus 15 percent). On the other hand, the advantage of conservative treatment is that the patient conserves a breast of nearly normal appearance. Physicians' experience has shown that in such circumstances, which are unfortunately frequent, decisions are not made by the patient on a rational basis. Moreoever, when the plain facts are presented to the patient, without help or advice, they create anxiety and confusion. The major responsibility for the decision falls on the physician who advises the patient, taking into account her age, as well as aesthetic, psychological, and family considerations. However, the arguments that may convince her to accept surgery, for instance, are not quantitative data or logical reasoning. Moreover, the final decision made by the patient is often difficult to interpret. A sixty-five year old widow might decide to keep a breast that nobody may ever see, whereas a good-looking young woman may choose mastectomy. Obviously, the decision cannot be explained by a cost versus benefit analysis.

This example underlines the limiting factors in such an analysis: risk and benefit are not of the same nature (in this case an increase in the probability of survival versus an unimpaired body). One cannot measure these two quantities with the same unit. Furthermore, individual evaluation of risk and benefit is influenced by subjective considerations. Any decision can, of course, be subsequently explained on a rational basis, but none can be predicted. It might be argued that examples taken from a physician's experience are biased because the emotional factors involved in disease and death are so strong that they interfere with objective analysis.

But, in fact, other examples lead to a similar conclusion. Important individual decisions are not made on a rational basis: the choice of a husband or a wife; the decision to enroll in the army or to desert during a war; the speed chosen by a driver on a highway—none of these are based on assessment of risk and benefit, but rather on the individual's subjectivity and fantasies. Although they are the result of "nonlogical" processes, this does not mean that they are illogical or impossible to understand. It does signify that they cannot be explained simply by a quantitative assessment of advantages and disadvantages.

Similarly, daily behavior is more influenced by beliefs, by fears, by custom, and by myths—or even by genetic information—than by solid data. For instance, in industrialized countries, people are frightened by cancer, the major health concern. Epidemiological data have shown that systematic vaginal smears and breast self-palpation can greatly reduce the mortality of breast and uterine cancers. Yet, only a relatively small percentage of women do these things regularly, and many women avoid them because they are afraid that they might have cancer. In the United States, $800 million were devoted to basic and clinical cancer research in 1978, but the American Cancer Society estimates that patients with cancer spend more than $2 billion every year to buy untested remedies or to consult charlatans. Moreoever, California, a state with a high standard of living and a high level of education but where all kinds of religious sects flourish, is renowned for its prosperous quacks. Furthermore, TV debates between cancer specialists and charlatans have been organized in the hope of better public information, and the results were clear: these encounters were favorable to the defenders of untested remedies. This is generally explained by considering that when a cancer patient has reached an advanced stage and when medical treatment has become ineffective, it is normal for the patient who is facing death, or for his or her family, to put their faith in anything that provides the faintest hope.

However, I am afraid that this optimistic explanation is not entirely valid. In Paris each year, we see hundreds of patients in cancer hospitals who could have been cured if they had not relied upon untested remedies during the preceding months. In fact, TV debates such as the ones I refer to have shown that there is no even balance between someone discussing facts and someone who is manipulating hopes and fears.

The paramount importance of unconscious motivations is indicated by other observations. Cancer research has demonstrated for twenty years that the three major carcinogenic factors are tobacco, alcohol, and overeating (animal fat); yet in industrialized countries, tobacco, alcohol, and animal fat consumption has increased, while public fears have been focused on substances whose role in cancer induction is dubious such as pesticides, coloring agents, and the like. Every adolescent knows that tobacco shortens life expectancy and causes cancer. Yet, logical reasoning does not prevent the use of tobacco. Unconscious imitation of adult models and other irrational motivations influence the adolescent's behavior—perhaps I should say "dictate it"—much more than the fear of death.

The risks are overlooked when they go against strong irrational motivations. Along the same line, it is well known that the fear of getting a speeding ticket is a much stronger influence than a speech on an increased probability of death when it comes to modifying driving habits. In France, speed limits on highways have reduced the annual death toll by 4,000, yet some associations are asking that the limits be done away with. In Germany, speed limits on highways were lifted in October 1978. Surprisingly, there has been practically no protest against this decision, which will cause thousands of deaths among careful drivers hit by careless drivers. More evidence, stressed by psychological studies, is that the fear caused by a risk is not related to the magnitude of the risk and is mostly influenced by subjective considerations. For example, in France, there are 3,500 vacation deaths a year (drowning, mountain climbing falls, sunstroke, etc.) and

2,600 deaths from professional diseases and work accidents. Nevertheless, we call attention to the professional risks, which is resented and raises fears, while vacation accidents are overlooked in spite of the fact they could easily be reduced. Many other examples could illustrate the subjective nature of a risk. For instance, E.E. Pochin[1] listed a number of human activities that create risks of the same magnitude (Table 24-1). Obviously, they do not have the same emotional impact and do not provoke the same reactions.

A last example will illustrate how fear and value judgment interact. Fear of cancer has become nearly hysterical in some countries during recent years, and it is often said that cancer is a consequence of industrialization and urban civilization. Yet the epidemiological data, and in particular those of the International Center for Cancer Research, do not substantiate these views. For people of the same age, the cancer incidence is no greater in hyperindustrialized areas than in nonindustrialized developing countries. In industrialized countries the cancer incidence has not increased during the last forty years and would even have decreased without the rapid rise in the incidence of lung cancer, which is solely due to tobacco.

To incriminate technical civilizations as a factor detrimental to health is relatively recent and contemporary with the hostility against science. People are now afraid to live in "polluted" cities; a few decades ago they feared a trip in uncivilized countries where food might be "polluted"—in other words, unhygienic.

Table 20-1. 10^{-6} Risk of Death From Various Causes.

0.8 cigarette	
Staying during two hours in a room with a smoker[a]	
One-half bottle of wine	
Traveling	100 km by car
	500 km by air
	15 km by bicycle
	6 km by motorcycle
	6 km by horse
1.5 minutes rock climbing	
6 minutes canoeing	
1 hour of sea fishing	
0.5 hour being a president of the United States	
One-half day irradiation at maximum permissible dose for professional worker[b]	
Living during three years in the vicinity of a nuclear plant[b]	

[a]Considering that the blood and urine concentrations of tobacco wastes are one-tenth in nonsmokers of what they are in smokers and assessing the risk by linear extrapolation from risks observed for smokers. Actual risk is probably lower.

[b]Upper limit of the possible risk as assessed by linear extrapolation from risks observed after high doses, assuming no threshold and including both carcinogenic and mutagenic effects. Actual risk is probably lower.

Table based in part on data from E. E. Pochin, "Colloquium on the Psycho-Social Implications of the Nuclear Industry" (Paris: Société Francaise de Radio protection, 1977), pp. 44-55.

If, after individual behavior or feelings, we consider collective value judgments, we find that phantasms and subjectivity also influence public opinion and therefore, in a democracy, government decisions. For instance, let us consider the case of saccharine. Epidemiological surveys have not demonstrated an increased cancer incidence among people who have used it for decades. Furthermore, the experimental evidence for some carcinogenic effects is extremely dubious. On the other hand, the benefits are of paramount importance both for diabetics and for nondiabetics in reducing obesity, which increases the probability of cancer. Yet under the pressure of public opinion in some countries, governments are considering a ban of saccharine despite the advice of experts. The example of DDT is even more striking. There is no proven risk for humans. The benefits are obvious: insecticides and especially DDT have helped to fight diseases transmitted by insects—in particular malaria, which causes millions of deaths each year, even in some industrialized countries. The search for insecticides that are less toxic for some animals, yet equally or more effective, was of great interest. But to ban DDT without offering good substitutes was an emotional decision, and it had detrimental consequences for health. Moreover, many people in the United States do not always realize how shocking it may seem to the inhabitants of the developing countries, where insects are a daily threat to their health, that so much emphasis is put on the protection of birds and insects and so little is said about human beings killed or starved by insects.

It would be easy to cite many other instances where emotional reactions of the public, or of small groups, have led, in recent years, to government decisions that are difficult to explain logically. For example, a greater and greater percentage of national income is put into the fight against pollution and into environmental protection. Each small increase in the stringency of environmental regulations requires progressively larger amounts of effort and money. Has anyone assessed the potential health benefits of these constantly increasing demand? Without even referring to the developing countries where a small percentage of the money spent by industrial countries in the fight against pollution would result in immense health and social benefits, it appears that even in the industrialized countries, the same amount of money, if it were spent in other fields, such as community medicine, medical research, or the fight against self-pollution (tobacco, alcohol, drug addiction), would be much more beneficial to public health. Moreover, while some environmentalists claim that the only permissible concentration of carcinogens should be zero, there is very little protest against smoking in public; bills that planned to forbid the presence of the most potent concentrations of carcinogens and mutagens known in our environment—namely, tobacco smoke—have been rejected. Furthermore in the United States, as in many other countries, tobacco growers are subsidized, and taxpayers pay twice for tobacco—once to subsidize growers and once for the diseases caused by tobacco, costs that in France are estimated at about $5 billion per year.

Collective phantasms in this example are leading to decisions different from those that would arise from a cost versus benefit analysis or from an optimization of the use of funds devoted to health.

Science does not have to judge, but to analyze the phenomena objectively. The importance of individual and collective myths in the present-day world and their influence on the future of civilization and of the species require that they be examined scientifically. Many scientists and technologists deceive themselves about the irrational nature of myths and believe that it is only necessary to expose them to the light of truth to make them fade away. But myths are firmly embedded in the human spirit and cannot easily be dispelled. Myths are, in fact, realities of the world that are as tangible and as real as material facts. Logical reason alone is not an efficient enough weapon to fight them. This does not mean that the human mind is insensitive to logical approaches, but that logic is only one of the many factors that may influence human decisionmaking and not one of the more powerful.

Furthermore, in times of crisis, numerous historical examples show that the human mind becomes more susceptible to myths. The victory of the Nazis in 1933, after a regular democratic poll, was obtained after a campaign based on a few myths such as the responsibility of the Jews in all of Germany's troubles and the hopes raised by the "infinite wisdom" of a guide, Hitler. This success would have been inconceivable without the social disruption caused in Germany by depression, unemployment, and inflation. Already, during the Middle Ages, thousands of Jews and "witches" were burned after each epidemic of the plague. The fears and anxieties caused by the disease had to crystalize around a group considered as responsible. The aggravation of irrationality caused by the present mental crisis is shown through many examples. There are three times more registered astrologers in Western Europe and in the United States than members of professional associations of physics and chemistry.[2] In France between 1960 and 1973, the annual number of pilgrims to Lourdes increased from two to three and a half million, while the proportion of the French population attending church services decreased 35 percent to 20 percent.[3]

What is the source of this present crisis? The fear of death and aging is stronger than a few decades ago. This is substantiated by the denial of death and the cult of youth, so evident in our society. I hypothesized recently[4] that these fears are due to social and family desintegration and a decline of all faiths. In turn they lead to anxiety and explain the need for protection by irrational beliefs. The fear of cancer may be a displacement of the fear of death; this could explain the ambiguity of the attitude toward the disease and why, in fact, so many people do so little to minimize the risk.

Another source of the mental crisis is the high rate of change that Western societies have undergone since the end of the Second World War. "It would seem that societies have a certain threshold of tolerance for rate of changes, which, if exceeded, tend to social desintegration. . . . from the standpoint of mental health irrational emotional states and unsatisfactory human relationships tend to be proportional to social desintegration. . . . "[5]

The nuclear debate should be analyzed in this prospect. The role of emotional stress in its genesis is highly probable. In the United States the antinuclear movement began during the end of the Viet Nam war. At this time the tensions and anxiety were transferred to the environmental movement and focused on nuclear energy. In France an analysis of what the press said about nuclear energy between 1945 and 1976 shows that the atomic age was heralded euphorically as the second technological revolution until the Bikini and Cuban crises. There is thus a time coincidence between the increase of terror and the probability of nuclear warfare and a change in the media's attitude toward peaceful uses of nuclear energy. Despite concerns about nuclear energy, the first nuclear stations were brought into service without arousing any controversy. Furthermore, the ecologists' campaigns had no impact on public opinion until the oil embargo in 1973-1974. Protests began at a time when oil shortage increased the usefulness of nuclear energy without intensifying its disadvantages.

As a member of the expert study group set up by the World Health Organization (WHO) in 1957 on "Mental Health Aspects of the Peaceful Uses of Atomic Energy."[6] I have since remained interested in those problems. In particular, as a radiotherapist and a radiobiologist, I have been fascinated by the discrepancy between the radiobiological risks, as assessed by the most reliable world expert committees (UNSCEAR 1977; ICRP), and the anxieties and fear so widespread in the public. I have already underlined the rarity of papers on this subject. The lack of interest on the part of administrators and business men is easy to understand. Mental processes evoke in their minds simply the need for more propaganda, advertisements, and public relations rather than scientific research. The necessity of immediate decisions leads in some way to shortsightedness and disinterest in long-term investigations. But how could one explain the relative silence of academic research on a so large and so deep a psychosociological reaction, which furthermore will have some impact on the world future. Another challenging question of comparable importance—Why do young people start smoking and, while they claim they want freedom, become deliberately slaves of this habit?—has also been the object of relatively little academic research. I do not think that this is only due to an absence of imagination. In fact, having been in academic research for more than thirty years, I have observed that there are fashions in research. Moreover, academic research is often afraid of subjects so large that they cannot be carried out in a relatively short amount of time; the number of papers devoted to a particular field of research is never proportional to its importance. For example, the first thorough paper on tobacco and cancer was published only in 1952, at a time when tens of millions had already died from lung cancer caused by tobacco and when each unfrequent, benign disease had already been studied in hundred of papers. The Nazi movement in Germany before the Second World War could have been the most fantastic source of information about the spread of myths and the influence of a nonrational propaganda upon human minds. All sociologists and psychologists were fascinated by this wave of deep irrationality but, as far as I know, only two books attempted to study it with a scientific methodology before the war.[7] Nevertheless, despite the small number of papers devoted to reactions to nuclear energy, some interesting data have been reported, mostly of European origin. In 1977 a colloquium was organized in Paris on the psycho-sociological implications of the development of nuclear industry. Many of the data to which I shall refer were reported during that meeting.[8]

FACTORS INFLUENCING PUBLIC OPINION OF NUCLEAR TECHNOLOGY

Two types of factors influence public opinion.[9] Some are nonspecific; for example, a latent hostility against any new technology has always been present, and the fights over railways at the beginning of their history provide good examples of this attitude.

These reactions are themselves due to two types of motivations. One is fear of the new, fear of a change. This is a widespread feeling, and the human attitude toward change has always been ambivalent–desire for innovation on the one hand; fear of the unknown on the other hand. Nowadays fear of change is also an unconsicous expression of satisfaction with the status quo. But it is mainly a protest against the too rapid rate of change that makes, for the first time in human history, the world a man knows as an adult different from the one he knew as a child. These feelings are expressed by "passeism" and also by a cult of nature, which is the symbol of something permanent. The second motivation is even more basic. As far in the history as one can go in all civilizations–Hebrew, Egyptian, Greek–some people have always considered knowledge dangerous because it is a sacrilege. This is expressed by the Greek legend of Prometheus.

Prometheus robbed fire from the gods and gave it to men. He is both a hero who challenged the gods and changed the human condition and at the same time a disrespectful thief punished by having his liver eaten by vultures for eternity. These old fears toward research never completely vanished. In the Roman world, Pliny the Younger wrote about the terrors provoked by the first attempts at vegetal grafts. In the middle of the nineteenth century, Pasteur's neighbors protested and petitioned against his laboratory work and asked for its interdiction because they thought it would cause epidemics.

Antitechnological and antiscience feelings were sufficiently strong in 1929 to be the subject of a book by Freud.[10] One of the conclusions was that technology (and science) is the most obvious symbol of the constraints of everyday life–for example, clock, subway. In opposition, for someone who lives in the city, "nature" is vacation, it is freedom. Historical data are, from this point of view, very illuminating: discontent with civilization and criticism of technology vanished from 1930 till 1950. When life is difficult, science and technology represent the hopes for a brighter future. But when life is easy and when there is plenty of everything, the social constraints become inacceptable, even if they are light. This is substantiated by geographical considerations. In the three-quarters *or* fourths of the world where the problem is protection against the environment, technology is highly praised. In the technological Western world, where there is no longer any need for protection against the environment, the emphasis is now put on the protection of the environment, hence technology is accused.

Second, there is also a widespread fear that science and technology are leading the world to catastrophes–atomic war, overpopulation, pollution, computer civilization, to name a few.

A third factor is political. Despite numerous historical examples that show that large oppressive empires existed long before modern technology (such as Egyptian pharaohs, pre-Colombian empires, Gengis-Khan's empire, etc.), science and technology are the preeminent targets of the revolt against large-scale social organization. This is, first, because a modern state, in the Western world as well as in the Eastern, is inextricably interdependent with technology. But it is also because some think, like Marcuse: "technology seems to institute new, more effective social control and social cohesion. . . . By virture of the way it has organized its technological base, contemporary society tends to be totalitarian."[11]

Among the specific factors, the following have been identified. (1) Nuclear power is associated with big science, big industry, and centralized technocracy. To fight nuclear energy is to initiate a movement toward "small self-governing communities." (2) Many characteristics of nuclear energy are felt to be magical or frightening–for example, the contrast between large amounts of energy and the small volumes from which they emanate; the invisibility of radiations, which can harm or kill without being perceived; and so forth. (3) However, all studies have in fact stressed that by far the most important factor is fear of the nuclear bomb. Psychoanalytical research suggests that public anxiety about nuclear reactors is, in reality, directed against the A-bomb. Fear of the bomb, toward which the public feels impotent and which it does not want to face, is displaced onto nuclear power, with which each individual can more readily express his anger.[12]

Another point that emerges from these psychological studies is that the hard core of nuclear opponents are not distributed randomly among various categories of the population.[13] Typically they are young, belong to middle or upper middle class, and often have a university degree but are without scientific education. Generally they are not well integrated into modern society. For instance, they belong to new professions not yet well recognized, or they feel that their social position is not what they deserve and thus resent it. Moreover, their psychological profile has a characteristic pattern. In contrast with proponents, who are self-confident, optimistic about the world future, and like to act–extroverts–the opponents are pessimistic about the future of the world, have diffuse anxieties and fears, and have more of a passive than of an active psychological attitude–introverts–and they are skeptical about technology's ability to solve the world's problems. Their resentment against the modern world is expressed by a nostalgia for the idealized past or the utopic side of the so-called "natural" or "wild" life. It has been said that one's attitude toward nuclear energy reflects or expresses the way the modern civilization is perceived–its acceptance or its rejection.

Greenhalgh[14] recently parallelled nuclear opponents and the group identified by Kaase and Marsch[15] as "protesters". In their study on political action, Kaase and Marsch distinguish five groups: those who are inactive, conformists, reformists, activists, and protesters. The protesters are young, generally well-educated, with a higher proportion of women; they are prone to a whole range of protest action from signing petitions to participating in street blockades and are easily persuaded to adopt such causes as the antinuclear campaign, which falls outside the conventional political sphere as long as it is "in" to be against.

STUDIES OF PUBLIC OPPOSITION TO NUCLEAR TECHNOLOGY

Two sets of data confirm the view that in the opposition to nuclear energy, rational beliefs or fears play a very small role. First, the issues taken up by the environmental opposition varied considerably during the past decade but still were adopted and propagated by the same groups of people. The first issue was thermal pollution. When it was widely known that all sources of thermal energy involve thermal pollution, the emphasis was put on the risks of low level radiation exposure, then on the storage and disposal of radioactive wastes, then on reactor safety, and more recently on proliferation of nuclear weapons. We begin now to hear questions regarding the "moral" aspects of a nuclear power program that could be a burden for future generations and about the need "to leave the options open." The variety of issues that have been successively raised by the same people suggests that these issues simply represent an attempt to rationalize their fears and their opposition rather than the source of their hostility.

The second argument is the failure of all information campaigns. In Austria,[16] Sweden,[17] Switzerland,[18] and Germany,[19] thorough, factual, educational campaigns not only have failed to give a better understanding but have heightened uncertainty and confusion. They have provided active opponents with the opportunity to express their criticism, have highlighted the fears instead of dissipating them.

A study carried out by A. Whyte in Canada, the United Kingdom, and United States, similarly concluded: "public information in nuclear risk assessment does not necessarily lead to more public knowledge and understanding. . . . in the U.S. experience of greater information in risk assessment resulted in greater distrust of those who make them"[20]

If motivations to be against—or for—nuclear energy are not rational, it can be easily understood that they cannot be influenced by data. A public information program dealing with purely nuclear issues cannot succeed because it does not meet the underlying social and psychological problems. This does not mean that information is harmful, but it shows that information restricted to nuclear energy alone can be unsufficient or detrimental.

The role of political opinion in the nuclear debate has often been discussed. In Europe governing parties have always been unequivocally in favor of nuclear energy whether they are Socialist, Social-Democrat, Labour, Center, or Conservative, and therefore there is a great temptation for opposition parties to be equivocal or against. The only temporary exception to that rule was in Sweden, where a Socialist government, strongly in favor of nuclear energy, was replaced by a coalition Center government whose prime minister was against nuclear energy; but the coalition broke on the nuclear issue, and the new government is again in favor of nuclear energy. Surveys performed in France have shown that opponents and proponents of nuclear energy are found in all conventional political parties. However, this does not mean that politics are absent from the nuclear debate. A few studies, and in particular those carried out in England during the Windscale hearing in 1977, have shown that the fears are deliberately fostered by very small groups of proponents of radical social changes who are exploiting the nuclear issue as a weapon against the modern world of technology.[21] This action is efficient because myths and unconscious fears are embedded in all segments of the population.

CONCLUSIONS

In conclusion, it appears that if, as indicated by these studies, hostility to nuclear energy is mainly based on myths and fears, active opponents and proponents of nuclear energy are not speaking the same language. Information by proponents of nuclear energy is based on factual data, on objective analysis of advantages and disadvantages. It cannot gain public acceptance on its own—the supporting facts are necessary but not sufficient. Fighting nuclear myths necessitates an understanding of them and an identification of those that are mobilizing the public and aggravating their fears. This battle is difficult because, as we have seen, the motivations and the mental processes involved are complex. But such a combat also requires facing the fact that myths and subjective considerations have more influence on our behavior and beliefs than plain factual data, even though this is unacceptable, and even scandalous, to a rational scientific mind. Marcel Proust, perhaps the greatest French novelist, wrote fifty years ago: "factual data do not penetrate the world of our beliefs, they have not inspired them, they cannot interfere with them."

In fact, it is highly inefficient to combat a "nonrational" fear by logical reasoning. Only faith, or other fears, can counterbalance fears. In a private conversation, Dr. Teller was citing Voltaire, who gave his name to the room in which we are meeting. During his life Voltaire faced myths and superstitions, and he fought them with a powerful weapon—satire. He stressed the absurdity of superstition. It is now up to us to find the right way to face the superstitions of the end of the twentieth century.

Rather than demonstrating how small the risk involved in nuclear energy is, the proponents of this method should give solid guarantees of their reliability. Much too often the scientific community has appeared to be split on the issue. When this problem was discussed a few years ago at a symposium in which I participated, Lederberg had an illuminating comment. He said that much too often the scientists who speak to the media are not the scientists who have something to say but scientists who have personal problems.[22] It should be stressed that, concerning factual data, there is very little disagreement within the competent scientific community—for instance, when radiobiologists are speaking about radiobiology. They may differ when they express value judgments, and one of the great mistakes of these past decades has been to ask scientific experts to give value judgments. As scientists, they should, in order to preserve their credibility, restrict their comments to scientific matters. Of course, they may express value judgments, but, then, it should be pointed out that they speak as citizens and not as scientists; the two should not be confused.

One other obvious conclusion is that it is not correct either from a scientific or technical point of view, or from an ethical point of view, to discuss only the risks versus benefits of nuclear energy. The debaters should enlarge the discussion to include the entire energy problem and show that a shortage of energy would create more hazardous changes than energy production. Nor can the energy problem be discussed at a national level. For example, the United States is using up a large share of the world oil reserves; it would be technically and, perhaps more important, ethically improper to consider the energy problem in the United States without considering simultaneously the world energy problem and in particular the energy problems of the developing countries.

NOTES

1. E.E. Pochin, in *Colloquium on the psycho-sociological implications of the development of the nuclear industry* (Paris: Société Française de Radioprotection, 1977), pp. 44–45.
2. G. Steiner, "Has truth a future?" (1st Bronowski Memorial lecture, London, January 1978).
3. Maurice Tubiana, Le Refus du Réel, (Paris: Laffont et Cie édit, 1977).
4. Ibid.
5. WHO, "Mental health aspects of the peaceful uses of Atomic Energy" Technical Report Series, 151 (Geneva, 1958).
6. Ibid.
7. S. Tchakotine, Le viol des foules par la propagande politique Paris: N.R.F., 1952; (orig. in German); W. Reigh, *La psychologie de masse du Fascisme* (Payot, 1972; orig. in German, 1933).

8. *Colloquium on the psycho-sociological implications of the development of the nuclear industry* (Paris: Société Française de Radioprotection, 1977).
9. Tubiana; *Colloquium.*
10. S. Freud, *Civilization and its discontents* (1930; rpt. London: Hogarth Press, 1969).
11. H. Marcuse, *One Dimensional Man* (London; Routledge and Kegan Paul, 1968 ed.).
12. C. Guedeney, and G. Mendel, *L'Angoisse Atomique* (Paris: Payot, 1973); *Colloquim.*
13. Colloquium.
14. G. Greenhalgh, *The opposition to nuclear power. An appreciation* (London: The Uranium Institute, 1978).
15. M. Kaase, and A. Marsh, *The Matrix of Political Protest and Participation in Five Nations* (10th World Congress of International Political Science Association, Edinburgh, August 1976).
16. H. Hirsch, "Information Campaign on Nuclear Energy," IAEA-CN-36/589, in *Nuclear Power and Its Fuel Cycle,* Proceedings of an International IAEA Conference, Salzburg, 1977 (Vienna: IAEA, 1977), vol. 7, p. 219.
17. M. Lonnroth, in *Colloquium on the psycho-sociological implications of the development of the nuclear industry* (Paris: Société Française de Radioprotection, 1977), p. 520.
18. Greenhalgh.
19. Ibid.
20. A. Whyte, in *Colloquium on the psycho-sociological implications of the development of the nuclear industry* (Paris: Société Francaise de Radioprotection, 1977), p. 438.
21. Greenhalgh; P. Taylor, Proof of Evidence on behalf of the Political Ecology Research Group Windscale Inquiry.
22. "The Challenge of Life–Biomedical progress and human values," *Experientia. suppl.* 17 (Basel, 1971).

REFERENCES

J.P. Delziani, and C. Carde. In *Colloquium on the psycho-sociological implications of the development of the nuclear industry,* pp. 354–74. (Paris: Société Française de Radioprotection, 1977).

J.M. Doderlein, The Nuclear Controversy International, March 1975.

A. Gauvenet, "Citizens agitation and political responsiveness." AIF/SVA Conference, Geneva, September 1977.

I.A.E.A. "Information Section Study on the Origins of the Nuclear Controversy in the U.S.A. In *Nuclear Power and Its Fuel Cycle,* vol. 7. Vienna, 1977.

I.A.E.A. "Nuclear Power and Public Opinion." In *Nuclear Power and Its Fuel Cycle,* Proceedings of an International IAEA Conference, Salzburg, May 1977, vol. 7, plenary sess. 5. Vienna, 1977.

S. Nealey, In *Colloquium on the psycho-sociological implications of the development of the nuclear industry,* p. 529. Paris: Société Française de Radioprotection, 1977).

Novotney, H. "Social Aspects of the Nuclear Power Controversy." Laxenburg, Austria: IIASA/IAEA, RM-76-33, 1976.

Otway, Harry J. "A Review of Research on the Identification of Factors Influencing Social Response to Technological Risk." IAEA-CN-36/4. In *Nuclear Power and Its Fuel Cycle*, p. 95. Vienna: IAEA, 1977.

J.F. Picard. In *Colloquium on the psycho-sociological implications of the development of the nuclear industry*," pp. 336–51. Paris: Société Française de Radioprotection, 1977.

R. Remond. "L'atome et la démocratie." *Revue Générale Nucléaire* 1 (1977): 6.

Sofres. Les Journalistes et le programme nucléaire français. Paris 1975. Rapport interne.

Yulish, C.B. Why the growing opposition to nuclear power?" AIF/SVA Conference, Geneva, September 1977.

Chapter 21

On the Methodology of Cost-Benefit Analysis and Risk Perception

Peter A. Engelmann
and
O. Renn
Jülich Nuclear Laboratory,
West Germany
(Presented by Peter A. Engelmann)

A CRITIQUE OF COST BENEFIT ANALYSIS

Cost-benefit analysis is a method to weigh the advantages of a project envisaged against its disadvantages. Generally four different types of analysis have been applied, predominantly:

- Cost efficiency analysis (a comparison of different benefit levels maintaining a constant cost structure);
- Benefit efficiency analysis (a comparison of different cost structures maintaining a constant benefit level);
- Quantitative risk-benefit analysis by applying a reference value to the positive and negative consequences of the event (prevalently in monetary units); and
- Revealed preference acceptance model (comparison of benefit and risk in analogy to historically accepted hazards and the extrapolation of the revealed values to innovative events).

Occasionally decision analysis has been included as a distinct method in cost-benefit analysis. But decision analysis is usually referred to as a general approach to rational decisionmaking in which cost-benefit analysis is one of the means.

There has been much criticism on cost-benefit analysis in the past years that is concentrated on the following points:

- Restricted number of possible input variables,
- Systematic biases in finding and evaluating input variables,
- Difficulty of assigning probabilities to rare events,
- Temporary and local inconsistencies of value judgments,
- Ignorance of input unit interaction and side effects, and
- Difficulty of converting qualitative consequences into quantitative measuring units (like dollars).

On a more normative base, cost-benefit analysis has been criticized as an undemocratic method to stabilize the existing power structure of a society and to leave the decisions to a privileged class of experts.

These general comments on cost-benefit analysis apply to all four types. But in particular the risk-benefit and revealed preference methods lead to severe methodological problems. The method of historical analysis by revealed preferences and its extrapolation to pending decisions on innovations does not comply with the weighing process prevailing in the public and at official institutions concerning the risk assessment. This approach appears to be questionable from the normative viewpoint as well (cumulation effect). The determination of an acceptance threshold on the grounds of historical risk comparisons is dependent on the requirement that for historical decisions on risks exact information about the import of risk was available prior to making the decisions, innovations were introduced as a result of rational decisions, and all innovations were brought about the social consent.

These conditions are not consistent with historical reality. The availability of an acceptance threshold requires:

1. A uniform standard for the evaluation of risk or hazards of different origin (for instance of natural or technological origin);
2. Universal standards of comparison for qualitatively varying benefit factors (e.g., shirt, newspaper, electricity, safety) and for cost factors different in quality (say monetary costs, injuries, fatalities); and
3. Restriction to the extent of benefit and risk without considering the sequence in time, the number and social structure of potential beneficiaries or cost payers, the varying identifiability of benefit and risk, and the possibility of individual or collective influencing control (qualitative risk properties).

All investigations in connection with risk perception result in the statement that these requirements are not fulfilled. Cost-benefit evaluations on the basis of monetary standards of comparison (for instance one human life = $400,000) are declined both among population and by decisionmaking agents as being incompatible with their personal formation of judgment. Thus, the direct comparison of cost-benefit functions is methodically inadmissible, because for both categories there is at present no universal yardstick for cardinal measurement.

THE PROBLEMS OF CALCULATING RISKS AND BENEFITS

At least for the moment, it is impossible to create any quantitative risk and benefit function for varying sources of hazards. Under methodical aspects exclusively, the cost efficiency or benefit efficiency analysis can be permitted, according to which potential alternatives are compared, contrasting their qualitative and quantitative risks with a constant benefit level and vice versa.

It is recommendable that calculation of the losses or benefits be made separately for each consequence, to avoid transferring distinct qualitative features to a base unit. By listing the costs of alternative production systems, for example, they should be separated into monetary losses; injuries; fatalities; redistribution effects; aesthetical, sociological and psychological effects; and so forth.

It is up to decisionmaking agencies (which in ideal circumstances are controlled democratically) to give a societal value to each consequence and to find a proper solution on the basis of value-oriented discussions and interaction with the relevant groups of society.

Even with this restriction on the calculation and interpretation of cost-benefit analysis, there remains one problem unsolved, which is being underestimated by the majority of critics—that of the divergence between individual, group, and societal perspectives of risk-benefit analysis. It has been emphasized in numerous papers that the ordinary layman forms his attitude on the basis of perceptive biases and postrationalized emotions. But it is hard to find a reference for the fact that it might be just as rational for an individual to be against something as for society to be for something.

The rational decision of the individual or of a group is not necessarily consistent with the rational decision of society. This is predominantly the case even in connection with technological large-scale projects, for:

- The direct benefit and the direct cost are of little relevance to the individual citizen (for example, he neither needs the lot of power nor is he obliged to finance the construction of the power station); however, as far as society is concerned, the cost and benefit amounts added up are the most important criteria;
- The indirect advantages and disadvantages are of immediate significance to the adjacent residents, whereas for the decisionmaking agencies, the same are averaged among the entire population and, consequently, related or modified (for example, the location of nuclear power stations in sparsely populated areas);
- As a rule, the indirect advantages and disadvantages are not equally distributed—that is, those who bear the risk will not with precedence profit by the benefit;
- The altruistic cost and benefit considerations of the individual citizen or of a

group will not necessarily be in conformity—and this is the normal case—with the political bearings established by the decisionmaking agencies. So even if the cost-benefit analysis by society is absolutely correct, logically sound, and consistent, special groups or individuals may reach completely opposite conclusions. Generally this behavior has been rated as a product of distorted perceptions. This is in fact partly true, but it is also possible that the deviant judgment is based on purely rational thinking.

It seems necessary in connection with risk-benefit analysis to distinguish among different segments of society—namely the individual level, the intermediate level and the societal level—and it is important to find the components of the risk-benefit analysis that pertain to each of the three levels. Naturally, all components have to be weighted by the (perceived) probability of the expected event and the subjective evaluation factor for each consequence. As a reference model we used the following concept:

1. *Individual cost-benefit analysis* (perception of the individual): direct cost, or benefit respectively (electricity rate, power supply); indirect cost, or benefit respectively (personal risk, improved local infrastructure); altruistic cost, or benefit respectively (perceived risk for society, benefit for society).
2. *Intermediate cost-benefit analysis* (group decision): total of direct cost and direct benefit for the group members (formal and informal groups); total of indirect cost and benefit for the group members and for the group in its capacity as institution and their distribution specific to the group (varying distribution of cost and benefit among the different groups); congruity with group-specific values and ideals, as well as with political and social functions and interdependencies (for instance, environmental protection or competition with other groups).
3. *Social cost-benefit evaluation* (decision model): total of individual beneficiaries or cost units respectively (power supply altogether, costs altogether); total of external effects (positive and negative production effects); congruity with political and social values and political programs.

There are special links among the three levels, forming a dynamic interactive system. The individual is a member of his or her reference group which influences the value and attitude commitment of each group member, his affiliation toward special perceptive patterns, and the selection of information. An exchange of functional support and control takes place between individuals, groups, and administration through political and economic institutions and processes. As to the question of nuclear power, both the general public and special groups have forced government agencies to react to protests and to look for new solutions.

AN EXPANDED COST BENEFIT ANALYSIS

The main characteristics of the enlarged cost benefit approach are:

- Separation between individual, intermediate, and social level and separate coverage of the cost and benefit considerations;
- Inclusion of the predispositive, dispositive, and situative coefficients of influence on the individual discernment;
- Inclusion of the dynamic structure of political and social institutions and of social processes of forming opinions and decisions at the level of intermediate and social analysis;
- Investigation of the links and connections among the three levels and of their relative importance for each perspective;
- Abandonment of any universal cost-benefit theory; and
- Abandonment of any universal reference unit for cost-benefit analysis. Instead of compound indexes, the qualitatively distinct consequences should be calculated separately.

The prospective advantages of this type of analysis are:

- The possibility of early prediction of conflicts among the levels concerned;
- The knowledge of the planning agencies about decisionmaking on the part of the public and special groups;
- A better understanding of attitude formation toward new projects;
- A higher sensitivity to temporary changes of values and opinions;
- An increased value-neutral procedure on the part of the planning agencies; and
- More evaluative power of the entitled decisionmaking organizations and of the public.

What has been done so far to implement this model by an empirical case study? At present, a study on the individual level of cost-benefit analysis is underway and will be completed by the end of 1979. For many reasons, nuclear power has been chosen to serve as the research model. The starting point for the research program is a functional model of the individual decision process. This model is described in Figure 21-1.

From right to left one finds a sequential order of variables starting with general internal and external characteristics of a person. These patterns influence the rationalization process of beliefs referring to individual and societal recognition of advantages and disadvantages. These perceived consequences are weighted by personal value and attitude commitments and the subjective probability of the events. These considerations result in the formation of cost and benefit. The balance between cost and benefit as well as general affects toward the object

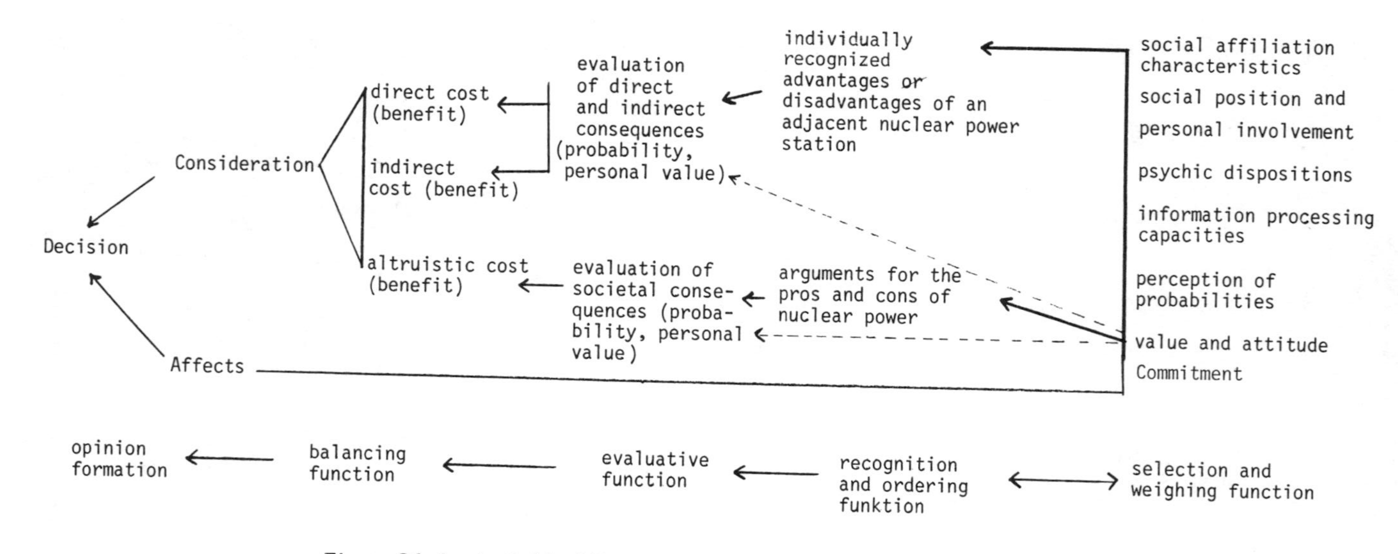

Figure 21-1. Individual Decision Model in the Case of Nuclear Energy.

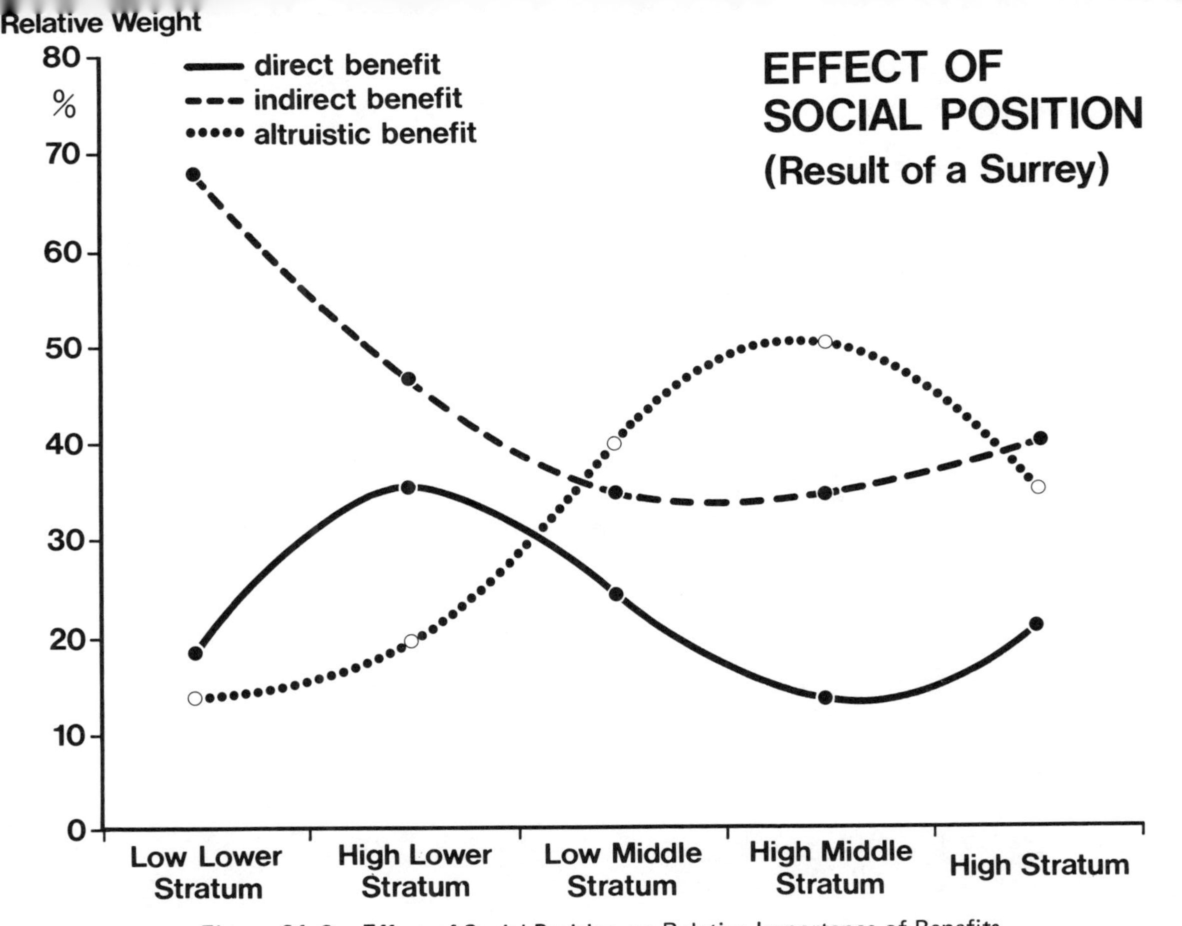

Figure 21-2. Effect of Social Position on Relative Importance of Benefits.

(both of course interfere) lead to the final decision. The testing of this model has not yet been completed, but first results of experiments and surveys show that there is much evidence for the validity of the concept, although the steps of decisionmaking are not consciously followed up by laymen.

Two results of interest become apparent already on first evaluation of the pretests:

- The probability of consequences to be expected is of little influence on personal decisionmaking, whereas the possibility that something might occur is much more important.
- The relative weight of the risk-benefit components differs significantly for the variable "stratification." Middle class persons predominantly reflect altruistic considerations, persons with low positions attribute utmost importance to indirect consequences, and upper class people evaluate the three components almost equally (Figure 21-2.).

These finding are not yet representative, but reassure us that our concept may be a meaningful attempt to improve the method of cost-benefit analysis.

Chapter 22

Bioethical Imperatives for Managing Energy Risks

Margaret N. Maxey
University of Detroit

The underlying theme of my remarks can best be introduced by recounting an ancient myth about a king who recognized his needs in a world of uncertainties, hazards, perils, and promises. He needed a risk manager.[1] To select the most qualified, he devised a test:

> The candidate manager could open either door he pleased. If he opened one of them, there would come forth from it a hungry tiger, the fiercest and most cruel that could be procured, which would immediately tear him to pieces. But if he opened the other door, there would come forth from it a lovely lady, the most suitable to his years and station that His Majesty could select among his fair subjects. Which door should he open?
>
> The first candidate refused to take the risk. He lived safe and died chaste.
>
> The second candidate hired risk assessment consultants. He collected all the available data on lady and tiger populations. He brought in sophisticated technology to listen for growling and to detect the faintest whiff of perfume. He completed checklists. He developed a utility function and assessed his risk averseness. Finally, realizing that in a few more years he would be in no condition to enjoy the lady anyway, he opened the optimal door. He was promptly eaten by a low probability tiger.
>
> The third candidate took the most direct route: he took a course in tiger taming. He opened a door at random—and was eaten by the lady.

My theme: the risks we perceive, scientifically analyze, and worry about may not be those that in the end are our undoing.

A fundamental premise of my remarks may be expressed in this form: any risk assessment methodology that fails to take into account the origin and widening dimensions of the political conflict over bringing an acceptable energy future into existence has little if anything to contribute to the institutional framework necessary for managing either perceived or actual risks. The same applies, for a stronger reason, to any ethical assessment of the social problem of managing alternative energy risks. I take seriously the admonition voiced by social scientist Peter Berger: unless an ethicist perceives the social realities of class struggle and class ideology as a preamble to the ethical consideration of societal problems, "ethics itself will become ideology."[2] That is to say, ethics can become used as an instrument or political weapon for bestowing moral legitimacy upon one or another political constituency to the extent that its practitioners remain uncircumspect about the uses and abuses of its legitimating function. The ethical task of primary importance, therefore, is to take into account some of the sources of political conflict over alternative energy sources as a preamble to any ethical consideration of managing energy risks.

The longer we live in an era shaped by such disparate national events as the fissioning of the atom, NEPA legislation, and the fallout from Seabrook demonstrations, the more reason we have to scrutinize and question the political process whereby two major social institutions are allegedly used to protect and enhance a time-honored ethical principle—namely, "the common good." (This principle has been translated, if not diluted, into "the public interest.") I refer specifically to our regulatory agency system and the judicial-legal system as institutions increasingly laden with the power and privilege that political struggles are all about.

With the advent of an "energy crisis" and the attendant public skepticism—plus a highly developed state of the art in measuring public risk perceptions about energy technologies—a new stage in the political arena has been set for a stirring psychodrama aptly entitled, "The Moral Equivalent of War." It has yet to be made clear whether the intent behind this descriptive phrase was to justify calling a state of affairs equivalent to "war" or whether it was intended to justify warlike energy policies as "moral." In either case, contending actors in our current psychodrama appear to regard the political struggle over energy technologies not simply as a matter of risk or benefits to individuals and groups, but as a matter of survival or extinction for our only habitable planet, "Spaceship Earth."

It is within this context that one of the major shortcomings of scientific risk assessments appears. The risks perceived and projected as probable consequences of "hard" versus "soft" technology are derived from qualitatively different universes of discourse. At the level of discourse familiar to scientists and engineers, the "risks" perceived and projected derive from considering technical options—that is, they are dictated by the physical nature and limits of things. But for the nonscientist or philosopher or generalist, the "risks" perceived emerge from increasingly powerful, complex, and uncontrollable energy systems with a potential for catastrophic consequences. The risks perceived are dictated by a philosophical vision of how things ought to be, quite apart from, or even in spite of, the physical nature of technical possibilities and constraints. It does not, therefore, seem accurate or productive to characterize either level of risk perception as "subjective" versus "objective" or as "imagined" versus "real." Continued use of these terms can only polarize still further those who subscribe to conflicting ideologies in the current politics of technology.

I would like to amplify the seriousness of this point by reviewing the results of an unsuccessful attempt to apply the concept of a science court in Minnesota.[3] Over the past year, national media coverage has spotlighted the opposition of militant farmers to the erection of 150 foot towers transmitting direct current high voltage transmission lines across 172 miles of farmland in western Minnesota. Since 1973, two "coop" utilities have been trying to find a transmission route from a coal-generating plant located in North Dakota near strip-mining facilities to a station on the outskirts of the Twin Cities. The Minnesota Highway Department refused to allow the route to follow interstate highways on aesthetic grounds—"it would be unsightly." Next, the state Department of Natural Resources refused to allow the route to cross wildlife areas on grounds that fauna and flora would suffer. Finally, exercising the right of eminent domain, the utilities chose paths that would cross many farmers' fields diagonally, thus interfering with common farming practices and, especially, with center pivot irrigation systems. When the utilities were asked to shift this route so as to skirt boundaries, they refused. According to their cost-benefit ratios, it would be too costly. Ensuing litigation up to the state supreme court, including public hearings in 1975, were all decided in favor of the utilities. Health and safety issues have not been absent; indeed, fears of exposure to electromagnetic fields have proved, as always, to galvanize the protest movement. However, the farmers have insisted that the dominant reasons behind their opposition are these: They are concerned for the impact of the powerline on valuable, increasingly needed farmland that they have a moral responsibility to improve and to hand down to their children. No existing social institutions or judicial procedures were available to them to deal with what they perceived to be long-term risks and hazards—namely, the sacrifice of their land without their consent, presumably for a greater social benefit and need they were not convinced was valid in the first place.

When a newly elected governor in 1977 made repeated attempts to resolve the rancorous dispute by setting up a science court procedure, the coop utilities insisted that the focus of the court be limited to technical issues of health and safety. But the farmers insisted that the focus be broadened to include all the issues of concern to them—findings of need, following alternate routes, possible alternative sources of power, and so forth. Additionally, since the court proceedings needed closure and the dispute was regarded as political rather than technical, the farmers' version of the science court called for an accountable public official to render a final judgment about nontechnical issues, rather than letting scientific experts acts as judges. The governor refused to act as judge or to extend the scope of the science court beyond health and safety issues. Subsequently, the science court has been consigned to limbo, the governor has just been defeated in his bid for reelection, and the transmission lines are going up.

I shall return to the science court concept later. The first point of this illustration is that it will not do to call either the utilities' or the farmers' risk perceptions "objective" and "real" and to dismiss the other's as "subjective" and "imagined." Second, the problem cannot be litigated away by some specious "due process." Neither can it be resolved merely by exhorting parties in the dispute to practice personal virtue, much less to adopt economic cost-benefit and technical risk-benefit methods of analysis. We must confront the seriousness of the problem about technology disputes by recognizing and remedying institutional deficiencies. A new kind of technosocial problem requires a new quality of intellectual analysis and institutionalized processes of dealing with it.

The widening dimensions of the political conflict over bringing an acceptable energy future into existence indicate that we must confront at least three ethical imperatives if we are to deal constructively with the politics of technology in general and with the politics of managing energy risks in particular. Tentatively, I propose these formulations as a first step toward our mutual understanding of the broader dimensions of the risk management problem.

FIRST IMPERATIVE

First, it has become an ethical imperative that the moral objections to certain energy sources and systems be made explicit, publicly debated, and resolved with some authoritative closure.

The underlying ethical principle can be stated generally thus: There are kinds of human actions that inflict risks and harm of such a nature that, no matter what benefits may also result, such actions cannot be justified. They cannot be justified by rational analysis and argumentation (they are "unethical") for the reason that they cannot be justified morally—that is to say, they violate basic norms of responsibility that human beings owe to each other (they are "immoral").

The fact that certain benefits, personal or social, can be shown to result from a policy that entails risk, hazards, or harm does not constitute an ethical justification for that policy. Doubtless the institution and practice of slavery resulted in tangible and essential benefits to slaveowners, to slave traders, and in some cases, to those enslaved. Nevertheless, the practice itself has merited moral condemnation because it violates fundamental values and norms—namely, the essential human conditions of freedom and well-being.

The current political debate over energy risks has become crystalized into a moral question: Are there any energy technologies that, by their nature and inherent consequences, risk inflicting a type of harm to present and future generations such that, no matter what benefits may also result, those technologies cannot be ethically or morally justified? It would appear that certain members of the scientific community, as well as influential political activists and a significant segment of the public, are persuaded that one technology—nuclear fission—is morally wrong, hence ethically unjustifiable. They argue that, because somatic and genetic effects from any level of radiation exposure are so deleterious to present and future progeny and because no technological method or institutional longevity can guarantee that long-lived radioactive wastes will remain safely isolated from the biosphere, nuclear technology is by its very nature immoral and unethical. These objections have been derived from allegedly sound scientific judgments of experts. Other objections are derived from a political philosophy: nuclear technology by its very nature sustains corporate, industrial, and economic systems that risk inflicting psychological and moral harm on unconsenting citizens today and in the future. Once again, nuclear technology is morally and ethically indefensible.

Because of time constraints, I cannot engage in any discussion of these moral objections. My only purpose here is to make them explicit and to suggest that they receive close inspection by the scientific and technical community. The validity of these objections, and the ethical or moral analyses made about them, not only can but must be just as "objective" and "empirically based" as a cost-benefit analysis. The quality of ethical and moral analyses is directly proportional to the quality of scientific and technical analyses that are made with their moral implications in mind.

Moreover, if a consideration and line of reasoning about energy risks is to be ethical, then it is unjustifiable for cost-risk-benefit experts to isolate out nuclear risks for some unique, unprecedented treatment. To protect the public from inequitable management of biohazards, it is an ethical imperative that nuclear risks be compared and evaluated—not only in relation to biospheric risks from other feasible alternatives, but also in relation to social risks of deprivation, civil disruptions, and serious distributional inequities if electricity shortages occur because nuclear technology becomes rejected or indefinitely delayed.

SECOND IMPERATIVE

The unwarranted stigmatization of nuclear risks and radiation hazards appears to have resulted from two institutional deficiencies:

1. The manner in which some professional experts are included and others excluded in the bureaucratic standards-setting process; and
2. The unfortunate fact that each regulatory agency both sets standards and enforces them by a process that is vulnerable to arbitrary revision and limitless litigation by career intervenors' use of the system.

Because the common good has become seriously jeopardized by the legislative ambiguities and their interpretation by self-serving regulatory agencies, it has become a second ethical imperative that the regulatory system be radically restructured by a more enlightened legislative mandate.

As presently functioning, our regulatory agency system is so chartered and mandated as to be compartmentalized, fragmented, and virtually unaccountable to any comprehensive guardian of the general welfare. Scientific risk assessments and economic cost-benefit ratios are forced to be piecemeal, ad hoc, haphazard, isolated for one at a time consideration. Each regulatory agency operates in such a way that one hazard is spotlighted for a time (because it is the current product of research projects), giving way to another in unending succession: DDT, lead, cyclamates, the Pill, red dye #2, PCBs and PVCs, triss, and now saccharin. Having completed a decade of concern about "the carcinogen of the week," we are entering a political climate that will doubtless force a decade of public concern over "the low level radiation source of the week."

Each regulatory agency has its own category of so-called hazards on which to conduct research, at the same time making a case for more federal funds to do more research in further risk reduction of units of hazards. Not only does this piecemeal, selective concentration magnify certain potential hazards at the data-gathering and risk assessment levels, but the public is misled into perceiving that, just because some risks are the more studied, they are by that very fact the more dangerous to public health and safety. But this is clearly not the case.

It may seem amusing to suggest that something in this nation has not yet been counted, but we have an urgent ethical imperative to conduct some undone business here. What we really need is a whole new field of numbers. We need to know, with the most comprehensive overview (1) how much public money is spent to reduce ordinary diseases and ordinary accidents that afflict major segments of the population, (2) the cost per capita that is being spent to reduce them effectively, and (3) precisely at what point huge amounts of money may be pouring into budgets that can assure only minor gains in the status of public health, if any at all. We have a surfeit of statistics on public health, but those data are not arranged by any responsible public institution so as to look at risks to the entire population relatively, to make comparisons, to maximize cost effectiveness so as to get the most public health protection for the many out of the expenditure of a finite amount of public money.

Instead of multiplying regulators and regulations in the advanced stages of social Parkinsonianism, our regulatory agency system must be profoundly altered. Its deficiencies must be taken seriously so as to assure that regulative standards for protecting health and safety actually consider the common good of the many and that finite amounts of public money are allocated in a just and equitable manner.

The time is long overdue for the institution of a separate cabinet level Department of Health and Safety. It would fall to this department's jurisdiction to make a comprehensive review of cost-effective health and safety standards, as well as justification for budgetary allocations. The Congress should charter and mandate such a department to consolidate and govern the following regulatory agencies:

Environmental Protection Agency,
Nuclear Regulatory Commission,
Federal Drug Administration,
Occupational Health and Safety Administration,
Federal Aviation Administration,
Public Health Service

—and any other agency currently engaged in setting standards and regulating conditions affecting public health and safety. Such a Department of Health and Safety should not be conceived as still another bureaucratic level but rather the contrary—a consolidating, streamlining, efficiency-centered governing organ to which regulatory agencies are answerable and accountable.

Properly mandated, this department could eliminate major jurisdictional disputes, duplicative standards, and piecemeal regulations that obstruct justice and equity in protecting the quality of our common life. Moreoever, if so chartered, this department would be set up to institutionalize a more enlightened process of having professionals in the sciences and engineering themselves set the standards for health and safety according to strict procedures of peer review, rather than the enforcing agencies. A proven model has been successfully operating over fifty years—The American Society for Testing and Materials' (ASTM) Voluntary Consensus System of professionally established standards. The enforcing agencies would, therefore, not be the arbiters of conflicts among the experts in any given profession; the professions themselves by their own peer review would be responsible to adjudicate conflicting judgments about scientific or engineering matters. Policymaking and standard setting would thus be derived from the best scientific judgment available at any given time, and haphazard or arbitrary revisions could be avoided.

THIRD IMPERATIVE

The Department of Health and Safety might also be the proper governmental arm for institutionalizing a method of dealing with a third ethical imperative—namely, the resolution of newly emerging technosocial issues by some authoritative closure for policymaking (beyond standard setting) with respect to public health and safety protection. The public disputes over fetal research, recombinant DNA research, and nuclear technology should be evidence enough that there is an urgent imperative to devise a new kind of social institution for establishing policy and guidelines in increasingly controverted technological developments.

Technosocial issues are of such a nature that they are seeking a policymaking end product, and traditional institutions are no longer adequate to the task. Policy has been set by the courts, reacting to individual cases, and decided by judicial fiat. Policy has also been set by legislatures whose members are responsive to a constituency with vested interests, and policy is decided by political tradeoffs. A new phenomenon in the sphere of public policymaking emerged in the aftermath of an outcry about fetal research at the National Institute of Health. In 1974, a National Commission for the Protection of Human Subjects was convened for the purpose of fact finding and policy formation. Constituted as a public commission, its members represented diverse backgrounds, competencies, and convictions in various disciplines–law, medicine, science, ethics. It conducted public discussion, deliberated openly and candidly, heard from each segment of public responsibility. This commission may well represent a precedent that could be emulated, amended with a broader objective in mind, and institutionalized for the purposes of policymaking in the Department of Health and Safety.

The application of a science court concept has many positive uses, yet its role in policymaking seems to have many drawbacks. According to its originator, Arthur Kantrowitz, the intent is to adjudicate and judge the preponderance of scientific fact on any given issue so as to settle a factual dispute. But the term suggests that a legal adversarial model would dominate the procedure. In practice, this model substitutes courtroom rhetoric, innuendo, dramatic overstatement, and prestructured questioning that seeks "an optimum resolution of conflict" profitable to the victor for an established scientific procedure that seeks the truth. Granted that courtroom adversary models are an appropriate method for adjudicating disputes over rights between individuals and groups, they fail miserably to provide a method for making enlightened policy about man's transactions with nature.

In any case, the ethical imperative remains to be confronted and resolved, if we are not going to be "eaten by the lady."

NOTES

1. This version is adapted from Frank Richard Stockton's text cited and freely translated by William C. Clark in "Managing the Unknown," in *Managing Technological Hazard: Research Needs and Opportunities,* Robert W. Kates et al., ed. (University of Colorado, Institute of Behavioral Science, Program on Technology, Environment and Man, Boulder, Colorado, Monograph #25, 1977), p. 111.

2. Peter Berger, "Ethics and the Present Class Struggle," *Worldview* 20 (April 1978): 6–11.

3. This summary is based upon an article by Barry M. Casper and Paul D. Wellstone, "The Science Court on Trial in Minnesota," *Hastings Center Report,* 8, no. 4 (August 1978): 5–7.

Part II

Limits to Growth of Due Process

 *Chapter 23*

A Statement on Limits to Growth of Due Process

John E. Gray
International Energy Associates Limited

My mission here is, first, to present a statement related to the limits to growth of due process. Subsequently, Pierre Zaleski had suggested that I also deal with the Cluff Lake, Ranger, and Windscale inquiries, as examples of due process. Consequently, I am going to comment on (1) some of the elements required for due process, (2) some examples of due and other process, and (3) factors that may limit the effectiveness of due process. I will then advance an opinion or two and retire. All of my remarks are, of course, directed to proceedings concerned with energy supply and related decisionmaking.

Webster defines due process of law as "a course of legal proceedings carried out regularly and in accordance with established rules and principles." Process is defined as "the whole course of proceedings in a legal action." Process is also defined as "something going on." I submit this latter definition as perhaps being quite relevant to current decisionmaking.

Turning to Black's Law Dictionary, due process of law is defined as

> Law in its regular course of administration through courts of justice. . . . "Due process of law in each particular case means such an exercise of the powers of the government as the settled maxims of law permit and sanction, and under such safeguards for the protection of individual rights as those maxims prescribe. . . ." To give such proceedings any validity, there must be a tribunal competent by its constitution—that is, by the law of its creation, to pass upon the subject-matter of the suit. . . . Due process of law implies the right of the parties affected thereby to be present before the tribunal which pronounces judgment upon the question of life, liberty, or property, in its most comprehensive sense; to be heard, by testimony or otherwise, and to have the right of controverting, by proof, every material

> fact which bears on the question of right in the matter involved. If any question of fact or liability be conclusively presumed . . . , this is not due process of law. . . .

Searching for a lay understanding of the major elements of due process, I come up with (1) the purpose being the enforcement and protection of private rights; (2) the mechanism being a legally constituted, competent tribunal that hears and weighs all pertinent material, facts, and arguments; and (3) the result being a lawful finding or judgment. Also, I presume that the tribunal must be able to determine the relevance of matters and parties to any inquiry, to limit them, and to determine the breadth, pace, and schedule of an inquiry or proceeding.

Given this background, I submit that the Ranger Inquiry in Australia, the Cluff Lake Inquiry in Canada, and the Windscale Inquiry in the United Kingdom appear to have satisfied those nations' requirements for, or perceptions of, an adequate process for evaluating and judging the rights and proper future conduct of the various parties to those inquiries. Also, those inquiries appear to have been conducted in accordance with the major elements of due process, as presented above. Legally constituted tribunals were given a job to do and did it. The results, although disputed by some, were accepted as lawful and as a basis for government and private decisions.

Turning to relevant inquiries or proceedings elsewhere, it is interesting to note the status of GESMO in the United States. It appears to me that a sound course of due process was initiated and then aborted. Query—Is it due process to abort due process?

The GESMO proceedings were intended to result in rulemaking and consideration of license applications for nuclear fuel reprocessing and a use of plutonium in nuclear reactors. For a period of time the properly constituted and competent hearing board conducted far-reaching but well-focused hearings. Then on December 23, 1977, the Nuclear Regulatory Commission terminated the GESMO and related licensing proceedings. That order has become the issue, in the United States Court of Appeals, with some of the various parties to the GESMO proceedings and other interested persons questioning, or supporting, the legality of the NRC's December 1977 action. I note that one of the arguments the Natural Resource Defense Council's in support of NRC's action is that the "[Nuclear Regulatory] Commission's decision gave proper weight to the President's views and is not arbitrary, capricious, or an abuse of discretion."

In effect, the process has become the issue, or else the initially germane issue has been obscured by the process. I do not question or pretend to know the legalities of this situation or the possible outcome. However, I do believe that the GESMO proceedings were downgraded from the level of careful, competent, productive due process to the level of that process defined earlier as "something going on."

Turning to the international arena, we find the International Fuel Cycle Evaluation Program (INFCE). Its conduct appears to have little to do with due process and a lot to do with the "something going on" process. INFCE may be a program in search of a process. It does not appear that it can find its home in any established competent legal, technical, or economic process, and thus it may be committed to being dealt with principally in the political process. I believe that the political process may come closest to the "something going on" category.

Turning to factors that would appear to limit the effectiveness of due process, I suggest the following: (1) lack of a competent, properly constituted tribunal; (2) excess of presumption relative to fact; (3) excess of parties or issues to be embraced in a single case; (4) excess of time required for the process; and (5) political (or policy) intervention or inherent conflict of interest derived from government organization and/or authority.

In closing, I would like to comment on the matter of limits of growth of due process. There do not appear to be any, inherently. However, in being applied to energy-related decisionmaking, due process demands both management and manageability, as well as a factual base that supports findings. And there must be limits. The interconnected technical-economic-legislative-administrative-judicial decisionmaking systems, which determine the economic utility and social acceptability of various energy supply and use options, should be limited to those that are manageable, fact oriented, timely, and conclusive.

Looked at in this light, the overall U.S. decisionmaking process on energy supply and use and on related environmental and health and safety issues may have become dysfunctional. Without some simplification and tightening up, it is likely to get worse. This applies especially to the proliferation of legislation, regulation, and adjudication.

Regulation, competently derived and administered, is no big deal as a problem. Regulation as an incentive to misuse of due process is another matter. The U.S. Supreme Court, in the Vermont Yankee case, stated that the "constant challenge of regulatory procedures is judicial license run riot." Due process, unlimited, is a threat, not a route, to fairness. My concern is that we have not yet reached a point where there is a broadly shared understanding of the dimension and impact of the unlimited growth of due process in national decisionmaking, much less of what to do about it.

 *Chapter 24*

The Energy Crisis and the Adjudicatory Process

Arthur W. Murphy
Columbia University

When lawyers and scientists embark on a discussion of the legal process it is well—at least my experience so suggests—to define very carefully the precise subject under discussion. An early description of the work of this session defined as its purpose the exploration of criteria to "define reasonable limits to the growth of *due process.*" I am more than a little uncomfortable with the use of the term "due process" in a pejorative sense. And I would like to make clear that the issue is not due process but "undue process." I take as my starting point that the constitutional requirement of due process (I speak, of course, only about the United States) commands no more than what is reasonable in the circumstances. Our problems with making decisions about energy and the environment—to the extent that they are procedurally based—do not stem from the implementation of constitutionally mandated procedures but from the use of procedures fashioned for one purpose in areas where they are inappropriate or, if you will, undue.

There is a second caveat that I would also state at the outset. My focus in this chapter is on the use and abuse of adjudicatory procedures in that part of the administrative process concerned with energy and the environment. There is a great deal wrong with the way we make decisions about energy and the environment in the United States, but to a considerable extent the difficulty is symptomatic of some very fundamental problems that our institutions of government are experiencing today. This is not the forum—and I am not the person—to discuss these problems, but the major elements are clear: a seriously weakened presidency; a legislature structured on the lines of party politics without party discipline; a public disenchanted with the political process and distrustful of all aspects of government—except, perhaps, the courts. I mention this because, important as procedure is, the best conceivable procedures can only facilitate the making of decisions that society wants made. Procedures will not make policy.

Procedures can on the other hand frustrate the effectuation of policies that have been made—and that does appear to be happening. Indeed, in the case of power plant licensing, differences in the procedures for licensing (e.g., nuclear as contrasted with coal) may dictate a choice of energy thought to be inferior on economic or on environmental grounds. One of the areas of concern is what I shall call, somewhat loosely, the judicialization of the decisionmaking process—or the push toward making decisions by adjudication.

THE USES AND ABUSES OF ADJUDICATION

What do I mean by adjudication (or adjudicatory procedures)? Generally, I mean the procedures followed by courts in the United States in hearing and deciding cases. Although all would not agree on all points, I think that most lawyers would agree that the essential elements of adjudication would include:

- Notice to persons interested in the pendency of a case;
- A "hearing" that affords an opportunity to the parties to present evidence, to rebut the evidence of other parties, and to test the other parties "case," preferably by cross-examination;
- The hearing to be held before, and the decision to be made by, a competent and impartial tribunal;
- The decision to be based on the "record" of the hearing, and the conclusions reached to be rationally explained;
- The decision (in the case of an administrative agency) to be subject to judicial review;
- The parties to be permitted representation by counsel.

This model, with variations, is essentially that developed by the Anglo-American courts for criminal and civil proceedings. Perhaps its most prominent characteristic is its adversarial nature; the tribunal making the decision is dependent on the evidence introduced by the parties. In the United States, moreover, there is a strong tradition of judicial restraint—some would call it passivity. The proceeding is dominated by counsel who are expected to—and expect to be allowed to—"make" their clients' cases and to destroy their opponents' cases by cross-examination.

A leading scholar in the field has stated the criteria for judging procedures in terms of the extent to which they promote *efficiency* in handling an agency's business, *accuracy* of its selection and determination of relevant issues, and *acceptability* of the agency decision to the participants (including the agency) and the general public.[1] Although not an inevitable characteristic of the judicial model, the judiciary in the United States has always been more concerned with acceptability—specifically with the rights of the parties to a fair hearing—than with efficiency of the system. To be sure, from time to time the bar pays elaborate lip service to efficiency, but the system nearly always leans over backwards to ensure procedural fairness even at the expense of efficient administration.

Adjudication works pretty well in the cases for which it was designed. In the ordinary criminal or civil case, the task of the tribunal is to ascertain the truth about specific past events: who did what, and with which, to whom. The facts at issue are primarily of concern to the individual parties and are likely to be best known to them; in the circumstances, their proof or disproof is perhaps best left to the adversarial process. In these cases, we are not, it should be remembered, primarily interested in truth. In the criminal case, we consciously subordinate truth to the protection of the accused by exclusion of some kinds of evidence, by requiring proof beyond a reasonable doubt, and so forth. In the civil case, it is often more important that a dispute be settled than that it be decided correctly. And, although efficiency is not a negligible factor in either civil or criminal cases, other factors are understandably given priority. In criminal cases, a primary concern with the rights of the person accused of a crime is properly the boast of our law. In civil cases, most of which involve essentially private controversies, convenience to the parties is given an appropriately high value. Finally, it should be noted that, in the ordinary civil suit, time is not of the essence, nor is it in a criminal action except to the extent that the accused is entitled to a speedy trial.

Satisfactory as the judicial model may be in its traditional sphere, it works much less well in some kinds of administrative proceedings. I stress "some kinds" because the administrative process covers a broad spectrum of proceedings of very different kinds. Where a proceeding is concerned with the imposition of a sanction—for example, the revocation of a license—it is only marginally different from a criminal proceeding in the courts, and the use of similar procedures in both situations is warranted. In other cases—for example, a refusal to bargain collectively—the tribunal is primarily interested in resolving a dispute between private parties about historical facts, and the use of the procedures of a civil action in the courts causes no difficulty. But many types of administrative proceedings are radically different from the typical civil or criminal case. In the kind of proceeding of interest to us here, for example, the differences include: (1) the nature of the facts at issue—they are not ordinarily historical facts peculiar to the parties, but generic "facts" about the future impact of a project; (2) the decision, which although of particular interest to the parties, will directly affect important public interests—for example, the satisfaction of energy demands; (3) that it is more important that a correct result be reached than that the parties be allowed to make their case; (4) that time is usually an important factor, and the consequences of delay more serious.

Despite the obvious differences, and not withstanding a considerable body of opinion that administrative procedures should be functional—that is, should be adapted to the mission of the agency and to the particular kinds of questions raised—administrative law has tended to follow the familiar model of the judiciary. In the words of one administrative scholar: "The obstacle [to development of a distinct administrative procedure] was and to some extent still is a perception among judges, lawyers and legislatures that the adversary system is the only legitimate way to make decisions in our society."[2] The devotion to the judicial model is not, it should be noted, explicable wholly in terms of lawyers' preference for the familiar. At least in part it reflects objection to the substance of regulation. It is no secret that adjudicatory procedures can be used to delay action. And it is no accident that in the 1930s the most bitter opponents of the New Deal programs in the United States were the strongest proponents of the judicial model for the administrative determination or that today many prominent adherents of the "pure" judicial model in the energy area are those who are opposed on environmental grounds to large-scale energy projects.

The victory of the judicial model was not complete. The federal Administrative Procedure Act (which became law in 1946) recognizes a basic distinction between an agency's rulemaking function and its function in adjudicating individual cases. The former was seen as quasi-legislative in nature, and the procedures to be followed are those thought to be appropriate to the legislative function—notice and an opportunity to make oral or written comment on the proposed rule. Adjudication on the other hand was to follow the judicial model. Although the dichotomy between rulemaking and adjudication was seen as basic, room was left for the development of procedures on functional lines. Rulemaking could be required to follow adjudicatory procedures, and at least by implication, adjudications need not follow the judicial model. Unfortunately, although the invitation to require formal rulemaking (i.e., using adjudicatory procedures) was often accepted, the invitation to use less formal adjudicatory procedures was not. Despite frequent criticism of the "judicialization" of the administrative process, little attention was paid to the need to develop different procedures for different situations. Interestingly enough, it was not only the courts and the legislatures who expressed a preference for the judicial model. The history of the last thirty years contains many examples of self-inflicted wounds—that is, situations in which an agency free to choose a less judicial model has nevertheless imposed upon itself the more onerous restrictions of that model.

Until fairly recently, although weaknesses in the process were frequently deplored, the costs of overjudicialization were by and large accepted. Why, then you may ask, is it now seen as a significant threat to our ability to make decisions? To some extent, the new concern reflects the fact that a different ox is being gored. Many lawyers who fiercely defend the judicial model when interests of their clients are at stake prophesy disaster when others adopt the same tactics. But there also have been very substantial changes in the picture that transform what was once largely a problem for the bar into a matter of national–even international–concern. One of these changes is a vast extension of government in areas previously left largely to private initiative. Particularly in the energy field, the trend has been to larger and larger projects, many of which need government support and all of which involve the government in some way or another. This latter extension has been greatly complicated by the multiplication of issues required to be considered by licensing agencies under the national and state environmental policy acts. At the same time there has been a significant relaxation of the barriers to intervention by members of the public in agency proceedings and a significant increase in the extent of public participation. Where once the active participants in a licensing action were likely to be the agency and the applicant (or competing applicants), today multiple parties representing a broad spectrum of interests tend to be the rule. Since intervenors, generally speaking, have all the rights of other parties, added complexity is inevitable. A final exacerbating condition, although not strictly speaking attributable to adjudication, is that under our federal system, many environmental issues must be decided more than once.

Most people concede that the adjudicatory process is a cumbersome method of making decisions and that it frequently causes delay. They argue, however, that its worst defects can be ameliorated by better management and that, properly managed, its virtues outweigh its benefits. No doubt there is room for improvement; a strong hearing officer can limit dilatory cross-examination; can prevent, within limits, duplicative testimony; and can eliminate peripheral issues. But I am skeptical about how much can be accomplished along these lines. The tradition of fairness to the parties and deference to lawyers, however frivolous their cases and however obvious their abuse of the process, is firmly established in our law. Moreover, there is considerable risk that the courts (with the benefit of hindsight) will find prejudicial error in rulings limiting the procedural rights of any party. For even a strong hearing officer there is a great inducement to err on the side of leniency if the alternative is the possibility that the whole proceeding may have to be gone through again. I do not mean to suggest that nothing can be done, but only that as a practical matter, the opportunities are limited.

Over the years the principal argument of the proponents of adjudication has been that it produces better–that is, more accurate–results. In recent years, this argument has been supplemented, and in some quarters superseded, by the argument that an opportunity to participate in an adjudicatory procedure is necessary to *acceptability* of the decision. Acceptability of decisions is, of course, an essential element of any legal process, but it has never been thought that a right to direct participation in the making of decisions was a prerequisite to acceptance. The new insistence on participation reflects the very different atmosphere toward government agencies that exists today. A large number of people have lost faith in the administrative process. The notion is widespread that administrative agencies have become captives of the industries they regulate and, therefore, no longer represent the public interest; the natural corollary of that proposition is that "true" representatives of the public interest must be allowed to participate in agency deliberations.

While I support public participation—in some form—in agency deliberations, I tend to be skeptical about the extent to which the argument based on *acceptability* can be pushed. Many of those who intervene in agency proceedings are not interested in the opportunity to be heard; they want to prevail. Surely no one could be afforded more opportunity to participate than the opponents of the nuclear power plant at Seabrook, New Hampshire. Yet when they lost, some "took to the streets," and there are veiled—and not so veiled—threats to do the same in other places. In my view, promotion of acceptability is not in itself a sufficient justification for affording procedural rights not otherwise warranted. The real questions should be the contribution to accuracy that present procedures make, whether that contribution could not be made in another way, and whether the contribution is worth its cost. The time is ripe, I believe, for a cost-benefit analysis of the process by which we make decisions.

On the benefit side, it seems beyond dispute that the agency perception of issues such as nuclear waste storage, thermal discharges from power plants, and the effects of pesticides have been materially broadened by the contribution of public groups. It is a matter of considerable dispute however, whether those contributions depended on the adjudicatory process. At least in my experience, the contributions were not so dependent and could as well have been made in rulemaking proceedings, with some adjustments. This should not be surprising. Most of the crucial issues in these matters involve matters of general fact or policy, rather than the specific historical facts for which adjudicatory procedures were designed.

If the benefits are questionable, the disbenefits are not. The problem of delay has been mentioned above. In the nuclear field, a great deal of debate has been expended on the question of whether delays in licensing are attributable to the adjudicatory process, as some contend, or to other factors. At best I find this debate not very useful. Case studies make clear that there are many causes of delay—unrealistic schedules, labor-management disputes, failure by suppliers to deliver key components when promised, design and construction failures, the regulatory process, difficulties in financing, and second thoughts by utilities about the need for a particular project are prominent, but by no means the only, causes. Whatever the cause may be, many licensing hearings have been the occasion for the use of dilatory tactics—for the precise purpose of causing delay—and the opportunity to use such tactics is inherent in the process.[3]

Delay in particular projects increases costs, in some cases to the point of making the project economically unviable. But there are, I suspect, more insidious effects of the adjudicatory process as applied to energy decisions.

Essentially, my suspicion is that the adjudicatory process operates in subtle ways to disserve the cause of accuracy—that is, in making the correct decision and, indeed, in addressing the correct issues. As noted above, surprisingly little attention has been paid to fashioning procedures appropriate to the occasion, and little empirical evidence has been collected on the disbenefits of adjudication except in terms of delay. Consequently, what I am about to say is based on intuition and, it must be confessed, a somewhat vague intuition at that. With that apology, let me just indicate some of the questions that I think need study.

QUESTIONS DESERVING FURTHER STUDY

What is the effect of the insistence on an "independent" tribunal and the requirement that decisions be based on the record of the hearing? One effect can be the isolation of the tribunal from the assistance of those who could be most helpful. Another may be a tendency to value all evidence uncritically and to credit testimony that a more knowledgeable person would disregard. Of course, no one likes to play cards with a stacked deck or to have a decision based on factors unknown to him, but the notion that everything is *tabula rasa*—that, like a jury, potential decisionmakers should be disqualified if they have any advance knowledge of the issues in dispute—is of doubtful validity with respect to complex scientific subjects.

What is the effect of the way in which issues in adjudication are traditionally phrased and answered? It seems to me that one of the serious drawbacks of a trial type proceeding is that it tends to harden positions on both sides. As a result, the person making the decision does not get the benefit of objective views; he is offered a choice between certainties instead of a range of alternatives. Moreover, the practice of answering issues in "yes" or "no" terms may create an illusion of certainty that can backfire. I have always felt, for example, that the Atomic Energy Commission would have been better off to state the issue of reactor safety in terms of accident probabilities, instead of requiring that a reactor be designed to contain the effects of maximum hypothetical accidents or design basis accidents.

What is the effect of such devices as presumptions or burden of proof? The categorical imperative of the civil or criminal suit is that a decision must be made. On each issue, one party must win and the other lose. To ensure this result, we assign to one party or another the "burden of proof." This has two effects. One is that in case of ties, we know how to decide the case; another is that the failure to establish any "essential" element of his or her case will cause the party with the burden of proof on that issue to lose. How does one discharge that burden on an issue such as the preferable site for a power plant or the relative safety of a nuclear plant and an (as yet unbuilt) solar alternative. The result in my view is to weight the scales heavily against known technologies and against taking affirmative risks (by which I mean the risk of doing something as compared to doing nothing). If that is true, the consequences to society could be enormous.

A somewhat different problem is caused by the sheer length of time taken up in the combination of administrative and judicial review. The licensing hearing in the case of the Midland, Michigan, reactors began in December 1970. The initial decision to grant a construction permit was rendered in December 1972. In July 1976, the court of appeals for the District of Columbia overturned the initial decision on grounds that called into question whether the project should be constructed. In April 1978, the United States Supreme Court reversed the court of appeals and reinstated the construction permit. While for the utility the ending of the legal saga may have been a happy one, the fact that the viability of the project remained uncertain for eight years after the decision was made that it was needed will not be lost on other companies.

THE OUTLOOK

What can be done? There is some evidence that the courts and the legislature are beginning to work out a pragmatic approach to procedures along functional lines. Some compromise between the minimum rulemaking procedures and the maximum adjudicatory procedures will probably be worked out in the long run, but I do not believe we can reach a satisfactory result unless we think more carefully than we have about what we want to achieve and about the effect of procedures in reaching these goals. In that process, we will have to work together; we will have to become more sophisticated about each other's fields. At least we have to be aware of what each means. Despite all of the talk about interdisciplinary cooperation, most of the time lawyers and scientists seem to be talking past one another. We can no longer afford to do that. Whether we like it or not, we must work out a way of working together.

One avenue to which we have paid insufficient attention is whether or not we can adapt the adversary process to technological issues. Scientists tend to associate the adversary process with cross-examination and particularly with the highly overrated, flamboyant pyrotechnics of some trial lawyers. But the adversary process need not mean only that. If I may be permitted one more quote from an advocate of a broader view of the adversary process:

> An effective consensus cannot be reached unless each party understands fully the position of the others. This understanding cannot be obtained unless each party is permitted to state fully what its own interest is and to urge with partisan zeal the vital importance of that interest to the enterprise as a whole. At the same time, since an effective consensus requires an understanding and willing cooperation of all concerned, no party should so abandon himself in advocacy that he loses the power to comprehend sympathetically the views of those different interests. What is required here is a spirit that can be called that of tolerant partisanship. This implies not only tolerance for opposing viewpoints, but tolerance for a partisan presentation of those viewpoints, since without that presentation they may easily be lost from sight.[4]

For the short run, I am somewhat gloomy about the prospects in the United States. The conventional wisdom—at least in a substantial segment of the legislature—seems to be that full participation in adjudicatory hearings is a necessary antidote to "captive agencies." For example, the recent attempt to rationalize the system of licensing nuclear power reactors has foundered at least in part on the insistence of some groups that the judicial model be preserved and indeed strengthened; in the case of the New Energy Act, the federal government has gone so far as to impose requirements of an evidentiary hearing upon the states and, moreover, to require funding of the intervenors in such proceedings. The picture is not uniform, and there are countervailing currents, but the perception that adjudication is the only legitimate method of making decisions does not seem to have changed much. Indeed, it is reinforced for some by the notion that the administrative process can be an effective substitute for the legislative process. With all sympathy for the feeling of frustration with government that many people are experiencing, it seems clear that it cannot. What it can do, unfortunately, is to give a virtual veto over the undertaking of affirmative programs to the opponents of those programs. Sooner or later the pendulum will swing, but in the meantime we may find it very hard to do what ought to be done.

NOTES

1. Cramton, "A Comment on Trial-Type Hearings in Nuclear Power Plant Licensing," *Va. L. Rev.* 58 (1972): 585–93.
2. Verkeuil, "The Emerging Concept of Administrative Procedure," *Colum. L. Rev.* 78 (1978): 258, 278.
3. The causes and effects of delay are discussed in Murphy, La Pierre, & Orloff, *The Licensing of Power Plants in the United States* (1978).
4. Lon Fuller, "The Adversary System," in *Talks on American Law,* a series of broadcasts to foreign audiences by members of the Harvard Law School faculty. Edited by Harold J. Berman (New York: Vintage Books, 1961), p. 32.

 *Chapter 25*

The Economics of Due Process

Kenneth W. Dam
University of Chicago

How would an economist look at the question of due process in the power-plant-siting field? In the first place, if he were a high-powered academic theoretician, he probably would not look at it at all because he would regard the questions involved as too mired in institutional and legal constraints to be of general interest to economists. Of course, most economists, following the insights of Adam Smith, will be prepared to analyze almost any problem if the compensation is right. But since this forum is *pro bono publico*, you shall have to content yourselves with a lawyer's impressions as to how due process can be subjected to economic analysis.

PRIVATE INTERVENTION DISCUSSED

Although I hope my long, sometimes pleasant associations with economists both in academia and in government provide some basis for my effort to combine legal and economic analysis, I would argue that a quick, hard look at the due process concept from the legal point of view will strip away some of the humbug that sometimes surrounds discussion of energy and environmental issues and will permit the economic discussion to remain at a relatively elementary level.

The term "due process" is used in the plant-siting field to mean something rather different from what it means in the lexicon of constitutional lawyers. To the latter, due process refers to constitutional rights—that is, rights to fair procedure (and sometimes substantive rights) rooted in the Constitution itself. To lawyers whose sights are limited to plant sites, due process is often taken in a broader political sense to refer to the practice of private intervention in plant-siting proceedings and to the purported undesirability and perhaps political impossibility of restricting private intervention.

What we have here, in short, is massive confusion of thought. And the confusion is not without consequences. By invoking a constitutional term, we suggest that private intervention is somehow rooted in the Constitution. Nothing could be further from the truth. Congress has the power to eliminate all private intervention in federal siting proceedings. Congress could, furthermore, eliminate judicial review of environmental impact statements. It could probably even eliminate third party review of Nuclear Regulatory Commission decisions, permitting only the applicant to appeal from the denial of a construction permit to operating license.

The growth of private intervention is a byproduct of the growth of regulation, particularly environmental regulation. Eliminate the underlying statutes and procedures, and one would eliminate private intervention. (This is not meant as a normative but rather purely as an analytical statement!) Moreover, private intervention is not a necessary byproduct of the underlying regulation. If Congress chose, it could—as indicated above—eliminate all private intervention both in agency proceedings and in the form of third party appeal from agency decisions. Indeed, the courts and agencies have been able to expand private intervention only because Congress has normally been silent, as it was, for example, in the Natural Environmental Policy Act (NEPA) of 1969.

In passing, let me point out that, for the sake of thinking straight, I use the adjective *private* rather than *public* to describe intervention because I do not accept the assertion that the intervenors in some recent nuclear plant disputes represent the public in fact. Moreover, I see no legal basis for believing that any intervenors represent the public in the same sense that the named plaintiff in a class action represents the class, thereby permitting a legally binding judgment against all class members.

One should not overlook the political impracticability of anyone asking Congress to eliminate private intervention at this time. The few examples (e.g., the 1973 Alaska pipeline legislation making an exception from NEPA) illustrate the controversy that would surround any such effort. Indeed, the perception that the administration's nuclear-power-siting legislation would somehow stand in the way of unfettered assertion of private rights was probably an important factor in stalling that legislation this past session. The fact is that the invocation of the due process terminology reflects a broad political consensus that citizens should have a right to have certain important issues, particularly in the environmental realm, decided in the courts or at least in an adjudicatory proceeding. The use of the term "due process" thus conveys the notion that the ability to intervene is a very highly ranked value in our way of making decisions. It is an illustration of the notion of participative democracy. If one fails to appreciate the strong hold that this value has on the polity (and particularly on those who come to maturity in the period of the segregation cases, the sit-in cases, and the first environmental cases), one cannot understand why the issue of delay in power plant siting is so controversial and emotional.

At the same time, we should be clear that raising private intervention to the level of a quasi-constitutional right is at odds with fundamental notions of representative democracy. The courts are an ademocratic, if perhaps not anti-democratic, institution. Private intervention is being used to fight out fundamental social and political choices better resolved in the legislature. Indeed, the rather broadly based belief that issues bearing directly on the energy future of our country should be the subject of litigation is a signal to the objective observor that the Congress has seriously defaulted on its responsibility of making fundamental, hard choices about tradeoffs in the energy area. That the courts have been so willing to fill the vacuum may surprise some lawyers and energy experts but would surprise no seasoned politician or bureaucrat.

If the invocation of constitutional concepts is unjustified, one is driven to take a cost-benefit approach to private intervention. Doubtless such intervention has benefits. There are certain kinds of factual issues, especially those pertaining to a specific site or to a specific plant component design, that yield best to the investment of the considerable time and expense associated with adjudicatory methods (oral testimony, cross-examination, etc.). At the end of the day (perhaps I should say at the end of the year) one may have learned a great deal more about a specific narrow issue of fact when lawyers pit their sometimes formidable skills against each other in front of a law officer and a legal stenographer than one would be likely to learn through ordinary bureaucratic processes. Let us call this benefit the "better decision effect." At the same time that we recognize the possibility of a better decision effect, let us concede that it does not extend to decisions involving a ranking of values. There is no reason, for example, to believe that an adjudicatory process will lead to a better decision about an energy-environmental tradeoff. Nor will it help us make the tradeoff between a rather likely benefit and an unlikely event with extremely high disutility (e.g., a nuclear plant accident). Indeed, I would argue that even if the adjudicatory process may not inhibit the making of such tradeoffs, decisions by judges will likely be less reflective of society's composite value rankings simply because judges are not politically responsible to the electorate. So while we recognize that there may be a better decision effect, we cannot be sure what percentage of private interventions actually lead to better decisions.

Another benefit is to be found in the feeling of justice that intervenors and much of society enjoy because the opportunity to intervene is there. Even if one is not an intervenor, one may feel better because one could intervene if one wanted to. Let us call this the "participation effect." The importance of the value of participation is precisely the reason that private intervention is associated with a word exuding such rectitude and tradition as "due process." To be sure, participation, particularly if allowed to run riot, has its costs, but let us defer that insight to our consideration of the cost side of the benefit-cost balance. More appropriate for consideration on the benefit side is whether most or all of the benefits associated with the participation effect cannot be achieved by notice and comment rulemaking proceedings in which private parties have the right to make written comments and perhaps even oral comments on the basis of a complete disclosure by the agency of the evidence it has already collected and of the theories of decision it is contemplating. I believe that adjudicatory processes add little to the value of the participation effect. To be sure, those out to stop a particular nuclear plant, and especially those out to stop nuclear power in general, will not be satisfied by rulemaking that grinds out an adverse rule, but they would probably not be satisfied by a full-scale trial ending in an adverse result either.

THE COSTS AND BENEFITS OF PRIVATE INTERVENTION

What then are the costs to balance against the benefits of the better decision and participation effects? Let us put to one side legal fees, though these are not insubstantial. It can hardly be doubted that private intervention takes time. The intervention proceeding will not always be "on the critical path," but when it is, the costs of delay are considerable. Putting aside economic accounting issues involving the distinction between nominal and real prices (for example, I do not believe that a higher nominal price level for inputs after a one year delay should itself be considered a cost of delay to the economy), one can in principle, nonetheless, make estimates of the cost of delay at different points in the plant construction cycle, and a number of such estimates have been published.

If the costs of delay are knowable, it is much more difficult to determine when a particular delay is attributable to private intervention and when to other causes. In many instances, the assignment of a particular *x* month period to private intervention rather than some other factor will be a subjective judgment on which reasonable men will differ. For example, during the period following the change in relative prices engendered by the 1973 oil crisis, many utilities reconsidered the desirability of continuing with their expansion plans, and many lenders and investors came to doubt the utilities' ability to pay for new plants (particularly given some state utility commissions' rate regulation policies). If private intervention delayed a plant where construction would have been held up for planning or financing reasons anyway, does one assign the delay to private intervention? Assigning to each factor the time period involved in it can easily lead to the absurdity of a total for delay exceeding the time from conception to actual operation.

As a matter of fact, there is even a question of defining delay. One could say that delay is the difference between the actual conception-to-operation period and some baseline period. But the baseline period depends upon a host of issues specific to a particular plant. Moreover, delay is both an *ex ante* and an *ex post* concept. Failure to appreciate this difference leads to much confusion. Ideally, one should speak of an optimal lead time and of the causes and costs of epartures from optimality. But the concept of delay is here to stay in policy discussions, and so one might as well use that term.

Assuming that a cost-benefit czar could determine the time period of delay attributable to private intervention, one could calculate the direct cost of delay for each project in the country, which could then be summed across all plant projects to arrive at some global direct cost of delay (which could be adjusted to give annual national direct delay costs or average delay costs per plant or per megawatt). But there are also indirect costs. Even if a utility decisionmaker knows the average period of delay per plant, there is always the risk that the period for any particular plant will be greater. Utility representatives assert that many of the financing difficulties that led to delay or canceling of power plant projects were in fact attributable to the risk that the plant would be seriously delayed or even blocked entirely (after large expenditures had already been made) as a result of private intervention and subsequent appellate litigation. It is a truism that uncertainty is costly, but it is unclear how to put a cost figure, for cost-benefit purposes, on what we may term the "uncertainty effect."

One way to value uncertainty, making use of the association between the concepts of risk and uncertainty, is to assign a risk premium to a known interest rate. Specifically, it is possible to explain the financing difficulties of utilities in the following way. The uncertainty as to the completion date of a plant created by private intervention (including the risk that the plant will never be completed, even after substantial costs are incurred) leads financial markets to require a risk premium on top of the normal interest rate required to sell long-term utility bonds. And if the utility will not, or cannot, pay that higher interest rate, then the plant will not be built or, if already in the early stages of construction, the project may be abandoned. The higher interest costs are surely a cost that is quite measureable in dollars. If the plant is not built because of the legal uncertainty, then this failure to add to capacity is a cost, though the method of calculation is troublesome. One can think of the cost as the cost to society of the next most costly method of providing the same electricity or other delivered energy. Or one can look to what the consumer would have been willing to pay for the foregone electricity as what is lost to the society.

An entirely different kind of indirect cost of private intervention is the equivalent of the cost of defensive medicine engendered by the medical malpractice tort system. Let us call this the "defensive regulation effect." Some of the additional requirements (ratcheting and backfitting) are clearly desirable, and the costs involved are offset by the better decision effect. But others are unnecessary or even affirmatively undesirable, and these are unmitigated net costs of private intervention. The defensive regulation effects is very hard to value, not because the cost of additional safety and health measures cannot be calculated, but because it is extremely difficult to know what the regulators would have done if there had not been the possibility of private intervention lurking in their future.

Thus, we have two principal kinds of benefits—the better decision effect and the participation effect. And we have both the direct costs of delay and the indirect costs, including the uncertainty effect and the defensive regulation effect.

The benefits are extremely difficult to value. This is especially true because the value of the participation effect tends to lie in the eyes of the beholder, and there is no market in citizen participation. More abstractly put, the participation effect is incommensurate with the other benefits and costs. This factor, of course, does not make it less important.

On the cost side, the direct costs of delay are relatively easy to measure. So too, in principle, is the uncertainty effect, though the actual calculation would be rather speculative. And the defensive regulation effect is very difficult to measure because we cannot know, with even modest confidence, what the regulators would have done in a hypothetical situation.

Some will say that using the cost-benefit approach on private intervention raises fewer problems than this approach does in analyzing many other public policy issues. And that may well be true. But some of the considerations I have suggested may explain why the utility industry has not tried to cost out, or even to collect statistics on, private intervention in a form that could be useful to an outside evaluator. The costs of private intervention might turn out to be small, and the calculation would certainly be controversial. Therefore, the industry has chosen to debate the issue not in the cost-benefit terms of the economist but rather by utilizing the anecdotal evidence techniques of the lawyer.

How should we, then, evaluate private intervention? I fear that we shall have to rely on anecdotal evidence to determine the impact of private intervention. We do know that today most nuclear plant proposals are challenged before the NRC at both the construction permit and operating license stages. (Knowledgeable insiders give guesstimates of the incidence of private challenge at 90 or even 99 percent.) But the extent to which the typical proceeding is delayed by private intervention is not at all clear. About all that we have to go on are anecdotes about particular cases in which private intervention clearly had a substantial delaying effect. Seabrook and Midland (Consumers Power) are the two principal examples.

Even after an intensive case study of particular proceedings, however, we would be left with the obvious—namely, that private intervention sometimes has delay costs. But proponents of private intervention have never denied that litigation sometimes causes delays. Indeed, the whole case for the use of administrative agencies (as opposed to conventional bureaucracies) and for judicial review rests on the belief that the additional time involved is justified by superior results. In short, I despair of being able to reach judgments that could be defended as being objective and rigorous about how much private intervention is worth on balance, or indeed whether the sign is positive or negative. Let us recall that private intervention is alive and well in plant licensing (though not in some other regulatory areas, such as Federal Aviation Administration licensing) and that therefore, as a practical matter, the burden is on those who would abolish it.

SOME RECOMMENDATIONS ON IMPLEMENTING INTERVENTION

My pessimism about being able to draw up a definitive balance on private intervention does not mean, however, that certain abuses of private intervention cannot be brought under control by better procedures. And there are some reforms that make a good deal of sense.

My first general recommendation would be to make much more generous use of notice and comment rulemaking and much less use of adjudicatory techniques. For many kinds of issues, adjudication makes no sense. This is particularly the case for issues such as whether we should have additional nuclear power at all or how much the utility should be required to specify, in its construction permit application, about the treatment of nuclear wastes that will not be produced for many years. For most "generic" issues, therefore, notice and comment rulemaking should be used, and the issues should therefore be barred in site-specific permit litigation. The court of appeals for the District of Columbia has put practical roadblocks in the way of this approach with its unpredictable and sporadic requirement of "hybrid" rulemaking (involving adjudicatory fact-finding methods for contested issues within the overall framework of notice and comment rulemaking). No agency head or general counsel wants to learn years later that he should have used hybrid techniques rather than pure notice and comment rulemaking, and hence his decisions will be biased toward adjudicatory methods. The Supreme Court's recent Vermont Yankee decision, however, seems to have vindicated and protected the notice and comment approach.

The second general recommendation would be to impose rigid limitations upon the time when private intervention would have to occur. My impression, which I cannot document in detail, is that some private intervention occurs quite late in the construction cycle and, in particular, later than one could reasonably require of intervenors. Nothing in the Constitution says that intervenors should be able to sit on their hands until construction is quite far along and then intervene with the hope of causing the sponsors to abort the project. Intervenors will, of course, protest that health and safety should come first, even if the intervention comes late. But the response is that intervenors should be encouraged, indeed required, to come forth early where important health and safety risks are involved. The concern that the health or safety risk is one that could not have been discovered or properly appreciated earlier (because of, say, later new knowledge) can be met by a provision along these lines: If the late intervenor can document in writing to the satisfaction of the agency or court that the intervention could not have been made within the specified period on the basis of knowledge than publicly available, the agency or court may waive the barrier to late intervention.

A recommendation in this form would take care of a third rather sticky problem. That problem is raised by the phenomenon of multiple intervention where one or more intervenors come in late. The constitutional rule is that an unsuccessful intervention by one public intervenor does not bind a second intervenor because the former does not represent the latter. In this sense there is something to the appellation "due process." Once a statutory right of intervention is created by Congress, one citizen's recourse to that right cannot under the due process clause bar another citizen's recourse to it. It is true, of course, that the second tribunal may dispose of the legal contentions of the second intervenor rather summarily under the principle of *stare decisis* (i.e., precedent). But if there are rigorous time limits on intervention, then all intervenors will have to come quite early into the regulatory or court proceeding. Once all intervenors are before the same tribunal at the same time, the costs and risks of multiple intervention are more manageable.

This is not the place to make more specific law reform recommendations. The point is simply that with some marginal legal improvements, private intervention would become a more attractive proposition (or some would say a less unattractive proposition) for the society as a whole.

❃ *Part III*

International Decisionmaking Processes

Chapter 26

Remarks on International Decisionmaking Processes

I. G. K. Williams
OECD Nuclear Energy Agency, Paris

The theme of this forum is "an acceptable world energy future," and we need, of course, to consider acceptability from various points of view. Many of the previous chapters have been devoted to technical feasibility and to economic acceptability. Earlier in this part of the book, we have been concerned with the factor of social acceptability. Social acceptability also has a political dimension, essentially in the domestic sense, but the larger political issues are by their very nature international. Political acceptability therefore involves consideration of international decisionmaking processes and the extent to which these are amenable to consensus relationships.

Certainly, it would make no sense at all to consider an acceptable world energy future without giving full weight to the profound international political elements that are involved. One has only to recall the events of late 1973, which conclusively demonstrated the fragility of the energy foundation upon which most industrialized nations had been conducting their economies and the crucial importance of energy policy and the security of energy supplies in any calculations of national self-interest. The impetus these events gave to the alternative of a massive dependence on nuclear energy also led to a heightened awareness of such problems as the need to insulate civilian nuclear development from weapons proliferation. Questions such as this provide an obvious potential for conflict between national interests and between differing concepts of international acceptability. While they provide an added complexity to international decisionmaking processes in the modern world, they can be satisfactorily resolved only by international agreement.

At the same time, it is also important to keep in mind the influence of domestic politics in the determination of international acceptability. For example, it would be foolish to ignore that an important part of the public resistance to modern technology is being fomented by movements that see the world as their constituency. These movements are noticeably less evident in countries having centrally planned economies, and for this reason among others, public opposition in those countries is much less apparent. In the long run, this difference in public attitudes could lead to an imbalance, with both domestic and international consequences, between countries having free market economies and those with centrally planned regimes.

In the international sense, political acceptability depends ultimately on the equilibrium reached between governments, and this is, of course, a variable phenomenon through time. It is fashionable to discuss international decisionmaking processes in terms of dominance relationships or the ability of the strong to impose their will on the less strong. There are, I think, two points to make about this: as man is an infinitely resourceful animal, the imposition of oppressive conditions encourages a search for more acceptable alternatives; and it follows from this that a durable equilibrium ultimately depends on consent, even from the weakest. Personally, I would doubt the realism, at least for another generation or two, of concluding from this that dominance relationships could in practice give way to genuine consensus relationships. What it does mean, I suggest, is that those with the strength in a dominance relationship must recognize the price to be paid through time for the inconsiderate use of their power. The notion that what is good for General Motors is good for the United States is a dubious foundation for harmonious international economic and political relations.

It is clear, of course, that political hegemony based on military force played a considerable role in the economic hegemony of the colonialist period. Although colonialism as such is largely a thing of the past, we see some of the same characteristics in the dominance relationships that prevail today. The weaker partners are usually more conscious of these than the stronger ones.

The political and economic equilibrium among the industrialized nations and between them and developing countries cannot, however, be determined by traditional political and military factors, which are no longer an appropriate means of obtaining control over such natural resources as raw materials and energy. This does not mean that the production and uses of these resources are now governed by market forces alone. On the contrary, as has been shown with particular clarity in the energy sector since 1973, regulation of the market for reasons that are at least as political as economic has led to the emergence of newly powerful developing countries whose relative increase in material wealth has given them an influence that bears no relation to the situation of only a few years ago. It is obviously a consideration for this forum that the balance in this respect between the industrialized and some developing countries has changed dramatically, while other developing countries without the same bargaining power have been left in an even more depressed condition. It is perhaps self-evident that there are dominance relationships in this new balance and that these are a key factor in determining the sense in which the world energy future will be regarded as acceptable and, if so, by whom. One's point of view is likely to be powerfully influenced by the position from which one starts.

These remarks are prompted, of course, by the situation concerning oil. It is noteworthy that, at least for the time being, the situation with respect to one of the principal alternatives to oil–namely, uranium–is equally dominated by political considerations. Though this could be an important factor in relation to the future availability of nuclear power, it is too early to say whether a small number of uranium producers will be able to maintain (and, if so, for how long) a dominance relationship over those needing long-term assurances of supply.

The international decisionmaking processes concerning nuclear energy are, in fact, currently concerned particularly with the search for an acceptable solution to the proliferation problem. The outcome can hardly expect to be successful if it involves an attempt to rewrite history. It has been said for example, no doubt with justification, that the Lilienthal-Baruch plan of 1946 was the last chance for humanity to live in a world without nuclear weapons. The notion of a supranational organization to manage a world nuclear industry in the interests of all nations failed because mankind is not yet ready to accept a system of world government. Instead, an equilibrium between nuclear weapon states and non-nuclear weapon states was established by the Non-Proliferation Treaty, which entered into force on March 5, 1970, and to which over a hundred states have become party. This equilibrium was founded on renunciation of nuclear weapons by the vast majority in exchange for an unquestioned right to the benefits of nuclear energy for peaceful purposes, a right that will be defended with great determination. Nevertheless, the NPT is an imperfect instrument, if only because several states with significant nuclear activities have stood aside.

The underlying thought behind the International Nuclear Fuel Cycle Evaluation (INFCE) is an attempt to remedy this imperfection. The idea that hitherto unchallenged technological strategies should be reexamined cannot be criticized, and the INFCE process seems likely to achieve its objective of defining alternatives. But INFCE is not a negotiation, and it does not address the question of how any particular course should be accepted and maintained. In line with what I was saying earlier, it is doubtful whether a durable long-term equilibrium would result from INFCE if it involved the imposition, through dominance relationships, of conditions that would be regarded as oppressive. For example, the acknowledgement in the NPT of their unquestioned right to the benefits of modern technology is of paramount importance to many countries, both industrialized and developing, and restriction or denial now would certainly be seen by these as discriminatory. It must also be doubted whether a voluntary renunciation by industrialized countries of technologies previously adopted for compelling economic reasons could last for long or be made universally applicable: the whole progress of economic history illustrates that interference with freedom of economic choice cannot be sustained because it induces an invariably successful search for alternatives. In the nuclear context, this could evidently involve some very unacceptable consequences. The more lasting significance of INFCE may therefore well prove to be that it is conditioning international opinion to a more receptive approach to institutional changes as the acceptable means of achieving its underlying objective.

Economic progress is based on continuous change. In the energy field, informed opinion is beginning to agree that the crucial period is more likely to be between 1985 and 2015 than in the very long term. The difficulties of facing up to this crucial period seem likely to be essentially political. For example, the next twenty-five years will have to be characterized by adaptation and transition to a very different pattern of usage, but public opinion is slow to grasp that the necessary lead times require measures to be taken even when there is no apparent scarcity of energy. The present economic recession provides a good illustration of this: insulation from some of the harsh realities that must be faced makes their recognition that much harder to achieve. At the international level, there are manifestly also divergences of interest between groups of countries even though all share the desire for a prosperous world economy.

In such a situation, the object of planning should be to achieve the required changes at a rate that can be assimilated by society without upheaval. In particular, an acceptable world energy future depends on a solution to the nuclear proliferation problem that does not itself induce the very situation that it is designed to avoid and on an accommodation concerning other energy resources that does not breed revolutionary change in economic and social structures. It needs recognition that the growing world economic interdependence requires political relations based on mutual respect. Such a future may sound utopian and certainly could not be achieved through successive equilibria based on crude dominance relationships: it would require that those in a position of strength use their power to achieve equilibria without oppression. This may often be irreconcilable with their own short-term self-interest.

As a disciple of this new cult, I suppose it falls to me to comment on the emergence of multilateralism as a factor in establishing this new climate in international economic and political relations. It is an interesting thought, though probably not one to be explored at this forum, that the postwar period has been notable for the proliferation of multilateral and multinational institutions at both governmental and nongovernmental levels. This is one of the characteristics of our age, although some of these bodies have, of course, been designed simply to elevate to a higher level a political or economic dominance relationship. Superficially, this phenomenon appears to support the thesis that we are moving toward consensus relationships in international affairs rather than the dominance relationships that were characterized in the nineteenth century by gunboat diplomacy. I say "superficially" because an evolution in this sense is certainly very gradual: multilateralism still means pursuit of national self-interest, though whenever possible in a conciliatory atmosphere. On the other hand, multilateralism enhances the relative influence of smaller countries. The increased ability of these, by participating actively in multilateral institutions, to gain a hearing for their particular interests is one of the newer facts of international life.

This is perhaps another way of saying that the existence of multilateralism has the inestimable merit of permitting, encouraging, or even requiring a dialogue between both sides in a dominance relationship and obliging each to recognize the factors influencing the other. In this sense, the emergence of multilateralism can evidently be a potent moderating influence in the equilibria that evolve. Although it is evidently an indispensable step in that direction, it would be unrealistic at this stage to claim that it represents the development of genuine consensus relationships. We still have a long way to go, because there can be no doubt that a world energy future acceptable to all will be possible only with a far greater measure of international accord than exists today.

Session E

Scrutiny of Future Energy Paths: Uncertainties, Long-term Policies, and Decisionmaking

❋ *Part I*

Major Uncertainties

Chapter 27

The Outlook for OPEC and World Oil Prices: Projections from World Energy Models for Three Decades

Dermot Gately
New York University

INTRODUCTION

At this time five years ago, the world was in the midst of a quadrupling of world oil prices by OPEC. There was sharp disagreement about the prospects for world oil prices, OPEC exports, and OPEC stability.

As we look back, we now see that many analysts underestimated OPEC's ability to restrict output and overestimated its need to do so. From 1973 to 1975 the demand for OPEC exports fell by less than 15 percent, with roughly proportional cutbacks by Arab and non-Arab members.[1] Following the recovery in world economic activity after 1975, OPEC exports have returned to about their 1973 level.

Yet if we restrict our attention to the formal modeling efforts of 1974 and 1975, ignoring the many opinions and intuitive assessments of that period, we see that some models have held up relatively well. These models, surveyed in Fischer-Gately-Kyle[2] although differing sharply on some issues, generally agreed that 1974 prices were too high to be sustained and that a price decline to the range of $7 to $10 (1974$) would be forthcoming. We now witness a real oil price near the upper end of this range, because inflation rates since 1974 have exceeded increases in nominal oil prices and because of the decline in the value of the dollar vis-à-vis other currencies.

Whether the current price will continue through the 1980s and beyond is one of the main concerns of this chapter. In the following section we review the projections of various modeling efforts with regard to the likely path of world oil prices and output levels. Next, we discuss some of the problems and uncertainties inherent in long-range modeling. One of the major uncertainties, the cost of "backstop technologies," is dealt with in the fourth section. Some broad conclusions are drawn in the final section.

PROJECTIONS FOR THE WORLD OIL MARKET OVER THE NEXT THREE DECADES

Most analyses agree that, over the next few years, real oil prices will remain roughly constant. (Throughout this chapter we will be discussing the real, rather than the nominal, price of oil.) But the outlook for the mid-1980s and beyond is clouded. Analyses generally take either of two positions. Some forecast that oil prices will continue at about the same level or increase by at most 2 percent per annum. Others predict increasingly severe upward pressure on price in the mid to late 1980s as OPEC's capacity becomes fully utilized.

Related to these different outlooks for world oil prices are different estimates of quantities to be demanded in world energy markets. Table 27-1 summarizes these estimates for a score of modeling efforts that have appeared since our earlier survey.[3] To simplify the comparisons, we focus on world demand for oil and gas from OPEC and on U.S. imports.

Before discussing Table 27-1, however, two caveats will be offered. First, the models represent a wide range of approaches, with different degrees of sophistication, different regional and energy product disaggregation, and different time horizons. Second, although it is noted in Table 27-1, some quantity projections are not directly comparable, because of different price path assumptions and because internal OPEC consumption and natural gas exports are not estimated uniformly. Yet even having made note of these considerations, the wide distribution of the projections is striking. Even for 1985, the estimates differ by a factor of two; for 1990 and beyond the range is even greater.

For 1985 most of the major analyses project a demand for OPEC exports to be about 35 MMBD of oil and the equivalent of 3 MMBD natural gas. This is about 15 to 20 percent above current levels. Only the CIA forecast is much higher, by about 25 percent; of those that are much lower, the most notable is Houthakker-Kennedy, lower by one-third. Projections of U.S. imports differ even more dramatically. Most cluster around 10-11 MMBD, with the Exxon and the CIA projections somewhat higher. By comparison, 1976 U.S. imports were about 7 MMBD. But Naill's projections are about 50 percent above that range, while the Houthakker-Kennedy forecast is only about half, with slightly more than 4 MMBD of oil.

For 1990 most projections involve small to moderate growth above 1985 levels, both for U.S. imports[4] and for world demand for OPEC exports.

Only a few of the models project to the year 2000 and beyond. Perhaps not too surprisingly, there is a wide range of estimates for both OPEC exports and U.S. imports. Under various WAES scenarios through the year 2000, the demand for OPEC exports could decline moderately (back to about current levels), could plateau at 1990 levels, or could continue to increase slightly. Likewise, Manne's projections of the demand facing OPEC in the year 2000 are moderately sensitive to the price assumptions—55.7 MMBD with a 2 percent annual price increase after 1980 and 38.7 for a 4 percent rate of increase.

Houthakker-Kennedy again differs dramatically, but now on the high side: assuming constant prices, they project an OPEC demand of 89 MMBD by 2005. Comparable differences exist for projections of U.S. imports, ranging from the CONAES estimates of less than 12 MMBD, which are perhaps artificially low (for reasons noted above), to estimates in the range of 26–28 MMBD (Houthakker-Kennedy with constant prices and Manne with 2 percent annual price increases).

Having reviewed these models there appears a strong, direct correlation between their projections of changes in the demand for OPEC exports and their estimates of likely changes in real oil prices. For example, the CIA report forecasts a surge in demand for OPEC oil and a "price break" in the early 1980s. In contrast, other models, such as Ben-Shahar, Cremer and Weitzman, and Pindyck, project no such increase in demand nor any substantial increase in real price over the next two decades. An exception to this correlation, of course, would be Houthakker-Kennedy; they project a tripling of demand for OPEC oil in the next three decades without any increase in price.

One particularly interesting comparison is Exxon's 1978 revision of its 1977 "World Energy Outlook." The 1977 version has significantly higher estimates of OPEC demand in 1985 and 1990, and it seems much more apprehensive about an energy "crunch" than the 1978 revision.

> The small projected "spare" oil producing capacity in the 1980's implies little supply flexibility to deal with normal operating variables. More importantly, the fragile overall energy balance suggests that capacity in place could be insufficient to cover unexpected growth in demand or slippage in supply. During the outlook period this is, to a major extent, a function of capacity expansion decisions by a few producing countries.[5]

> Resources in the OPEC nations appear adequate to supply the levels of imports shown through 1990, provided producing-country governments are willing to develop and produce them.[6]

Both of these quotes point to an additional uncertainty: the discretionary power of OPEC over capacity expansion as well as over price. This uncertainty is reflected in several models, such as Gately-Kyle-Fischer[7] and Manne (see Chapter 13, above), who observes: "Through the year 2000, OPEC has considerable flexibility in its pricing or production policy. This may be determined either by economic or political objectives and is not constrained by petroleum resource limits."

UNCERTAINTIES AND PROBLEMS IN LONG-RANGE MODELING

As the discussion in the previous Section indicates, there is substantial agreement among the various analyses about the world oil market of the coming

Table 27-1. Projected U.S. Imports and Demand for OPEC Exports of Oil and Gas (in MMBD oil equivalent).

	World Demand for OPEC Exports			*U.S. Imports*		
Model	*1985*	*1990*	*2000*	*1985*	*1990*	*2010*
CIA [16] [a,b]	43–47			11	14	
WAES [17] [a,b]	35–38	35–40	25–45			
Exxon [6] [a]	38	44				
Exxon [7] [a]	34	36		13	14	
OECD [14] [c]	35			11		
Moran [12]	40					
Gately-Kyle-Fischer [8] [a,d]	35	45				
Eckbo [5] [a,c]	30	51				
Pindyck [15] [a,e]	26	28				
Hnylicza-Pindyck [9] [a,f]	30	28				
Ben-Shahar [1] [a]						
$10 constant price (1974$)	32	32				
$13 constant price (1974$)	17	3				
Charles River Association [2] [a]	26	40		6	11	
Cremer-Weitzman [9] [a,g]	16	16				
Naill [13] [h]				17	20	28
Houthakker-Kennedy [10] [c,i]	25		89	6		26
CONAES [3] [j]						
SRI					7	9
Nordhaus					7	11
DESOM (Brookhaven)					9	12
FEA					10	
ETA (Manne)					11	6
Manne [11] [k]						
2% p. a. price increase after 1980	32	38	52	11	16	30
4% p. a. price increase after 1980	30	30	57	10	10	13

Notes:

[a] These quantities exclude natural gas. Most other estimates of the demand for OPEC exports of natural gas in 1985–1990 are equivalent to about 3–4 MMBD.

[b] Due to internal OPEC demand, required OPEC production would be about 4 MMBD greater than the CIA range, about 4.5 MMBD greater for WAES in 1985, about 7 MMBD in 1990, and about 10 MMBD in 1990 [ref. 17: 280].

[c] "base case," constant price projection.

[d] This corresponds to one of several price-output paths that were simulated; it is not a "base case" [ref. 8: 223].

[e] This corresponds to an "optimal," quasi-U-shaped price path, starting at $13 in 1975, declining to about $10 in 1978, and rising to about $13 by 1995.

[f] This corresponds to the Nash cooperative solution for a two part OPEC with variable market shares.

[g] This includes production only from Persian Gulf and North African members.

[h] These are "reference" projections, assuming no major policy changes. The effects of various demand and supply policies are investigated; little reduction in imports is possible by 1985, but dramatic reduction is possible by 2000.

[i] The natural gas components of these amounts (in oil equivalents) are as follows: for 1985, OPEC exports 3 MMBD and for U.S. imports 2 MMBD (half from Canada); in 2005, of the 89 MMBD exported by OPEC, 4 MMBD are natural gas and of the 26 MMBD imported by the United States, 3 MMBD are natural gas (one-third from Canada).

Table 27-1. (continued)

---

[j]CONAES fixed the price of imported oil at $13 and imposed two constraints: one limited U.S. imports of oil and gas in each year to one-third of domestic consumption, the other limited cumulative U.S. imports to 1,000 quads (172 billion barrels). The latter constraint was binding for all models. The former was binding for Nordhaus and ETA in 1990 and for all models in 2010.

[k]We assume that half of Manne's "non-OPEC LDC consumption" comes from OPEC exports.

Sources:

1. H. Ben-Shahar, *Oil: Prices and Capital* (Lexington, Massachusetts: Lexington Books, 1976), p. 54.

2. Charles River Associates, *Policy Implications of Producer Country Supply Restrictions: The World Energy Market* (Washington, D.C.: National Bureau of Standards, NBS-GCR-ETIP 76-33, 1976), p. 254.

3. CONAES (Committee on Nuclear and Alternative Energy Systems, National Research Council), *Energy Modeling for an Uncertain Future,* Supporting Paper #2 (Washington, D.C.: National Academy of Sciences, 1978), pp. 47, 52, 57.

4. J. Cremer and M. L. Weitzman, "OPEC and the Monopoly Price of World Oil," *European Economic Review 8,* no. 2 (August 1976): 161.

5. P. L. Eckbo, *The Future of World Oil* (Cambridge, Massachusetts: Ballinger Publishing Company, 1976), p. 93.

6. Exxon Corporation, Public Affairs Department, "World Energy Outlook," January 1977, p. 39.

7. Exxon Corporation, Public Affairs Department, "World Energy Outlook," April 1978, p. 39.

8. D. Gately and J. F. Kyle, in association with D. Fischer, "Strategies for OPEC's Pricing Decisions," *European Economic Review* (December 1977).

9. E. Hnylicza and R. S. Pincdyck, "Pricing Policies for a Two-Part Exhaustible Resource Cartel: The Case of OPEC," *European Economic Review* no. 2 (August 1976): 16.

10. H. S. Houthakker and M. Kennedy, "Long Range Energy Prospects," Harvard Institute of Economic Research Discussion Paper No. 634 (July 1978), pp. 11, 16.

11. A. Manne, "Energy Transition Strategies for the Industrialized Nations," see Chapter 13 of this book pp.

12. T. H. Moran, "Oil Prices and the Future of OPEC" (Washington, D.C.: Resources for the Future, Research Paper No. R-8 1978), p. 72.

13. R. F. Naill, *Managing the Energy Transition* (Cambridge, Massachusetts: Ballinger Publishing Company, 1977), ch. 7.

14. Organization for Economic Cooperation and Development (OECD), *World Energy Outlook* (Paris, 1977), p. 39.

15. R. S. Pindyck, "Gains to Producers for the Cartelization of Exhaustible Resources," *Review of Economics and Statistics* (May 1978), p. 242.

16. United States Central Intelligence Agency, "The International Energy Situation Outlook to 1985" (Washington, D.C.: April 1977), p. 15.

17. Workshop on Alternative Energy Strategies, *Energy: Global Prospects 1985–2000* (New York: McGraw-Hill, 1977), p. 135.

decade. It is difficult, however, to narrow the range of uncertainty regarding world oil prices and the demand for OPEC exports much beyond that.

But this should not be surprising. Projections for the coming decade largely involve evaluating the implications of what is happening in world energy markets and of what has already happened, such as the adjustments to the 1973–1974 price quadrupling. In contrast, projections much beyond 1990 involve an evaluation of trends and events yet to happen. Not only is there the difficulty of predicting the effects of certain assumptions, there is the more fundamental problem of knowing which assumptions will turn out to be correct. That is, one level of uncertainty involves the magnitudes of response of world demands and supplies of various types of energy to changes in price and in world income and also the substitutability among fuels and the effects of various government policies. But even more fundamentally, there are uncertainties about the level of world economic growth, technological changes and the cost of "backstop" technologies, additions to world oil reserves, and changes in the concentration of power within OPEC, as well as the policies (either anticipatory or reactive) to be adopted by various governments in the future.

The most we can realistically expect from such long-term projections is to narrow the range of uncertainty, in the sense of a roughly defined confidence interval. Any assessment of alternative national and international energy policies must deal explicitly with this range of uncertainty. It would, of course, be desirable that important policy decisions be robustly satisfactory over the range of uncertainty.

One attempt to come to grips with some of these uncertainties is our own model, Gately-Kyle-Fischer.[8] We assume that OPEC gropes its way toward a satisfactory price path by employing some rule of thumb pricing strategy in which prices are changed in response to changes in OPEC export demand and OPEC capacity utilization. A strategy that is relatively cautious about further major price increases serves OPEC relatively well in comparison with other strategies, but there exists a real possibility of major price increases over the next fifteen years.

ON BACKSTOP PRICES

One of the major unknowns in long-term world energy analysis is the cost of "backstop technologies." After a review of some recent estimates, we offer two general comments about this concept.

In the years immediately following the 1973–1974 price quadrupling it was sometimes observed, not entirely facetiously, that the only prices rising faster than world oil prices were the estimated costs of shale oil, coal liquifaction, and other backstop technologies. And it was often said that the cost of shale oil would always be $2 above the world price of oil.

Indeed, the "backstop" prices used in some of the models surveyed in Fischer-Gately-Kyle[9] are low in comparison with some recent estimates. Bohi and Russell[10] assumed that the demand for OPEC oil would disappear at prices above its long-run equilibrium prices of $10 (in 1974$). Kalymon[11] used $15 (in 1974$) as the price at which the demand for OPEC oil would be zero. Using a slightly different notion of an upper limit, Blitzer-Meeraus-Stoutjesdijk[12] employed various estimates from $8 to $12 (in 1974$) of the price at which non-OPEC supply will satisfy demand growth.

Somewhat higher estimates were used in the recent joint modeling effort of CONAES, which used two assumptions about the cost of synthetic fuels (in 1975$)–a base case estimate of about $21.50 and lower estimate of about $14.80.[13] Still higher estimates, in the range of $20 to $30, have been reported, going even as high as $60, which is the estimated cost of a commercial scale coal liquefaction plant in South Africa. Clearly, until there is more agreement about the price at which substantial supplies of such "backstop fuels" would be forthcoming, this will continue as a major uncertainty in the analysis of world energy markets.

Finally, whatever is the true price of the backstop technology, two points should be noted. First, the backstop holds as an upper limit on price only in the long run, when there is sufficient time for capacity expansion. In the short run, the price could go well above the backstop price and come down either when the backstop capacity has been installed or in order to deter its installation.

Second, it is important to take account of the backstop technology's variable costs of production–namely, those costs that must continue to be incurred after the capacity is in place and that depend on the level of output. These costs would be relatively low for solar power and relatively high for shale oil. They become particularly important after the capacity has been installed, inasmuch as they determine a price below which the capacity (at least theoretically) would not be used. However, the practical relevance of this would be limited insofar as governments would be pressured to protect such producers.

SOME CONCLUSIONS

Having reviewed all these analyses, what can we say about OPEC exports and world oil prices over the next three decades? Over the next five years we do not foresee any significant change in real oil prices, and we expect OPEC spare capacity to remain at about its current level (6 MMBD). While there may be continuing tension within OPEC over price and output shares, as several members may be unable to satisfy their perceived revenue needs, we expect Saudi Arabia and its Arab allies on the Persian Gulf to continue their stabilizing role.

But the outlook for the late 1980s and beyond is fairly uncertain. There could be a continuation of current market conditions, with OPEC's exports and price roughly constant or increasing slowly. But there could also be a moderately strong tightening of the world oil market. If that appears likely—and we ought to know by the early 1980s—we expect several gradual price increases to anticipate such a tightening, warding off another crunch like 1973-1974.

OPEC exports are perhaps easier to project than OPEC prices, because we expect exports to be relatively stable and price to be the adjusting variable. Our expectation is that OPEC will export no less than 25 MMBD nor more than 45 MMBD over the next three decades. While this is a fairly wide range, it is not as wide as our comparable confidence interval for prices—from about $10 to $30 (1978$).

We noted earlier the shale oil cynic's forecast that the cost of shale oil would always be $2 more than the world price. But turning this equation around, we have an important element of truth: the world price of oil will be kept $2 below the cost of shale oil (or that of some better backstop technology). This idea is formalized in the dynamic limit pricing literature, particularly the recent article by Gilbert.[14]

Hence we expect the world price of oil to be determined by the demand for OPEC exports, not by its cost of production (even at the margin). OPEC's price ought to change in response to changes in the demand it faces and expects to face.

This has an important corollary for present and future evaluations of programs to reduce energy demand growth and to increase non-OPEC supply. While large-scale efforts may in the future appear to have been unnecessary if world oil prices remain at or near current levels, it will only be due to such efforts that world oil prices did not increase.

NOTES

1. Arab output fell from 886 to 787 million metric tons, while non-Arab output fell from 655 to 562; see United Nations; United Nations, *World Energy Supplies* (New York: Statistical Office, U.N. Department of Economics and Social Affairs, Series J, No. 20, (1977).

2. D. Fischer, D. Gately, and J. F. Kyle, "The Prospects for OPEC: A Critical Survey of Models of the World Oil Market," *Journal of Development Economics* 2 (1975): 363-86. (Included were C. Blitzer, A. Meeraus, and A. Stroutjesdijk, "A Dynamic Model of OPEC Trade and Production," *Journal of Development Economics* 2 (1975): 319-35; D. R. Bohi and M. Russell, *U.S. Energy Policy* (Baltimore: Resources for the Future, Johns Hopkins University Press, 1975); United States Federal Energy Administration, *Project Independence Report* (Washington, D.C.: U.S. Government Printing Office, 1974); B. A. Kalymon, "Economic Incentives in OPEC Oil Pricing Policy,"

Journal of Development Economics 2 (1975): 337–62; M. Kennedy, "An Economic Model of the World Oil Market," *Bell Journal of Economics and Management Science* vol. 5, no. 2 (1974): 540–77; W. J. Levy Consultants, "Implications of World Oil Austerity," a private communique for clients (1974) (Mimeo); and W. D. Nordhaus, "The Allocation of Energy Resources," *Brookings Papers on Economic Activity* vol. 4, no. 3 (1973): 529–76.

3. Fisher, Gately, and Kyle.

4. The relatively low projections of U.S. imports in the CONAES report are attributable in part to the imposition of two somewhat artificial constraints, described in a footnote to Table 27-1.

5. Exxon Corporation, Public Affairs Department, "World Energy Outlook" (New York, January 1977), p. 37.

6. Exxon Corporation, Public Affairs Department, "World Energy Outlook" (New York, April 1978), p. 38.

7. D. Gately and J. F. Kyle, in association with D. Fischer, "Strategies for OPEC's Pricing Decisions," *European Economic Review* (December 1977).

8. Gately, Kyle, and Fisher.

9. Fisher, Gately, and Kyle.

10. Bohi and Russell.

11. Kalymon.

12. Blitzer, Meeraus, and Stoutjesdijk.

13. CONAES (Committee on Nuclear and Alternative Energy Systems, National Research Council), *Energy Modeling for an Uncertain Future.* Supporting Paper #2 (Washington, D.C.: National Academy of Sciences, 1978), p. 74. Supplementing these estimates, a recent study by N. R. Ericsson and P. Morgan, "The Economic Feasibility of Shale Oil: An Activity Analysis," *Bell Journal of Economics* vol. 9, no. 2 (Autumn 1978): 457–87 yields an output of U.S. shale oil of 15 MMBD at a long-run price of $18 (in 1975$).

14. R. J. Gilbert, "Dominant Firm Pricing Policy in a Market for an Exhaustible Resource," *Bell Journal of Economics,* vol. 9, no. 2 (Autumn 1978): 385–95.

REFERENCES

Eckbo, P. L. *The Future of World Oil.* Cambridge, Mass.: Ballinger Publishing Company, 1976.

Rockefeller Foundation. "Working Paper on International Energy Supply: A Perspective from the Industrial World." May 1978.

Rustow, D. A. "U.S.–Saudi Relations and the Oil Crises of the 1980s." *Foreign Affairs* (1977): 494–516.

United States Central Intelligence Agency. "The International Energy Situation Outlook to 1985." April 1977.

❋ *Chapter 28*

Energy Demand and Energy Policy: What Have We Learned?

Robert S. Pindyck
Massachusetts Institute of Technology

Unfortunately, discussions of future energy availability and use, whether in the context of the United States or in an international context, often revolve around questions of the physical limits of various resource supplies and the technical feasibilities of producing alternative supplies at different prices. The notion seems to be that the evolution of energy markets should in some way be managed, with government controls and subsidies used as instruments to control the supplies and consumptions of alternative forms of energy. It seems to me that in most cases, market mechanisms can bring about a more efficient use of resources. In the United States, for example, much of the recent failure of energy policy has been a failure to allow markets to work.

To what extent will rising energy prices dampen the growth of energy demand and stimulate the production of new energy supplies? Critics of the market often argue that too little is known about the likely impact of changing energy prices—that there is simply too much uncertainty about demand and supply elasticities. While our uncertainty is no doubt considerable, we have recently learned a good deal about the magnitudes and time response of these elasticities.

I will focus here on elasticities of energy demand, and I will attempt to make two basic points. First, our uncertainty over the characteristics of energy demand (including the magnitudes of demand elasticities) has been considerably reduced as a result of statistical studies conducted over the past few years. Second, these elasticities are much larger than we had previously thought to be the case, and this, of course, has important implications for the design of energy policy.

I begin with a survey of recent statistical studies, including my own, of energy demand elasticities. I will discuss the reasons for differences in the reported estimates of demand elasticities, and I will argue that in fact there are "consensus" estimates that can be derived and used for policy analysis. I will also discuss the extent of our current uncertainty over these elasticities, as well as the implications for policy.

A SURVEY OF RECENT ENERGY DEMAND ELASTICITY ESTIMATES

A survey of recent statistical studies of energy demand elasticities is presented in Table 28-1. This survey is by no means exhaustive, nor is it completely up to date. Numerous other studies have been conducted in the past, and the current rate of output of econometric studies of the energy sector is at an all time high. However, those studies reported in the table are relatively recent and give a good overall representation of the existing econometric work on energy demand. The studies and estimates cited in the table are grouped by sector of energy use (the characteristics of energy demand–and thus elasticities–differ considerably across sector) and by type of elasticity within each sector.

It is important to emphasize that the studies surveyed here differ considerably in scope, model formulation, and type of data used. Some of these studies are based on simple ad hoc models (e.g., static or dynamic double logarithmic equations) that impose severe restrictions on the structure of demand but that require relatively little data for estimation. Other studies are based on the estimation of expenditure share equations generated from a generalized (and unrestrictive) utility or cost function. And some studies are based on time series data for a single country, while others use pooled data covering a range of countries.[1]

A casual scanning of Table 28-1 might lead to the conclusion that there is wide disagreement about energy demand elasticities and that it would be impossible to reach a consensus on a set of "working" estimates that could be used for policy analysis. Indeed, looking at the own price elasticity of aggregate energy use in the residential sector, the long-run estimates range from –0.28 to –1.10. However, let us examine these estimates and the studies behind them more closely.

We can begin with the own price elasticity of aggregate energy use in the residential sector. The first five studies cited in Table 28-1 (three for the United States, one for Canada, and one for Norway) actually show rather close agreement on the estimate for this elasticity; all of the estimates are within the range of about –0.3 to –0.5. The last two studies cited, however, indicate an own price elaticity in the range of –0.7 to –1.1. Why have these two studies produced elasticity estimates that are so much larger (about twice as large) as the others? I will argue that the estimates produced by the first set of studies are too low–although it is worth noting that they are larger than the typical "consensus"

Table 28-1. Alternative Estimates of Energy Demand Elasticities (all estimates for the United States unless otherwise indicated).

Elasticity	*Study*	*Estimate*
	A. Residential Sector	
Aggregate energy use – own price elasticity	Joskow and Baughman [15]	S.R.: –0.12, L.R.: –0.50
	Nelson [19]	–0.28
	Jorgenson [14]	–0.40
	Fuss and Waverman [5] (Canada)	–0.33 to –0.56
	Rødseth and Strøm [26] (Norway)	–0.30
	Nordhaus [20] (6 countries pooled)	–0.71
	Pindyck [24] (9 countries pooled)	–1.10
Aggregate energy use – income elasticity	Joskow and Baughman [15]	S.R.: 0.10, L.R.: 0.60
	Nelson [19]	0.27
	Fuss and Waverman [5] (Canada)	0.83 to 1.26
	Rødseth and Strøm [26] (Norway)	1.08
	Nordhaus [20] (6 countries pooled)	1.09
	Pindyck [24] (9 countries pooled)	1.0
Fuel consumption – own price elasticity	Joskow and Baughman [15]	gas, oil, coal: all between –1.0 and –1.1 in long run
	Halvorsen [10]	electricity: –1.0 to –1.2
	Liew [17]	natural gas: –1.28 to –1.77
		electricity: –0.40
	Hirst, Lin, and Cope [12]	gas, oil and electricity: all between –0.84 and –0.91
	Pindyck, 24 (9 countries pooled)	oil: –1.1 to –1.3, gas: –1.3 to –2.1, electricity: –0.3 to –0.7
	B. Industrial Sector	
Price elasticities for factor inputs – K = capital L = labor E = energy	Berndt and Wood [3]	η_{KK}:–0.44, η_{LL}:–0.45, η_{EE}:–0.49, η_{KE}:–0.15, η_{LE}:0.03
	Halvorsen and Ford [11] (2 digit industries)	η_{KK}:–0.67 to –1.16, η_{LL}:–0.28 to –1.55, η_{EE}:–0.66 to –2.56
	Fuss [4] (Canada)	η_{KK}:–0.76, η_{LL}:–0.49, η_{EE}:–0.49, η_{KE}:–0.05, η_{LE}:0.55
	Magnus [18] (Netherlands)	η_{KK}:0.05, η_{LL}:–0.26, η_{EE}:–0.90
	Griffin and Gregory [7] (9 countries pooled)	η_{KK}:–0.28, η_{LL}:–0.20, η_{EE}:–0.80, η_{KE}:0.13, η_{LE}:0.11
	Pindyck [21,24] (10 countries pooled)	η_{KK}:–0.29 to –0.78, η_{LL}:–0.23 to –0.66, η_{EE}:–0.83 to –0.85, η_{KE}:0.02 to 0.08, η_{LE}:0.0 to 0.09

Table 28-1. (continued)

Fuel consumption – own price elasticity	Halvorsen [9]	oil: –2.82, gas: –1.47, coal: –1.52, electricity: –0.92
	Fuss [4] (Canada)	oil: –1.30, gas: –1.30, coal: –0.48, electricity: –0.74
	Pindyck [21,24] (10 countries pooled)	oil: –0.2 to –1.2, gas: –0.4 to –2.3, coal: –1.3 to –2.2, electric: –0.5 to –0.6
	C. Transportation Sector	
Motor gasoline – own price elasticity	Houthakker et al. [13]	S.R.: –0.07, L.R.: –0.24
	Ramsey, Rasche, and Allen [25]	–0.70
	Fuss and Waverman [5] (Canada)	–0.22 to –0.45
	Adams, Graham, and Griffin [1] (20 countries pooled)	–0.92
	Pindyck [24] (11 countries pooled)	–1.3
Motor gasoline – income elasticity	Houthakker et al. [13]	S.R.: 0.30, L.R.: 0.98
	Ramsey, Rasche, and Allen [25]	1.15
	Adams, Graham, and Griffin [1] (20 countries pooled)	0.54
	Pindyck [24] (11 countries pooled)	0.8
	D. Electric Generation Sector	
Electricity generation – fuel price elasticities	Griffin [6]	oil: –1.0 to –4.0, gas: –0.8 to –1.2, coal: –0.5 to –0.8, cross-price elasticities between 0.2 and 1.2.
	Atkinson and Halvorsen [2]	oil: –1.5 to –1.6, gas: –1.4, coal: –0.4 to 1.2, cross-price elasticities between 0.4 and 1.0.
	Joskow and Mishkin [16]	Elasticities vary greatly across utilities. Fuel price elasticities will be very large when expected fuel prices are roughly the same, but very small when fuel prices differ considerably.

Sources:

1. F. G. Adams, H. Graham, and J. M. Griffin, "Demand Elasticities for Gasoline: Another View," Discussion Paper No. 279, Department of Economics, University of Pennsylvania, June 1974.

2. S. E. Atkinson and R. Halvorsen, "Interfuel Substitution in Steam Electric Power Generation," *Journal of Political Economy,* October 1976.

3. E. R. Berndt, and D. Wood, "Technology, Prices, and the Desired Demand for Energy," *Review of Economics and Statistics,* August 1975.

4. M. A. Fuss, "The Demand for Energy in Canadian Manufacturing," *Journal of Econometrics* 5, (1977).

5. M. Fuss, and L. Waverman, "The Demand for Energy in Canada," Working Paper, Institute for Policy Analysis, University of Toronto, 1975.

6. J. M. Griffin, "Interfuel Substitution Possibilities: A Translog Application to Pooled Data," *International Economic Review,* October 1977.

7. J. M. Griffin, and P. R. Gregory, "An Intercountry Translog Model of Energy Substitution Responses," *American Economic Review,* December 1976.

8. R. E. Hall and R. S. Pindyck, "The Conflicting Goals of National Energy Policy," *The Public Interest,* Spring 1977.

9. R. Halvorsen, "Energy Substitution in U.S. Manufacturing," (Unpublished paper, 1976).

Table 28-1. (continued)

10. R. Halvorsen, "Residential Demand for Electric Energy," *Review of Economics and Statistics,* February 1975.

11. R. Halvorsen, and J. Ford, "Substitution Among Energy, Capital, and Labor Inputs in U.S. Manufacturing," in R. S. Pindyck, ed., *Advances in the Economics of Energy and Resources,* vol. I (Greenwich, Connecticut: J.A.I. Press, 1978).

12. E. Hirst, W. Lin, and J. Cope, "An Engineering-Economic Model of Residential Energy Use," Oak Ridge National Laboratory, Technical Report #TM-5470, July 1976.

13. H. S. Houthakker, P. K. Verleger, and D. P. Sheehan, "Dynamic Demand Analyses for Gasoline and Residential Electricity," *American Journal of Agricultural Economics* 56, no. 2, (May 1974).

14. D. W. Jorgenson, "Consumer Demand for Energy," Harvard Institute of Economic Research, Discussion Paper 386, November 1974.

15. P. L. Joskow, and M. L. Baughman, "The Future of the U.S. Nuclear Energy Industry," *Bell Journal of Economics,* 7, no. 1, (Spring 1976).

16. P. L. Joskow, and F. S. Mishkin, "Electric Utility Fuel Choice Behavior in the United States," *International Economic Review,* October 1977.

17. C. K. Liew, "Measuring the Substitutability of Energy Consumption" (Unpublished paper, December 1974).

18. J. R. Magnus, "Substitution Between Energy and Non-Energy Inputs in the Netherlands: 1950–1974," *International Economic Review,* in press.

19. J. P. Nelson, "The Demand for Space Heating Energy," *Review of Economics and Statistics,* November 1975.

20. W. D. Nordhaus, "The Demand for Energy: An International Perspective" (Unpublished paper, September 1975).

21. R. S. Pindyck, "Interfuel Substitution and the Industrial Demand for Energy: An International Comparison," *Review of Economics and Statistics,* to appear May 1979.

22. R. S. Pindyck, "The Characteristics of the Demand for Energy," in J. Sawhill, ed., *Improving the Energy Productivity of the American Economy,* (Englewood Cliffs, New Jersey: Prentice-Hall, forthcoming).

23. R. S. Pindyck, "Prices and Shortages: Policy Options for the Natural Gas Industry," in *Options for U.S. Energy Policy* (San Francisco: Institute for Contemporary Studies, 1977).

24. R. S. Pindyck, *The Structure of World Energy Demand* (Cambridge, Massachusetts: M.I.T. Press, 1979).

25. J. Ramsey, R. Rasche, and B. Allen, "An Analysis of the Private and Commercial Demand for Gasoline," *Review of Economics and Statistics,* November 1975.

26. A. Rødseth, and S. Strøm, "The Demand for Energy in Norwegian Households with Special Emphasis on the Demand for Electricity," University of Oslo, Institute of Economics, Research Memorandum, April 1976.

estimates of this long-run elasticity that have been applied to policy analysis.

The first five studies cited in Table 28-1 are all based on the use of time series data for a single country—and as a result are likely to have underestimated the true long-run price elasticity demand. There are two important problems with the use of such time series data. First, some of the demand, price, or income variables under study may not have changed very much over the historical time horizon for which data are available. Second, unless the historical time horizon is very long, the approach is like to capture short-run rather than long-run elasticities. The reason is simply that the measurement of successive changes in demand that follow successive changes in price or income precludes the possibility that enough time had elapsed for a full adjustment to those prices and for income changes to have taken place.

In order to estimate long-run elasticities of demand, it is necessary to compare the equilibrium demands for energy corresponding to prices that are significantly different from each other. By equilibrium demand we mean the demand that would prevail after sufficient time had elapsed for consumers to completely adapt to a new price or set of prices. How much time is sufficient will depend on the particular sector or subsector of energy use, but it might be anywhere from five to twenty years. Given the limited time horizon for which data are available for any one country, it is unlikely that we can compare equilibrium prices and demands by using data for only a single country. On the other hand, energy prices are and have been quite different across countries, so that the use of pooled international data may indeed allow us to elicit long-run elasticity estimates.

The studies by Nordhaus and this author were both based on the use of pooled cross-section times series data spanning a number of different countries. Because energy prices and per capita consumption levels differed considerably across these countries, it is more likely that these studies would have captured long-run elasticities. It is true that these two studies use very different structural formulations to represent demand, but it is likely that their use of pooled international data rather than time series data for a single country is more critical. Thus, while it is debatable whether the true long-run own price elasticity of aggregate energy use in the residential sector is closer to -0.7 or -1.1, it is quite likely that it is somewhere within this range and about two or three times as large as the typical estimates that people have used earlier.

Turning back to Table 28-1, there is little disagreement about the income elasticity of energy demand in the residential sector. With the exception of the Joskow-Baughman study and the Nelson study, all of the other studies show estimates in the range of 1.0. The lower estimates obtained by Joskow-Baughman and Nelson might have again resulted from the use of time series data.

Estimates for price elasticities of demand for individual fuels used in the residential sector would seem to indicate that those for oil and natural gas are at least -1.0 and possibly larger in magnitude, so that there should be considerable room for interfuel substitution between these fuels. There seems to be more disagreement about the elasticity of electricity demand, with estimates varying from -0.3 to -1.2. We would expect electricity demand, however, to be less elastic than fuel oil or natural gas demands since for many uses (e.g., household appliances) there is simply no substitute for electricity.[2]

Let us now turn to the industrial sector, and the elasticities of demand for energy as well as other factors of production. Three of the studies cited in Table 28-1 used aggregate time series data for a single country–Berndt-Wood (the United States), Fuss (Canada), and Magnus (the Netherlands). All of these studies use generalized cost functions to estimate derived demands for energy (translog functions in the cases of Berndt-Wood and Fuss, and generalized Cobb-Douglas functions in the case of Magnus). The first two of these studies find own price elasticities for energy demand similar to those obtained from time series studies of the residential sector, about −0.49. In addition, they both indicate that capital and energy are complementary rather than substitutable. (Note that the cross-price elasticity of capital and energy is negative, so that an increase in the price of energy results not only in the use of less energy, but also in the use of less capital.) These results have disturbing implications for the use of policies to increase energy productivity, but again, they are based on time series data for a single country and may well represent short-run, rather than long-run, characteristics of industrial energy demand.

The studies by Griffin and Gregory and by Pindyck are both based on pooled international data, and both use generalized cost functions (translog) to avoid imposing a priori restrictions on the structure of demand. Both of these studies indicate a much larger own price elasticity for energy demand (−0.8 or greater), as well as substitutability between capital and energy (note that the cross-price elasticities between these factors are now positive, so that an increase in the price of energy results in some substitution toward capital). These results are encouraging and are consistent with our intuitive expectation that capital and energy should be at least to some extent substitutable, given that enough time is allowed to pass. One would expect that as machines eventually wear out, they might be replaced with more energy-efficient ones if the price of energy has indeed increased significantly.[3]

Finally, it is worth noting the interesting results of the Halvorsen-Ford study. This study estimated elasticities for eight individual industries on the grounds that the potential for factor substitution might vary widely across industries. Indeed, they obtained estimates for the own price elasticity of energy demand that range from −0.66 to −2.56.[4] This reinforces our belief that energy demand is indeed quite price elastic in the industrial sector, but also indicates that more detailed analyses are needed at the level of individual industries. In fact, not only are the magnitudes of energy demand elasticities likely to differ considerably across industries, but because of differences in capital depreciation rates across industries, the speeds of response to price changes are likely to differ as well.

There seems to be little agreement about price elasticities for individual fuels used in the industrial sector, except that again electricity demand seems to be less elastic (as we would expect) than the demands for other fuels. Otherwise, it would seem difficult to pinpoint single numbers or "consensus" estimates of elasticities for individual fuels. The reason for this is probably that the demands for individual fuels vary considerably across industries; an industry that uses a fuel as a chemical feed stock or for some special application would exhibit almost no flexibility in demand, while industries that use fuel simply to generate steam would probably have considerable flexibility in choosing the fuel they use. Again, detailed industry-specific analyses are needed.

We turn next to the transportation sector and, in particular, to the demand for motor gasoline. Again, there seem to be a basic difference between the results obtained through the use of data for a single country (the first three studies cited) and those that use data spanning a number of countries. The first three studies find the long-run own price elasticity for gasoline demand to be in the range of –0.22 to –0.70. It is in fact the lowest end of this range that has been most widely used in policy analyses involving gasoline taxes or other incentives to reduce demand. Even the high end of this range, however, may be too low. As indicated from the last two studies cited, which use data that span a number of countries (and that are more likely to elicit long-run elasticities), this elasticity might be much closer to about –1.0. This would indicate that higher gasoline prices could be quite effective as a means of reducing demand, but that one must wait a number of years for the effects of price increases to take place.

Finally, we turn to the choice of fuels used in the electric generation sector. Here little work has been done, and we can cite only three studies. The first two indicate considerable room for interfuel substitution in electricity generation, which would indicate that financial incentives might indeed be effective as a means of shifting from one fuel to another. These results are in fact reinforced by the third study cited, which indicates that the elasticities vary greatly across utilities, but will be large when expected fuel prices are roughly the same in terms of thermal content. In other words, if fuel prices differ considerably (e.g., natural gas is much cheaper on a Btu basis than oil or coal), the cheap fuel will almost always be chosen, even if minor shifts occur in its price. As one would expect, small price changes will affect the choice of fuel only when the relative fuel prices are already quite close to each other.

In summary, we find that the elasticity of energy demand is a dynamic concept and depends critically on how much time is allowed to elapse after the price of energy (or income) has changed. We find, however, that if enough time is allowed to elapse, the demand response can be considerable–that is, long-run price elasticities appear to be quite large. How much time is enough is unfortunately difficult to say and will depend not only on the sector of use, but also on the particular use within each sector. Unfortunately, most uses of energy involve expensive capital equipment, and in many cases it is not easy to convert this capital quickly. On the other hand, to the extent that much of the debate over energy policy is concerned with how best to achieve what are essentially long-run targets, we can argue that price should be an extremely potent policy instrument.

THE IMPLICATIONS FOR ENERGY POLICY AND THE EVOLUTION OF ENERGY MARKETS

As we have seen, there is considerable evidence that higher prices (through deregulation, taxes, or other measures) can indeed reduce energy demand significantly, but it will take some time, perhaps seven to ten years or even longer, to see the full effect of these price increases. While demand elasticities certainly vary across individual industries, in the aggregate this price responsiveness holds across all sectors of use. If we were to pick "working" estimates for price elasticities of energy demand, we might choose a number on the order of −0.8 to −1.0 for the residential sector, a number like −0.8 for the industrial sector, and a number like −0.9 for gasoline demand. Of course, these working estimates should be accompanied by confidence intervals that express our degree of uncertainty, and intervals on the order of ±25 percent would not be unreasonable. On the other hand, it is interesting to note that all of these working estimates are at least twice as large, and in some cases three times or four times as large, as those that have often been used previously for policy analysis.

We also find that there seems to be considerable room for interfuel substitution in the residential, industrial, and electric generation sectors, although again, particular elasticities may vary considerably across industries and across individual electric utilities. It would appear, therefore, that taxes or other financial incentives to encourage "fuel switching" should work, at least if they are applied to the particular industries where there is indeed room for switching. (Here we should stress that individual industry studies are needed, so that we can better understand where and how to apply fuel switching incentives.)

We have seen that in the long run, energy demand in the industrial sector is indeed reasonably elastic. The responsiveness of energy demand—as well as the demands for labor and capital—to price also becomes apparent if one considers the fact that between the end of World War II and 1972 a slow but steady shift occurred in the structure of industrial production in the United States and in most of the other advanced economies. During this period, two factors of production—energy and capital—became significantly cheaper in real terms relative to a third important factor, labor. This shift in relative prices occurred for a number of reasons. Reserves of energy resources, and energy production, were increasing worldwide, which drove down the real cost of energy. Tax policies in many countries (e.g., the investment tax credit in the United States), designed to encourge new capital investment as a spur to economic expansion, helped to reduce the growth in the price of capital services. Finally, tax and social welfare policies, combined with greater wage demands on the part of workers, tended to greatly increase the effective cost of labor services for production. The result of these changing prices was a shift in the factor mix used in production. Gradually, producers replaced labor with less expensive capital and energy.

The secular shift away from labor and toward energy and capital helped to exacerbate the impact of the increases in energy prices that were brought about by the OPEC cartel. When energy prices rose, industries in many countries were unable to achieve a significant shift away from energy-intensive production. In the short term at least, energy and capital were complementary inputs, and the only substitutable alternative–labor–was already very expensive. Thus, increases in energy prices were translated into an increase in the cost of industrial output–an increase in cost nearly as large as the percentage increase in the price of energy times energy's share in the total cost of output. But with labor and capital fixed in the short term, this had to mean a drop in the level of real output.

On the more encouraging side, the increase in energy prices should help to reverse the long-term shift that we have observed in the factor input mix (albeit slowly). And, to the extent that capital and energy are substitutable in the long run, the macroeconomic impact of energy price increases should be ameliorated after a number of years have gone by.

To the extent that limiting the growth of energy consumption is a goal of energy policy, it is clear that the most effective policy instrument available is the price of energy itself. As we have shown, we now have good reason to expect that our consumption of energy can indeed be reduced if the price of energy is allowed to rise. This is particularly important for the United States, where for a long time the goal of low energy prices has dominated national energy policy.[5] Allowing the price of energy in the United States to climb to the world level (energy in the United States is now about 25 percent cheaper than it is in the rest of the world) would effectively reduce American energy consumption, would stimulate energy production, and would thereby reduce the growing level of American energy imports.

NOTES

1. For a general discussion of the determinants of energy demand and the issues involved in constructing and estimating energy demand models, see R.S. Pindyck, "The Characteristics of the Demand for Energy," in J. Sawhill, ed., *Improving the Energy Productivity of the American Economy* (Englewood Cliffs, New Jersey: Prentice-Hall, forthcoming); R.S. Pindyck, *The Structure of World Energy Demand* (Cambridge, Massachusetts: M.I.T. Press, 1979).

2. There is a problem in modeling electricity demand, however, since with most available data it is difficult to differentiate between the use of electricity for heating (where there is considerable room for substitution) and the use for lighting (where there is no substitution possibility).

3. We are arguing that the larger elasticities obtained by Griffin and Gregory and by Pindyck can be attributed to the use of pooled international data. It should be pointed out, however, that these studies work with only three factors of production (capital, labor, and energy), while the Berndt-Wood, Fuss, and Magnus studies include a fourth factor (nonenergy raw materials), and this is another possible explanation for the difference in the results. It is possible for two factors that are net complements in a four factor production space to be net substitutes in a three factor subspace.

4. This is supported by a research study currently being conducted by Fuss and Waverman at the University of Toronto using industry data for Canada.

Although that study is incomplete, initial results also show wide elasticity variation across industries.

5. For a discussion of some of the problems in U.S. Energy policy, see R.E. Hall and R.S. Pindyck, "The Conflicting Goals of National Energy Policy; *The Public Interest*, Spring 1977.

 *Chapter 29*

Long-term Prospects for Energy Conservation

John H. Gibbons
University of Tennessee

INTRODUCTION

Even a cursory examination of energy utilization today anywhere in the world discloses the fact that the efficiency of use is generally very remote from theoretical limits. There are good reasons for this situation to exist today but equally good reasons to embark on a long-term course to very much higher energy productivity. The purpose of this chapter is to briefly review the long-term opportunities for higher energy productivity and the impedimenta or limitations to the extent and rate at which this higher productivity can be reached. While the focus is on the United States, the general principles can also be applied to other societies.

In addressing the need and likelihood for more productive energy use in the future, several assumptions held by the author, but shared by many others, should be made explicit:

1. The price of energy, compared to other goods and services, will be substantially higher than present. However, except for relatively short-term and emergency situations, real energy price will probably never be more than about four times higher than present price, because at that price level, given time, extremely large quantities of renewable and long-term sustainable energy resources are economic. In the long run, we will not run out of energy—but we are rapidly running out of cheap energy. It is not unlikely that gas and oil prices (in real terms) will be double present prices by 2000.
2. The dominant forms of energy will be the same as today—electricity, gas, and liquids (albeit derived from a variety of sources, increasingly nonfossil in origin).

3. As energy prices rise, it will pay to alter patterns and efficiencies of energy use, substituting technological ingenuity and institutional innovation for energy resource consumption.

LIMITATIONS TO ENERGY PRODUCTIVITY

Natural law

While energy is always conserved, it becomes less useful after each time we use it. For example, when natural gas is used to heat water, no calories are lost, but the inherently very useful calories stored in the gas are converted to hot water and waste heat, and in their new form they are far less "available" to do work. As energy gets more expensive, the technologist designs machines that cause smaller amounts of high availability energy to be used for a given service rendered. The technologist also tries to match the quality of the energy source to the quality of energy needed to perform a service so that minimum "availability" is lost.

The maximum efficiency of any cyclical process (e.g., a heat engine such as a steam boiler) is related to the combustion temperature T_H and to the coolant (discharge) temperature T_C. Well over a century ago Carnot showed theoretically that the maximum efficiency, E, is: $E = 1 - (T_C/T_H)$. Clearly, unless the combustion temperature is extremely high and/or the discharge temperature extremely low, efficiencies are relatively small. It is possible using fossil fuels to generate combustion temperatures that result in very high efficiencies (fossil fuels, with air combustion, can burn as hot as 3,000°F), but there are no materials that can readily be used to contain the steam above about 1,300°F. Such limitations in a power plant can be overcome by using a combined cycle wherein a "topping" cycle consisting, for example, of a gas turbine followed by a steam cycle is used. Even by combining such methods one is still practically constrained to maximum Carnot efficiencies of less than 70 percent, with 30 to 40 percent actually attained for modern electricity generation plants.

While most heat pumps and air conditioners provide about twice as much heating or cooling energy as input energy by extracting heat from outside air or dumping heat into the outside air, they could theoretically supply about twenty times as much (hence they have an actual efficiency of less than 10 percent of Carnot efficiency). Equipment could be made with much improved energy efficiency and still be much lower than Carnot's limit but each improvement in efficiency requires more investment (e.g., a more efficient heat radiator). Some typical Carnot efficiencies are listed in Table 29-1.

Since a large fraction of uses of energy actually only require relatively low temperatures,* it is clear that thermodynamic matching between energy sources

*Thirty percent of all U.S. fuel and electricity is used for tasks that change temperature by less than 80°C; another 15 percent produces steam at less than 400°C.

Table 29-1. Second Law Efficiencies for Some Energy-consuming Activities.

Sector	*Carnot Efficiency (percent)*
Residential	
Space Conditioning	
Furnace, heat pump	5
Electric Resistance	2.5
Air Conditioning	5
Water Heating	
Gas	3
Electric	1.5
Refrigeration	4
Transportation	
Automobile	10
Industry	
Electric Power Generation	33
Process Steam Production	34
Steel Production	23
Aluminum Production	13
Average Manufacturing	13

Selected sources: M. H. Ross and R. H. Williams, "Energy Efficiency: Our Most Underrated Energy Resource," *Bulletin of the Atomic Scientists,* 32 (November 1976): 33ff; G. N. Hatsopoulos, E. F. Gyftopoulos, R. W. Sant, and T. F. Widner, "Capital Investment to Save Energy," *Harvard Business Review,* March-April 1978, pp. 111ff.

and energy uses holds much promise. In other words, much better thermodynamic utilization can be obtained through a variety of techniques, including total energy systems, cascaded energy use such as cogeneration, and low temperature storage.

It is conceivable that some of the same goods and services now provided by heat engines can be provided by some other mechanism. If so, the Carnot efficiency limits could be bypassed. For example, an electric motor powered by a chemical battery might be used to replace an internal combustion engine to power a car. These electrochemical reactions are not limited by Carnot's heat engine laws and can operate with very high efficiency. However, in thinking about such substitutes, we must remember that other constraints will operate. For example, the electricity to recharge the battery must be generated somewhere, likely in a Carnot-limited device. Of course, it is possible that electricity could be generated from fuels via a hydrogen-oxygen fuel cell and again avoid the limitations of efficiency imposed on a thermal process. But where do the hydrogen and oxygen come from?

Other limitations to efficiency of energy-consuming devices that are imposed by natural law include properties of materials such as heat conductivity, friction, and high temperature strength. For example:

- A large fraction of energy required to power a car or truck above 60 mph is due to wind resistance and rolling friction between the tires and the road;

- The efficiency of a refrigerator, freezer, or air conditioner is strongly dependent upon the heat transfer properties of the heat exchanger, which dumps heat into the outside air;
- The efficiency of high temperature, high pressure combustion devices is limited by the loss of mechanical strength of materials at high temperature;
- The finite resistance to heat transfer through insulating materials limits the conservation design in appliances, homes, and industrial processes where we wish to hold heat in—or out.

Technology

One of the most fundamental challenges to improving energy productivity lies in improvement of various properties of existing materials and processes and in development of new materials and processes with desired special characteristics. As scientific understanding of the properties of materials grows, we have already witnessed profound impacts on technology and, in turn, on the amount of energy required to provide a given amenity (examples are given later in the chapter).

In virtually all instances of improved energy productivity the notion of saving energy has been important, but not necessarily dominant, in spurring development of the processes. Indeed, many of these developments occurred while energy prices were falling. More important in inducing the developments has been the drive to improve overall quality, cost, and reliability of product. Another very significant factor has been the desire to improve the yield of final product compared to the consumption of raw materials. Major advances in energy efficiency have occurred over the past several decades but some (e.g., conversion of coal-fired steam locomotives to diesel electric drive) were achieved at the cost of switching to less plentiful fuels. Nevertheless, close consideration of virtually every major energy-consuming activity reveals major opportunities for energy to be replaced by technological advances.

Economics, Complexity, and Interdependence

As nations industrialize and urbanize, their citizens seem to accept conditions of increasing socioeconomic and technical complexity in exchange for material benefits. The United States economy is now remarkably complex and interdependent, in large measure because increased scale of operations (e.g., farming, communications, power production, transportation) tends to decrease unit production costs. Of course, at the same time, this trend tends to have undesired but "external" side effects such as requiring workers to commute farther (at their own expense).

In the case of energy, the "waste" from one user (e.g., a power plant) can be used as the "input" energy for another (e.g., heat for an apartment building). Similarly, the purchased energy required to provide space conditioning in a house can be considerably decreased by installing a complex automatic electronic sensing and control system; or, transportation energy for urban commuter traffic can be drastically lowered by extensive use of carpools and vanpools. There are five conditions that tend to govern the extent to which these kinds of conservation tactics are attractive:

Energy Price. As energy gets relatively more expensive, increased complexity and interdependence, if a large improvement in energy productivity results, become more tolerable.

Reliability. We tend to accept complexity and interdependence more readily when the system can be counted on to operate without unwanted interruption. An apartment owner is not interested in being heated by "waste heat" from a power plant that cannot be counted on to operate reliably throughout the heating system. Improvement in reliability can be expected to continue to result from investments in science and technology; however, consumers are increasingly suspicious of complexity and big technology.

Proximity and Scale. Because of the past performance of energy price, economies of scale in industry, and land use patterns of residential and industrial development, there are not many major opportunities for energy savings through increased interdependence in our existing system. Present sources of low grade and waste heat, sufficient to heat and cool buildings, are generally sited too far from existing buildings to justify construction and transport costs of the heat. The extremely large size of most power plants cause so much waste heat to be generated that markets within plausible range are oversaturated. As new capital investments are made, much improvement could be gained from energy-conscious siting and sizing of energy production and other industrial plants.

Convenience. Citizens, be they individual or corporate, value independence. A petrochemical plant manager might save fuel bills by accepting waste heat from a power plant, but his hidden costs in doing so include threat of regulation, unscheduled interruptions, and inflexibility in future negotiations for fuel purchases. An urban commuter can save money in joining a vanpool, but in doing so must forego the freedom that would otherwise be available in driving his own car. An auto driven at 55 mph might get 20 percent better mileage than one driven at 70, but the occupant who highly values his time may choose the faster speed despite the losses in fuel and safety. A refrigerator with a certain color may be more important to a buyer than the fact that it uses half again as much electricity as one of another color. Thus "convenience" and independence can be strong factors in limiting energy productivity.

Economics. While the U.S. markets of energy production and energy-utilizing equipment (producers and consumers) are not "free," they are competitive.

Therefore, as energy prices and availability change, all of these actors adjust to maximize their welfare. In the past, and still to a large extent today, the energy characteristics of consumer goods are either invisible or poorly visible at point of sale, even for items (e.g., refrigerators) where the energy costs to operate over its lifetime will considerably exceed the purchase price.

With upward trending energy prices, the wise energy user will make purchase decisions that take account of projected future cost of energy as well as current prices. There are at least three reasons why this situation generally does not obtain. First, energy costs are poorly visible, especially on new goods. Second, future energy prices are uncertain and subject to political and cartel manipulation. Third, in a period of high inflation, we tend to pay less attention to future expenditures than to present expenditures. This last situation is particularly important for poor people because this effective discount rate used in capital investment decisions (e.g., 50 percent per year for those with incomes of less than $5,000 per year) is several times higher than the rate used by the middle class.

Thus, the relative lack of access to capital markets by many energy consumers, combined with lack of clear signals about present and future prices and a time of substantial inflation, all tend to frustrate a shift from energy purchases over to capital investment in more energy-productive stock.

WHAT CAN BE DONE?

There are now ample examples of the reality of substitutibility of technological and managerial sophistication for energy consumption throughout the world's economics and especially in the United States. Therefore, rather than a comprehensive review we relate only a few examples to illustrate not only what can be done but also what is being done.

Transportation

Automobile mileage can be doubled for less than a 10 percent increase in purchase price.[2] However, considerable time (e.g., seven years or more) is reasonably required for a production changeover. While the new cars will have less capability to leave rubber on the road from high acceleration they will otherwise provide the same comfort, safety, and performance. Similar major improvements in energy productivity can accrue in new production of trucks and airlines. Beyond these purely technological changes, intermodal shifts of freight and urban/commuter mass transportation can provide substantial further improvements.

Buildings and Appliances

As for the automobile, there are a lengthy string of options to radially improve energy productivity in this sector. Energy consciousness in design (e.g., orientation of building, location of windows, use of internal thermal mass, proper overhanging eaves) can make a major impact of itself. The construction of the thermal envelope (shell) can cut heating requirements dramatically for minor changes in construction cost (Table 29-2, Figures 29-1, 29-2). Finally, the design and operation of the space and water heating, cooling, ventilation, and lighting systems can have a major impact upon requirements for purchased energy (Figure 29-3). A caveat must be added: As the thermal envelope and passive solar effectiveness are improved, the amount of purchased energy required becomes so small that one cannot justify capital-intensive, highly energy-efficient heating and cooling systems.

In short, the application of only modestly sophisticated energy-sensitive design and engineering principles can cut overall purchased energy requirements in half or less. Numerous instances of such improvements have occurred over the past five years. Results are especially impressive for larger buildings. Eight years ago a typical commercial building required 180,000 Btu/ft^2 per year. Nowadays the (achieved) design figure is more like 50,000,[3] and the opportunity to make further improvements appears to be substantial. Because of these facts, it appears to be quite possible that despite the continued expansion of residential and commercial space in the United States, total energy requirements in this sector could roughly level out over the next thirty years or at least not grow faster than abour 1 percent per year despite a 30 percent population increase and a doubling of real GNP.[4]

Table 29-2. List of Design Options and Associated Additional Features for Increasingly Energy-efficient Thermal Envelope of a Residence (key to Figures 29-1 and 29-2).

Design Option	*Added Feature*
1	Baseline
2	R-11 attic insulation
3	R-11 wall insulation
4	R-19 attic insulation
5	R-11 floor insulation
6	Storm windows
7	R-30 attic insulation
8	R-19 floor insulation
9	2-paned SGD
10	R-13 wall insulation
11	R-38 attic insulation
12	R-19 wall insulation
13	3-paned windows
14	R-49 attic insulation
15	Storm door
16	R-23 wall insulation

Source: E. Hirst and P. Hutchins, *Engineering-Economic Analysis of Single-Family Dwelling Thermal Performance* (Oak Ridge, Tennessee: National Laboratory, Report ORNL-Con-35, 1978).

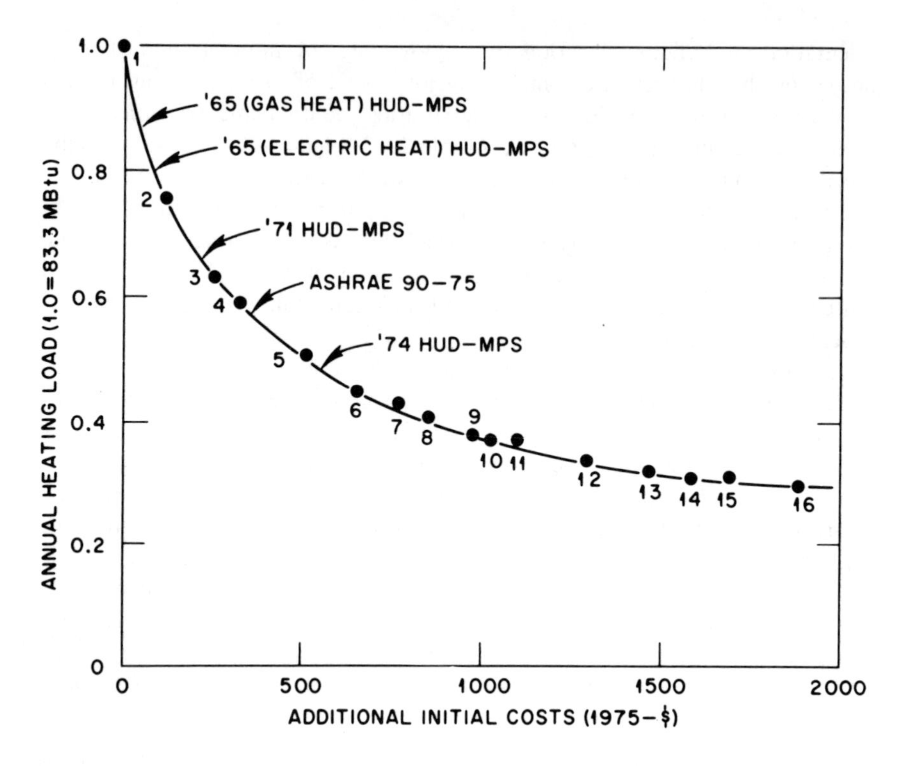

Source: E. Hirst and P. Hutchins, *Engineering-Economic Analysis of Single-Family Dwelling Thermal Performance* (Oak Ridge, Tennessee: Oak Ridge National Laboratory, Report ORNL-Con-35, 1978)

Figure 29-1. Heating-Cooling Load versus Incremental Cost for a New Single Family Residence in Weather Conditions Equivalent to Kansas City. The numbers refer to options described in Table 29-2. The arrows point to loads that correspond to various indicated codes and standards.

Industry

Carnot efficiencies presently achieved in the industrial sector range from low to impressively high (Table 29-1). Besides many opportunities for improvements in energy productivity using existing processes, there are important additional opportunities to develop new processes that can markedly improve productivity. Here are a few examples of recent advances that typify the sorts of things that can continue to happen as we further substitute technological innovation for energy consumption:[5]

- The progression from vacuum tubes to transistors and thence to large-scale integrated circuits has enabled the energy required for various functions in communications and computers to be reduced about a millionfold in two decades.

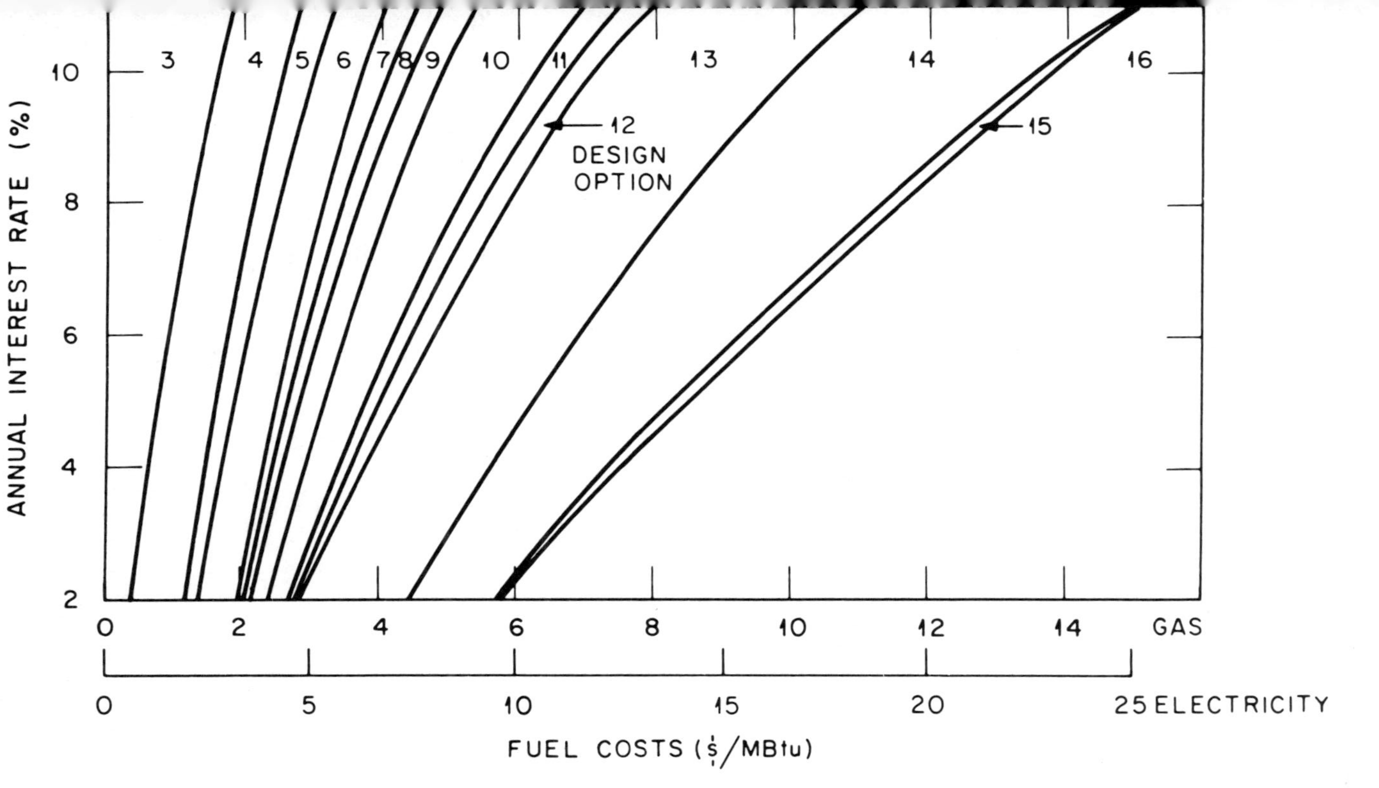

Source: E. Hirst and P. Hutchins, *Engineering-Economic Analysis of Single-Family Dwelling Thermal Performance* (Oak Ridge, Tennessee: Oak Ridge National Laboratory, Report ORNL-Con-35 1978).

Figure 29–2. Sensitivity of Option Selection to Assumed Annual Discount (interest) Rate and Fuel Costs. See Table 29–2 for description of options and Figure 29–1 for energy intensiveness.

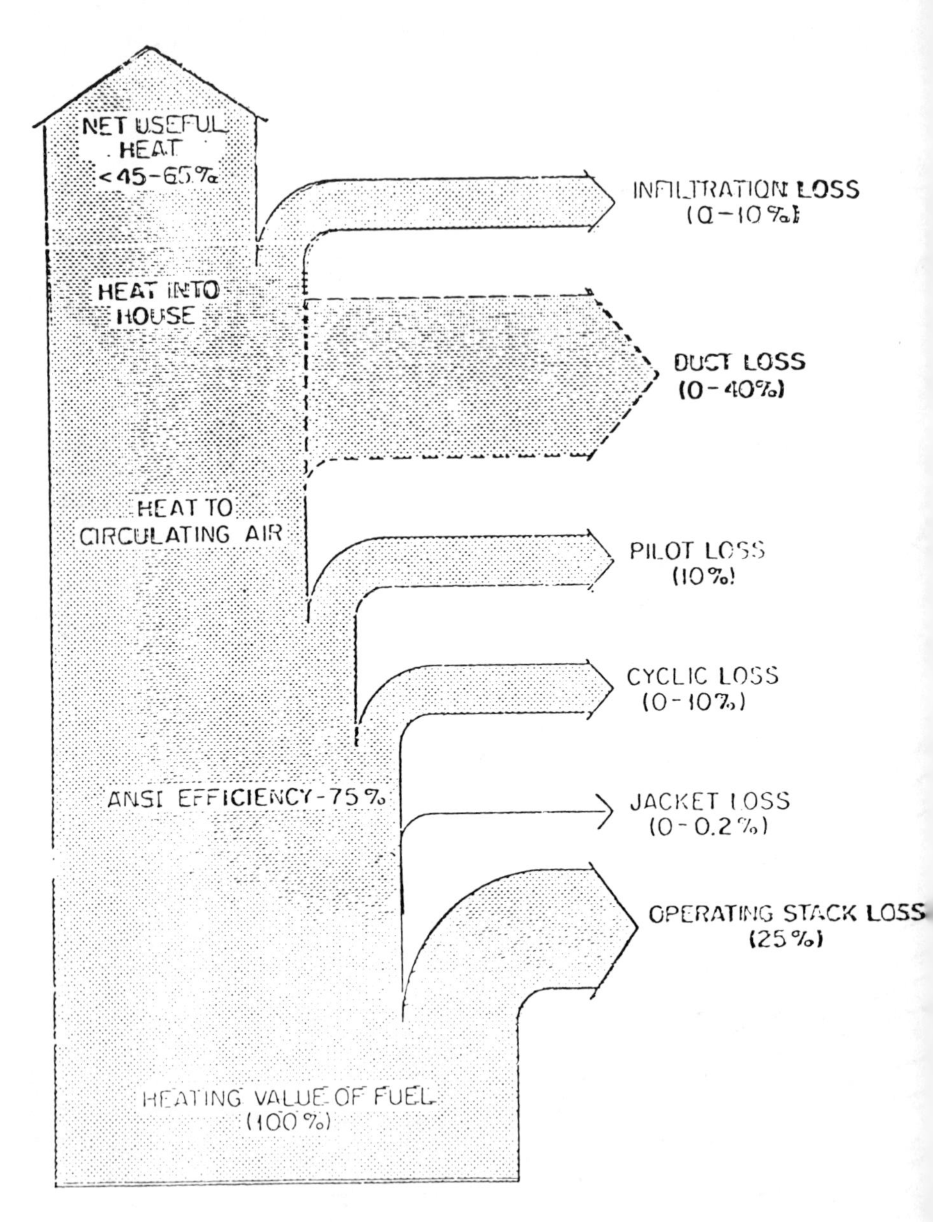

Source: E. C. Hise and A. S. Holman, *Heat Balance and Efficiency Measurements of Central forced Air, Residential Gas Furnaces* (Oak Ridge, Tennessee: Oak Ridge National Laboratory, Report ORNL-NSF-EP-88, October 1975).

Figure 29-3. Energy Flow for a Gas Furnace System.

- The efficiency of electricity conversion into lighting has increased from less than 5 percent (incandescent) to more than 30 percent (alkali-halide) since 1950.
- The energy required to produce a pound of aluminum from alumina has fallen from more than 12 kWh/lb in 1950 to less than 5 kWh/lb using the new Alcoa molten chloride process.
- The energy required to produce low density polyethylene has been halved due to a new process developed by Union Carbide.
- Research on producing a fire-safe diesel fuel accidentally resulted in a fuel that burns with 10 percent better economy and 50 percent less exhaust smoke.
- Energy consumed per unit output of product has fallen substantially for all industry over the past decade, as much as 50 percent for many petrochemical products.

Behavioral Change

We have contrained the previous discussion to technological innovations and applications. Clearly there are a host of plausible changes in consumer priorities that could have marked impact on energy demand over the next several decades. Such priority changes could push energy demand up or down. One thing is certain: lifestyles are always changing. Even in the low GNP growth scenarios studied by the CONAES Panel on Demand and Conservation[6] real income doubles over thirty-five years, and that fact alone implies substantial lifestyle change, especially when combined with major demographic shifts due to lowered birth rates. If energy prices increase, it is likely that consumers, seeking maximum welfare, will shift their behavior toward less energy-intensive activities (e.g., purchase a night setback thermostat, alter indoor space temperatures). Another form of behavior change could (hopefully) become more important in future years in the emergence of an ethic that results in energy savings as a means to share energy resources with other people who live in other places and in future times. The extent to which these accommodations will occur is problematic, but even minor changes could have substantial impacts on energy demand.

Research Opportunities[7]

The areas of energy utilization are so diverse that the richness of the opportunity to improve energy productivity through research is difficult to overestimate. Major public support should be provided not only for basic science, engineering, and social science research relevant to energy utilization, but also for developing the technological capability to use energy at an efficiency that corresponds to energy prices that would obtain from such sources as gas synthesized from coal. It is naturally difficult to anticipate how much this research will contribute to higher energy productivity (that is, economically feasible for a given energy price), but my guess is that such research will be likely to be very productive, in part because the field has received little attention until very recently. The fact that current practice is far removed from theoretical limits also gives strong support to this notion.

LONG-TERM PROSPECTS

After examining in detail the specific prospects to achieve major increases in energy productivity, given several decades of time, it can be concluded that for the United States, a real GNP of about twice current level can be achieved without a significant increase in (purchased) energy.[8] The biggest uncertainty is not so much the extent to which energy productivity can increase but rather the time over which such improvements can accrue. Such a situation could result from the vigorous application of human ingenuity, mostly through technology and management. Further shifts toward higher energy productivity could occur as a result of research and/or by altered consumer behavior patterns and life-styles.

In view of the manifold uncertainties about our energy supply future and the relative ease–if not attractiveness (e.g., capital requirements, employment and inflation implications)–of effecting much higher levels of energy productivity, it seems only prudent as national insurance to move ahead in this area with dispatch, since decades will be required once we commence.

The things we have concentrated on in this chapter probably are more readily associated with "soft" energy paths than with "hard" paths. For example, the long-term route to higher energy productivity shifts a lot of capital investment from the energy production sector to the energy-consuming sectors. It opts for dispersed energy conversion, so that productive uses of "waste" heat are easier to capture; consequently, it tends to push technology toward smaller scale economy. But it does not seek higher productivity via denial of amenities. Rather, it rests upon the reality that human ingenuity, applied through both technological and institutional innovations, can substitute for energy to a major degree.

In this light the "soft" path can be interpreted as a new level of technological and institutional sophistication, adapted by a society that seeks its affluence in a way that is more harmonious with the rest of nature. Such an application of human wisdom could translate the specific issue of "prospects for energy conservation" into a more general one: To what extent can mankind substitute ingenuity for resource consumption? As a consequence, our measure of progress could become less that of counting how much material and energy resources we consume and more that of how little we require to provide a given level of amenities.

NOTES

1. See J. H. Gibbons, "Long Term Research Opportunities," in *Improving the Energy Productivity of the American Economy*, ed. John Sawhill (Englewood Cliffs, New Jersey: Prentice-Hall, forthcoming).

2. Gibbons et al, *Science* 200 (April 14, 1978): 142.
3. Charles Ince, AIA Research Institute (private communication, 1978).
4. E. Hirst and J. Jackson, *Energy* 2 (1977): 131.
5. Gibbons et al.; C. Berg, *Science* 199 (February 10, 1978): 608; G. N. Hotsopoulos, E. P. Gyftopoulos, R. W. Sant, and T. F. Widner, *Harvard Business Review*, March-April 1978, pp. 111 ff.
6. Gibbons et al.
7. Gibbons
8. Gibbons et al.

Part II

Policy Directions and Decisionmaking

❀ *Chapter 30*

Long-term Strategic Analysis (Toward an Energy Doctrine)

Kenneth C. Hoffman
and
Steven C. Carhart
Brookhaven National Laboratory
(Presented by K.C. Hoffman)

ABSTRACT

The formulation of long-term energy strategies requires the synthesis of a large number of economic, social, technical, and environmental factors with due regard to regional and international needs and constraints. It is also clear that the long-term uncertainties are great and that any strategy must be adaptive in being able to respond to new circumstances and events. There is a need to debate and develop the strategic basis of energy policies and to test policies against the uncertainties and contingencies that are identified in the strategic basis.

Policy options that would lead to government involvement in energy markets should be analyzed and evaluated in terms of clearly defined objectives with means established to monitor their effectiveness. The orderly functioning of the decision process in industry and government requires a base of analysis and information on strategic issues.

This chapter outlines an approach to strategic analysis in the energy sector. Emphasis is placed on the logic of the approach to strategic planning and analysis. The process begins with the definition of overall goals and objectives. A range of policy options may be defined to achieve the objectives, and these must be integrated into a coherent strategy to achieve the objectives. Supporting analysis can indicate the effectiveness of the policy and the continuing need

*Under Contract No. EY-76-C-02-0016 with the United States Department of Energy.

to modify objectives and policies as events unfold, uncertainties are resolved, or new situations are developed.

INTRODUCTION

Long-range energy strategy must provide the basis of coordinated government action to implement energy policies consistent with national goals and objectives. This chapter outlines a structured approach to the development of such a national energy strategy. It points toward the establishment of an "energy doctrine" that will provide a record of analysis and debate of issues and possible solutions as a basis for future government policy. The emphasis in this chapter is on defining an appropriate scope and structural approach to policy analysis rather than on specific policy recommendations. Any policies that are defined are in the nature of options to be studied and evaluated.

In view of the need to provide maximum freedom of choice for Americans consistent with the values and goals of present and future generations, a long-range energy strategy must be consistent with current knowledge and adaptive as uncertainties become resolved and new options arise. At any given time, the strategy must be sufficiently specific to provide a stable climate for decision by the principal decisionmakers in the energy market, industry, and citizens. Involvement in the market by government should have clearly stated objectives, and its effects should be monitored and measured against those objectives. Examples of the appropriate role of government in the energy market include regulation of natural monopolies; regulation to correct market imperfections in dealing with environmental, conservation, and national security issues; basic research; long-term or high-risk development; and demonstration.

Energy is a pervasive element in the economic and social fabric of a nation. Energy strategy and policy have not yet been woven into our political and decisionmaking infrastructure to the extent of, say, defense and environmental strategy. Debate on energy is sometimes unstructured, with too much emphasis on specific detail before the overall strategic objectives have been articulated. Work is in progress at the federal and state levels to establish a structure for analysis and debate of the issues in order to identify policy elements and focus debate at the appropriate policy level. The energy system is a physical system involving resources and technologies. The primary energy policy goals of providing sufficient energy for economic and social development with maximum security and minimum environmental impact can be represented in physical terms and monitored through parameters herein referred to as "energy system indicators."

In formulating and analyzing a long-range strategy, attention must be given to the specific goals and objectives of the strategy and to monitoring or measurement of the effectiveness of the policies employed to attain those objectives. Particular attention must be paid to criteria for the escalation or withdrawal of a policy option should circumstances and resolution of uncertainties warrant.

The physical energy system operates in a complex institutional and regulatory framework. These aspects of energy must be dealt with in energy strategy. The logic of energy strategy followed in this chapter begins with the definition of urgent physical needs of society that comprise the goals and objectives of the strategy, followed by a review of potential physical solutions. Policy levers that can affect the physical aspects of the energy system in the desired direction must be identified, and a spectrum of consistent policy options must be laid out. Criteria that govern the selection of the appropriate level of policy action and of the escalation of policy responses in response to new information and emergencies may then be defined. Finally, recommendations can be made of new policy options to be studied in support of the future development of a consistent and comprehensive long-range energy strategy.

An essential feature of this approach to long-range strategy is the definition of a "prudent planning basis." This basis identifies the specific objectives to be achieved by the energy system and the specific contingencies that must be prepared for. A "prudent planning basis" must contain sufficient margin for error. Government must be "risk averse" in its long-term planning and must seek to avoid pitfalls and problems that may be perceived by the private sector but that are beyond its ability or mission to deal with. Factors that cause indecision on the part of industry and consumers must also be dealt with. In any event, the strategy must reflect the appropriate roles of federal and state government and of industry and recognize that these appropriate roles will change in response to future events.

The inclusion of a margin for error in planning raises the "self-fulfilling prophesy" problem. If, for example, a technical or administrative system is developed to prepare for some contingency, there is often considerable pressure to employ that system whether or not the specific contingency arises. Safeguards must be established in advance to the maximum possible extent through the definition of decision points and of the criteria that will apply at those decision point in the future; not that the criteria cannot be modified if necessary, but a base decision point and criteria must be established.

Any strategy requires the immediate initiation of some policies to resolve future difficulties that are most certain and preparation for contingencies that pose serious threats to the United States. Steps must be taken to further develop information on those problems and contingencies that are less certain and, where sufficient lead time is available, for gathering further information. This procedure takes into account the sequential nature of the decision process, whereby it is not necessary to make every decision right now. There is an opportunity to keep some options open, to develop new options, and to reduce uncertainty; however, there are also some problems and contingencies that must be dealt with or prepared for immediately.

Current policy difficulties result primarily from transition problems and equity issues that arise when policies must be reversed—for example, moving from regulation of some energy forms to deregulation and taxation. We must avoid future traps of this kind in providing an overall policy structure.

In summary, the recommended logic of a long-term strategy is based on the following sequential steps.

1. Identify and define the urgent energy system problems that we can foresee and the major contingencies for which we must prepare. These must be defined in relation to national goals and objectives as determined by the political process.
2. Specify energy system indicators that represent the status of the energy system. Characterize a prudent planning basis in terms of the energy system indicators. This procedure provides the conceptual basis for energy strategy and policy and exhibits energy targets, key policy assumptions, and implications. Policy debate can be focused on the primary issues of needs, objectives, capabilities, and means.
3. Evaluate alternative means of achieving the prudent planning basis that will address the physical elements of the nation's urgent energy system needs and contingencies.
4. Design a set of policies consistent with an overall strategy to employ political, economic, and technical capabilities to achieve the prudent planning basis. Analyze and monitor impacts of policies on energy system indicators.
5. Define backstop policies and measures with implementation or "trigger action" keyed to tracking or monitoring of key energy system indicators and comparison with previously defined objectives. Include measures for termination or phaseout of policy in response to indicators.

A successful strategy must be implemented through the effective meshing of regulatory, technological, and political measures. Regulatory policy must be designed to address specific problems in decisionmaking and institutions and to alleviate any regional or other inequities that may arise. R&D will serve to develop and deploy the needed systems, to develop new options to support future policy, and to reduce uncertainties about future issues and options.

It is recognized that the philosophical approach to energy policy outlined here emphasizes physical attributes and requirements, expressed whenever possible in quantitative terms. It must also be recognized that many nonquantifiable objectives and needs often dominate energy policy. The emphasis here on physical and quantifiable attributes is deliberate in order to ensure that these factors receive appropriate attention in the broader policy discussions.

METHODS OF STRATEGIC ANALYSIS

A variety of energy models and data bases have been developed and are available for general use. The energy models used by government are maintained by the Energy Information Administration and have been documented.[3] Models developed and in use outside of the government have been reviewed and summarized by Hoffman and Wood.[4] Most of the models described in these two compilations are deterministic in nature. The sensitivity of recommended strategy, policy, or options to uncertain events must be analyzed explicitly in individual solutions of the deterministic models.

Decision theory approaches have been applied to structure analyses and to deal with the effect of uncertainty on decision in a direct way. This approach was used by the Modeling Resources Group of the Committee on Nuclear and Alternative Energy Systems (CONAES)[5] to evaluate alternative nuclear development strategies. The uncertainties that were dealt with in that analysis are listed in Table 30-1, and the decision tree developed for the analysis is shown in a truncated form in Figure 30-1. The decision theory approach was used to calculate the expected value of the net benefits of developing alternative nuclear reactor systems and the probability distribution of those benefits given a set of subjective probability estimates regarding the uncertain events that were defined.

Clearly, the results obtained using the decision theory approach may be sensitive to subjective probabilities assigned to uncertain events. There is often considerable objection to the thought of estimating probabilities for future events, because such estimates generally use a poor information base and sometimes involve events that are inherently "unknowable." While these estimates are difficult, it is generally quite useful, and indeed essential, that they be revealed. When any individual or group makes a decision, by whatever process, there is an implicit weighting and selection among uncertainties that bear on that decision. The formal decision theory approach supports this type of decision-making by making the weightings and selection explicit and visible. The decision itself is still in the hands of the decisionmaker, but hopefully the process will be improved by a more complete definition of uncertainties and a more complete and open discussion of their likelihood and impacts.

Baseline and alternative projections of future energy use have been developed by the Energy Information Administration.[6] These provide the general information background for decisions and policy. Based on a survey of these forecasts and others developed outside the government, a number of energy issues and problems have been identified that, if real, would require policy attention at different stages in time.

Near Time Issues and Contingencies (0-5 years)

- Possible oil embargo, related to Middle East political situation may develop. This requires standby rationing policy and strategic petroleum reserve.
- Dollar devaluation will be linked to even higher level of energy imports as a major part of trade.
- Emerging problems of environmental and health nature such as sulfate emissions can be anticipated.
- Increased possibility of proliferation of nuclear weapons will be linked to growth in nuclear power.

Table 30-1. Definition of Uncertainties, Dates of Resolution, and Judgmental Probability Estimate in CONAES-MRG Decision Analysis.

1986		*1996*		*Later*	
1. Demand	H (0.29) M (0.31) L[a] (0.4)	5. Uranium resources	low (0.06) intermediate (0.17) high (0.77)	6. 6 GW solar electric in 2000	2.7 × LWR (0.78) 1.25 × LWR (0.22)
2. Coal and shale constraints	50 Quads (0.62) (0.38) None[a]				
3. Cost of nonelectric AES	\$8. (0.27) \$5. (0.55) \$2.[a] (0.18)				
4. Public acceptance of breeder	yes (0.54) no[a] (0.46)				

[a]Resulted in zero breeder benefits in analysis.

Source: National Academy of Sciences, "Energy Modeling for an Uncertain Future," Modeling Resources Group of Committee on Nuclear and Alternative Energy Systems (Washington, D.C., 1978).

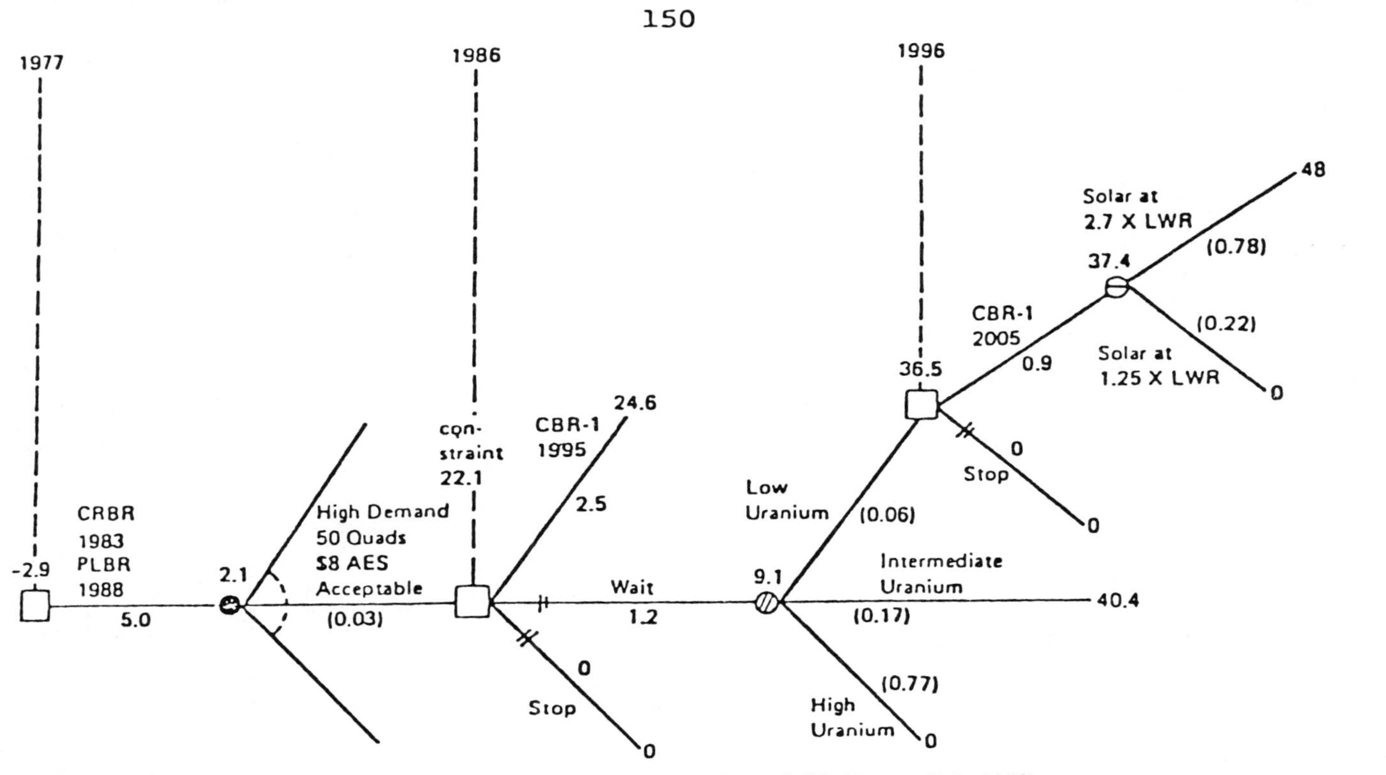

Source: National Academy of Sciences, "Energy Modeling for an Uncertain Future" (Washington, D.C., 1978).

Figure 30–1. Truncated Decision Flow Diagram Showing Benefits of Concurrent Development of Demonstration and Commercial Breeder Reactor ($billion discounted to 1975 at 6 percent per year).

- Severe weather may cause seasonal shortages of gas and abnormally high electric demand.
- Deferral of major investment decisions by industry because of economic and regulatory uncertainties will lead to increased future demand for oil.

Midterm Issues (5-15 years)

- If short-term issues listed above are not resolved, they may be anticipated to affect the midterm outlook.
- Major increases in oil prices may emerge in this time period based on resource limitations.
- Due in part to possible major increases in the price of oil, many LDCs may be expected to be unable to obtain the energy required for growth targets.
- If this situation develops abrutply, the shock to the international financial system may be so severe as to cripple growth or cause political instability in much of the world.
- New sources of oil in Mexico, offshore areas, and/or new discoveries may ease the midterm transition problems and allow longer term solutions to be effective.

Long-term Issues

- The continued growth of world population and energy demand will require aggressive development of sources other than oil and gas in this period.
- Various long lead time environmental issues may be expected to emerge. These could include climatic effects from CO_2, particulates, and other fossil fuel emissions and long-term nuclear issues such as waste disposal and plant decommissioning.

THE BASIS OF STRATEGY: URGENT ENERGY SYSTEM PROBLEMS AND CONTINGENCIES

As the basis for an energy strategy, it is necessary to describe the nature of the energy problem in clear terms. Behind any set of policies, there are generally subjective estimates of the magnitude of anticipated problems and of the likelihood of the problem arising. It is extremely important to make these estimates and descriptions as explicit as possible, so that they may be analyzed and discussed in an objective manner. This section represents a format for the definition of urgent anticipated problems that we must begin to resolve now and contingencies that are of lesser likelihood but have serious potential consequences that must be prepared for.

For the purpose of this chapter, urgent problems are defined as those events having a likelihood of occurrence of 0.5 or greater. Contingencies are defined as events having a likelihood of occurrence of less than 0.5. Some sample problems and contingencies have been developed in a complete format (although even here only in brief summary), while other areas are simply identified for further development. The categorization of an event as a contingency or urgent problem is, of course, a subjective estimate. The categorization presented here is that of the author.

URGENT PROBLEMS

1. Energy System
   a. Shortfall or disruptive price changes of liquid fuels for transportation and other uses requiring a clean convenient fuel.

      Indicators and levels of concern: Indicator is projected shortfall or price changes of domestic production, plus imports likely to be available, compared with desired use. Moderate problem exists when projected shortfall reaches one million barrels per day. Serious problem exists when projected shortfall reaches three million barrels per day.

      Time Frame: Potential shortfalls are projected reaching five million barrels per day by 1985.

      Lead time: Trends in domestic production and desired use can be projected with reliability over five year period. Availability of imports can probably be projected over five year period with some downside risk. Based on estimate that serious problem exists, action is required now.

      Corollary problems: Significant price increase (factor of 2 or 3) possible by mid-1980s.

      Options: Stringent efficiency standards in use of liquids, large-scale synthetic fuels program, large-scale exploration and development of frontier areas.

      Emergency measures: Rationing program, end use allocation.

      Ability to influence likelihood: Political action can improve stability of global supply.

   b. Scarcity or high price of secure fuels for industrial use. Need to provide secure low cost fuels to industry to provide regional stability and ability to meet overseas competition.

      Indicators and levels of concern: Indicators are extent and frequency of curtailments, price of fuel, and environmental restrictions. Risk of significant curtailment of supply in excess of once in five years represents serious problems.

      Time frame: Problem exists now for gas supply, possibility of similar problem in the 1980–1985 time frame for electricity. Further deterioration of situation expected.

      Lead time: Two to five year warning possible, immediate action required.

      Options: Interfuel substitution (coal and/or electrification).

      Emergency measures: End use allocation of scarce fuels.

   c. Increased vulnerability of urban areas to supply scarcity and higher prices. Need for appropriate technologies for urban areas based on abundant energy forms and stringent environmental control.

Time frame: Vulnerability is large now, will increase over period to 1985 in view of limited options from environmental viewpoint.

Options: Incorporate new energy supply infrastructure (cogeneration, district heat) in conjunction with urban reconstruction over next ten to fifteen years.

Emergency measures: End use allocation, rationing, load shedding.

d. Need for increased range of siting options regarding appropriate scale technology (decentralized options).

e. Vulnerability of energy system to periodic supply or demand perturbations (strikes, bad weather, etc.). Increased storage of fuels and flexibility to use alternate sources is needed.

2. Decision making

a. An atmosphere conducive to open debate of issues; balancing of economic, security, and environmental objectives; and effective regional and national communication is urgently needed. Issues should be defined as far in advance of crisis period as possible, allowing time for research and analysis.

3. International Problems

a. Need for increased cooperation and collaboration with oil producers, consumers, and developing countries.

b. Global effects of policies must be ascertained in advance to maintain spirit of international collaboration on major problems.

4. Special Regional Needs

a. The heavy reliance upon oil existing in the Northeast, and thus its susceptibility to economic and social disruption, mandates that emphasis be placed upon conservation and substitution of alternate fuels.

b. The urban problem has been discussed previously, but the problem of supplying clean fuels at reasonable cost to urban areas bears repeating as a special regional problem.

c. Storage capability at various regional sites is needed. Prudent planning of energy requirements, regional storage capabilities, and an assessment of the delivery infrastructure can serve to provide a system more resilient to disruption. Policy should be implemented that can smooth out perturbations during periods of fuel shortage or emergency.

d. A study of the delivery infrastructure is a necessary part of a long-term strategy. Will the delivery system serve the demographic patterns projected in the future; which railroads need improving to carry the vast coal shipments envisioned, and where are new delivery systems required?

CONTINGENCIES

1. Embargo of oil imports

Indicators and levels of concern: Indicators are total fuel imports, fraction of imports from any single political bloc. Serious problem exists if reduction of 40 percent of import or 20 percent of total fuel usage is realized.

Time frame: Problem exists over full period of significant imports. Likelihood will increase over time related to political developments and vulnerability of consumers.

Lead time: Embargo may be effected with advance warning of two to six months. Sufficient fuel is in "pipeline" for three to four months.

Options: Storage, political action, military action.

Emergency measures: Rationing program, end use allocation.

Ability to influence likelihood: Political action, vulnerability to embargo affects likelihood.

2. Nuclear disruption related to incidents of sabotage, reactor accident, waste or reprocessing accident. United States is sensitive to incident occurring anywhere in the world.
3. CO_2 crisis. Could require, for example, immediate leveling off of fossil fuel use as soon as crisis is apparent, followed by 10 percent annual use reduction beginning three to five years later.
4. Serious exacerbation of trade deficit. If current deficit level continues or deteriorates over three year period and energy accounts for significant fraction of deficit, immediate emergency situation will exist.
5. Serious financial disruption of world economy or of major friendly nation related to energy (scarcity or high prices).
6. Significant increase in leisure time desired by public.
7. Adverse political development in major exporting country.
8. Increased regional independence with reduced interregional cooperation.
9. Major oil discovery in United States or major technological innovation that would resolve United States problem.
10. Major oil discovery in other than United States or major technological innovation that resolves the global energy problem.

POLICY OPTIONS AVAILABLE TO IMPLEMENT STRATEGY

A nation has a variety of policy options at its disposal to address the urgent energy problems and contingencies that it faces. The problems and contingencies identified previously may be characterized by inadequate system performance represented by one or more of the general categories of energy system indicators: (1) material and social development, (2) environmental quality, (3) efficiency, (4) security, and/or (5) equity.

A common denominator of most policies is their effect on energy supply and demand. Policies may be grouped according to whether they affect supply or demand and their degree of intensity. Using this approach, a policy space can be identified as in Figure 30-2 according to whether policies affect supply or demand and their degree of intensity. Different cells in the policy matrix

ENERGY POLICY TAXONOMY

		Demand-Oriented				
		Strong Efficiency Improvement	Moderate Efficiency Improvement	Neutral	Moderate Efficiency Discouragement	Strong Efficiency Discouragement
Supply-Oriented	Strong Incentive					
	Moderate Incentive					
	Neutral					
	Moderate Disincentive					
	Strong Disincentive					

Figure 30–2. Policy Space in Terms of Impacts on Energy Supply and Demand.

represent combinations of supply and demand policies. Energy is likely to be in surplus in the upper left corner combinations and is constrained in the lower right. Some policies may not be readily classified because their effects are hard to predict—for example, horizontal divestiture. Other policies may be difficult to classify because they represent preference within supply or demand categories—for example, a tax on fossil fuels but not on renewables.

Each step to a more intensive supply or demand policy represents a level of escalation in the effort to affect directly either supply or demand. Note that preferences among alternative approaches to supply- and demand-oriented options exist on the basis of other social objectives, as discussed later in the chapter. Although the policy option space has been characterized in terms of impact on supply and demand, it is also necessary to estimate the impact of policy option on the critical energy system indicators (development, environmental quality, efficiency, security, and equity). Comprehensive analysis is required to estimate these impacts.

To illustrate the approach toward characterizing levels of policy escalation, the following are offered as examples of each type of policy. This list is illustrative only and is not exhaustive.

Strong Supply Encouragement

Eminent domain for energy facilities siting
Return equalization taxes to producers
Suspension of environmental regulations
Subsidies for accelerated developments
Decontrol of crude oil prices
Decontrol of natural gas prices
Subsidies for commercial prototype supply plants
New coal leasing program
Sale of entitlements to import oil

Moderate Supply Encouragement

Accelerated OCS leasing
Renew Price-Anderson Act
Accelerated nuclear licensing
Higher (regulated) natural gas prices
Major RD&D on supply technologies
Full production of NPR1-3, development of NPR 4
Tax incentives for secondary and tertiary recovery

Moderate Supply Disincentive

Regulation of natural gas prices
Clean Air Act limitations on expansion of coal assumption

Strong Supply Disincentive

Level or lower (regulated) natural gas prices

Strong Efficiency Improvement
Federal reconstruction of older urban areas to increase efficiency

Moderate Efficiency Improvement
Marginal cost pricing
Regulations on consumer items requiring best available conservation equipment (BACE)
Crude oil equalization tax
Investment tax credit for energy-saving equipment
Extension of tax credits to building owners who improve efficiency
Federal action to reduce institutional obstacles to cogeneration
Extention of automobile efficiency standards

Moderate Efficiency Disincentive
Oil depletion allowance
Low regulated natural gas prices

Strong Efficiency Discouragement
Extension of natural gas regulation to intrastate markets and to other fuels

ANALYTIC BASIS FOR STRATEGY ANALYSIS

The effectiveness of any particular set of policies may be monitored and measured in terms of impacts on the physical energy system. The configuration of the energy supply system and the energy utilization equipment stock is a function of energy policy and a variety of other factors beyond the influence of public or private policy, such as geology and demography.

The success of any particular strategy of energy supply and utilization must be in terms of how successfully it serves energy-related social and economic objectives. Those that are amenable to quantification may be grouped into energy system indicators within the general categories of societal objectives.

The ultimate determination of energy strategy and policies that balance security, environmental, and economic objectives with appropriate regional equity is arrived at through the political process. It is essential that this process be supported by the best available base of information and analysis of the consequences of alternative actions.

Analysis of strategy and policy options must be done in the context of sequential decisions under uncertainty in response to the evaluation of energy system indicators. Issues and alternatives must be analyzed in a consistent and comparable manner using a variety of methods and capabilities.

A measurable set of performance criteria that characterize objectives must be developed. In this chapter they are referred to as energy system indicators—for example, oil imports, regional emissions and air quality, and regional employment as a partial measure of allocative efficiency. Equity must be considered in all analyses. In the final analysis, however, the issue of equity is not solely a DOE function, but must be dealt with in the context of overall federal policies.

Energy System Indicators

Material and Social Objectives. Energy system indicators that represent the level of energy services that are delivered and satisfied by the energy system are listed below. These can be related through analysis to macroeconomic forecasts including consideration of social and demographic trends and lifestyle considerations.

- Transportation: Passenger miles by mode of travel; freight ton miles by mode of travel.
- Industry: Physical output and/or output measured in physical and monetary terms. This needs to capture both the quantity and quality of basic materials and manufactured products.
- Residential-Commercial: Degree day–square feet of heating; degree-day–square feet of cooling; gallons of hot water; fuel and electricity per commercial square feet or per household for miscellaneous appliances and services.

Security. To a first approximation, security can be measured in terms of import levels on an annual and seasonal basis: imports of oil; imports of gas; imports of electricty. At a greater level of sophistication, we would measure only the levels of imports not covered by the strategic petroleum reserve or other backup supplies and the distribution of those imports among various countries.

Allocation Efficiency. The principal indexes of allocative efficiency are prices. Provided they include appropriate externalities but no subsidies, the lower that prices are, the greater the system efficiency in economic terms. One may wish to measure both delivered fuel prices and the price of energy services as measured by the material and social indicators.

Environmental Quality. A first approximation of environmental quality is contained in an air and water emission and land use vector reported at the regional level. This is only a first approximation, as we are ultimately interested in morbidity and mortality impacts that require the sequences of analyses shown in Figure 30-3.

Emission ⟶ (Transport Phenomena) ⟶ Ambient Concentrations
⟶ (Damage Functions) ⟶ Morbidity and Mortality

Figure 30-3. Sequences of Analyses for Morbidity and Mortality Impacts

Presently our knowledge of transport phenomena and damage functions is limited, so at present we use the emissions vector as a proxy for morbidity and mortality. One may wish to weigh the various emissions relative to one another in the performance of multiobjective analyses. Increased understanding of transport phenomena and damage functions represent key research needs to improve the integration of health effects and other environmental analysis into energy policy analysis.

Equity. Equity is probably the most difficult of the objectives to treat quantitatively. At the macrolevel, income distribution is conventionally described by the Lorenz curve, Figure 30-4, plotting percentage of income versus percentage of families. The degree to which actual income distribution is skewed from equality is expressed by the Gini coefficient, which characterizes the mathematical formulation of the Lorenz curve. In principle, energy policy impacts on equity could be measured by changes in the Gini coefficient. In practice, we may not be able to make meaningful quantitative estimates of these impacts.

Other dimensions of equity that may be important are the interregional and intergenerational dimensions. Typical consumer units (families) may be identified as standard units for the measurement of impacts.

Energy System Configuration. The physical stocks and flows in various points of the energy system represent the starting point for indicators of energy system configuration. The energy system has been represented in many regional, national, and international studies in the format of the Reference Energy System (RES), shown in Figure 30-5. The RES shows the sequence of activities required

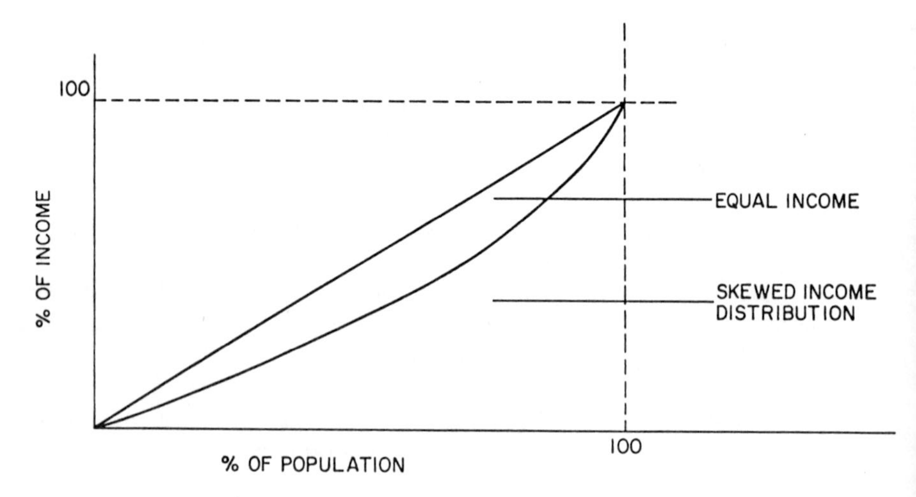

Figure 30-4. Lorenz Curve of Income Distribution.

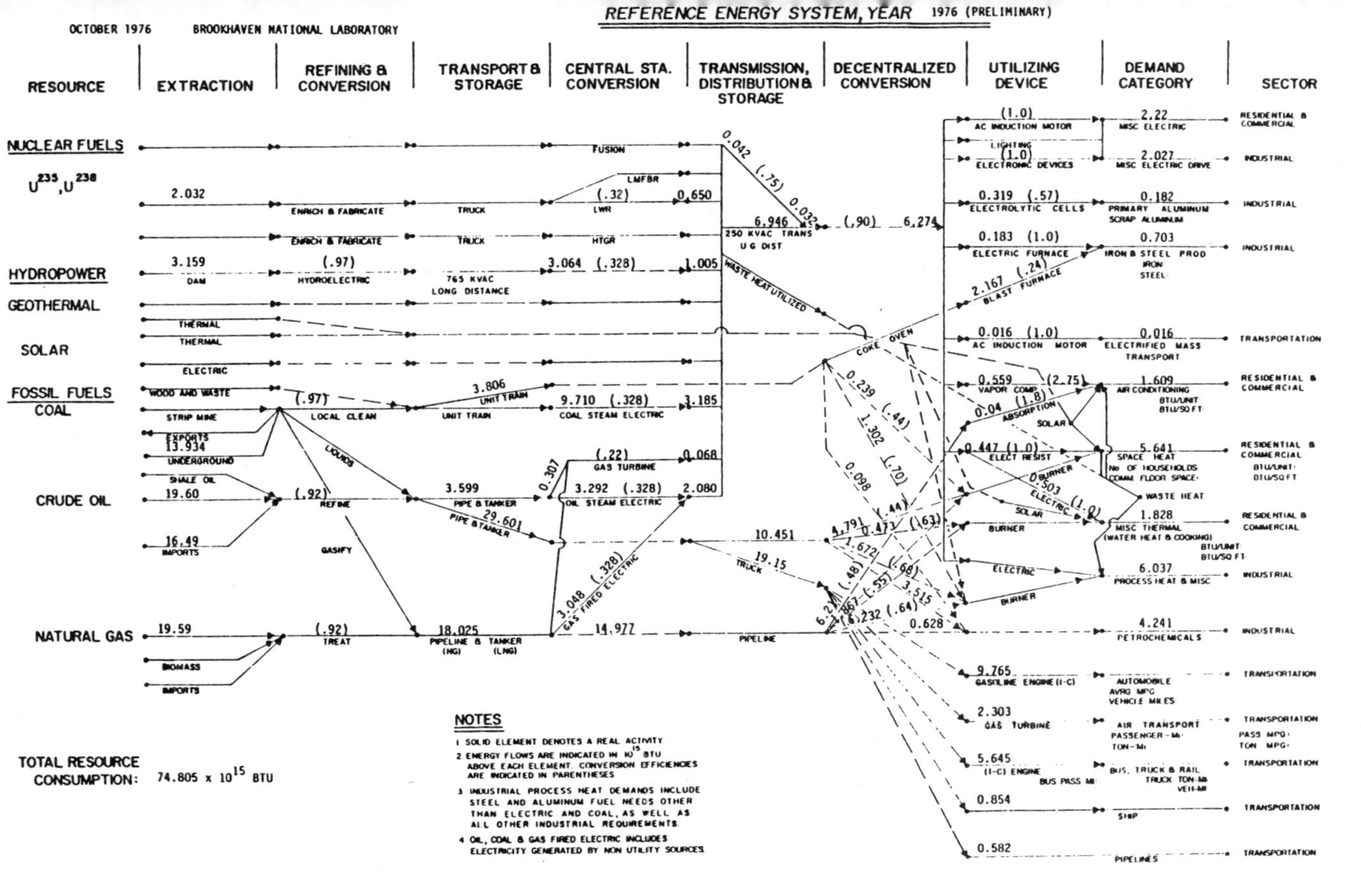

Figure 30-5. Reference Energy System, Year 1976 (preliminary).

to employ resources, convert them to useable secondary energy forms, and utilize them to deliver the energy services identified in the first group of material and social objective indicators. The physical stock and flow indicators shown in the Reference Energy System may be augmented by other indicators of the institutional, financial, and environmental characteristics of the elements of the energy system.

Based on measurements or perceptions of success or failure in these areas the political process reevaluates policies in order to find a more satisfactory pattern of energy system operation. This sequence is shown in Figure 30-6. The indicators are used as a basis for identification of urgent system needs in the following context:

1. Economic System. Energy must be considered in a general equilibrium framework in a way that long-term transition effects can be estimated along with short-run impacts.
2. Energy System. The physical nature of the energy system must be represented with balanced treatment of the supply and end use sectors.
3. Energy Processes. The characteristics of industrial processes and technology must be described. Tradeoff possibilities between cost, efficiency, and emissions must be represented. Analysis must include existing technologies as well as those under development.
4. Resources. The energy and material resources as well as the labor resources on which the energy system is based must be defined.

The major results of analysis must be the estimation of the quantitative and qualitative impact of strategy recommendations on the energy system

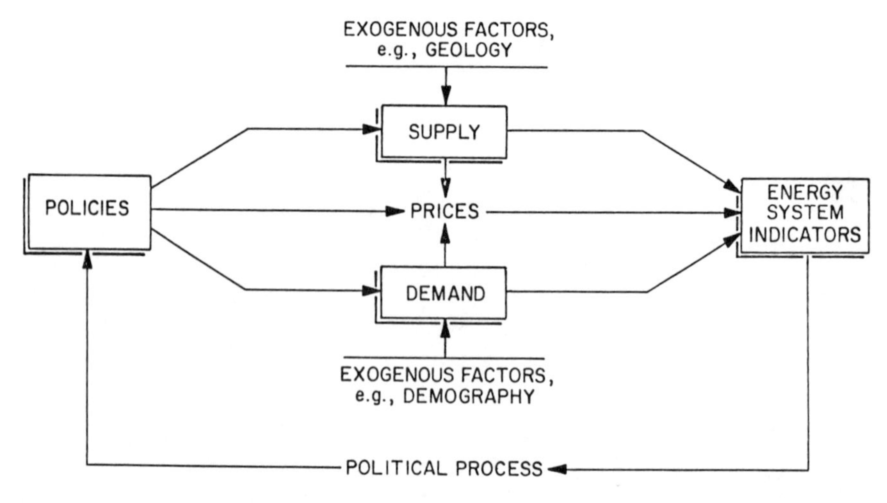

Figure 30-6. Logic of Energy Analysis to Support the Political Process.

problems and contingencies. The objectives of the government policies adopted as part of a comprehensive strategy may be summarized in three categories:

- The gathering of additional information on future uncertainties to reduce the uncertainties faced by consumers and producers;
- The reduction of the likelihood of future events identified as problems or contingencies; and
- Preparations to ameliorate the effects of problems or contingencies should they occur.

LONG-TERM STRATEGY

Long-term strategy recommendations must be responsive to the energy needs of urban, suburban, and rural consumers; industries with severe overseas competition; transportation needs of the nation; national security; and protection of the environment. One possible set of recommendations would comprise a moderate efficiency improvement–moderate supply incentive set of policies in terms of the energy policy space described previously. Major reasons for holding the supply and demand incentives at moderate levels involve the transition problem from previous demand policies that are best characterized as moderate efficiency discouragement. In view of the contingencies that must be prepared for and the significant damage to the nation's economy that could result, preparatory measures would also be needed to lay the basis for escalation to strong efficiency improvement and/or strong supply incentive policies should future circumstances as measured by the energy system indicators dictate those actions.

Planning Basis

The uncertainties of future economic growth and world trade make it difficult to relate future energy requirements directly to these factors. It is important that the planning basis strategy provide for sufficient supply and conservation so that energy does not act as a constraint on economic growth and other social policies. Indeed, energy strategy should support social objectives.

As an alternative to specific objectives expressed in terms of economic growth, a basis for prudent planning could provide for social upward mobility such that every citizen of the United States could have access to a level of energy services (square feet of living space, miles of travel, etc.) equal to, for example, that of the average of the upper quartile income group of 1978. Similar upward mobility goals on a world scale, but with different targets, could serve as the basis for U.S. cooperation with an assistance to other nations.

It might also be appropriate to establish "Principles of Regional and Sectoral Equity" as a basis of national strategy. All options should be evaluated for impact on census regions—urban, suburban and rural areas, as well as across income groups.

The planning basis for long-term strategy would be based on several levels of activism or intervention in markets. Discrete levels of activism or intervention might involve the following:

1. Laissez-Faire. Allow markets to function with minimal regulation and control. Regulate and correct serious market failures where they occur and where action is dictated by political process.
2. Selective Intervention (market oriented). Adopt a more active stance in identification and resolution of market imperfections (environmental externalities, competition, conservation, pricing of resources, capital formation) and of potential longer range problems, but still rely on market solutions.
3. Active Intervention. Define potential solutions to near and long-term energy problems and develop technological and policy options to correct those. Still rely on market to choose among options and initiate them. Continue to identify and resolve market imperfections.
4. Government Control. Develop options to solve problems and actively force implementation, either through market system or by government.

Current energy policies are somewhere between Steps 2 and 3. It is clear that circumstances would be defined that would lead to reduced government involvement (Step 1) such as the discovery of abundant, low cost, clean resources in the United States (possibly a gas bonanza?). Also, a national emergency could dictate escalation of policy to Step 4, as in wartime economy.

Supply Strategy Option

A triad supply strategy, based on solar energy, coal and unconventional fossil, and nuclear power seems to be most reasonable as a guide to long-term R&D policy. There are several unique and appropriate applications of these energy sources that must be encouraged—for example, solar for water and space heating, liquids from coal and biomass for transportation, nuclear power for electricity. A possible long-term prudent planning basis for the triad supply strategy could be delivery of, say 150×10^{15} Btu of energy (in primary resource equivalents) equally divided among the three elements. About two-thirds of this energy could come from centralized sources and one-third from decentralized sources. A possible mix of resources and supply options is shown in Figure 30-7 to implement such a strategy. Specific initiatives would be needed to demonstrate, in the 1985-1990 time frame, the major technological elements of the long-term strategy. For example:

- Synthetic liquid fuels demonstration plants with capacity of 100,000 bbl/day, in each of the nine census regions (using shale, biomass, coal, and so forth and feedstock as appropriate to region and in markets appropriate to region);
- Combined electric-gas plant based on geopressured gas;

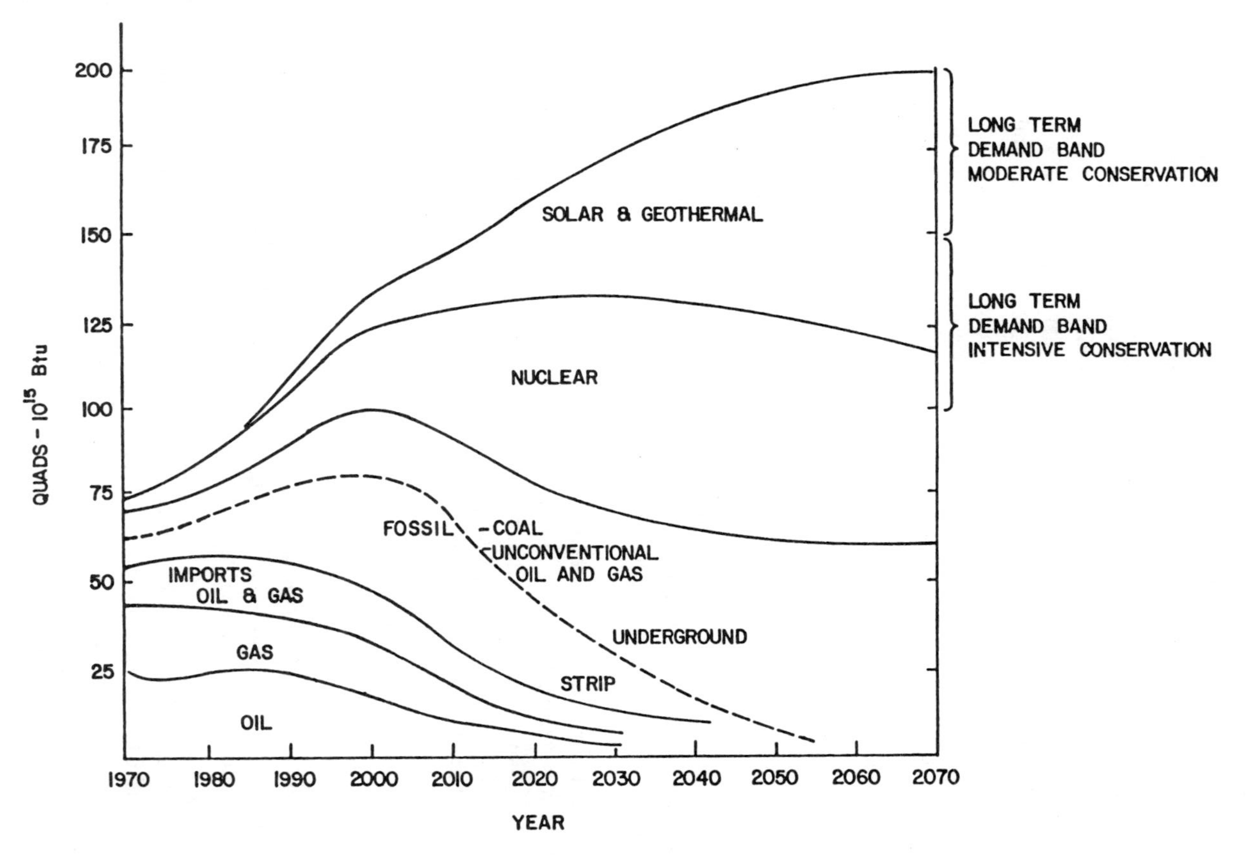

Figure 30-7. Long-range Energy Supply Strategy.

- Advanced coal-based total energy conversion systems suitable for urban siting;
- Variety of solar conversion plants including wood and agricultural wastes;
- Closed nuclear fuel cycle and waste disposal with maximum safeguards.

Attention would be directed toward the complete energy delivery system, involving technology appropriate to urban, suburban, and/or rural areas.

Consistent with such a strategy, research and development policy over this time period would aim toward the development of base technology and a wide range of options having significantly improved technical, economic, and environmental characteristics compared with those systems to be demonstrated. Other options include:

1. A specific option consistent with the supply strategy, including expansion of coal production for purpose of increasing exports.
2. Establishment of a program for the sale at auction of entitlements to import oil. This will provide an economic incentive to reduce imports and increase domestic supply.
3. Preparatory action to implement stronger supply- and demand-oriented policies, should future trends of energy system indicators dictate that action. Included in this preparatory action would be:

    a. Identification and dedication of national energy reserve sites in each census region consisting of sites with adequate water, transportation, and the like that could be utilized for energy supply facilities in the event that strong supply policy is needed. At time of any future determination to utilize these sites, they would be leased to utilities and/or private industry for further development and construction.
    b. Government incentives provided to upgrade energy transportation facilities and storage facilities.

End Use and Fuel Substitution Option

A moderate conservation policy might consist of the following new elements added to the National Energy Plan.

1. Definition of best available conservation equipment (BACE) technology to be established as efficiency standard for utilizing devices over a seven year period. The definition of best available conservation equipment should be based on achieving maximum energy efficiency with no significant increase in life cycle cost and no significant deterioration in the level of service delivered by the device.
2. Marginal cost pricing of oil and gas with appropriate return of excess revenues to suppliers and consumers.

3. Regulations for building, processes, and so forth to ensure life cycle optimal investment decisions.
4. Encouragement to energy supply industries to market fuels and power to appropriate end uses as indicated by analyses.
5. Research and development through both incentives and federal funding for new series of end use devices that promise significant efficiency improvement compared with currently defined best available conservation technology.

It is important to increase the capability for interfuel substitution in the future. Devices in residential, commercial, and industrial service must have improved capability for multifuel capability and for manual operation. Standards for these devices must incorporate requirements for these features in order to make the future energy system more adaptable to a variety of energy sources and more reliable in the event of fuel or power shortages.

Preparatory action should be taken in the following areas to prepare for further conservation measures should they be required. Items 2-5 cover the basic issues, while items 6-9 detail some policy traps and unanswered questions.

1. The U.S. Department of Energy should estimate and publish forecasts of future average and marginal energy prices under specific resource and technological assumptions. Recommendations for future regulatory, tax, and rationing programs that would be triggered by trends in key energy system indicators should be developed (examples of pertinent indicators that would trigger action are oil imports of more than 50 percent of total consumption, fraction of balance of payment deficit attributable to oil imports in excess of 0.5, electric generation reserve margins of less than 10 percent, reserve-to-production ratio for domestic oil and gas below 10).
2. Government energy offices, state and federal, should identify and raise public awareness of the five most urgent national and regional energy needs anticipated over the next twenty-year period.
3. Government should identify and raise public awareness of the five most serious energy-related contingencies that the United States must prepare for in its energy policy.
4. The nation's research capabilities must be motivated to address the critical social, political, technical, and economic issues that complicate the nation's energy situation. A complete doctrine of energy strategy must be developed to guide future policy in a way that is responsive, effective, and adaptive to changing circumstances.
5. Government should improve coordination of energy policy with other policy areas that affect the energy system. Using the analytic base defined previously, all federal policies must be evaluated in terms of their impact on key energy system indicators by means of an energy impact assessment.

6. Some propose that high frontend cost options are desirable because future inflation will wipe out the incurred debt. To the extent that inflation is driven by the need to offset large debt commitment, such deliberate actions will further strengthen inflationary forces.
7. Some propose that we need not worry about a world oil crisis because lower economic growth will hold down the demand for oil. This rosy view attempts to paint the problem as a solution.
8. Be careful in the use of GNP and employment impacts as a basis of policy. When analyzed in terms of a less than fully employed economy, the least efficient project can have the greatest positive impact. Comparative analyses should be done in a context of full employment to seek the most efficient policy. Consider policies that address specific problems–for example, urban unemployment–separately.
9. What are the relationships between energy policy and other policy areas such as economics, transportation, defense, trade, and so forth? Instances where energy policy may be used as an instrument of social change should be identified so that this aspect of energy policy is explicit–for example, energy stamps, lifeline rates, rationing, load shedding on selective basis, and price regulation in general.

NOTES

1. Webster's Third New Internaional Dictionary Unabridged (Springfield, Mass.: C.C. Merriam Company, 1966).

2. Ibid.

3. U.S. Department of Energy, Energy Information Administration, *Models of the Energy Information Administration* (Washington, D.C., May 1978).

4. K.C. Hoffman and D.O. Wood, "Energy System Modeling and Forecasting," *Annual Review of Energy*, vol. 1 (Palo Alto: Annual Review, Inc., 1976).

5. National Academy of Sciences, Energy Modeling for an Uncertain Future," Modeling Resources Group of Committee on Nuclear and Alternative Energy Systems (Washington, D.C., 1978).

6. U.S. Department of Energy, Energy Information Administration, *Annual Report to Congress*, DOE/EIA-0036/2 (Washington, D.C., 1978).

Chapter 31

Energy Decisions

Thomas B. Neville
U.S. Department of Energy

The Department of Energy faces a wide range of options from which to select energy policies. The process by which the department develops and evaluates options is made up of a number of steps. These steps, each of which is discussed below, include:

- Employing a decision criterion,
- Justifying government actions,
- Developing options,
- Identifying relevant considerations,
- Analyzing the benefits and costs of the economic considerations, and
- Making a decision—the balancing process.

EMPLOYING A DECISION CRITERION

Deciding whether the government should adopt any particular energy policy option requires an appropriate decision criterion: Is the nation better off with or without that particular energy policy option? Deciding whether the nation will be better off with than without a particular policy option requires a careful evaluation of the benefits and the costs of that option and its comparison with other policy options.

JUSTIFYING GOVERNMENT ACTIONS

Government actions may be generally justified when the private sector left to itself either does not or cannot act to realize the greatest net benefits potentially available to the nation as a whole. Likewise, energy policy options that require government action may be justified when the private sector does not or cannot act to realize the greatest net energy benefits potentially available to the nation. Since the nation is a larger importer of oil, there are a number of categories of government actions that can realize benefits that will not be obtained by the private sector alone. Several examples are discussed below.

First, the government can adopt policy options to reduce imports that are desirable for the nation as a whole but are uneconomic for any single producer or consumer. Any increases in reduction in imports tends to mitigate world oil prices. All oil importers benefit from a decrease in the price of imports, not just the producers or consumers who cause the reduction; yet no individual producer or consumer has any incentive to consider this benefit to the nation of additional production or conservation. Therefore, government actions to reduce imports may be justified by the benefit to the nation from the mitigation of world oil prices. Policy options should be evaluated by using the world oil price plus a premium that reflects the benefit of a reduction in imports.

Second, the government can develop new technologies earlier than the private sector would. To the extent that OPEC and other oil producers base production decisions or their expectations about the future value of oil, RD&D can influence their decisions. For if these producers believe that new technologies may limit the value of their oil in the future, they have less incentive to delay production. Oil prices will tend to be mitigated if more oil is produced rather than saved for the future. But private sector producers and consumers do not capture the benefit to the nation of downward pressure on world oil prices caused by successful RD&D. In fact, the mitigation of world oil prices delays the date at which new technologies will become profitable. In other words, some of the potential gain to the nation represents a potential loss to the firm undertaking the RD&D.

Third, the government can also act to minimize the likelihood, extent, or duration of an oil import disruption. Private firms have little incentive to take into account the deterrent benefit of reducing imports or of storing oil.

DEVELOPING OPTIONS

Most energy policy options are designed to deal with two principal economic aspects of the energy problem.

- World oil prices are high today, and future prices are uncertain but may increase substantially over time.
- Oil imports from insecure sources pose the continuing threat of import disruptions.

Previous chapters have discussed areas of concern that provide targets of opportunity for energy policy options. For example, conservation or increased domestic oil production can reduce oil imports and their associated costs; and storing large amounts of oil can minimize the effects and, probably, the likelihood of import disruptions.

Once a target of opportunity is identified, the appropriate instruments for achieving desired outcomes must be considered. The relevant question is whether the government should undertake direct action or rely on the private sector (to some extent) to achieve the desired outcomes. More precisely, the question usually turns out to be how the government can induce the private sector to undertake the desired action. The appropriate instruments range from government purchases and regulations to incentives such as tax credits.

IDENTIFYING RELEVANT CONSIDERATIONS

The consequences of a particular policy option must be identified in order to evaluate that option. These are of two basic kinds. First, there are the energy outcomes that are related to the principal economic aspects of the energy problem. Second, there are the associated nonenergy outcomes, which are also important to the nation.

The energy outcomes relate to two principal economic aspects of the energy problem. First, the energy outcomes of a particular policy option must be measured in economic terms. These are the benefits and costs for the economy of the nation as a whole. Second, the benefits of protecting the nation from import disruptions and in deterring future import disruptions must be identified. In short, the appropriate questions are how the specific instrument achieves the desired outcomes and how these outcomes relate to the principal economic aspects of the energy problem.

Associated nonenergy outcomes that may affect the choice of energy policy options include the protection of the environment and equity. Since some energy policy options have unfavorable environmental outcomes, these outcomes should be considered in the evaluation. In addition, any consideration of equity—that is, who gains and who loses—may substantially affect the choice of energy policy options and also should be involved in an evaluation of energy policy options.

ANALYZING THE BENEFITS AND COSTS OF THE ECONOMIC CONSIDERATIONS

The evaluation of benefits and costs of an energy policy option must balance energy and associated nonenergy outcomes. It is important in this evaluation to make all of the outcomes explicit and to quantify as many of them as possible. It is particularly important to quantify the economic benefits and costs of an energy policy option. The evaluation of economic benefits and costs should address three questions.

First, have all the relevant options been considered and evaluated? For example, have the likely private sector actions in the absence of an energy policy option been considered correctly? Or has an incentive program, as well as a regulatory program, been considered?

Second, have the benefits and costs of implementation been fully considered? On the one hand, many government programs often fall short of achieving all of the benefits promised by optimistic projections. On the other hand, they involve administrative costs both to the government and to the private sector.

Third, do the benefits exceed the costs when uncertainty is considered? The energy future is uncertain. Estimates of future oil prices may span a wide range, and the costs of most new technologies cannot be reliably assessed at this time. The environmental impacts of current and proposed technologies are usually not well known, and the regulations that govern them may change over time.

These benefits and costs should be evaluated over a wide range of uncertainty, particularly over a wide range of oil prices, since the best policy option at one oil price may be a very bad option at another oil price. As a result, the most desirable policy options are often the ones that are satisfactory over a wide range of uncertainty.

The most desirable policy options are the ones that can be adapted to new information about oil prices and other uncertainties. Such policy options are often like insurance; they have a very high payoff only at high oil prices and no payoff at low oil prices.

RD&D will have a high payoff if the new technologies and the energy produced by them turn out to be low in cost compared to future oil prices. Conversely, RD&D will have no payoff if the new technologies and the energy produced by them turn out to be high in cost compared to future oil prices. RD&D policy options can increase the likelihood of high payoffs by pursuing several new technologies simultaneously to ascertain which are most economically promising. This approach is adaptive because later decisions can defer or stop the development of new technologies or the production of energy if either is relatively high in cost.

MAKING A DECISION—THE BALANCING PROCESS

The energy policy options easiest to choose are the ones that have net positive economic benefits over a wide range of uncertainty and do not have any undesirable environmental or equity consequences. However, almost all energy policy options to involve important environmental and equity considerations that must be involved in making a decision. In these instances, making a decision requires the careful balancing of the economic and other considerations involved. The steps in the process of evaluating energy policy options are designed to make all of these considerations explicit and to quantify as many of them as possible.

 *Chapter 32*

Long-term Research and Development Policy

Robert W. Fri

The topic of long-term research and development policy is not an easy one. Certainly it is too broad to capture, or perhaps even to define, in a short time. And, occupying the last position on the last panel of the last session of a five day conference, it is entirely possible that anything I have to say you may have already been said. So, both to contain the topic and to guard against massive repetition, I intend to speak as a practitioner of policymaking and to discuss what seemed to me relevant when I had to make decisions on long-term research and development policy.

Most relevant perhaps to the policymaker is to acknowledge what he does not know. We cannot, for example, predict the future—surely not for the long term and probably not very well for the short term, either. A few years ago, few of us thought that energy demand in the United States would grow at a rate of much less than 3 or 4 percent per year. We were skeptical of forecasts of 2 percent annual growth, although we realized something like that rate was highly desirable. Yet it turns out that energy demand grew at only a 0.6 percent rate from 1973 to early 1978—a fact that should humble us when we try to foresee even longer range trends.

Similarly, we need to be careful about our understanding of the options for the future. By definition, we do not know all the good options, or even the most promising paths for research. Contrary to earlier views, for example, coal gasification now looks more like a good way to use coal and a poor way of making more gas; in only a few years, more promising sources of natural gas have appeared or at least have sunk into our thinking. Longer term, who knows? I spent a day last week with an inventor who works in the chemistry of atomic hydrogen and who, if he is right, could open an entirely new line of energy development.

We deal, in other words, in uncertainty, and the wrong basis for long-term research and development policy is to suppose that we now know what options to pursue toward an already decided future. But the cause is not lost, for we can characterize the dynamics of the setting to which policy must respond and some of the common sense imperatives with which sound policy must abide. These are the topics I want to discuss, together with a few notes on the substance of long-term research and development policy.

POLICY DYNAMICS

Energy scientific and technical activities move along a spectrum from basic research to the deployment of new forms of energy consumption and production in the real world. The end result of these activities may be hardware or social change, but in either case the product has gone through the winnowing of possibility, practicality, and acceptability in the eyes of a user, not a researcher. And so as we come closer to this product, research and development policy deals increasingly with the tangible, the observable, and the desirable. Nearer term policy thus seems easy to make, if hard to carry out. We need not comment here on this illusion.

Longer term policy responds to other forces, and it is useful to understand the dynamics with which such policy must deal, for those forces are no less real and powerful than those that shape the more immediate end of the research spectrum. There seem to me to be four such forces.

First, long-term research policy must incorporate social and political goals, for such goals are often attainable only in the longer term. In the nearer future, we have little choice but to burn coal, employ nuclear power, and attempt conservation in a fundamentally wasteful energy system. We may not like the resulting pattern of environmental damage, expensive technology, and excessive demand, but it is ours for the time being.

Longer term, however, we have choice in these matters. We can opt for less environmental damage, for more dispersed power generation, for both technological and social change that restructures energy demand, for higher or lower fractions of our wealth spent on energy, and for more or less economic development. We can opt for these things or, more likely, our children can. Our duty, in shaping long-term research and development policy, is at a minimum to ensure that future generations have not only the technological options, but also the social and economic options, from which they can choose their own life. The only error more devastating than failing to serve up the full range of choice would be consciously to foreclose future social and economic choices by making up our technological minds today.

The second force at work in long-range research and development is that attention focuses on the resource base rather than the product. Like the first, this aspect of policy arises from the flexibility of the longer time horizon.

When dealing with more immediate technology, it is crucial to produce energy in a form that existing markets and infrastructure can absorb. This important constraint requires that we worry about keeping gas pipelines full, having liquid fuels for transportation, cleaning up pollution rather than not creating it, and the like.

With more time before us, however, we can think of the resources we would like to use for the long haul. The desire to move to inexhaustible or renewable resources is an obvious consideration, and it is equally clear that fusion, solar, and the breeder represent different outcomes. Less obvious, perhaps, is how this force shapes our thinking about somewhat closer in technologies. For example, gasification is probably a cleaner and cheaper way to use coal then electrification. Yet it is only with a longer term perspective that we would dare now to pursue this technology and the related problem of substituting gas for electricity in our energy system.

Third, long-range energy planning must maintain continuity with, and feedback to, nearer term activities. Future energy technologies cannot, in other words, be a discontinuity in some future world. Fine new ways to generate electricity will be of little use if our automobiles still run on liquid fuels.

To avoid such discontinuities—a polite name for a technological white elephant—requires feedback between present directions and future possibilities. Near term policy must be nudged toward future options; electric vehicles should be tried out now if an all electric future looks promising. And long-term options need to be rethought if the world persists in moving in other directions. Thus, if no one seems to like electric cars, options that produce only electricity need to be scaled back.

This feedback is a constraint on the social, economic, and resource breadth otherwise inherent in long-term research and development planning. Some desirable energy options require changes in markets and other social behavior and, as already noted, the full range of choice should be available. But when presented with choice, people choose; and when people choose, the planner's options shrink. A wise planner will try to test his nontechnological options early and to abide by the results.

Last, even long-range energy research has time limits. There is an earliest time before which no long-term technology can be deployed. More important, there is a latest time after which our descendants will have selected an energy system. For example, it is altogether conceivable that, in the absence of timely alternatives, we could be locked into a high cost, coal-based energy economy. And because the coal resource is very large, the incentive to move to solar or fusion would be correspondingly small.

I would judge the window for new energy systems to close in the next fifty or so years. Lest this seem like a long time, however, remember that no option—technological, social, economic, or resource—is viable unless it is widely deployed by that time. Even fifty years is precious little time to deploy fusion, alter our transportation systems, or turnover our housing stock—all to use technology that is only now being invented. Time disciplines the long term as firmly, but with dangerously less urgency, as the short.

POLICY IMPERATIVES

The dynamics of energy policy guarantee that the content of long-range energy research and development will shift over time. But at any given time, the policy-maker must cope with a few hard nuts, and they always seem to be the same ones. It may be useful to enumerate these imperatives of policy, the tough choices that seem to abide.

First, the long-range research program requires stability. There are always budget stringencies and immediate demands for funds from near term projects. Programs with less operational content always seem good places to seek money. Yet it is just these long-range programs that suffer most from yo-yo budgeting, and resisting this procedure is among the decisionmaker's hardest tasks.

Second, there is a limit on the cost of energy from new production or conservation systems, and it is pointless to pursue approaches likely to cost more. Such a limit clearly exists, although there are two ways to get at it.

The first approach to the cost limit is simply to observe that all technologies that do the same thing do not cost the same. If there are four ways to generate electricty—fusion, breeders, geothermal, and solar thermal electric, for example—then there is some ordering of the energy cost among them. Prudence dictates following several courses, but does not argue for running as hard toward the highest cost option as toward the lowest.

The limit may also be viewed as a more absolute one. It can be argued that a perfectly respectable energy future can be had at a real cost of not much more than twice that of today's energy. If that is true, as I am inclined to believe, then there is a fairly clear cutoff beyond which higher cost energy is not likely to succeed. In any case, policymakers need to accept the notion of a cost limit if they are ever to make the hard choices to pursue some options and not others.

Third, early demonstration is a mistake. By its nature, long-range research is a quest for better ideas, and a good idea is far more valuable than a demonstrable failure. Fortunately, it is also cheaper.

But the policymaker will find that the pressures are all in the other direction. Despite the expense, and despite the fact the demonstration freezes technological progress, program managers want to demonstrate. Resisting the impulse is the hard core of long-term planning.

Finally, nonhardware research must be sustained, usually by force of will. What we call basic research is in this category, but so too is social and environmental research. I suspect that neither the need for, nor the dangers of inaction in, these areas requires elaboration here.

NOTES ON POLICY

Thus far, my purpose has been to outline from my experience the nature of the policymaker's job—the forces with which he or she must content and the imperatives he or she must face. Some grasp of these factors is a necessary, if insufficient, basis of wise policy and seems a more appropriate contribution for this conference than a dissertation on policy content. However, it is hard to resist a few observations of a substantive nature, and I have yielded to the temptation. The notes that follow are incomplete and possibly wrong, but may provide some insight into the working of the forces that are shaping research and development policy for the long term.

It first seems to me that energy research must be more international in scope. Energy is not a problem confined by national boundaries. Not only must we loosen the hold of OPEC oil, we must also seek security of supply for all countries. Only collectively can we reduce global demand for depletable energy resources. It is simply enlightened self-interest to proceed cooperatively.

Our responsibilities extend especially to the developing nations. Unlike the wealthy of the world, these nations must increase their energy consumption per unit of output. And since developing economies generally run a current account deficit in trade, they can least afford the burden of energy imports. Research must increasingly be concerned with their needs.

Second, we should not, I think, anticipate an all electric future, especially in the United States. Electricity is expensive energy, for one thing. Furthermore, there may be an alternative. All the current evidence suggests that both the supply and the conservation potential for gas are large. Given the convenience of using the pipeline system we have, gas may well be one of those continuities that we should recognize now.

More important, perhaps, is our devotion to stored energy for transportation. The social change required to move decisively to other transportation energy sources—electrified rail, for example—is so large as to be improbable. And the costs of making and storing electricity are substantial.

None of this means that we should not pursue new ways of generating electricity. But the principle of cost limits suggests that we need not pursue them all, as we now seem to be doing. Rather, observing the feedback between present and future, it would seem useful to concentrate more heavily on long-range research into transportable fuels.

Third, it appears that we are in danger of conducting unnecessary and premature demonstrations of longer range technology. It is hard to believe that we have thought enough about solar power towers, ocean thermal energy conversion, solar power satellites, or large fusion machines to be comfortable in contemplating their demonstration. We do not have so many new ideas in any of these technologies that we can now afford to choke off thinking.

Finally, we need to worry about nuclear proliferation, but we must also stop paralyzing research in the face of it. This is probably not the place, and certainly not the time, to go into the complexities of proliferation. But perhaps a sketch of my thoughts will serve.

Proliferation happens when nations get their hands on plutonium. Plutonium is available when it is used to fuel reactors or when nations have plants that produce it. Setting aside clandestine or avowedly warlike activities, commercial nuclear power creates the possibility of proliferation when spent fuel is reprocessed for recycle or for breeders.

However, plutonium recycle is not an attractive activity. Almost no country will soon have a power program large enough to justify recycle. And in the few that do, recycle is economically marginal and always less attractive than saving plutonium for breeders. With some exercise of good sense and political skill, recycle need not be a credible avenue to proliferation.

We are thus left with breeders, and breeders can be a credible proliferation threat. But it will certainly be decades before breeders are widely deployed outside a few developed countries. Time is thus with us, at least in the sense of affording room to do the research to answer the technical and other questions about breeders. What a tragedy it would be if we failed now to take the time to do our best work and later find that time is too short to do anything but our worst.

The subject of proliferation amply illustrates my final point. Precisely because it is long range, long-range energy research and development policy must grapple with controversial social and political issues as well as with different technical ones. And precisely because it concerns research, the same policy must unfailingly seek the facts to inform decision on these issues. These two goals are not always compatible, whether one's political mind is made up in advance or whether one's technology advocacy overcomes good sense. Mustering the courage to seek the facts, and to let them influence judgment, is sometimes the policymaker's hardest job.

Glossary

ACR – Advanced Converter Reactor
ANISN – Anisotropic S_n (a neutron transport code prepared by ORNL)
Alcator – M.I.T. experimental Tokomak device
ATC – A small version of Tokomak
AES – Unconventional nonelectric alternative energy systems
ASTM – American Society for Testing and Materials

Btu – British thermal unit
BOE – Barrels of oil equivalent
bbl – barrel
BWR – Boiling Water Reactor
Be – beryllium
BACT – Best Available Control Technology
BAT – Best Available Control Technology
BACE – Best Available Conservation Equipment
BESOM – Brookhaven Energy System Optimization Model

CONAES – Committee for Nuclear and Alternative Energy Systems (of the National Research Council)
CPE – Centrally Planned Economies
CIA – Central Intelligence Agency
cm – centimeter
CO_2 – carbon dioxide

DOE – Department of Energy
$/kW – Dollars per kilowatt
$/HP – Dollars per horsepower
D-T – deuterium-tritium (reaction)
D-D – deuterium-deuterium (reaction)

ERDA – Energy Research and Development Administration (de funct)
E/GNP – Energy use per gross national product (energy intensity)
EJ – 10^{18} joules
EJ/y – EJ per year
ETA – Energy Technology Assessment
EIA – Energy Information Administration

FBR – fast breeder reactor

GAO – General Accounting Office
GSA – General Services Administration
GJ – gigajoule ≡ 10^9 joule
GJ/y – gigajoule per year
g – gram
GDP – Gross Domestic Product
GW – gigawatt ≡ 10^9 watt
GWe – gigawatt electric
GNP – gross national product

HTR – High temperature reactor
HTGR – high temperature gas-cooled reactor
HEU – highly enriched uranium
HWR – heavy water reactor
HF – hydrogen fluoride
Hz – hertz ≡ cycle per second

IAEA – International Atomic Energy Agency
INFCE – International Fuel Cycle Evaluation
IIASA – International Institute for Applied Systems Analysis

J – joule

kW – kilowatt ≡ 10^3 watts
KT – kiloton
kg/cap. y – kilogram per capita per year
kWh – kilowatt hour = 10^3 watt-hours
kgoe/cap. y – kilograms oil equivalent per capita per year
kg – kilogram ≡ 10^3 grams
kG – kilo Gauss ≡ 10^3 gauss
kWe – kilowatt electric ≡ 10^3 watt electric
°K – degree Kelvin (absolute temperature)
keV – kiloelectron volt ≡ 10^3 eV

kA – kiloampere ≡ 10^3 A
kT – Boltzmann energy (T = Kelvin temperature)
km – kilometer ≡ 10^3 meters

LDC – less developed countries
LWR – light water reactor
LNG – liquified natural gas
LITM – Long Term Interindustry Transactions Model
LMFBR – liquid metal fast breeder reactor
Li – lithium

MJ – megajoule = 10^6 joules
MBOE/day – million barrels oil equivalent per day
MW – Megawatt ≡ 10^6 watts
MRG – Modelling Resource Group
MBDOE – Million barrels per day oil equivalent
MWD – Megawatt days
MEU –moderately enriched uranium
MeV – megaelectron volt ≡ 10^6 eV
MWD/MT – Megawatt days per megaton
MWe – megawatt electric ≡ 10^6 watts electric
MWt – megawatt thermal = 10^6 watts thermal
m – meter
mm – millimeter ≡ 10^{-3} meter
mph – miles per hour
MMBD – million barrels per day
MBOE – Million barrels of oil equivalent
MMscfd – Million standard cubic feet per day

NNP – net national product
NASAP – Nonproliferation alternative systems assessment program
NPT – Nonproliferation treaty
ns – nanosecond ≡ 10^{-9} second
Na – sodium
nsec – nanosecond ≡ 10^{-9} second
NRC – National Research Council
Nuclear Regulatory Commission
NEPA – National Environmental Policy Act
NRDC – Natural Resources Defense Council

OMB – Office of Management and Budget
OECD – Organization for Economic Cooperation and Development
OPEC – Organization of the Petroleum Exporting Countries

ORNL – Oak Ridge National Laboratory
ORMAX – a small version of Tokomak

/cap. y – per capita per year
PIRINC – Petroleum Industry Research Foundation Inc.
Pu – plutonium
ppm – part per million
Pb – lead
PPPL – Princeton Plasma Physics Laboratory
PLT – Princeton Large Torus
PDX – Princeton experimental device using magnetic "divertor"
PCB – polychlorinated biphenyl
PVC – polyvinyl chloride

quad – 10^{15} Btu

R&D – research and development
ROE – the rest of the economy
RES – Reference Energy System

SWU – Separative work unit (reactor fuel enrichment)
SOLASE-H – fusion-fission hybrid reactor study (U. of Wisconsin)
s – second

t – ton
TESOM – Time-Phased Energy Systems Optimization Model
TCF – Trillion cubic feet
Th-232 – Thorium isotope 232
to – tonne
TW – terawatt $\equiv 10^{12}$ watts
TFTR – Tokomak Fusion Test Reactor
TIP – TFTR Improvements Project
torr – 1 mm Hg (pressure unit)

U-233 – uranium isotope 233
U-235 – uranium isotope 235
UBR – uranium breeding ratio

WOCA – World Areas Outside the Centrally Planned Economies
WAES – Workshop on Alternative Energy Strategies
WEC – World Energy Conference

Program

Second in a Series

INTERNATIONAL SCIENTIFIC FORUM ON AN ACCEPTABLE WORLD ENERGY FUTURE
(An Interdisciplinary Approach)

November 27 - December 1, 1978

Voltaire Room

FONTAINEBLEAU HILTON
4441 Collins Avenue
Miami Beach, Florida

CENTER FOR THEORETICAL STUDIES
University of Miami
Coral Gables, Florida 33124

SUNDAY, NOVEMBER 26, 1978

9:00 A.M.	Registration and Forum Check In Information (Free day for participants)
8:00 P.M.	Meeting of Forum Planning Committee and Session Moderators

MONDAY, NOVEMBER 27, 1978

8:00 A.M. Registration completion

8:30 A.M. PROLOGUE

PATRICK J. CESARANO
Chairman, University of Miami Board of Trustees
Welcoming Address

BEHRAM KURSUNOGLU
University of Miami
Opening Remarks

9:00 A.M. SESSION A
Finding an Evolving Balance of Energy Technologies

SESSION A I
Evolving Global Energy Balances and Constraints

Moderator: BENT ELBEK
Niels Bohr Institutet, Denmark

Dissertators: CARL D. PURSELL
Congress of the United States
"Political Fusion: Scientific Research and the Legislative Process"

BENT ELBEK
Niels Bohr Institutet, Denmark
"World Energy Outlook and Options"

10:15-10:30 A.M. COFFEE BREAK

JAY B. KOPELMAN
Electric Power Research Institute, Palo Alto
"A Global Energy Balance to Year 2000"

EDWARD TELLER
Stanford University
" 'Renewable' Resources"

Annotators: ERNEST BAUMEISTER
Atomics International (Rockwell)

J.F. BLACK
Exxon Research & Engineering Company

PETER FORTESCUE
General Atomic Company

PAULO ROBERTO KRAHE
Brazilian National Research Council

T.N. SRINIVASAN
World Bank Development Research Center

12:00 NOON — LUNCH BREAK (End of Session A I)

2:00 P.M. — SESSION A II
Living Through the Transition Period

Moderator: KARL COHEN
Stanford University

Dissertators: JOHN C. FISHER
General Electric Company, Schenectady, New York
"Transition Topography"

JOHN GIBBONS
University of Tennessee
"Some Alternative Demand Paths Through the Transition to a Long-term, Sustainable Energy Future"

3:15-3:30 P.M. — COFFEE BREAK

THOMAS H. LEE
General Electric Company, Fairfield, Connecticut
"Realities of the Transition to New Energy Sources—A Manufacturer's View"

C.C. BURWELL
Institute for Energy Analysis, Oak Ridge
"Feasibility of a Nuclear Siting Policy Based on the Expansion of Existing Sites"

Annotators: CHARLES F. COOK
Phillips Petroleum Company

JEAN COUTURE
Societe Generale, France

RUSSELL L. CROWTHER
General Electric Company

H.J. DAGER
Florida Power & Light Company

HERMAN DIECKAMP
General Public Utilities

Rapporteur: WILLEM VEDDER
General Electric Corporate Research & Development

5:00 P.M. FORUM ADJOURNS FOR THE DAY (End of Sessions A)

6:00 P.M. Welcoming Cocktails (by invitation)
Fontaine Room

TUESDAY, NOVEMBER 28, 1978

8:30 A.M. SESSION B
Energy Models for Exploring Alternative Evolving Balances

SESSION B I
Impacts of Energy Strategies on the U.S. Economy

Moderator: TJALLING C. KOOPMANS
Yale University

Dissertators: TJALLING C. KOOPMANS
Yale University
"Alternative Futures With or Without Constraints on the Energy Technology Mix"

10:15-10:30 A.M. COFFEE BREAK

JAMES L. SWEENEY
Stanford University
"Energy and Economic Growth: A Conceptual Framework"

	DALE JORGENSON Harvard University "The Economic Impact of Policies to Reduce U.S. Energy Growth"
Annotators:	JOSEPH R. DIETRICH Combustion Engineering, Inc.
	MAURICE F. DURET Atomic Energy of Canada Limited
	PETER ENGELMANN Jülich Nuclear Laboratory, West Germany
	ROBERT LITAN Council of Economic Advisors to the President
12:00 NOON	LUNCH BREAK (End of Session B I)
2:00 P.M.	SESSION B II World Models
Moderator:	JAMES SWEENEY Stanford University
Dissertators:	HENDRIK S. HOUTHAKKER Harvard University "A Long-run Model of World Energy Demands, Supplies, and Prices"
3:15-3:30 P.M.	COFFEE BREAK
	ALAN S. MANNE Stanford University "Energy Transition Strategies for the Industrialized Nations"
	HENRY S. ROWEN Stanford University "The Underestimated Potential of World Natural Gas"
Annotators:	MARKUS FRITZ Max-Planck-Institut, West Germany

ROLFDIETER GERHARDT
Nukem GmbH, West Germany

H.H. HASIBA
Gulf Oil Corporation

ARTHUR C. JOHNSON
Ontario Ministry of Energy

Rapporteur: HARRY DAVITIAN
Brookhaven National Laboratory

5:00 P.M. FORUM ADJOURNS FOR THE DAY (End of Sessions B)

WEDNESDAY, NOVEMBER 29, 1978

8:30 A.M. SESSION C
Identifying Critical Parameters of Resource Availability and Technology (Thorium and Uranium Availability and Cost, Response of Consumers to Energy Prices, Discount Rates, Etc.)

SESSION C I
Fission and Solar Parameters

Moderator: EUGENE P. WIGNER
Princeton University

Dissertators: KARL COHEN
Stanford University
"High Capital Cost: A Deterrent to Technological Change"

PETER J. JANSEN
Nuclear Research Center, West Germany
"The Need of the Plutonium-fueled LMFBR"

10:15–10:30 A.M. COFFEE BREAK

EDWARD TELLER
Stanford University
"On the Thorium Economy"

EUGENE P. WIGNER
Princeton University
"What We Have Learned"

Annotators: JOSEPH KESTIN
Brown University

W.D. KREBS
Kraftwerk Union, c/o Combustion Engineering, Inc.

SIDNEY H. LAW
Northeast Utilities Service Company

PETER MURRAY
Westinghouse Electric Corporation

FEDERICK TAPPERT
University of Miami

12:00 NOON LUNCH BREAK (End of Session C I)

2:00 P.M. SESSION C II
Fusion and Other Parameters

Moderator: ROBERT HOFSTADTER
Stanford University

Dissertators: GREGORY A. MOSES
University of Wisconsin
"Laser Fusion Hybrids–Technical, Economic, and Proliferation Considerations"

HAROLD P. FURTH
Princeton University
"Progress in the Toroidal Approach to a Fusion Reactor"

3:15–3:30 P.M. COFFEE BREAK

THOMAS GOLD
Cornell University
"Terrestrial Sources of Carbon and Earthquake Outgassing"

Annotators: W.W. BRANDFON
Sargent & Lundy, Chicago

LARS-AKE NOJD
Studsvik Energiteknik AB, Sweden

TIHIRO OHKAWA
General Atomic Company

ROBERT L. RINNE
Sandia Laboratories, Livermore

COLLEEN SEN
Institute of Gas Technology

5:00 P.M. FORUM ADJOURNS FOR THE DAY (End of Sessions C)

7:00 P.M. Cocktails and Buffet Dinner (by invitation)
at the home of
DR. and MRS. MAXWELL DAUER
5930 North Bay Road, Miami Beach

THURSDAY, NOVEMBER 30, 1978

8:30 A.M. SESSION D
Progress in Resolving Critical Environmental Problems, Issues of Risk-Benefit, Perceptions of Public Acceptance, and Institutional Constraints

SESSION D I
Risk-Benefit Ethics and Public Perception

Moderator: EDWIN L. ZEBROSKI
Electric Power Research Institute

Dissertators: ROBERT LATTES
French Atomic Energy Commission, Paris
"Two Basic Aspects of the World Energy Prolematique"

CHAUNCEY STARR
Electric Power Research Institute, Palo Alto
"Risk-Benefit Analysis and Its Relation to the Energy-Environment Debate"

MAURICE TUBIANA
Institut Gustave-Roussy, Villejuif, France
"One Approach to the Study of Public Acceptance"

PAUL SLOVIC
Decision Research, Eugene, Oregon
"Images of Disaster: Perception and Acceptance of Risks from Nuclear Power"

10:15–10:30 A.M. COFFEE BREAK

HERBERT INHABER
Atomic Energy Control Board, Canada
"The Paper Chase: Evaluating the Risk of Energy Systems"

PETER ENGELMANN
Jülich Nuclear Laboratory, West Germany
"On the Methodology of Cost-Benefit Analysis and Risk Perception"

MARGARET N. MAXEY
University of Detroit
"Bioethical Imperatives for Managing Energy Risks"

Annotators: MIRO M. TODOROVICH
City University of New York

ERSEL A. EVANS
Westinghouse Hanford Company

WM. CORNELIUS HALL
Chemtree Corporation, New York

JOSEPH W. STRALEY
University of North Carolina at Chapel Hill

ALLEN K. MERRILL
U. S. Department of State

Rapporteur: STEVEN B. SHANTZIS
International Energy Associates Limited, Washington

12:00 NOON LUNCH BREAK (End of Session D I)

2:00 P.M. SESSION D II
Limits to Growth of Due Process

Moderator: MARCUS ROWDEN
Fried, Frank, Harris Shriver & Kampelman, Washington

Dissertators: JOHN GRAY
International Energy Associates Limited, Washington
"A Statement on Limits to Growth of Due Process"

	ARTHUR W. MURPHY Columbia University "The Energy Crisis and the Adjudicatory Process"
3:15–3:30 P.M.	COFFEE BREAK
	KENNETH W. DAM University of Chicago "The Economics of Due Process"
Annotators:	VICTOR GILINSKY U.S. Nuclear Regulatory Commission
	HENRY LINDEN Gas Research Institute, Chicago
	JAMES M. WILLIAMS Los Alamos Scientific Laboratory, New Mexico
	DAVID S. SMITH Environmental Protection Agency, Washington
Rapporteur:	JAMES P. McGRANERY, JR. LeBoeuf, Lamb, Leiby & MacRae, Washington
5:00 P.M.	FORUM ADJOURNS DAY SESSIONS (End of Sessions D I and D II)
8:00 P.M.	SESSION D III International Decisionmaking Processes
Moderator:	LINCOLN GORDON Resources for the Future, Inc., Washington
Dissertator:	I.G.K. WILLIAMS O.E.C.D. Nuclear Energy Agency, Paris "Remarks on International Decisionmaking Processes"
Annotators:	WILLIAM V. SAUNDERS Ministry of Mining and Natural Resources, Jamaica
	FRANK L. HUBAND National Science Foundation, Washington

JACK M. HOLLANDER
Lawrence Berkeley Laboratory

DALE JORGENSON
Harvard University

JOHN P. HOWE
University of California, San Diego

Rapporteur: ACHILLES ADAMANTIADES
Electric Power Research Institute, Washington

10:00 P.M. FORUM ADJOURNS FOR THE DAY (End of Sessions D)
[These sessions were organized jointly by
C. PIERRE L. ZALESKI, Embassy of France, Washington, D.C., and
EDWIN L. ZEBROSKI, Electric Power Research Institute, Palo Alto]

FRIDAY, DECEMBER 1, 1978

8:30 P.M. SESSION E
Scrutiny of Future Energy Paths: Uncertainties, Long-term Policies, and Decisionmaking

SESSION E I
Major Uncertainties

Moderator: KENNETH C. HOFFMAN
Brookhaven National Laboratory

Dissertators: DERMOT GATELY
New York University
"'The Outlook for OPEC and World Oil Prices: Projections from World Energy Models for Three Decades"

ROBERT PINDYCK
Massachusetts Institute of Technology
"Energy Demand and Energy Policy: What Have We Learned?"

10:15-10:30 A.M. COFFEE BREAK

JOHN H. GIBBONS
University of Tennessee
"Long-term Prospects for Energy Conservation"

Annotators:	JOSEPH W. STRALEY University of North Carolina at Chapel Hill
	EDWARD A. AITKEN General Electric Company
	ROLLON O. BONDELID U.S. Naval Research Laboratory, Washington
	MARY D. SCHROT Lawrence Livermore Laboratory
Rapporteur:	STEVEN CARHART Brookhaven National Laboratory
12:00 NOON	LUNCH BREAK
2:00 P.M.	SESSION E II Policy Directions and Decisionmaking
Moderator:	WILLIAM HOGAN Harvard University
Dissertators:	KENNETH C. HOFFMAN Brookhaven National Laboratory "Long-term Strategic Analysis (Toward an Energy Doctrine)"
	THOMAS B. NEVILLE U.S. Department of Energy "Energy Decisions"
	ROBERT W. FRI Cambridge, Massachusetts "Long-term Research and Development Policy"
	VICTOR GILINSKY Nuclear Regulatory Commission " 'Delay', Public Participation, and Public Confidence"
Annotators:	MICHAEL TELSON U.S. House of Representatives
	JOSEPH W. STRALEY University of North Carolina at Chapel Hill

RICHARD C. PERRY
Union Carbide Corporation

H.H. INSTON
United Kingdom Atomic Energy Authority

ERSEL A. EVANS
Westinghouse Hanford Company

5:00 P.M. FORUM OFFICIALLY ADJOURNS

EPILOGUE

The Center for Theoretical Studies of the University of Miami wishes to extend their gratitude to all members of the planning committee, forum moderators, dissertators, annotators, rapporteurs, and other participants for their contributions to what this energy forum may have hoped to achieve on this most critical problem of all time. We hope, at the end of this assembly, we are wiser than before!

(This international scientific forum was supported in part by a grant from the Exxon Education Foundation. Additional support was obtained from registration fees paid by participants.)

Forum Moderators and Organizers

Marcelo Alonso
Organization of American States, Washington

Nikolai Basov
USSR Academy of Sciences, Moscow

Thomas H. Burbank
Edison Electric Institute, New York

Karl Cohen
Stanford University

Bent Elbek
Niels Bohr Institutet, Denmark

Kenneth C. Hoffman
Brookhaven National Laboratory

Robert Hofstadter
Stanford University

William W. Hogan
Harvard University

Tjalling C. Koopmans
Yale University

Behram Kursunoglu (Forum Chairman)
University of Miami

Marcus Rowden
Fried, Frank, Harris, Shriver, & Kampelman, Washington

Sam H. Schurr
Resources for the Future, Inc., Washington

Frederick Seitz
Rockefeller University

James Sweeney
Stanford University

Eugene P. Wigner
Princeton University

C. Pierre L. Zaleski
Embassy of France, Washington

Edwin L. Zebroski
Electric Power Research Institute, Palo Alto

Arnold Perlmutter (Forum Editor)
University of Miami

Osman K. Kadiroglu (Forum Editor)
University of Miami

Linda Scott (Forum Secretary)
University of Miami

Forum Secretariat

Helga S. Billings
University of Miami

Sandra M. Soto
University of Miami

List of Participants

Achilles G. Adamantiades
Electric Power Research Institute
Washington, D.C.

Edward A. Aitken
General Electric Company
Sunnyvale, California

Marcelo Alonso
Organization of American States
Washington, D.C.

Angel Alvarado
Ministerio de Energia y Minas de Venezuela
Caracas, Venezuela

Ernest B. Baumeister
Rockwell International
Canoga Park, California

Sebastian Bernstein
Comision Nacional de Energia
Santiago, Chile

J.F. Black
Exxon Research and Engineering Company
Linden, New Jersey

Albert W. Blackburn
BKW Associates
Washington, D.C.

Luciano N. Blanco
University of Miami
Coral Gables, Florida

Lucio M. Bolivar
Instituto Ecutoriano de Electrificacion
Quito, Ecuador

Rollon O. Bondelid
Naval Research Laboratory
Washington, D.C.

R.T. Bowles
Florida Power Corporation
St. Petersburg, Florida

W.W. Brandfon
Sargent & Lundy
Chicago, Illinois

Calvin C. Burwell
Institute for Energy Analysis
Oak Ridge, Tennessee

Steven Carhart
Carnegie-Mellon Institute of Research
Arlington, Virginia

Eric Casamiquela
Comision Ejecutiva Hidroelectrica del Rio Lempa
San Salvador, El Salvador

J.J. Chacon
Consejo Nacional de Investigaciones Cientificas y Tecnologicas de Costa Rica
Costa Rica

Mou-shan Chen
University of Miami
Coral Gables, Florida

Karl Cohen
Stanford University
Stanford, California

Jean Couture
Societe Generale
Paris, France

Russell L. Crowther
General Electric Company
San Jose, California

Benjamin F. Crump
Union Carbide
Oak Ridge, Tennessee

H.J. Dager
Florida Power & Light Company
Miami, Florida

Kenneth W. Dam
University of Chicago
Chicago, Illinois

Harry Davitian
Brookhaven National Laboratory
Upton, New York

Herman Dieckamp
General Public Utilities Corporation
Parsippany, New Jersey

Joseph R. Dietrich
Combustion Engineering, Inc.
Windsor, Connecticut

John A. Dillon, Jr.
University of Louisville
Louisville, Kentucky

P.A.M. Dirac
Florida State University
Tallahassee, Florida

E.A. Doryan
Consejo Nacional de Investigaciones Cientificas y Tecnologicas de Costa Rica
Costa Rica

John W. Duane
Consumers Power Company
Jackson, Michigan

Herbert G. Duggan
Union Carbide Corporation
Oak Ridge, Tennessee

Maurice F. Duret
Atomic Energy of Canada Limited
Chalk River, Ontario, Canada

Bent Elbek
Niels Bohr Institutet
Roskilde, Denmark

P. Engelmann
Nuclear Research Center
Jülich
Jülich, West Germany

Noel Espinoza Chavarria
Comision Ejecutiva Hidroelectrica
del Rio Lempa
San Salvador, El Salvador

Ersel A. Evans
Westinghouse Hanford Company
Richland, Washington

Anthony J. Favale
Grumman Aerospace
Corporation
Bethpage, New York

John C. Fisher
General Electric Company
Schenectady, New York

William M. Fitzgerald
Exxon Research and Engineering
Company
Florham Park, New Jersey

Peter Fortescue
General Atomic Company
San Diego, California

Robert W. Fri
Harvard University
Cambridge, Massachusetts

Markus Fritz
Max-Planck-Institut
Starnberg, West Germany

Harold P. Furth
Princeton University
Princeton, New Jersey

Edmund P. Gaines, Jr.
Vermont Yankee Nuclear Power Corp.
Rutland, Vermont

Dermot Gately
New York University
New York, New York

Rolfdieter Gerhardt
Nukem GmbH
Hanau, West Germany

John H. Gibbons
University of Tennessee
Knoxville, Tennessee

Victor Gilinsky
Nuclear Regulatory Commission
Washington, D.C.

C. Robert Glassey
U.S. Department of Energy
Washington, D.C.

Thomas Gold
Cornell University
Ithaca, New York

Henry J. Gomberg
KMS Fusion, Inc.
Ann Arbor, Michigan

Lincoln Gordon
Resources for the Future, Inc.
Washington, D.C.

John E. Gray
International Energy Associates, Ltd.
Washington, D.C.

Jules Gueron
Framatome
Paris, France

Michael Guhin
Department of State
Washington, D.C.

J. Curtis Haire
EG&G Idaho, Inc.
Idaho Falls, Idaho

William Cornelius Hall
Chemtree Corporation
Central Valley, New York

H.H. Hasiba
Gulf Oil Corporation
Pittsburgh, Pennsylvania

Kenneth C. Hoffman
Brookhaven National Laboratory
Upton, New York

Robert Hofstadter
Stanford University
Stanford, California

William W. Hogan
Harvard University
Cambridge, Massachusetts

Hendrik S. Houthakker
Harvard University
Cambridge, Massachusetts

John P. Howe
University of California–San Diego
La Jolla, California

Frank L. Huband
National Science Foundation
Washington, D.C.

Bruce Hutchins
General Electric Company
Sunnyvale, California

Herbert Inhaber
Atomic Energy Control Board
Ottawa, Canada

H. Inston
U.K. Atomic Energy Authority
London, England

Henri Jammet
Commissariat a l'Energie Atomique
Fontenay-aux-Roses, France

Peter J. Jansen
Nuclear Research Center
Karsruhe, West Germany

Arthur C. Johnson
Ontario Ministry of Energy
Toronto, Ontario, Canada

Dale W. Jorgenson
Harvard University
Cambridge, Massachusetts

Osman K. Kadiroglu
University of Miami
Coral Gables, Florida

Joseph Kestin
Brown University
Providence, Rhode Island

Tarek M. Khalil
University of Miami
Coral Gables, Florida

Chihiro Kikuchi
University of Michigan
Ann Arbor, Michigan

Tjalling C. Koopmans
Yale University
New Haven, Connecticut

Jay B. Kopelman
Electric Power Research Institute
Palo Alto, California

Paulo Roberto Krahe
Conselho Nacional de Desenvolvimento Cientifico e Tecnologico
Brasilia, Brasil

Frank Krausz
University of Miami
Coral Gables, Florida

Myron Kratzer
International Energy Associates, Ltd.
Washington, D.C.

W.D. Krebs
Kraftwerk Union AG
Combustion Engineering, Inc.
Windsor, Connecticut

Behram Kursunoglu
University of Miami
Coral Gables, Florida

Jack A. Kyger
Argonne National Laboratory
Argonne, Illinois

Gerald C. Lalor
University of the West Indies
Mona, Kingston, Jamaica

Robert Lattes
French Atomic Energy Commission
Paris, France

Sidney H. Law
Northeast Utilities Service Company
Hartford, Connecticut

Thomas H. Lee
General Electric Company
Fairfield, Connecticut

Lars Leine
ASEA-ATOM
Vaestreaas, Sweden

Henry R. Linden
Gas Research Institute
Chicago, Illinois

Robert Litan
Council of Economic Advisors to the President
Washington, D.C.

Alan S. Manne
Stanford University
Stanford, California

Luigi Massimo
European Communities
Washington, D.C.

Margaret N. Maxey
University of Detroit
Detroit, Michigan

James P. McGranery, Jr.
LeBoeuf, Lamb, Leiby & MacRae
Washington, D.C.

Carolyn Meinel Henson
L-5 Society
Tucson, Arizona

Allen K. Merrill
U.S. Department of State
Washington, D.C.

A.J. Meyer II
The Chase Manhattan Bank, N.A.
New York, New York

Gregory A. Moses
Universtiy of Wisconsin
Madison, Wisconsin

Arthur W. Murphy
Columbia University
New York, New York

Peter Murray
Westinghouse Electric Corporation
Madison, Pennsylvania

Thomas B. Neville
U.S. Department of Energy
Washington, D.C.

Lars-Ake Nojd
Studsvik Energiteknik AB
Nykoping, Sweden

Paul A. Northrop
University of Miami
Coral Gables, Florida

Uwe Parpart
Fusion Energy Foundation
New York, New York

Arnold Perlmutter
University of Miami
Coral Gables, Florida

Richard C. Perry
Union Carbide Corporation
New York, New York

Bruno Philippi
Comision Nacional de Energia
Santiago, Chile

George W. Pickering
University of Detroit
Detroit, Michigan

Guido Pincheira
University de Chile
Santiago, Chile

Robert S. Pindyck
Massachusetts Institute of Technology
Cambridge, Massachusetts

Carl D. Pursell
United States Congress
Washington, D.C.

Robert L. Rinne
Sandia Laboratories
Livermore, California

Dennis Ripley
Phillips Petroleum Co.
Bartlesville, Oklahoma

Marcus A. Rowden
Fried, Frank, Harris, Shriver, & Kampelman
Washington, D.C.

Henry S. Rowen
Stanford University
Stanford, California

Mary D. Schrot
Lawrence Livermore Laboratory
Livermore, California

Colleen Sen
Institute of Gas Technology
Chicago, Illinois

Steve Shantzis
International Energy Associates, Ltd.
Washington, D.C.

Paul Slovic
Decision Research
Eugene, Oregon

Paul Smith-Fontana
Comision Chilena de Energia Nuclear
Santiago, Chile

T.N. Srinivasan
World Bank
Washington, D.C.

Chauncey Starr
Electric Power Research Institute
Palo Alto, California

Joseph W. Straley
University of North Carolina
Chapel Hill, North Carolina

James L. Sweeney
Stanford University
Stanford, California

Frederick Tappert
University of Miami
Coral Gables, Florida

Edward Teller
Lawrence Livermore Laboratory
Livermore, California

Michael L. Telson
U.S. House of Representatives
Washington, D.C.

Micro M. Todorovich
City University of New York
New York, New York

Maurice Tubiana
Institut Gustave-Roussy
Villejuif, France

Willem Vedder
General Electric Company
Schenectady, New York

E.P. Wigner
Princeton University
Princeton, New Jersey

I.G.K. Williams
O.E.C.D.
Paris, France

James M. Williams
Los Alamos Scientific Laboratory
Los Alamos, New Mexico

C. Pierre Zaleski
Embassy of France
Washington, D.C.

E.L. Zebroski
Electric Power Research Institute
Palo Alto, California

Index